Antimicrobial Resistance
in Developing Countries

Aníbal de J. Sosa · Denis K. Byarugaba ·
Carlos F. Amábile-Cuevas · Po-Ren Hsueh ·
Samuel Kariuki · Iruka N. Okeke
Editors

Antimicrobial Resistance in Developing Countries

Foreword by Thomas F. O'Brien
Introductory Preface by the Editors
Guest Preface by Stuart B. Levy

 Springer

Editors

Aníbal de J. Sosa
Alliance for the Prudent Use
  of Antibiotics (APUA)
Tufts University
Boston, MA USA
anibal.sosa@tufts.edu

Denis K. Byarugaba
Department of Veterinary Medicine
Makerere University
Kampala, Uganda
dkb@vetmed.mak.ac.ug

Carlos F. Amábile-Cuevas
Fundación Lusara para la
  Investigación Científica
Mexico City, Mexico
carlos.amabile@lusara.org

Po-Ren Hsueh
Department of Laboratory Medicine and
  Internal Medicine
National Taiwan University Hospital
Taipei, Taiwan, R.O.C
hsporen@ha.mc.ntu.edu.tw

Samuel Kariuki
Kenya Medical Research
  Institute (KEMRI), Nairobi,
Kenya
SKariuki@kemri.org

Iruka N. Okeke
Department of Biology
Haverford College
Haverford, PA, USA
iokeke@haverford.edu

ISBN 978-0-387-89369-3   e-ISBN 978-0-387-89370-9
DOI 10.1007/978-0-387-89370-9
Springer New York Dordrecht Heidelberg London

Library of Congress Control Number: 2009920952

Printed on acid-free paper

Springer is part of Springer Science+Business Media (www.springer.com)

Foreword

Avoiding infection has always been expensive. Some human populations escaped tropical infections by migrating into cold climates but then had to procure fuel, warm clothing, durable housing, and crops from a short growing season. Waterborne infections were averted by owning your own well or supporting a community reservoir. Everyone got vaccines in rich countries, while people in others got them later if at all.

Antimicrobial agents seemed at first to be an exception. They did not need to be delivered through a cold chain and to everyone, as vaccines did. They had to be given only to infected patients and often then as relatively cheap injectables or pills off a shelf for only a few days to get astonishing cures. Antimicrobials not only were better than most other innovations but also reached more of the world's people sooner.

The problem appeared later. After each new antimicrobial became widely used, genes expressing resistance to it began to emerge and spread through bacterial populations. Patients infected with bacteria expressing such resistance genes then failed treatment and remained infected or died. Growing resistance to antimicrobial agents began to take away more and more of the cures that the agents had brought.

It then proved to be much more resource-intensive to keep patients from becoming infected with and failing treatment for drug-resistant bacteria than it had been to deliver the drugs that had caused the problem. Resource-limited countries that had managed to make antimicrobials available to their infected patients could not afford to do all the things that were then needed to manage the antimicrobial resistance that resulted.

Antimicrobial resistance seems a function of how many bacteria have been exposed to antimicrobials, for example, so treat only infections that antimicrobials cure for as long, but only as long, as needed. Treating an infecting germ with a drug it resists not only fails but also makes that resistant germ spread, so treat only with the drug that can still kill it. Resistant germs spread to others, so identify them and interrupt their spread.

Each of these ways to control resistance costs much more than it had cost to distribute the boxes of pills and injectables that had begun the resistance. Expensive microbiology laboratories in rich countries test whether the germ infecting any patient is of a kind that antimicrobial agents kill, and if so which agent could still kill it. Those countries then make that agent promptly available by keeping ubiquitous costly stocks of all agents.

Adequate housing and support for personal hygiene may also minimize interpersonal exchange of resistant bacteria in communities of developed countries, and clean water limits their ingestion. Many developed countries feed large amounts of resistance-selecting antimicrobials to food animals, however, but they do ban their residuals in food or their use to preserve food. Less is known about these in developing countries

In both rich and poor nations, resistant bacteria cause their most frequent, costly, and deadly infections in hospitals and intensive care units. In the rich nations, however, disposable items come in truckloads from warehouses to help nurses, and well-organized infection control teams slow the spread of resistant bacteria between private rooms, while the poorer struggle with shortages of reusable items and one hand-washing sink for an open ward.

Further promoting resistance in resource-limited countries is a cruel underlying inequity. They have more of the infections that richer countries have, e.g., pneumococcal, AIDS, meningococcal, trauma-related, tuberculosis, shigellosis, plus many they do not have, such as typhoid, malaria, and cholera. These require more valid antimicrobial use and also elicit more inadvertent misuse by complicating diagnosis, thus making resistance worse.

In a developed country, a sick febrile patient has prompt laboratory testing and compiled local test results to predict diagnosis and best therapy immediately, with confirmation or adjustment in a few days. In an undeveloped country, there will be fewer or no tests or compiled results, fewer antimicrobials available for often blind therapy of more possible diagnoses and so more chance of treatment failure and further spread of resistance.

For all of these reasons, the management of antimicrobial resistance in the resource-limited world faces special challenges and appears to need tools that are less resource-intensive than those that have evolved in the developed world. Affordable strategies and tools to manage antimicrobial resistance in less-resourced regions may need special effort to develop, but possible examples can be considered.

Microbiology laboratories in developed countries, for example, using supply-intensive instruments and highly trained and salaried staff for a huge menu of tests, become the model of excellence. Attempts to duplicate them in a developing country, however, may deplete the pool of trained professionals, outrun supply sources, and price tests so high that few patients can afford them and they become, after all, largely unused.

An alternative model might be developed in which a limited set of inexpensive essential tests using a common list of supplies could be performed by less-extensively but specifically trained workers supported by a web-based support and oversight system. This would not replace any currently functioning laboratories but supplement them at the next lower tier of medical facilities that now have no laboratories.

Similarly, growth of the Internet may provide cheap ways to provide updated information on currently prevalent infections and their drug resistances, as locally compiled data now do in developed countries, and use it to update and disseminate treatment guidelines. Every microbial test result that can be produced in a resource-limited country should be captured and analyzed to help overview its problems and update responses to them.

Each of the tools now used to control antimicrobial resistance might thus be reviewed with developing world caregivers to explore ways in which they could be modified to be cost-effective in their circumstances. Might not hospitals that cannot afford a team of infection control nurses, for example, have access to special training materials and information support to help their existing nursing staff carry out some of their functions?

Such a rethinking of the tools and strategies for controlling antimicrobial resistance to improve their application in the resource-limited world will not happen spontaneously. It will require concerted and funded effort by specialists from both worlds. Gaining support for it may prove difficult, moreover, since the whole problem of antimicrobial resistance has repeatedly slipped from the attention of public health and other funding agencies.

The first step in such a needed initiative is to recognize that antimicrobial resistance is a different and more deadly problem in the developing world and to elaborate in detail all of the aspects of that problem. That is what this book does and why it is important. It can be seen as both an informative work and a basis for action.

Boston, Massachusetts   Thomas F. O'Brien

Introduction

This book was mainly compiled while we quietly celebrated the 80th anniversary of the discovery of penicillin, a fact that formally inaugurated the "era of antibiotics" and it will become available during the celebration of the 150th anniversary of Darwin's *On the Origin of Species* (and the 200th anniversary of Darwin himself). These two milestones remind us, on the one hand, that we have been witnesses to the evolution of bacteria and other microorganisms from mostly susceptible to mostly resistant to drugs and, on the other, how little to heart even those of us who are not blinded by superstition have taken the basic notions of genetic change and selection. But in addition to the formidable evolutionary forces behind resistance, and the equally formidable neglect in putting our knowledge of it to work, developing countries have to deal with a wide variety of peculiar conditions that foster the emergence and spread of resistant germs. These aggravating factors range from malnutrition to lack of medical services and inadequate medical training and then to counterfeit drugs and incompetent governments. These traits are not exclusive to poor countries, but coalesce in the worst possible ways here. There is an exceptional need for rigorous data on the scale and spread of antimicrobial resistance, as well as for effective means for sharing the data and using it as an evidence base for effective containment strategies. In developing countries where resistance is a prime issue, data are least available, evidence is rarely collated, and containment interventions have been poorly implemented. Indeed, the validity of many proposed interventions in developing countries remains untested.

In putting together this book, we tried to assemble an overview of the magnitude, causes, consequences, and possible actions on microbial resistance in developing countries. If the book appears brief, it is because we know little about this particular side of the problem, as one of the main features of poor countries is insufficient scientific and medical research, and of opportunities to publish the scarce findings in "international" scientific journals, which routinely dismiss papers from developing countries because results are "only of local interest". Therefore, in addition to presenting information and ideas, the book explicitly highlights gaps that represent opportunities for research and policy innovation. High among our priorities has been to obtain data and critique from scholars who work not *on* but *in* developing countries and to include input from a variety of geographic regions. The picture that emerges, although incomplete, allows the reader to assess the current and emerging threats, the distinct issues that influence the evolution of resistance, the main problems caused by resistance, and the potential avenues to tackle at least some of this complex panorama.

Preface

The problem of antimicrobial resistance knows no boundaries. Drug-resistant microbes of all kinds can move among people and animals, from one country to another—without notice. From the early stages of identifying and discovering antibiotic resistance, the problem was clearly severe in developing countries where drug availability was limited and resistance was high. However, it has been in the developed world, with its abundant resources, where resistance has been more vigorously studied. Therefore, it is of some interest that out of a 1981 meeting in the Dominican Republic, where representatives from developing as well as industrialized countries assembled, came an Antibiotic Misuse Statement declaring the consequences of inappropriate use of antibiotics, namely the emergence and spread of antibiotic resistance. The response to the wide circulation of the statement led to the establishment of the Alliance for the Prudent Use of Antibiotics (APUA). This international organization continues today, 27 years later, to champion increased awareness and appropriate antimicrobial use so as to curtail drug resistance and "preserve the power of antibiotics." APUA fosters partnerships and communications among people in both developed and developing countries to improve antibiotic accessibility and reverse resistance.

This book, edited by Drs. Anibal Sosa and Denis Byarugaba and their associate editors, is unique in focusing on antimicrobial resistance as it relates to and threatens developing countries. It is curious that it has taken this long to produce a book dedicated to antibiotic resistance in developing parts of the world. One can ask "why?" since resistance is and has been so common there. In fact, whereas resistance has been addressed for the past four decades by experts in the industrialized world, studies describing the problem and the public health situation in the developing world have lagged behind. Although we have learned much from studies of the genetics and molecular biology of the problem from investigations in industrialized countries, it is in developing countries where more studies and efforts are needed. With travel encouraging the transport of microbes, the information in this book will have wide-sweeping benefit, not only for developing countries but also for the world at large. Surveillance of resistance and the prevention of resistance need attention on a worldwide basis. Improving antibiotic use requires a global effort.

One hopes that bringing an organized focus to the problem in developing nations will help efforts to improve accessibility to effective antibiotics and reduce resistance in previously neglected regions and countries of the globe. And so it is relevant that there are chapters in this book devoted to a particular country or region in which resistance poses a life-threatening challenge. We read about treatment failures and resistance challenges in Asia, Africa, and Latin America. The microbes under discussion in the book are not confined to bacteria, but encompass HIV, fungi, and parasites, including the agents of malaria and trypanosomiasis.

Lessons learned in one country can help others. What is needed is a clear idea of what the magnitude of the problem is and what efforts are being made to address it. The focus needs to be on the antimicrobial—its use and its availability—as well as the presence of resistant organisms and their resistance genes. Spread of resistance traits and resistant organisms is a further complicating feature of the resistance problem.

The chapter on the economics of resistance opens the potential for important cost analysis comparisons in this part of the world with studies in hospitals and health-care systems in industrialized countries. Cost is an important obstacle to change but needs to be assessed if we are to see change. Other chapters include discussions of the pivotal positive and negative roles of the pharmaceutical industry in delivering and marketing drugs in the developing world. Quite clearly, greater recognition of the needs and objectives of the stakeholders is critical to an understanding of and a cooperation in improving antibiotic availability and the decreasing frequency of resistance in these parts of the world.

The authors of these chapters are each distinguished in their own right and internationally recognized. The subjects are broadly inclusive of different infectious diseases, including those of the respiratory, urinary, and gastrointestinal tracts. The role of vaccines in helping to control organisms and avoid the overuse of antibiotics is critically important and discussed.

It is time, and timely, to focus attention on the developing countries in terms of helping people understand their role in reversing the global resistance problem. This book is an important step and will join other efforts, at both the government and nongovernment levels, including those of such organizations as the Alliance for Prudent Use of Antibiotics, the World Health Organization, the Pan American Health Organization, and others, to bring attention and potential solutions to the antimicrobial resistance problem as it presents in developing countries. The book shows how the problem has similar causes and the solution has similar goals as those in the industrialized countries. Drug resistance has no geographic preference—it compromises infectious disease treatments in countries throughout the world.

Boston, Massachusetts   Stuart B. Levy

Acknowledgment

Thanks to like-minded scientists from all over the world who dedicated many hours to research to provide evidence that will inform policy. Thanks to all those who relentlessly took the task to review and improve the scientific content of many manuscript drafts. Thanks to Jessica Restucci, APUA Executive Assistant for her valuable assistance throughout the editing and proof-reading of selected chapters, as well as to Andrea Macaluso, Springer Editorial Director, Life Sciences & Biomedicine, and Melanie Wilichinsky, Editorial Assistant, for their support throughout the process of producing and copy-editing.

Contents

Contributors

Lorena Abadía-Patiño Departamento de Biomedicina, Instituto de Investigaciones en Biomedicina y Ciencias Aplicadas, Universidad de Oriente, Cumaná. Edo. Sucre, Venezuela, abalor@movistar.net.ve

Richard A. Adegbola Medical Research Council Laboratories (UK), Fajara, Banjul, The Gambia, West Africa, radegbola@mrc.gm

Pedro L. Alonso Barcelona Center for International Health Research (CRESIB), Hospital Clínic/Institut d'Investigacions Biomèdiques August Pi i Sunyer, Universitat de Barcelona, Barcelona, Spain; Centro de Investigação em Saúde de Manhiça (CISM), Manhiça, Maputo, Mozambique, palonso@clinic.ub.es

Celia M. Alpuche Aranda Laboratorio de Infectología y Microbiología, Medicina Experimental, Facultad de Medicina UNAM – Hospital General de México, Instituto Nacional de Referencia Epidemiológica de Mexico, México, DF, México, calpuche@salud.gob.mx

Carlos F. Amábile-Cuevas Fundación Lusara para la Investigación Científica, Mexico City, Mexico, carlos.amabile@lusara.org

Soraya Sgambatti de Andrade Infectious Diseases Division, Laboratório Especial de Microbiologia Clínica, Department of Medicine, Universidade Federal de São Paulo, São Paulo, Brazil, soraya.andrade@lemc.com.br

Rebecca F. Baggaley Department of Infectious Disease Epidemiology, Faculty of Medicine, Imperial College, London, UK; Département de pathologie et microbiologie, Faculté de médecine vétérinaire, Université de Montréal, Montréal, Québec, Canada, r.baggaley@imperial.ac.uk

Abhijit M. Bal Department of Medical Microbiology, Aberdeen Royal Infirmary, Foresterhill, Aberdeen, Scotland, abhijit.bal@nhs.net

Alessandro Bartoloni Infectious Diseases Unit, Department of Critical Care Medicine and Surgery, Careggi Hospital, University of Florence, Florence, Italy, bartoloni@unifi.it

Quique Bassat Barcelona Center for International Health Research (CRESIB), Hospital Clínic/Institut d'Investigacions Biomèdiques August Pi i Sunyer, Universitat de Barcelona, Barcelona, Spain; Centro de Investigação em Saúde de Manhiça (CISM), Manhiça, Maputo, Mozambique, quique.bassat@cresib.cat

Francisco Inácio P. M. Bastos Laboratory of Health Information, Institute for Communication and Information on Science & Technology, Oswaldo Cruz Foundation, Rio de Janeiro, Brazil, francisco.inacio.bastos@hotmail.com

Michael L. Bennish Mpilonhle, Mtubatuba, South Africa; Department of Population, Family and Reproductive Health, Bloomberg School of Public Health, Johns Hopkins University, Baltimore, MD, USA, michael@mpilonhle.org

Marie-Claude Boily Department of Infectious Disease Epidemiology, Faculty of Medicine, Imperial College, London, UK, mc.boily@imperial.ac.uk

Freddie Bwanga Department of Medical Microbiology, Faculty of Medicine, Makerere University Medical School, Kampala, Uganda, fbwanga@med.mak.ac.ug

Denis K. Byarugaba Department of Veterinary Microbiology and Parasitology, Faculty of Veterinary Medicine, Makerere University, Kampala, Uganda, dkb@vetmed.mak.ac.ug

Béatrice Doizé Département de pathologie et microbiologie, Faculté de médecine vétérinaire, Université de Montréal, Montréal, Québec, Canada; Faculté de médecine vétérinaire – Université de Montréal, Montréal, Québec, Canada, beatrice.doize@umontreal.ca

Michael Feldgarden The Alliance for the Prudent Use of Antibiotics; Genome Sequencing and Analysis Program, The Broad Institute, Boston, MA, USA, feldgard@broad.mit.edu

Facundo M. Fernández School of Chemistry and Biochemistry, Georgia Institute of Technology, Atlanta, GA, USA, facundo.fernandez@chemistry.gatech.edu

Susan D. Foster Alliance for the Prudent Use of Antibiotics (APUA); Boston University, School of Public Health, Boston, MA, USA, susan.foster@tufts.edu

Carlos Franco-Paredes Hospital Infantil de México, Federico Gómez, México, DF, México; Emory University School of Medicine, Atlanta GA, USA, cfranco@himfg.edu.mx; cfranco@sph.emory.edu

Ana Cristina Gales Infectious Diseases Division, Laboratório Especial de Microbiologia Clínica, Department of Medicine, Universidade Federal de São Paulo, São Paulo, Brazil, ana.gales@gmail.com

Eduardo Gotuzzo Instituto de Medicina Tropical "Alexander von Humboldt", Universidad Peruana Cayetano Heredia, Lima, Perú, egh@upch.edu.pe

Ian M. Gould Department of Medical Microbiology, Aberdeen Royal Infirmary, Foresterhill, Aberdeen, Scotland, i.m.gould@abdn.ac.uk

Michael D. Green Division of Parasitic Diseases, Centers for Disease Control & Prevention, Atlanta, Georgia, USA, mgreen@cdc.gov

Manuel Guzmán-Blanco Hospital Vargas de Caracas; Department of Medicine, Infectious Diseases, Centro Médico de Caracas, San Bernardino, Caracas, Venezuela, mibeli@cantv.net

Hilbrand Haak Consultants for Health and Development, Leiden, The Netherlands, haakh@chd-consultants.nl

Po-Ren Hsueh Divisions of Clinical Microbiology and Infectious Diseases, Departments of Laboratory Medicine and Internal Medicine, National Taiwan University Hospital, National Taiwan University College of Medicine, Taipei, Taiwan, hsporen@ntu.edu.tw

Yu-Tsung Huang Divisions of Clinical Microbiology and Infectious Diseases, Department of Laboratory Medicine and Internal Medicine, National Taiwan University Hospital, Taipei, Taiwan, yutsunghuang@ntu.edu.tw

Raul E. Istúriz Department of Medicine, Infectious Diseases, Centro Médico de Caracas, Centro Medico Docente La Trinidad, San Bernardino, Caracas, Venezuela, mgrijm@cantv.net

Moses Joloba Department of Medical Microbiology, Faculty of Medicine, Makerere University Medical School, Kampala, Uganda, mlj10@cwru.edu

Samuel Kariuki Kenya Medical Research Institute (KEMRI), Centre for Microbiology Research, Nairobi, Kenya, skariuki@kemri.org

Wasif Ali Khan CSD, International Centre for Diarrheal Disease Research, Mohakhali, Dhaka, Bangladesh, wakhan@icddrb.org

Pascal P. Mäser University of Bern, Institute of Cell Biology, Bern, Switzerland, pascal.maeser@izb.unibe.ch

Enock Matovu Department of Veterinary Parasitology and Microbiology, Faculty of Veterinary Medicine, Makerere University, Kampala Uganda, matovue@vetmed.mak.ac.ug

Fatima Mir Department of Paediatrics and Child Health, Aga Khan University Medical College, Karachi, Pakistan, fatima.mir@aku.edu

Maria A. Miralles Center for Pharmaceutical Management, Management Sciences for Health, Arlington, VA, USA, mmiralles@msh.org

Eric S. Mitema Department of Public Health, Pharmacology and Toxicology, Faculty of Veterinary Medicine, University of Nairobi, Nairobi, Kenya, esmitema@uonbi.ac.ke

Paul N. Newton Wellcome Trust–Mahosot Hospital–Oxford Tropical Medicine Research Collaboration, Microbiology Laboratory, Mahosot Hospital, Vientiane, Lao PDR; Centre for Tropical Medicine, Churchill Hospital, University of Oxford, UK, paul@tropmedres.ac

Kayode K. Ojo Division of Allergy and Infectious Diseases, Department of Medicine, University of Washington, Seattle, WA, ojo67kk@u.washington.edu

Iruka N. Okeke Department of Biology, Haverford College, Haverford, PA, USA, iokeke@haverford.edu

Catherine Olivier Programmes de bioéthique, Département de médicine sociale et préventive, Université de Montréal, Montréal, Québec, Canada, catherine.olivier@umontreal.ca

Vural Ozdemir Bioethics Programs, Department of Social and Preventive Medicine, Faculty of Medicine, University of Montreal, Montréal, Québec, Canada, vural.ozdemir@umontreal.ca

David S. Perlin Public Health Research Institute, New Jersey Medical School / UMDNJ at the International Center for Public Health (ICPH), Newark, NJ, USA, perlinds@umdnj.edu

Maya L. Petersen School of Public Health, University of California, Berkeley, CA, USA, mayaliv@berkeley.edu

Joyce Primo-Carpenter U.S. Pharmacopeia, Drug Quality and Information Program, Rockville, MD, USA, joypcar@yahoo.com

Aryanti Radyowijati Consultants for Health and Development, Leiden, The Netherlands, aryanti@chd-consultants.nl

Luis Romano Mazzotti Infectious Diseases Department, Hospital Infantil de México Federico Gómez, México DF, México, luisromano@mac.com

Helio Silva Sader Infectious Diseases Division, Laboratório Especial de Microbiologia Clínica, Department of Medicine, Universidade Federal de São Paulo, São Paulo, Brazil; JMI Laboratories, North Liberty, IA, USA, helio-sader@jmilabs.com

Debasish Saha Medical Research Council (UK) Laboratories, Fajara, Banjul, The Gambia, West Africa. dsaha@mrc.gm

Marisabel Sánchez Links Media, LLC, Gaithersburg, MD, USA, msanchez@linksmedia.net

Jose Ignacio Santos-Preciado Facultad de Medicina, Departamento de Medicina Experimental, Universidad Nacional Autonoma de Maxico, jisantosp@gmail.com

Satya Sivaraman Action on Antibiotic Resistance (ReAct), New Delhi, India, satyasagar@gmail.com

Marcelo A. Soares Department of Genetics, Federal University of Rio de Janeiro; Division of Genetics, Brazilian Cancer Institute, Cidade Universitaria – Ilha do Fundao, Rio de Janeiro, RJ, Brazil, masoares@biologia.ufrj.br

Aníbal de J. Sosa Alliance for the Prudent Use of Antibiotics, Boston, MA, USA, anibal.sosa@tufts.edu

Erika Vlieghe Department of Clinical Sciences, Institute of Tropical Medicine, Antwerp, Belgium, evlieghe@itg.be

Nicholas J. White Wellcome Trust–Mahosot Hospital–Oxford Tropical Medicine Research Collaboration, Microbiology Laboratory, Mahosot Hospital, Vientiane, Lao PDR; Centre for Tropical Medicine, Churchill Hospital, University of Oxford, UK; Faculty of Tropical Medicine, Mahidol University, Bangkok, Thailand, nickw@tropmedres.ac

Bryn Williams-Jones Programmes de bioéthique, Département de médicine sociale et préventive, Faculté de médicine, Université de Montréal, Montréal, Québec, Canada, bryn.williams-jones@umontreal.ca

Anita K.M. Zaidi Department of Pediatric and Child Health, Aga Khan University, Karachi, Pakistan, anita.zaidi@aku.edu

Part I
General Issues in Antimicrobial Resistance

Chapter 1
Global Perspectives of Antibiotic Resistance

Carlos F. Amábile-Cuevas

Abstract The threat of antibiotic resistance is growing at an alarming pace, perhaps more rapidly in developing countries. Aside from the abuse of antibiotics, a number of circumstances converge to this rapid growth and spread, ranging from the biological traits that bacteria deploy to face antibiotics, which we are still trying to understand, to regulatory and financial issues behind antibiotic abuse. And, aside from the more-difficult-to-cure side of resistance, which is bad enough, bacteria are evolving to be more competent to face environmental stress, which includes antibiotics, present and future ones. In developing countries, we face peculiarities that go from antibiotic self-prescription to poor sanitary conditions, even at hospitals, that foster the threat of particular multi-resistant pathogens that are not common in developed countries and against which no new antibiotics are being investigated. In addition to the local repercussions of these peculiarities upon resistance trends, it is important to realize that these can easily cross borders in this era of globalization. A worldwide strategy must be developed and enforced if we are to have a good start at the post-antibiotic era.

1.1 Introduction

Bacterial resistance to antimicrobial drugs is one of the most serious jeopardies to global public health. While writing this chapter, newspapers commented during weeks that, at least in the United States, deaths caused by a single multi-resistant species, MRSA, might be more than those caused by AIDS. This explosive growth of resistance is caused by a host of factors, ranging from the peculiar genetic traits of bacteria to the complex and often corrupt relationships among big pharma companies, physicians, and governments; and its consequences go from "the usual" increase in sickness and deaths to the excess expenditure of billions of dollars and the somber perspective of an accelerating

C.F. Amábile-Cuevas (✉)
Fundación Lusara para la Investigación Científica, Mexico City, Mexico
e-mail: carlos.amabile@lusara.org

A. de J. Sosa et al. (eds.), *Antimicrobial Resistance in Developing Countries*,
DOI 10.1007/978-0-387-89370-9_1, © Springer Science+Business Media, LLC 2010

shrinkage of our antibacterial arsenal. Although we have successfully made people, both lay and not, aware of the problem, as well as of one of the main causes (i.e., antibiotic abuse), this is yet to become an integral strategy to harness resistance and to develop new ways to fight bacterial infections.

This book is about microbial resistance, which includes viruses, fungi, and protozoa along with bacteria. Resistance is, of course, the evolutionary consequence of the deployment of a selective pressure; therefore, it has been documented among all organisms against which we have declared biological wars, from viruses to insects. However, resistance among bacteria poses a distinct threat because of various reasons: (a) the abuse of antibacterial drugs is much higher than that of antifungal or antiviral agents; the later ones are seldom self-prescribed, wrongfully used as prophylaxis, or have agricultural usage; (b) bacterial genetic characteristics and abilities enable a rapid evolution toward resistance in ways that exceed by far those of viruses, fungi, and protozoa: haploidy, horizontal gene transfer mechanisms, extrachromosomal elements are all features that foster resistance and that are almost unique to bacteria; (c) bacteria appear to be much more abundant than viruses, fungi, and protozoa as microbiota of humans, which increases exponentially the exposure of the former to antibiotics each time they are used clinically, creating more chances of resistance to emerge and be selected; (d) bacterial diseases are also more abundant, at least for treatment purposes, increasing also the exposure to antibacterial drugs, perhaps with the exception of malaria. Therefore, although microbial resistance in general is posing grave problems for public health, it is not possible to view all resistance from the same perspective. This chapter deals with bacterial resistance only, and further chapters will explore the details of other kinds of resistant microbes.

In the following sections, I would try to assemble a view of the biological, pharmacological, educational, regulatory, and financial sides of bacterial resistance, along with an overview of its biological and clinical consequences. As this volume is devoted to the problem as we see it from the – so-called – developing countries, I will discuss the nuances that make antibiotic abuse, increasing resistance and particular pathogens, a distinctive threat to such countries. However, it is important to highlight that, in the end, and thanks to globalization, the dangers posed by resistance in poor countries will inevitably reach the developed world. If walls have proven to be insufficient to stop migrating people, it will certainly be so for multi-resistant microbes and their consequences.

1.2 The Biological Side of Resistance

At the same time antibiotics were being discovered and developed for clinical use, some of the particular traits of bacteria were also being discovered, mainly their ability to exchange genetic information (Amábile-Cuevas 2003). As the antibiotic era evolved, so did our understanding of bacteria, and bacteria themselves. A striking example of this simultaneous evolution is provided by quinolones and its associated resistance. Introduced in the 1960s, starting with nalidixic acid, quinolones had an explosive growth with the advent of fluorinated derivatives in the 1980s. Resistance emerged but was supposed to be restrained to vertical inheritance (Fuchs et al. 1996), and a mini-review paper in the early 1990s posed the question of the – so far – missing plasmid-mediated quinolone resistance (Courvalin 1990). Shortly before, a remarkable stress defense mechanism of several gram-negative bacteria, the *mar* regulon, was discovered among fluoroquinolone-resistant clinical isolates (Hooper et al. 1989). And shortly after, the first of a series of plasmid-borne fluoroquinolone genes was reported, possibly evolved from microcin-resistant traits (Tran and Jacoby 2002), soon followed by a variety of similar genes. And then, an aminoglycoside-modifying enzyme evolved to modify also a fluoroquinolone (Robicsek et al. 2006), an entirely different molecule, both chemically and pharmacologically. Meanwhile, in the gram-positive side, where plasmid-mediated quinolone resistance is yet to be found, the supposed restriction to vertical inheritance was also bypassed by transformation, allowing resistant alleles of target genes to be mobilized among different streptococcal species (Balsalobre et al. 2003). And this is just an example.

   In addition to the quinolone story, we have learned during the past 20 years or so about the many ways bacteria can successfully face the threat of antibiotics. Some of these include (a) global responses to environmental stress, which include the *mar* regulon earlier mentioned, and the somehow related oxidative-stress response system, the *soxRS* regulon of *Escherichia coli* and *Salmonella enterica* (Demple and Amábile-Cuevas 2003), along with several other efflux-mediated resistance mechanisms (Davin-Regli and Pagès 2007), but also including the SOS response (Miller et al. 2004), which can, additionally, promote the horizontal dissemination of resistance genes (Beaber et al. 2003); (b) biofilm growth and its conferred resistance, or tolerance, the mechanism of which is still a matter of debate (Gilbert et al. 2007), but that is clearly causing antibiotic treatment failure. Also, biofilms are playgrounds for bacteria in terms of exchanging and "concentrating" canonical resistance traits (Delissalde and Amábile-Cuevas 2004); (c) plasmid-antibiotic resistance interdependency, in a way similar to the toxin–antitoxin systems that prevent the emergence of plasmid-less cells within a resistant bacterial population (Heinemann and Silby 2003), and that is but one cause of the overall failure of antibiotic cycling strategies within hospitals; (d) resistance gene mobility, mainly mediated by plasmids, which has been known for more than 60 years (Amábile-Cuevas and Chicurel 1992) but whose components we are still trying to understand: integrons and gene cassettes, transposons including insertion sequences and conjugative transposons, conjugative and mobilizable plasmids, etc. (Amábile-Cuevas 1993). All these features, and perhaps many more that we are yet to discover, make resistance very resistant to elimination and even control. Additionally, antibiotic usage is not the only selective pressure that we are applying to foster antibiotic resistance: disinfectants (Aiello and Larson 2003),

environmental pollutants (Jiménez-Arribas et al. 2001), etc. can keep the prevalence of resistance high, even if we manage to restrain the abuse of antibiotics.

1.3 The Pharmacological Side of Resistance

Two pharmacological issues have particular relevance to the resistance problem: (a) spectrum and (b) compliance. Wide-spectrum antibiotics do have an
important role in fighting infection, either those caused by several different
bacterial species or those for which assessing the etiology is too difficult or takes
too much time. However, wide spectrum has been an obvious goal in the R&D
of antibiotics, as it ensures also a wide variety of clinical uses and, of course, of
sales. Wide spectrum has also been presented to the medical community as a
general advantage so that the physician need not worry about the etiology of an
infectious disease to start treatment. Of course, the likelihood of this strategy to
succeed in the short term and the individual patient is high. But this shotgun
notion contributes to resistance as it applies selective pressure, not only upon
the etiological agent of the infectious episode but also upon a larger fraction of
the patient's microbiota. Some reports indicate that, although worldwide use of
antibiotics is receding, the use of wide-spectrum ones is dramatically increasing.
Along with increasing resistance trends, marketing efforts are aimed at posi-
tioning newer fluoroquinolones as choice drugs against lower respiratory tract
infections, instead of aminopenicillins and macrolides, and even oral, third-
generation cephalosporins against diseases as minor as pharyngotonsillitis. It is
of course not surprising to see among community-acquired uropathogens up to
one-third of them highly resistant to fluoroquinolones, and even a high prevalence of ESBL enzymes, previously confined to hospital settings (Casellas and
Quinteros 2007).

Compliance is a pharmacological issue that has many repercussions upon
bacterial resistance. Patients often miss drug doses, both by mistake and deliberately (there is, for instance, the urban legend that antibiotics and alcohol
interact dangerously so that people under antibiotic therapy, but attending a
party, would rather miss a dose than avoid alcohol at the party). Other patients
decide to suspend the treatment prematurely when they feel well enough. Both
instances result in the exposure of surviving pathogens to subinhibitory concentrations of antibiotics and, consequently, to increased chances of acquiring
resistance. When a patient gets antibiotics through self-prescription, the patient
is more likely to take shorter treatments (Calva et al. 1996). A side effect of a
lack of compliance is to have remaining doses of antibiotics that are often stored
"just in case"; this favors future self-prescription, using antibiotics even beyond
their expiration dates, which could expose bacteria to further subinhibitory
concentrations. A list of suggestions for improving patient compliance included
in *Goodman & Gilman's Pharmacological Basis of Therapeutics* (Buxton 2006),
which is sound for the United States and other developed countries, is simply

unrealistic for poor countries where overworked physicians cannot devote the time needed for approaches such as "developing satisfactory, collaborative relationships between doctor and patient..." and "using behavioral techniques such as goal setting, self-monitoring, cognitive restructuring..." and where "easy-to-read written information" is not useful for an illiterate patient, while "mechanical compliance aids" are simply not available. However, it might be useful to try to dispel wrong notions (such as the alcohol–antibiotic interaction; but, then again, the notion might have been acquired from misinformed physicians) and to teach the patient that the risk of not taking each and every dose is not only – or perhaps not even – for him/her but for his/her family, friends, and coworkers, in the mid- and long term (but, then again, physicians would need to be convinced of these themselves).

1.4 The Educational Side of Resistance

If we assume that antibiotic abuse is one or perhaps *the* most important cause of the high prevalence of resistance among bacteria, then there must be something very wrong in the way physicians are trained to use antibiotics. Of course, not all abuse is related to the medical environment, but it is often referred that about half of antibiotic prescriptions are wrong (i.e., wrong antibiotic, wrong doses, or antibiotics not necessary at all). A number of factors contribute to this problem: (a) the hard-to-kill notion that antibiotics, if not useful, are not harmful, which is often true in the very short term but misses the grave danger imposed in the mid- and long term; (b) the lack of legal consequences of wrongful prescription of antibiotics, contrasted with the risk of actual consequences if antibiotics are not prescribed; (c) a sort of "prescription addiction," which makes prescription a satisfactory experience; this explanation was presented to me by an infectious disease specialist and an expert in nosocomial infections, for whom it seems an acceptable behavior; and (d) plain laziness. There is little that can be done to change all of the above, so I left aside the one educational issue that can and actually must be changed: the very few time devoted to microbiology, infectious diseases, and antimicrobial pharmacology during medical training. Considering that most causes of medical consultation, illness, and deaths come from infections, it is only inappropriate that medical students get only one course on microbiology and about a third of the one on pharmacology devoted to antibiotics. Additionally, the infectious diseases speciality is rightfully perceived as a low-income, low-profile one, contrasting to cardiology or plastic surgery, even (or, perhaps, particularly) in developing countries. Treating physicians try to avoid, and often disregard, the opinion of the ID specialist, and many hospitals do not even have one in the staff. Therefore, the handling of infectious diseases is left in the hands of undertrained physicians; all other consequences aside, antibiotic abuse is rampant and so is the growing bacterial resistance.

1.5 The Regulatory Side of Resistance

Regulations are the responsibility of governments, and for some of us, that clearly indicates that something inevitably is going to be completely wrong, as governments seldom do something right. However, on the one hand, some issues must be regulated regarding antibiotic usage, production, and marketing; on the other hand, we pay handsomely to governments and it is only fair to expect them to do their work every once in a while. The main question is, What should be regulated?

In the so-called developing countries, one of the first things that must be regulated is the sale of antibiotics: antibiotics, as most other drugs, can be purchased without medical prescription, on over-the-counter basis. This leads to self-prescription, which we can make it inclusive of all non-medically prescribed drugs, often from the advice of a relative, a friend, or even the sales clerk at the drug store, all of whom lack any kind of medical training. It is difficult to assess how much antibiotics are sold and used in this way; other chapters of this books will analyze this in detail. As an example, a survey in Mexico City revealed that at selected drug stores, about 30% of antibiotics were sold without medical prescription (Amábile-Cuevas et al. 1998). It is much more difficult to assess the burden this practice poses upon resistance, but we can assume that self-prescription would be wrong more often than medical one, leading to inadequate choice of drugs, inadequate doses and/or length of treatments, and, of course, a more common use of antibiotics where they were not necessary. A simple, draconian approach to this problem would be simple: to ban the sale of antibiotics without medical prescription. This has been made in a few Latin American countries, which we can only hope would attempt to record the impact of the measure. However, a trait of the "developing" countries is that a huge fraction of their populations lack access to medical care; for them, self-prescription might very well be the only means of access to antibiotics. Since we do not know how many lives are actually being saved by self-prescribed antibiotics, banning the free sale of such drugs could prevent a potential jeopardy (rising resistance) by causing an actual damage, denying access to life-saving drugs. It is my opinion that we must wait to see the impact of a prohibition, before extending it to all countries. Besides, and based only on the Mexican results cited before, medical prescription still accounts for more antibiotic abuse than self-prescription: if 70% of antibiotics are prescribed by a physician, and such prescriptions are 50% wrong, then this makes for 35% of all antibiotics sold at drugstores, more than the 30% accountable by self-prescription.

The mere suggestion of the regulation of the clinical use of antibiotics almost always results in the angry opposition of clinicians: they definitely know better than a simple rule, and are trained to understand signs and symptoms, and to decide when, where and which drug to use. Unfortunately, that is simply not true. Clinicians often use antibiotics to treat viral infections or even ailments that are not infectious at all; have difficulties to cope with the variety of microorganisms, drugs, mechanisms of action, pharmacokinetics and bioavailabilities, and resistance trends that must be considered before selecting a drug; and do not know what to expect from and what to do with a clinical microbiology report. For those who acknowledge some or all of these problems, the solution should lie upon educational efforts, perhaps only aided by some of the ubiquitous "guidelines" that pharmaceutical companies are only too kind to support; regulation, they say, hinders good medical practice. Although, as mentioned before, education should and must be improved, it is my belief that some basic rules must be set up and enforced. At least while further educational efforts are established and proven to work. For those physicians that do not adhere to these rules, legal consequences should be established; we are well beyond the point where just admonishing would do. These rules should include, but not be limited to, the use of antibiotics when not needed, as in non-infectious, viral, or self-limited diseases; the reckless use of wide-spectrum antibiotics; the unwarranted use of antimicrobial prophylaxis; and errors in dosing.

Another important issue calling for stronger regulatory efforts is drug production and marketing. Although this is important for all drugs, it is particularly relevant for antibiotics, as low-quality antimicrobials do not "only" affect the patient, but can also include effects upon microbial populations. Taking again Mexico as an example, we do have three drug categories that affect the quality of drugs: those manufactured by the original patent holder, invariably a transnational big pharma company; the "exchangeable generic," which is often manufactured overseas and only packed here but which must comply with a minimal set of tests to prove equivalency (tests that must be performed by licensed private pharmacology laboratories that work under obscure regulations); and the "similar" drugs that are beyond any regulation and that are manufactured by a single company owned by a politically untouchable subject. So a system that should provide cheap but reliable drugs to the ill turned out to be a source of questionable business (the private laboratories), bribery (the office inspecting those laboratories), and immense wealth (the single "similar" manufacturer), while most physicians and patients know that the only reliable drug is the one from big pharma companies. And this is the legal side of the pharmaceutical business: drugs stolen from public hospitals (and most likely stored inappropriately), and even counterfeits, are also sold to patients, physicians, hospitals, and drug stores. The counterfeit side of this problem will be reviewed in other chapters.

Final issues regarding antibiotic usage from the regulatory point of view include their non-clinical use, namely for agricultural purposes. Also, strict surveillance upon marketing strategies from pharmaceutical companies is necessary to prevent them to push for inadequate uses of their products.

1.6 The Financial Side of Resistance

There are at least two financial elements that foster the threat of resistance: (a) the pressure to recover the investment in R&D on new antibiotics, which results in aggressive marketing campaigns designed to have new drugs used as much as possible, and (b) the perception that investing in antimicrobial R&D is finally bad for business, since these drugs face great competition, are only used for a limited time period, and have a host of experts predicating to reduce the use of them. Measures to counteract these factors have been proposed, from extending the span of patent protection for antimicrobial drugs (Outterson et al. 2007), allowing for a reduced pressure to recover the investment, to fiscal incentives to pharmaceutical companies that engage in antibiotic R&D. Although these measures might achieve the reactivation of the search for new drugs, the underlying problem is left untouched: the social and political factors that shape the very way antibiotics are discovered, tested, and used (Heinemann and Goven 2007). First, most if not all drug research was left in the hands of private companies, which would rate each product not in terms of the public health benefits that can be derived from them but in terms of the income that they would generate. Even university researchers that engage in drug discovery end up selling their patents to pharmaceutical companies. This is not meant to be an attack on the big pharma but on our political will to leave our health decided by financial interests. We can only hope that Francis Fukuyama and his proposed end of history are dead wrong.

1.7 The Biological Consequences of Resistance

Resistance is often considered from the biological point of view as an interesting evolutionary model, one that we must thank as an incontrovertible proof of evolution by natural – or not-so-natural – selection, in this age of growing superstition. But the biological consequences of antibiotic deployment have seldom been explored outside the obvious field of clinical microbiology, despite clear indications that antibiotic usage is selecting a variety of traits other than resistance, which could have important biological impact. Unfortunately, the supply of pre-antibiotic-era microbes is short, so we cannot draw relevant conclusions as to the biological effects that antibiotic usage had, by comparing the "before" and "after" bugs. The recent analysis of class 1 integron integrases showed that these gene mobilization machineries are kept only in *E. coli* under selective pressure, likely to be antibiotics (Díaz-Mejía et al. 2008); although, on the one hand, it is encouraging to know that by diminishing antibiotic pressure we could reduce the presence of these elements, on the other hand, it is a reminder that, by using antibiotics, we are exerting pressure to keep integrons into clinically relevant bacteria. Error-prone DNA repair mechanisms were present in plasmids in the "pre-antibiotic" era (Amábile-Cuevas 1993); such plasmids acquired resistance genes and are now being co-selected by antibiotics so that after antibiotic exposure, surviving germs will be resistant and will have increased mutagenic rates. If other elements and genes, such as the horizontally transferable *mutS* gene, that accelerate the mutation rate of bacteria (Brown et al. 2001) are also being co-selected by antibiotics, the legacy of this "antibiotic era" would be not only resistant bacteria but also bacteria that are much more competent to face environmental threats, by means of increased prevalence of determinants that increase genomic plasticity.

1.8 The Clinical Consequences of Resistance

In the end, our primary concern about resistance is that resistant germs are more difficult to get rid off and that complications and deaths resulting from infections caused by them will only increase in time. Very few really new antibiotics will be developed in the short- and mid-term, and perhaps we already have as many antibiotics as possible, judging from recent genomic evidences (Becker et al. 2006). From a global point of view, this means that infections will impose a growing toll on humans, despite the dramatic reduction that we accomplished during the first 80 years of the twentieth century. But it will be different between regions. In the first paragraph of this chapter, I mentioned the alarm caused by rising MRSA infections and deaths in the United States; but in "developing" countries, MRSA is not as big a problem as multi-resistant salmonellosis, shigellosis, and tuberculosis in the community setting and enteric bacteria and *Pseudomonas aeruginosa* infections in hospitals. Some may think that resistance is affecting equally the developed and developing countries, if only through different bugs. However, developing countries do not have pharmaceutical research of their own, so we are dependent on what developed countries do develop; and their worries are different from ours. So while new anti-staphylococcal drugs are actually being deployed (linezolid, daptomycin, newer glycopeptides), drugs that can be used against MDR- and XDR-TB, multi-resistant *Shigella* and *Salmonella* strains, or multi-resistant enteric bacteria causing nosocomial infections much more often in our deficient hospitals are not being explored.

1.9 Final Considerations

"War is too important to be left to the generals," Clemenceau famously said. In this context, war against microbes is too important to be left to the clinicians. This is not meant to be offensive; there are too many aspects of infection management that are entirely out of the scope and training of the clinician. Epidemiologists, pharmacologists (and, most importantly, experts in pharmacoeconomics), microbiologists, evolutionary, population, and molecular biologists all must participate in a concerted effort to design sound strategies to handle infectious diseases, including the use of antibiotics. Unfortunately, in developing countries, we do have only a few qualified experts in each field; and, on the other hand, the political control of most of these issues relies only upon physicians (or bureaucrats with medical training) who tend to think of all other disciplines as lesser fields. In any case, non-clinical experts need to get involved in this area, despite their natural aversion to clinical things; and clinicians need to allow others to join the fight against resistance. Building bridges is the first task.

Antibiotic resistance is the consequence of a variety of biological, pharmacological, and societal variables that occur worldwide but that present themselves in the worst possible combinations in developing countries. Then we need to consider that infectious diseases are much more common here, as poor sanitary and work-safety conditions, starvation and malnutrition, lack of medical services (and an excess of "alternative medicine" options), and larger exposure to environmental agents that increase the likelihood of infection (e.g., weather changes, arthropod vectors) affect much more and a much larger fraction of the population than in developed countries. Conditions are only likely to get worse, as the divide between rich and poor countries widens, as it does between rich and poor people within poor countries, and also as climate change, war, and migration introduce entirely new variables to systems that were in an equilibrium of sorts for many years. To revert or at least stabilize things as they are now, we must start addressing the problem that affects particularly the poor countries: poverty itself. All other approaches are palliative at best.

References

Aiello, A.E. and Larson, E.L. (2003) "Antibacterial cleaning and hygiene products as emerging risk factor for antibiotic resistance in the community." *Lancet Infectious Diseases* 3, 501–506.

Amábile-Cuevas, C.F. (1993) *Origin, evolution and spread of antibiotic resistance genes.* RG Landes, Austin.

Amábile-Cuevas, C.F. (2003) "Gathering of resistance genes in gram-negative bacteria: an overview". In: C.F. Amábile-Cuevas (Ed.), *Multiple drug resistant bacteria.* Horizon Scientific Press, Wymondham, UK, pp. 9–31.

Amábile-Cuevas, C.F., Cabrera, R., Fuchs, L.Y. and Valenzuela, F. (1998) "Antibiotic resistance and prescription practices in developing countries." *Methods in Microbiology* 27, 587–594.

Amábile-Cuevas, C.F. and Chicurel, M.E. (1992) "Bacterial plasmids and gene flux." *Cell* 70, 189–199.

Balsalobre, L., Ferrándiz, M.J., Liñares, J., Tubau, F. and de la Campa, A.G. (2003) "Viridans group streptococci are donors in horizontal transfer of topoisomerase IV genes to *Streptococcus pneumoniae.*" *Antimicrobial Agents and Chemotherapy* 47, 2072–2081.

Beaber, J.W., Hochhut, B. and Waldor, M.K. (2003) "SOS response promotes horizontal dissemination of antibiotic resistance genes." *Nature* 427, 72–74.

Becker, D., Selbach, M., Rollenhagen, C., Ballmaier, M., Meyer, T.F., Mann, M. and Bumann, D. (2006) "Robust *Salmonella* metabolism limits possibilities for new antimicrobials." *Nature* 440, 303–307.

Brown, E.W., LeClerc, J.E., Li, B., Payne, W.L. and Cebula, T.S. (2001) "Phylogenetic evidence for horizontal transfer of *mutS* alleles among naturally occurring *Escherichia coli* strains." *Journal of Bacteriology* 183, 1631–1644.

Buxton, I.L.O. (2006) "Principles of prescription order writing and patient compliance." In: L.L. Brunton, J.S. Lazo and K.L. Parker (Ed.), *The pharmacological basis of therapeutics.* McGraw-Hill, New York, pp. 1777–1786.

Calva, J.J., Niebla-Pérez, A., Rodríguez-Lemoine, V., Santos, J.I. and Amábile-Cuevas, C.F. (1996) "Antibiotic usage and antibiotic resistance in Latin America." In: C.F. Amábile-Cuevas (Ed.), *Antibiotic resistance: from molecular basics to therapeutic options*. R.G. Landes/ Chapman & Hall, Austin/New York, pp. 73–97.

Casellas, J.M. and Quinteros, M.G. (2007) "A Latin American "point de vue" on the epidemiology, control, and treatment options of infections caused by extended-spectrum beta-lactamase producers." In: C.F. Amábile-Cuevas (Ed.), *Antimicrobial resistance in bacteria*. Horizon Bioscience, Wymondham, pp. 99–122.

Courvalin, P. (1990) "Plasmid-mediated 4-quinolone resistance: a real or apparent absence?" *Antimicrobial Agents and Chemotherapy* **34**, 681–684.

Davin-Regli, A. and Pagès, J.-M. (2007) "Regulation of efflux pumps in Enterobacteriaceae: genetic and chemical effectors." In: C.F. Amábile-Cuevas (Ed.), *Antimicrobial resistance in bacteria*. Horizon Bioscience, Wymondham, UK, pp. 55–75.

Delissalde, F. and Amábile-Cuevas, C.F. (2004) "Comparison of antibiotic susceptibility and plasmid content, between biofilm producing and non-producing clinical isolates of *Pseudomonas aeruginosa*." *International Journal of Antimicrobial Agents* **24**, 405–408.

Demple, B. and Amábile-Cuevas, C.F. (2003) "Multiple resistance mediated by individual genetic loci." In: C.F. Amábile-Cuevas (Ed.), *Multiple drug resistant bacteria*. Horizon Scientific Press, Wymondham, UK, pp. 61–80.

Díaz-Mejía, J.J., Amábile-Cuevas, C.F., Rosas, I. and Souza, V. (2008) "An analysis of the evolutionary relationships of integron integrases, with emphasis on the prevalence of class 1 integron in *Escherichia coli* isolates from clinical and environmental origins." *Microbiology* **154**, 94–102.

Fuchs, L.Y., Reyna, F., Chihu, L. and Carrillo, B. (1996) "Molecular aspects of fluoroquinolone resistance." In: C.F. Amábile-Cuevas (Ed.), *Antibiotic resistance: from molecular basics to therapeutic options*. Landes/Chapman & Hall, Austin/New York, pp. 147–174.

Gilbert, P., McBain, A. and Lindsay, S. (2007) "Biofilms, multi-resistance, and persistence." In: C.F. Amábile-Cuevas (Ed.), *Antimicrobial resistance in bacteria*. Horizon Bioscience, Wymondham, UK, pp. 77–98.

Heinemann, J.A. and Goven, J. (2007) "The social context of drug discovery and safety testing." In: C.F. Amábile-Cuevas (Ed.), *Antimicrobial resistance in bacteria*. Horizon Bioscience, Wymondham, pp. 179–196.

Heinemann, J.A. and Silby, M.W. (2003) "Horizontal gene transfer and the selection of antibiotic resistance." In: C.F. Amábile-Cuevas (Ed.), *Multiple drug resistant bacteria*. Horizon, Wymondham, pp. 161–178.

Hooper, D.C., Wolfson, J.S., Souza, K.S., Ng, E.Y., McHugh, G.L. and Swartz, M.N. (1989) "Mechanisms of quinolone resistance in *Escherichia coli*: characterization of *nfxB* and *cfxB*, two mutant resistance loci decreasing norfloxacin accumulation." *Antimicrobial Agents and Chemotherapy* **33**, 283–290.

Jiménez-Arribas, G., Léautaud, V. and Amábile-Cuevas, C.F. (2001) "Regulatory locus soxRS partially protects Escherichia coli against ozone." *FEMS Microbiology Letters* **195**, 175–177.

Miller, C., Thomsen, L.E., Gaggero, C., Mosseri, R., Ingmer, H. and Cohen, S.N. (2004) "SOS response induction by β-lactams and bacterial defense against antibiotic lethality." *Science* **305**, 1629–1631.

Outterson, K., Samora, J.B. and Keller-Cuda, K. (2007) "Will longer antimicrobial patents improve global public health?" *Lancet Infectious Diseases* **7**, 559–566.

Robicsek, A., Strahilevitz, J., Jacoby, G.A., Macielag, M., Abbanat, D., Park, C.H., Bush, K. and Hooper, D.C. (2006) "Fluoroquinolone-modifying enzyme: a new adaptation of a common aminoglycoside acetyltransferase." *Nature Medicine* **12**, 83–88.

Tran, J.H. and Jacoby, G.A. (2002) "Mechanism of plasmid-mediated quinolone resistance." *Proceedings of the National Academy of Sciences USA* **99**, 5638–5642.

Chapter 2
Mechanisms of Antimicrobial Resistance

Denis K. Byarugaba

Abstract There is no doubt that antimicrobial agents have saved the human race from a lot of suffering due to infectious disease burden. Without antimicrobial agents, millions of people would have succumbed to infectious diseases. Man has survived the accidental wrath of microorganisms using antimicrobial agents and other mechanisms that keep them at bay. Hardly years after the discovery and use of the first antibiotics was observation made of organisms that still survived the effects of the antimicrobial agents. That was the beginning of the suspicion that different microorganisms were getting a way around previously harmful agents that is known today as antimicrobial resistance. Microbial resistance to antimicrobial agents was not a new phenomenon for it had been constantly used as competitive/survival mechanisms by microorganisms against others. These mechanisms have been well documented. This chapter therefore gives a brief overview of the mechanisms of resistance by bacteria against antimicrobial agents, and the mechanisms, levels, and patterns of resistance to the different microorganisms in developing countries are dealt with in detail elsewhere in the book. Understanding the mechanisms of resistance is important in order to define better ways to keep existing agents useful for a little longer but also to help in the design of better antimicrobial agents that are not affected by the currently known, predicted, or unknown mechanisms of resistance.

2.1 Introduction

Microorganisms have existed on the earth for more than 3.8 billion years and exhibit the greatest genetic and metabolic diversity. They are an essential component of the biosphere and serve an important role in the maintenance and sustainability of ecosystems. It is believed that they compose about 50% of

D.K. Byarugaba (✉)
Department of Veterinary Microbiology and Parasitology, Faculty of Veterinary Medicine, Makerere University, Kampala, Uganda
e-mail: dkb@vetmed.mak.ac.ug

A. de J. Sosa et al. (eds.), *Antimicrobial Resistance in Developing Countries*,
DOI 10.1007/978-0-387-89370-9_2, © Springer Science+Business Media, LLC 2010

the living biomass. In order to survive, they have evolved mechanisms that enable them to respond to selective pressure exerted by various environments and competitive challenges. The disease-causing microorganisms have particularly been vulnerable to man's selfishness for survival who has sought to deprive them of their habitat using antimicrobial agents. These microorganisms have responded by developing resistance mechanisms to fight off this offensive. Currently antimicrobial resistance among bacteria, viruses, parasites, and other disease-causing organisms is a serious threat to infectious disease management globally.

Antibiotics were discovered in the middle of the nineteenth century and brought down the threat of infectious diseases which had devastated the human race. However, soon after the discovery of penicillin in 1940, a number of treatment failures and occurrence of some bacteria such as staphylococci which were no longer sensitive to penicillin started being noticed. This marked the beginning of the error of antimicrobial resistance. Scientific antibiotic discovery started in the early 1900s by Alexander Fleming, who observed inhibition of growth on his agar plate on which he was growing *Staphylococcus* spp. It was later found that a microorganism that was later to be called *Penicillium notatum* was the cause of the inhibition of the *Staphylococcus* around it as a result of excreting some chemical into the media. That marked the beginning of the discovery of penicillin which together with several other different antimicrobial agents was later to save millions of humans and animals from infectious disease-causing organisms. The detailed history and documentation of man's search for agents to cure infectious disease has been described extensively elsewhere.

The observation of *Staphylococci* spp. that could still grow in the presence of penicillin was the beginning of the era of antimicrobial resistance and the realization that after all the drugs that were described as "magical bullets" were not to last for long due to the selective pressure that was being exerted by the use of these agents. However, the complacency between the 1940s and the 1970s that infectious microorganisms had been dealt a blow was later proved to be a misplaced belief that available antibiotics would always effectively treat all infections. Nevertheless, antimicrobial agents have improved the management of infectious diseases up to date.

Increasing prevalence of resistance has been reported in many pathogens over the years in different regions of the world including developing countries (Byarugaba, 2005). This has been attributed to changing microbial characteristics, selective pressures of antimicrobial use, and societal and technological changes that enhance the development and transmission of drug-resistant organisms. Although antimicrobial resistance is a natural biological phenomenon, it often enhanced as a consequence of infectious agents' adaptation to exposure to antimicrobials used in humans or agriculture and the widespread use of disinfectants at the farm and the household levels (Walsh, 2000). It is now accepted that antimicrobial use is the single most important factor responsible for increased antimicrobial resistance (Aarestrup et al., 2001; Byarugaba, 2004).

In general, the reasons for increasing resistance levels include the following:

- suboptimal use of antimicrobials for prophylaxis and treatment of infection,
- noncompliance with infection-control practices,
- prolonged hospitalization, increased number and duration of intensive-care-unit stays,
- multiple comorbidities in hospitalized patients,
- increased use of invasive devices and catheters,
- ineffective infection-control practices, transfer of colonized patients from hospital to hospital,
- grouping of colonized patients in long-term-care facilities,
- antibiotic use in agriculture and household chores, and
- increasing national and international travel.

The level of antibiotic resistance is dependent on the following:

- the population of organisms that spontaneously acquire resistance mechanisms as a result of selective pressure either from antibiotic use or otherwise,
- the rate of introduction from the community of those resistant organisms into health care settings, and
- the proportion that is spread from person to person.

All of these factors must be addressed in order to control the spread of antimicrobial-resistant organisms within health care settings. Community-acquired antimicrobial resistance is increasing in large part because of the widespread suboptimal use of antibiotics in the outpatient settings and the use of antibiotics in animal husbandry and agriculture.

2.2 Mechanisms of Action of Antimicrobial Agents

In order to appreciate the mechanisms of resistance, it is important to understand how antimicrobial agents act. Antimicrobial agents act selectively on vital microbial functions with minimal effects or without affecting host functions. Different antimicrobial agents act in different ways. The understanding of these mechanisms as well as the chemical nature of the antimicrobial agents is crucial in the understanding of the ways how resistance against them develops. Broadly, antimicrobial agents may be described as either bacteriostatic or bactericidal. Bacteriostatic antimicrobial agents only inhibit the growth or multiplication of the bacteria giving the immune system of the host time to clear them from the system. Complete elimination of the bacteria in this case therefore is dependent on the competence of the immune system. Bactericidal agents kill the bacteria and therefore with or without a competent immune system of the host, the bacteria will be dead. However, the mechanism of action

of antimicrobial agents can be categorized further based on the structure of the
bacteria or the function that is affected by the agents. These include generally
the following:

- Inhibition of the cell wall synthesis
- Inhibition of ribosome function
- Inhibition of nucleic acid synthesis
- Inhibition of folate metabolism
- Inhibition of cell membrane function

The chemical structure and details of these mechanisms have been described
in several literature elsewhere and a summary of the mode of action for the
major classes is provided in Table 2.1.

2.3 Mechanisms of Antimicrobial Resistance

Prior to the 1990s, the problem of antimicrobial resistance was never taken to be
such a threat to the management of infectious diseases. But gradually treatment
failures were increasingly being seen in health care settings against first-line
drugs and second-line drugs or more. Microorganisms were increasingly
becoming resistant to ensure their survival against the arsenal of antimicrobial
agents to which they were being bombarded. They achieved this through
different means but primarily based on the chemical structure of the antimicrobial agent and the mechanisms through which the agents acted. The resistance
mechanisms therefore depend on which specific pathways are inhibited by the
drugs and the alternative ways available for those pathways that the organisms
can modify to get a way around in order to survive.

Table 2.1 Summary of mechanisms of action of antimicrobial agents

Group of antimicrobial agents	Effect on bacteria	Mode of action in general
Penicillins	Bactericidal	Inhibition of cell wall synthesis
Cephalosporins	Bactericidal	Inhibition of cell wall synthesis
Carbanepems	Bactericidal	Inhibition of cell wall synthesis
Polypeptide antibiotics	Bactericidal	Inhibition of cell wall synthesis
Quinolones	Bactericidal	Inhibits DNA synthesis
Metronidazole	Bactericidal	Inhibits DNA synthesis
Rifamycins	Bactericidal	Inhibitions of RNA transcription
Lincosamides	Bactericidal	Inhibition of protein synthesis
Aminoglycosides	Bactericidal	Inhibition of protein synthesis
Macrolides	Bacteriostatic	Inhibition of protein synthesis
Tetracyclines	Bacteriostatic	Inhibition of protein synthesis
Chloramphenicol	Bacteriostatic	Inhibition of protein synthesis
Sulfonamides	Bacteriostatic	Competitive inhibition

Resistance can be described in two ways:

a) intrinsic or natural whereby microorganisms naturally do not posses target sites for the drugs and therefore the drug does not affect them or they naturally have low permeability to those agents because of the differences in the chemical nature of the drug and the microbial membrane structures especially for those that require entry into the microbial cell in order to effect their action or
b) acquired resistance whereby a naturally susceptible microorganism acquires ways of not being affected by the drug.

Acquired resistance mechanisms can occur through various ways as described by Fluit et al. (2001) summarized in Box 2.1 and illustrated in Fig. 2.1.

Box 2.1 Mechanisms for acquired resistance
- the presence of an enzyme that inactivates the antimicrobial agent
- the presence of an alternative enzyme for the enzyme that is inhibited by the antimicrobial agent
- a mutation in the antimicrobial agent's target, which reduces the binding of the antimicrobial agent
- post-transcriptional or post-translational modification of the antimicrobial agent's target, which reduces binding of the antimicrobial agent
- reduced uptake of the antimicrobial agent
- active efflux of the antimicrobial agent
- overproduction of the target of the antimicrobial agent
- expression or suppression of a gene in vivo in contrast to the situation in vitro
- previously unrecognized mechanisms

2.3.1 Resistance to β-Lactam Antibiotics

β-Lactam antibiotics are a group of antibiotics characterized by possession of a β-lactam ring and they include penicillins, cephalosporins, carbapenems, oxapenams, and cephamycins. The penicillins are one of the most commonly used antibiotics in developing countries because of their ready availability and relatively low cost. The β-lactam ring is important for the activity of these antibiotics which results in the inactivation of a set of transpeptidases that catalyze the final cross-linking reactions of peptidoglycan synthesis in bacteria. The effectiveness of these antibiotics relies on their ability to reach the penicillin-binding protein (PBP) intact and their ability to bind to the PBPs.

Resistance to β-lactams in many bacteria is usually due to the hydrolysis of the antibiotic by a β-lactamase or the modification of PBPs or cellular permeability. β-Lactamases constitute a heterogenous group of enzymes which are classified according to different ways including their hydrolytic spectrum,

Fig. 2.1 Illustration of how
some antimicrobial agents
are rendered ineffective
(Adopted from http://
www.chembio.uoguelph.ca)

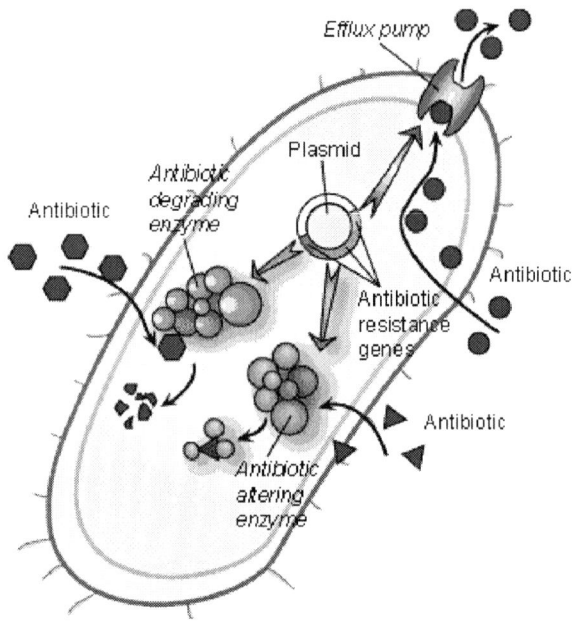

susceptibility to inhibitors, genetic localization (plasmidic or chromosomal), and gene or amino acid protein sequence. The functional classification scheme of β-lactamases proposed by Bush, Jacoby and Medeiros (1995) defines four groups according to their substrate and inhibitor profiles:

- Group 1 are cephalosporinases that are not well inhibited by clavulanic acid;
- Group 2 are penicillinases, cephalosporinases, and broad-spectrum β-lacta-mases that are generally inhibited by active site-directed β-lactamase inhibitors;
- Group 3 are metallo-β-lactamases that hydrolyze penicillins, cephalosporins, and carbapenems and that are poorly inhibited by almost all β-lactam-containing molecules;
- Group 4 are penicillinases that are not well inhibited by clavulanic acid.

2.3.2 Tetracycline Resistance

Tetracyclines are another of the very commonly used antimicrobial agents in both human and veterinary medicine in developing countries because of their availability and low cost as well as low toxicity and broad spectrum of activity. The tetracyclines were discovered in the 1940s. They inhibit protein synthesis by preventing the attachment of aminoacyl-tRNA to the ribosomal acceptor (A) site. They are broad-spectrum agents, exhibiting activity against a wide range of gram-positive and gram-negative bacteria, atypical organisms such as chlamydiae, mycoplasmas, and rickettsiae, and protozoan parasites. Examples of these include drugs such as tetracycline, doxycycline, minocycline, and oxtetracycline. Resistance to these agents occurs mainly through three mechanisms (Roberts, 1996), namely

- Efflux of the antibiotics,
- Ribosome protection, and
- Modification of the antibiotic.

These tetracycline resistance determinants are widespread in different microorganisms (Levy, 1988).

Efflux of the drug occurs through an export protein from the major facilitator superfamily (MFS). These export proteins are membrane-associated proteins which are coded for by *tet* efflux genes and export tetracycline from the cell. Export of tetracycline reduces the intracellular drug concentration and thus protects the ribosomes within the cell. Tetracycline efflux proteins have amino acid and protein structure similarities with other efflux proteins involved in multiple-drug resistance, quaternary ammonium resistance, and chloramphenicol and quinolone resistance. The gram-negative efflux genes are widely distributed and normally associated with large plasmids, most of which are conjugative.

Ribosome protection occurs through ribosome protection proteins that protect the ribosomes from the action of tetracyclines (Taylor and Chau, 1996). Ribosome protection proteins are cytoplasmic proteins that bind to the ribosome and cause an alteration in ribosomal conformation which prevents tetracycline from binding to the ribosome, without altering or stopping protein synthesis. They confer resistance mainly to doxycycline and minocycline and confer a wider spectrum of resistance to tetracyclines than is seen with bacteria that carry tetracycline efflux proteins.

Modification of the antibiotic on the other hand occurs through enzymatic alteration of the drugs. Some of these genes are coded for by tet(*X*) genes.

2.3.3 Chloramphenicol Resistance

Chloramphenicol binds to the 50S ribosomal subunit and inhibits the peptidyl transferase step in protein synthesis. Resistance to chloramphenicol is generally due to inactivation of the antibiotic by a chloramphenicol acetyltransferase (Traced et al., 1993). Various enzymes have been described and are coded for by the *cat* genes found in gram-negative and gram-positive bacteria and usually show little homology (Kehrenberg et al., 2001). Sometimes decreased outer membrane permeability or active efflux is responsible for the resistance in gram-negative bacteria (Butaye et al., 2003).

2.3.4 Aminoglycoside Resistance

Aminoglycosides include a group of drugs which are characterized by the presence of an aminocyclitol ring linked to amino sugars in their structure and have a broad spectrum of activity against bacteria. Examples of these drugs include streptomycin, kanamycin, gentamycin, tobramycin, and amikacin, which are commonly used in the treatment of infections by both gram-negative and gram-positive organisms. Their bactericidal activity is attributed to the irreversible binding to the ribosomes but effects resulting from interaction with other cellular structures and metabolic processes are also known.

Resistance to aminoglycosides such as gentamicin, tobramycin, amikacin, and streptomycin is widespread, with more than 50 aminoglycoside-modifying enzymes described (Schmitz and Fluit, 1999). Most of these genes are associated with gram-negative bacteria. Depending on their type of modification, these enzymes are classified as aminoglycoside acetyltransferases (AAC), aminoglycoside adenyltransferases (also named aminoglycoside nucleotidyltransferases [ANT]), and aminoglycoside phosphotransferases (APH) (Shaw et al., 1993). Aminoglycosides modified at amino groups by AAC enzymes or at hydroxyl groups by ANT or APH enzymes lose their ribosome-binding ability and thus no longer inhibit protein synthesis. Besides aminoglycoside-modifying enzymes, efflux systems and rRNA mutations have been described (Quintiliani and Courvalin, 1995).

2.3.5 Quinolone Resistance

The first quinolone with antibacterial activity (nalidixic acid) was discovered in 1962 during the process of synthesis and purification of chloroquine (an antimalarial agent). Since then several derivatives have been made available on the market, with the most important ones being fluoroquinolones which contain a substitution of a fluorine atom at position 6 of the quinolone molecule. This greatly enhanced their activity against gram-positive and gram-negative bacteria as well as anaerobes. These agents exert their antibacterial effects by inhibition of certain bacterial topoisomerase enzymes, namely DNA gyrase (bacterial topoisomerase II) and topoisomerase IV.

These essential bacterial enzymes alter the topology of double-stranded DNA (dsDNA) within the cell. DNA gyrase and topoisomerase IV are heterotetrameric proteins composed of two subunits, designated A and B.

Mechanisms of bacterial resistance to quinolones as described by Hooper (1999) fall into two principal categories:

- alterations in drug target enzymes and
- alterations that limit the permeability of the drug to the target.

The target enzymes are most commonly altered in domains near the enzyme-active sites, and in some cases reduced drug-binding affinity. In gram-negative organisms, DNA gyrase seems to be the primary target for all quinolones. In gram-positive organisms, topoisomerase IV or DNA gyrase is the primary target depending on the fluoroquinolones considered. In almost all instances, amino acid substitutions within the quinolone resistance-determining region (QRDR) involve the replacement of a hydroxyl group with a bulky hydrophobic residue. Mutations in *gyrA* induce changes in the binding-site conformation and/or charge that may be important for quinolone–DNA gyrase interaction (Everett and Piddock, 1998). Changes in the cell envelope of gram-negative bacteria, particularly in the outer membrane, have been associated with decreased uptake and increased resistance to fluoroquinolones, and this has not been demonstrated in gram-positive bacteria.

2.3.6 Macrolide, Lincosamide, and Streptogramin (MLS) Resistance

MLS antibiotics are chemically distinct inhibitors of bacterial protein synthesis. Intrinsic resistance to MLS_B (including streptogramin B) antibiotics in gram-negative bacilli is due to low permeability of the outer membrane to these hydrophobic compounds. Three different mechanisms of acquired MLS resistance have been found in gram-positive bacteria (Johnston et al., 1998). These include the following:

- Post-transcriptional modifications of the 23S rRNA by the adenine-N^6-methyltransferase which alters a site in 23S rRNA common to the binding of MLS_B antibiotics which also confers cross-resistance to MLS_B antibiotics (MLS_B-resistant phenotype) and remains the most frequent mechanism of resistance. In general, genes encoding these methylases have been designated *erm* (erythromycin ribosome methylation).
- Efflux proteins, which pump these antibiotics out of the cell or the cellular membrane, keeping intracellular concentrations low and ribosomes free from antibiotic, and these have become more frequent in gram-positive populations and often coded for by *mef, msr,* and *vga* genes.
- Hydrolytic enzymes which hydrolyze streptogramin B or modify the antibiotic by adding an acetyl group (acetyltransferases) to streptogramin A have also been described and these confer resistance to structurally related drugs.

2.3.7 Glycopeptide Resistance

Glycopeptides comprise peptide antibiotics of clinical interest such as vancomycin and teicoplanin. Their antimicrobial activity is due to binding to D-alanyl-D-alanine side chains of peptidoglycan or its precursors, thereby preventing cross-linking of the peptidoglycan chain and thus are largely effective against gram-positive microorganisms which poses a bigger layer of the peptidoglycan although not all gram-positive organisms are susceptible to these agents. High-level resistance to vancomycin is encoded by the *vanA* gene that results in the production of VanA, a novel D-Ala-D-Ala ligase resulting in the rebuilding of the peptidoglycan side chain to express D-alanyl-D-lactate type which has less affinity for glycopeptides (Leclerq and Courvalin, 1997). There are also other proteins in this gene cluster that are necessary for resistance including VanH and VanX, as well as VanB which confers moderate levels of resistance to vancomycin and susceptibility to teicoplanin. Vancomycin gained clinical importance because it was traditionally reserved as a last resort treatment for resistant infections especially of methicillin-resistant *Staphylococcus aureus* (MRSA). The emergency of vancomycin-resistant organisms has deprived the usefulness of this drug.

2.3.8 Sulfonamides and Trimethoprim Resistance

Resistance in sulfonamides is commonly mediated by alternative, drug-resistant forms of dihydropteroate synthase (DHPS). Sulfonamide resistance in gram-negative bacilli generally arises from the acquisition of either of the two genes *sul1* and *sul2*, encoding forms of dihydropteroate synthase that are not inhibited by the drug (Enne et al., 2001). The *sul1* gene is normally found linked to other resistance genes in class 1 integrons, while *sul2* is usually located on small nonconjugative plasmids or large transmissible multi-resistance plasmids.

Trimethoprim is an analog of dihydrofolic acid, an essential component in the synthesis of amino acid and nucleotides that competitively inhibits the enzyme dihydrofolate reductase (DHFR). Trimethoprim resistance is caused by a number of mechanisms (Thomson, 1993) including

- overproduction of the host DHFR,
- mutations in the structural gene for DHFR, and
- acquisition of a gene (*dfr*) encoding a resistant DHFR enzyme which is the most resistant mechanism in clinical isolates.

At least 15 DHFR enzyme types are known based on their properties and sequence homology (Schmitz and Fluit, 1999).

2.3.9 Multidrug Resistance

Multidrug resistance among many organisms has become a big challenge to infectious disease management. It is increasingly being reported in bacteria and is often mediated by genetic mobile elements such as plasmids, transposons, and integrons (Dessen et al., 2001). Integrons are mobile DNA elements with the ability to capture genes, notably those encoding antibiotic resistance, by site-specific recombination, and they have an intergrase gene (*int*), a nearby recombination site (*attI*), and a promoter, *Pant* (Hall, 1997). Integrons seem to have a major role in the spread of multidrug resistance in gram-negative bacteria but integrons in gram-positive bacteria have also been described (Dessen et al., 2001). Class 1 integrons are often associated with the sulfonamide resistance gene *sulI* and are the most common integrons. Class 2 integrons are associated with Tn7. The majority of genes encode antibiotic disinfectant resistance, including resistance to aminoglycosides, penicillins, cephalosporins, trimethoprim, tetracycline, erythromycin, and chloramphenicol.

References

Aarestrup, F. M., Seyfarth, A. M., Emborg, H. D., Pedersen, K., Hendriksen, R. S., and Bager, F. 2001. Effect of abolishment of the use of antimicrobial agents for growth promotion on occurrence of antimicrobial resistance in fecal enterococci from food animals in Denmark. Antimicrob. Agents Chemother. 45: 2054–2059.

Bush, K., Jacoby, G. A., and Medeiros A. A. 1995. A functional classification scheme for beta lactamases and its correlation with molecular structure. Antimicrob. Agents Chemother 39(6): 1211–1233.

Butaye, P., Cloeckaert, A., and Schwarz, S. 2003. Mobile genes coding for efflux-mediated antimicrobial resistance in Gram-positive and Gram-negative bacteria. Int. J. Antimicrob. Agents 22: 205–210.

Byarugaba, D. K. 2004. A view on antmicrobial resistance in developing countries and responsible risk factors. Int. J. Antimicrob. Agents 24: 105–110.

Byarugaba, D. K. (2005). Antimicrobial resistance and its containment in developing countries. In Antibiotic Policies: Theory and Practice, ed. I. Gould and V. Meer, pp 617–646. New York: Springer.

Dessen, A., Di Guilmi, A. M., Vernet, T., and Dideberg, O. 2001. Molecular mechanisms of antibiotic resistance in gram-positive pathogens. Curr. Drug Targets Infect. Dis.. 1:63–77

Enne, V. I., Livermore, D. M., Stephens, P., and Hall, L. M. C. 2001. Persistence of sulfonamide resistance in *Escherichia coli* in the UK despite national prescribing restriction. Lancet 357:1325–1328

Everett, M. J. and L. J. V. Piddock. 1998. Mechanisms of resistance to fluoroquinolones. In Quinolone Antibacterials ed. J. Kuhlmann, A. Dahlhoff, and H. J. Zeiler pp. 259–297. Berlin:Springer-Verlag KG.

Fluit, A. C., Visser, M. R., and Schmitz, F. J. 2001. Molecular detection of antimicrobial resistance. Clin. Microbiol. Rev. 14:836–71.

Hall, R. M. 1997. Mobile gene cassettes and integrons: moving antibiotic resistance genes in Gram-negative bacteria. Ciba Found. Symp. 207: 192–205

Hooper, D. C. 1999. Mechanisms of fluoroquinolone resistance. Drug Resist. Updates 2:38–55

Johnston, N. J., de Azavedo, J. C., Kellner, J. D., and Low, D. E. 1998. Prevalence and characterization of the mechanisms of macrolide, lincosamide and streptogramin resistance in isolates of *Streptococcus pneumoniae*. Antimicrob. Agents Chemother. 42:2425–2426

Kehrenberg, C., Schulze-Tanzil, G., Martel, J. L., Chaslus-Dancla, E., and Schwarz, S. 2001. Antimicrobial resistance in *Pasteurella* and *Mannheimia*: epidemiology and genetic basis. Vet. Res. 32(3–4): 323–339.

Leclerq, R. and Courvalin, P.. 1997. Resistance to glycopeptides in enterococci. Clin. Infect. Dis. 24: 545–556

Levy, S. B. 1988. Tetracycline resistance determinants are widespread. ASM News. 54:418–421.

Quintiliani, R. and Courvalin, P. 1995. Mechanisms of resistance to antimicrobial agents, In Manual of Clinical Microbiology, ed P. R. Murray, E. J. Baron, M. A. Pfaller, F. R. Tenover, and R. H. Yolken, pp. 1308–1326. Washington, D.C: ASM Press.

Roberts, M. C. 1996. Tetracycline resistance determinants: mechanisms of action, regulation of expression, genetic mobility, and distribution. FEMS Microbiol. Rev. 19:1–24

Schmitz, F. J. and Fluit. A. C. 1999. Mechanisms of resistance. In Infectious Diseases. ed D. Armstrong, and S. Cohen, pp. 7.2.1–7.2.14 London: Mosby, Ltd.,

Shaw, K. J., Rather, P. N., Have, R. S., and Miller, G. M. 1993. Molecular genetics of aminoglycoside resistance genes and familial relationships of the aminoglycoside modifying enzymes. Microbiol. Rev. 57: 138–163

Taylor, D. E. and Chau, A. 1996. Tetracycline resistance mediated by ribosomal protection. Antimicrob. Agents Chemother. 40: 1–5

Thomson, C. J. 1993. Trimethoprim and brodimoprim resistance of gram-positive and gram-negative bacteria. J. Chemother. 5: 458–464

Traced, P., de Cespédès, G., Bentorcha, F., Delbos, F., Gaspar, E., and Horaud, T. 1993. Study of heterogeneity of chloramphenicol acetyltransferase (CAT) genes in streptococci and enterococci by polymerase chain reaction: characterization of a new CAT determinant. Antimicrob. Agents Chemother. 37: 2593–2598

Walsh, C. 2000. Molecular mechanisms that confer antibacterial drug resistance. Nature 406: 775–781

Chapter 3
Poverty and Root Causes of Resistance in Developing Countries

Iruka N. Okeke

Abstract Antimicrobial use provides selective pressure for resistant strains but there are other factors that combine with use to promote the emergence and spread of resistant bacteria, mobile elements, and genes. Many of these factors are only superficially understood, if at all, and are complicated to study. Comparative antimicrobial resistance data from populations with different risk factors for resistance are hard to come by and inevitably suffer from bias. Nonetheless, a few studies document regional variations in resistance and some provide clues about factors that might exacerbate resistance. Data from these studies appear to suggest that antimicrobial misuse, prophylactic use, diagnostic imprecision, and interpersonal spread are key factors in the selection and dissemination of resistant strains. All these factors are promoted by poverty at the individual patient, health system, and national levels.

A 1990 report examined resistance among urinary *Escherichia coli* isolated from Tikur Anbessa Hospital, Addis Ababa, Ethiopia, with resistance in isolates from Karolinska Hospital, Stockholm in Sweden (Ringertz et al. 1990). There was a four- to eightfold higher rate of resistance to ampicillin, cephalotin, tetracycline, chloramphenicol, sulphonamides, and trimethoprim in the Addis Ababa isolates as compared to Stockholm strains isolated during the study. Moreover, resistance rates were higher at the Tikur Anbessa Hospital for most agents in the period between 1986 and 1987, as compared to 1976–1981. Interestingly, even though resistance rates were significantly greater in the Ethiopian hospital, total antimicrobial use at both institutions was similar. These and other data support the idea that antimicrobial pressure combines with other factors to exacerbate resistance and/or it is the type rather than the volume of use that contributes to the resistance problem. Ringertz et al. (1990) made two observations, which they hypothesized contributed to the differences in resistance at the two institutions: antimicrobials were more likely to be used

I.N. Okeke (✉)

Department of Biology, Haverford College, Haverford, PA, USA

e-mail: iokeke@haverford.edu

A. de J. Sosa et al. (eds.), *Antimicrobial Resistance in Developing Countries*,
DOI 10.1007/978-0-387-89370-9_3, © Springer Science+Business Media, LLC 2010

27

prophylactically at the Ethiopian institution and they were more likely to be selected following antimicrobial susceptibility testing at the Swedish hospital.

Outside hospitals, a number of studies have compared the carriage of resistant commensals in individuals residing in high-, medium-, and low-income countries (Lester et al. 1990; Nys et al. 2004). Studies comparing resistance among commensal *E. coli* from urban and provincial residents in developing countries have found that the urban residents were more likely to harbor resistant strains (Lamikanra and Okeke 1997; Nys et al. 2004). Similarly, carriage of resistant *Streptococcus pneumoniae* was significantly more common in urban than in rural children in Asia, the Middle East, and Lesotho (Lee et al. 2001; Mthwalo et al. 1998; Quagliarello et al. 2003). These and other studies point to within-country factors that exacerbate resistance, including access to medicines as well as opportunities for person-to-person transfer of resistant clones. In a Mexican study, a risk factor for carriage of quinolone-resistant *E. coli* was hospitalization of a family member. In the same study, carriage of quinolone-resistant *E. coli* was associated with *Salmonella* carriage (Zaidi et al. 2003). The data appear to suggest that resistant organisms are spread from person to person much the same way as pathogens.

Health disparities between rich and poor countries are very striking (Murray and Lopez, 1997). Within-country disparities can be even more polarized so that separation of public health problems along geographical and regional lines alone is overly simplistic. It is not entirely clear how these disparities play out in the epidemiology of antimicrobial resistance. Resistance is an externality so that the resistance-promoting activities of some may have consequences for others. People living in poverty are least able to buffer themselves from the consequences of resistance, even though they may contribute less to the problem in terms of selective pressure.

The burden of resistant infection is disproportionately borne by the less privileged. Poor people are typically at a greater risk of becoming infected but they are also less likely to be able to access health care, including antimicrobials. Evidence is accumulating to suggest that the emergence and spread of resistance is strongly influenced by socioeconomic factors at the individual and national levels. Data are limited and, while poverty appears to be associated with resistance in compelling studies, it must be emphasized that there is no evidence to suggest that poverty *causes* resistance. Rather, poverty influences downstream factors that could alter selective pressure for resistance and promote the dissemination of resistant strains. It is also important to acknowledge that the income variable overlaps to a considerable degree with other factors (such as climate or location) that may not be associated with poverty or even underdevelopment but may also impact resistance.

3.1 Susceptibility to Infection

Poor countries, and the poorest people within them, bear the predominant infectious disease burden (Murray and Lopez, 1997). Diseases are more efficiently transmitted to people whose immune systems have been made less robust by malnutrition and chronic or repeated infection and who are less likely to have access to uncontaminated food and water. These hypersusceptible individuals become reservoirs for onward transmission to others in the population. In South Africa, an outbreak of extensively drug-resistant tuberculosis (XDR-TB) began in the rural town of Tugela Ferry and took hold by infecting patients immunocompromised by HIV (Singh et al. 2007; XDR-TB strains are resistant to at least five anti-tuberculosis drugs, including isoniazid and rifampicin). HIV-positive patients succumbed so rapidly to infection that it very well might have burned out if it did not also spread to HIV-negative individuals, in particular health workers.

Some pathogens produce different syndromes in low- and high-income patient populations. In industrialized countries, multidrug-resistant non-typhoidal *Salmonella* are typically associated with foodborne gastroenteritis. Among children residing in the poor slums of Kenya and some other parts of Africa, however, non-typhoidal *Salmonella* are predominant causes of life-threatening bacteremia in infants and young children (Berkley et al. 2005; Kariuki et al. 2006). Gastroenteritis is typically self-limiting; bacteremia by contrast requires prompt and effective antimicrobial therapy. Moreover, expensive drugs such as cefuroxime, to which Kenyan non-typhoidal *Salmonella* are susceptible, are beyond the reach of many children for whom it should be prescribed empirically. Thus, poor people are not only more likely to succumb to resistant infection but also less likely to be able to address it.

In less affluent countries, people who live in urban centers have, on average, better access to amenities, including health care, and higher incomes than rural residents. Averages are, however, misleading. Health indicators as well as access to safe water and sanitation among urban residents in the lowest income quartile are as bad as, or worse than, poor people in rural areas (Songsore 2004). The urban poor often reside in cramped, overcrowded accommodations, which facilitate the dissemination of airborne disease. A rising number of newly emerged poor are the victims of natural disaster and civil unrest. They reside in camps for displaced persons, which are on occasion the site for disease outbreaks. The most common causes of mortality in complex emergencies have been diarrheal diseases, malaria, acute respiratory infection, and measles (Toole and Waldman, 1997). In recent times, outbreaks of cholera and bacillary dysentery in refugee camps have featured resistant pathogens (Engels et al. 1995; Goma Epidemiology Group, 1995; Siddique et al. 1995). Inadequate security in urban slums and refugee camps also places women and children at risk of acquiring sexually transmitted diseases through non-consensual sex.

Poverty is also a prime instigation to involvement in prostitution. In all these cases, when the transmitted pathogen is resistant, its spread among the poor is enhanced.

3.2 Access to Appropriate Care and Antimicrobials

Health systems in poor countries often cannot offer the most optimal therapies or the medical technology required to appropriately deliver such medicines. The lack of routine drug sensitivity testing and surveillance in many poor countries means that antibiotics and other antimicrobial drugs are not rationally used (Okeke 2006; Petti et al. 2006). Consequent overuse of cheap, broad-spectrum drugs has fostered widespread resistance. When therapeutic failure occurs, it can arise from misdiagnosis, drug resistance, or poor drug quality. At the very least, diagnostic laboratories would make it possible to resolve this type of confounding (Okeke 2006).

Poor people often cannot afford to consult properly qualified diagnosticians and prescribers. Even when consultation fees are waived, low-income patients have higher patient:health worker ratios so that they must wait longer for medical attention and loose more in forfeited income. Poor patients also typically pay a proportionately higher fraction of their income to reach a health center and procure care. Patients from low-income economic brackets are less likely to have the resources to fill an antimicrobial prescription and more likely to acquire and consume partial doses. In countries where prescription drugs are ineffectively regulated, such as many developing nations, purchasing partial regimen is a practical option. The patient who chooses to do so typically pays more per unit dose than they might have for a complete regimen but is under no obligation to purchase more than a few tablets or capsules.

Antimalarials are recommended for home management of malaria in hyper-endemic areas (Schapira 1994) and community health workers assist in ensuring that patients receive appropriate regimen in remote areas. However, many patients cannot access correct information or choose not to follow it. When home management is implemented, poor patients choose the cheapest, rather than the most appropriate antimicrobial (Onwujekwe et al. 2007). They are therefore more vulnerable to extended morbidity that could arise from resistance. Misuse of antimalarials in the health sector as well as inappropriate self-medication has contributed to the emergence and spread of resistance to the effective, affordable drugs chloroquine and sulfadoxine-pyrimethamine in Africa and Asia (D'Alessandro and Buttiens, 2001). As detailed by Okeke and Ojo in this volume, antibacterials are subjected to even more abuse, being available from unsanctioned distributors who are often not trained as to their storage, dispensing, and use.

Patients who cannot afford the time or expense to consult health practitioners or even dispensers will self-medicate and are assisted by the unregulated availability of antimicrobials from unofficial outlets. Up to a third of patients in developing countries choose to self-medicate even though most of them are likely to be suffering from a disease that requires medical attention. Self-medication often includes antimicrobials and may even include parenteral drugs (Bi et al. 2000). In Jimma, southwest Ethiopia, 15 of 42 self-medicating patients reported that they chose this mode of care because of its low cost (Worku and Abebe 2003). Fourteen others gave the need to avoid long clinic waiting times as their reason for self-medication. Self-medication is, however, only in part spurred by poverty. Low-income patients with poor access to medicines cannot self-medicate, as was seen in rural Ethiopia, where self-medication rates were 8.6% (Tsegaye 1998). Conversely, a combination of familiarity with medicines in underprivileged cultures, access to unregulated drugs, and mistrust of health systems means that more affluent patients may also self-medicate (Obadare 2005; Okeke 2003; Okeke and Lamikanra 2003; Whyte et al. 2002).

3.3 The Challenge of Antimicrobial Quality Assurance in Low-Income Countries

The poor obtain their medicines through less secure supply chains and are therefore more likely to receive sub-standard drugs. High-quality dispensed medicines are more likely to be improperly stored en route to and in the hands of the poor. Thus among low-income patients, the potential for insufficient use or for intentional or unintentional misuse of antimicrobials, predisposing to resistance, rises.

Low-income countries are least able to regulate drug distribution and drug quality. The responsibility of ensuring that drug products dispensed to patients will provide the necessary activity and will not harm lies with the health system and the government. Quality assurance begins with licensing and monitoring manufacturers and importers. The supply chain must also be certified and secured and dispensers must be properly trained to deliver drugs in a way that best serves the patient. In less affluent countries, it is difficult to avoid compromise at any or all of these levels. This is particularly true for antimicrobials which are in high demand in countries with high infectious disease burdens and which are often heat and moisture labile. Rigorous standards and strict laws govern the quality and distribution of medicines. Standards are expensive to develop and maintain; therefore, poor countries adopt pharmacopeias from abroad. However, implementing quality assurance standards written in these books may be difficult or impossible in under-equipped laboratories and enforcing laws has been a challenge in many parts of Africa.

As detailed in Chapter 24, counterfeits may influence resistance by providing more conducive selective pressure or by producing the perception of resistance due to therapeutic failure. Most clinical antimicrobials are administered to humans in a country that is less wealthy than the country that produced the drug. While many low-income countries do have some local capacity for dosage form production, that is to convert active ingredients into tablets, capsules, suspension or injectables, almost all must import a large proportion of finished drugs. Local dosage form producers also import active ingredient and most of the formulation ingredients or excipients they employ. Thus efficacious regulation of the manufacturing industry contributes little to the overall drug quality status. The quality of imported products contributes more but is difficult to regulate because the channels through which these products enter and move through low-income drug markets are often obscure and highly variable. Ballereau et al. (1997) observed that "chance methods" are the norm for drug procurement in Burkina Faso. This means that drug distributors constantly modify their sources according to availability, and also accept unpredictable, and even unjustified, drug donations. "Chance" procurement is unavoidable wherever there is chronic drug scarcity and economic means limited. Infiltration of this tenuous supply-web is easy, indeed almost certain; thus low-income countries are vulnerable to the dissemination of counterfeits (Kindermans et al. 2007). Undocumented drug supply systems also make it impossible to quantify drug consumption, which is an important prerequisite for many interventions to contain resistance.

3.4 Resistance Containment

The evidence is insufficient but points to higher rates of resistance in developing countries. Moreover the rate at which resistance is rising appears to be faster in developing countries (Okeke et al. 2000; Zhang et al. 2006). Therefore although resistance containment interventions have mostly been implemented in northern Europe and North America, there is a more pressing need to intervene in the resistance pandemic in developing countries.

It has recently been suggested that integrated interventions be applied to address the infectious disease burden of the poor (Ehrenberg and Ault 2005). In view of the inseparability of the problem of resistance from mitigating factors like poverty, which also impact the overall infectious disease burden, it makes sense to pursue resistance containment as part of an overall health package for the poor. Many interventions including national disease control and management programs and vaccination, which improve health care in general, also assist in containing resistance (Okeke et al. 2005). One reason why the poor bear more of the burden from resistance is because interventions designed to contain resistance, or to improve health, are not necessarily accessible to the people who need them most. For example, the Integrated Management of Childhood Illnesses (IMCI) program has been implemented in many poor countries, but poor people in these countries are least likely to benefit from it because it is less successfully implemented in remote areas (Victora et al. 2006). Paradoxically,

IMCI is most needed in areas that are distant from secondary and tertiary health care institutions.

The conclusions of Ringertz et al. (1990), highlighted at the beginning of this chapter, were that prophylactic and irrational antimicrobial use were exacerbating resistance in an Ethiopian hospital. The alternatives, stellar infection control and routine susceptibility testing, are by no means impossible, if more difficult, to instigate in poor countries. By contrast, the failure to contain resistance results in greater losses from infections and increased costs in overcoming them. Therefore, while many of the root causes of resistance arise from poverty and poverty is a roadblock to combating resistance, all, and especially the poor, will benefit from resistance containment.

Acknowledgment INO is a Branco Weiss Fellow of the Society-in-Science, ETHZ, Zürich, Switzerland.

References

Ballereau, F., Prazuck, T., Schrive, I., Lafleuriel, M., Rozec, D., Fisch, A., and Lafaix, C. 1997. Stability of essential drugs in the field: results of a study conducted over a two-year period in Burkina Faso. Am. J. Trop. Med. Hyg. 57:31–36.

Berkley, J. A., Lowe, B. S., Mwangi, I., Williams, T., Bauni, E., Mwarumba, S., Ngetsa, C., Slack, M. P., Njenga, S., Hart, C. A., Maitland, K., English, M., Marsh, K., and Scott, J. A. 2005. Bacteremia among children admitted to a rural hospital in Kenya. N. Engl. J. Med. 352:39–47.

Bi, P., Tong, S., and Parton, K. A. 2000. Family self-medication and antibiotics abuse for children and juveniles in a Chinese city. Soc. Sci. Med. 50:1445–1450.

D'Alessandro, U. and Buttiens, H. 2001. History and importance of antimalarial drug resistance. Trop. Med. Int. Health 6:845–848.

Ehrenberg, J. P. and Ault, S. K. 2005. Neglected diseases of neglected populations: thinking to reshape the determinants of health in Latin America and the Caribbean. BMC Public Health 5:119.

Engels, D., Madaras, T., Nyandwi, S., and Murray, J. 1995. Epidemic dysentery caused by *Shigella dysenteriae* type 1: a sentinel site surveillance of antimicrobial resistance patterns in Burundi. Bull. World Health Organ. 73:787–791.

Goma Epidemiology Group. 1995. Public health impact of Rwandan refugee crisis: what happened in Goma, Zaire, in July, 1994. Lancet 345:339–344.

Kariuki, S., Revathi, G., Kariuki, N., Kiiru, J., Mwituria, J., and Hart, C. A. 2006. Characterisation of community acquired non-typhoidal *Salmonella* from bacteraemia and diarrhoeal infections in children admitted to hospital in Nairobi, Kenya. BMC Microbiol. 6:101.

Kindermans, J. M., Vandenbergh, D., Vreeke, E., Olliaro, P., and D'Altilia, J. P. 2007. Estimating antimalarial drugs consumption in Africa before the switch to artemisinin-based combination therapies (ACTs). Malar. J. 6:91.

Lamikanra, A. and Okeke, I. N. 1997. A study of the effect of the urban/rural divide on the incidence of antibiotic resistance in *Escherichia coli*. Biomed. Lett. 55:91–97.

Lee, N. Y., Song, J. H., Kim, S., Peck, K. R., Ahn, K. M., Lee, S. I., Yang, Y., Li, J., Chongthaleong, A., Tiengrim, S., Aswapokee, N., Lin, T. Y., Wu, J. L., Chiu, C. H., Lalitha, M. K., Thomas, K., Cherian, T., Perera, J., Yee, T. T., Jamal, F., Warsa, U. C., Van, P. H., Carlos, C.C., Shibl, A. M., Jacobs, M. R., and Appelbaum, P. C. 2001.

Carriage of antibiotic-resistant pneumococci among Asian children: a multinational surveillance by the Asian Network for Surveillance of Resistant Pathogens (ANSORP). Clin. Infect. Dis. 32:1463–1469.

Lester, S. C., del Pilar Pla, M., Wang, F., Perez Schael, I., Jiang, H., and O'Brien, T. F. 1990. The carriage of *Escherichia coli* resistant to antimicrobial agents by healthy children in Boston, in Caracas, Venezuela, and in Qin Pu, China. N. Engl. J. Med. 323:285–289.

Mthwalo, M., Wasas, A., Huebner, R., Koornhof, H. J., and Klugman, K. P. 1998. Antibiotic resistance of nasopharyngeal isolates of *Streptococcus pneumoniae* from children in Lesotho. Bull. World Health Organ. 76:641–650.

Murray, C. J. and Lopez, A. D. 1997. Mortality by cause for eight regions of the world: Global Burden of Disease Study. Lancet 349:1269–1276.

Nys, S., Okeke, I. N., Kariuki, S., Dinant, G. J., Driessen, C., and Stobberingh, E. E. 2004. Antibiotic resistance of faecal *Escherichia coli* from healthy volunteers from eight developing countries. J. Antimicrob. Chemother. 54:952–955.

Obadare, E. 2005. A crisis of trust: history, politics, religion and the polio controversy in Northern Nigeria. Patterns Prejudice 39:265–266.

Okeke, I. 2003. Antibiotic use and resistance in developing countries. In the resistance phenomenon in microbes and infectious disease vectors: implications for human health and strategies for containment – Workshop Summary, ed. S. Knobler, S. Lemon, M. Najafi, and T. Burroughs, pp. 132–139. Washington DC: Institute of Medicine, National Academy of Science.

Okeke, I. N., Fayinka, S. T., and Lamikanra, A. 2000. Antibiotic resistance in *Escherichia coli* from Nigerian students, 1986 1998. Emerg. Infect. Dis. 6:393–396.

Okeke, I. N., and Lamikanra, A. 2003. Export of antimicrobial drugs by West African travelers. J. Travel. Med. 10:133–135.

Okeke, I. N., Klugman, K. P., Bhutta, Z. A., Duse, A. G., Jenkins, P., O'Brien, T. F., Pablos-Mendez, A., and Laxminarayan, R. 2005. Antimicrobial resistance in developing countries. Part II: strategies for containment. Lancet Infect. Dis. 5:568–580.

Okeke, I. N. 2006. Diagnostic insufficiency in Africa. Clin. Infect. Dis. 42:1501–1503.

Onwujekwe, O., Ojukwu, J., Shu, E., and Uzochukwu, B. 2007. Inequities in valuation of benefits, choice of drugs, and mode of payment for malaria treatment services provided by Community Health Workers in Nigeria. Am. J. Trop. Med. Hyg. 77:16–21.

Petti, C. A., Polage, C. R., Quinn, T. C., Ronald, A. R., and Sande, M. A. 2006. Laboratory medicine in Africa: a barrier to effective health care. Clin. Infect. Dis. 42:377–382.

Quagliarello, A. B., Parry, C. M., Hien, T. T., and Farrar, J. J. 2003. Factors associated with carriage of penicillin-resistant *Streptococcus pneumoniae* among Vietnamese children: a rural-urban divide. J. Health Popul. Nutr. 21:316–324.

Ringertz, S., Bellete, B., Karlsson, I., Ohman, G., Gedebou, M., and Kronvall, G. 1990. Antibiotic susceptibility of *Escherichia coli* isolates from inpatients with urinary tract infections in hospitals in Addis Ababa and Stockholm. Bull. World Health Organ. 68:61–68.

Schapira, A. 1994. A standard protocol for assessing the proportion of children presenting with febrile disease who suffer from malarial disease. Geneva: World Health Organization.

Siddique, A. K., Salam, A., Islam, M. S., Akram, K., Majumdar, R. N., Zaman, K., Fronczak, N., and Laston, S. 1995. Why treatment centres failed to prevent cholera deaths among Rwandan refugees in Goma, Zaire. Lancet 345:359–361.

Singh, J. A., Upshur, R., and Padayatchi, N. 2007. XDR-TB in South Africa: No time for denial or complacency. PLoS Med. 4:e50.

Songsore, J. 2004. Urbanization and health in Africa: exploring the interconnections between poverty, inequality and the burden of disease. Heath Clark Lecture delivered on June 2002 at the London School of Tropical Medicine and Hygiene. Accra: Ghana Universities Press.

Toole, M. J., and Waldman, R. J. 1997. The public health aspects of complex emergencies and refugee situations. Annu. Rev. Public Health 18:283–312.

Tsegaye, G. 1998. Assessment of knowledge and practice in urban and rural communities in Jimma zone. J. Health Sci. 8:92–93.

Victora, C. G., Huicho, L., Amaral, J. J., Armstrong-Schellenberg, J., Manzi, F., Mason, E., and Scherpbier, R. 2006. Are health interventions implemented where they are most needed? District uptake of the integrated management of childhood illness strategy in Brazil, Peru and the United Republic of Tanzania. Bull. World Health Organ. 84:792–801.

Whyte, S. R., Geest, S. V. D., and Hardon, A. 2002. Social lives of medicines. Cambridge: Cambridge University Press

Worku, S. and Abebe, G. 2003. Practice of self medication in Jimma town. Ethiop J. Health Dev. 17:111–116.

Zaidi, M. B., Zamora, E., Diaz, P., Tollefson, L., Fedorka-Cray, P. J., and Headrick, M. L. 2003. Risk factors for fecal quinolone-resistant *Escherichia coli* in Mexican children. Antimicrob. Agents Chemother. 47:1999–2001.

Zhang, R., Eggleston, K., Rotimi, V., and Zeckhauser, R. J. 2006. Antibiotic resistance as a global threat: evidence from China, Kuwait and the United States. Global Health 2:6.

Chapter 4
What the Future Holds for Resistance in Developing Countries

Michael L. Bennish and Wasif Ali Khan

Abstract The challenge to controlling antimicrobial resistance in coming years is to put into practice recent policy and programmatic advances. Increasing attention to the problem of antimicrobial resistance, and how resistance in developing countries can affect industrialized countries, has led to increased attention to the problem of resistance in developing countries.

Efforts, however, have lagged behind good intentions. Inappropriate use of antimicrobials, thought to be the major driver of resistance, remains the norm rather than the exception in most of the developing world, as well as in many industrialized countries. Enhanced educational outreach to both consumers and health-care providers to change the pattern of antimicrobial use is crucial and methods to do this effectively have been developed. There is an urgent need for greater regulation of antimicrobial distribution and sale so that private shops staffed by untrained owners and employees are no longer a common source of antimicrobials. Greater management capacity is required to ensure adherence to regulations, to audit prescribing in both the public and the private sectors, and to control corrupt practices and the proliferation of counterfeit or sub-standard drugs. Implementing and sustaining resistance surveillance systems that will alert the medical and public health communities to changes in resistance is also crucial. Development and introduction of rapid techniques to determine infecting pathogens and their susceptibility should enhance both surveillance and care.

The substantial funding that is now flowing to targeted diseases – HIV and AIDS, tuberculosis, and malaria, all three of which have substantial problems with antimicrobial resistance – can in the coming years be both a boon (if funds are used to enhance infrastructure to manage all diseases of public health concern) and a detriment (if efforts remain narrowly focused on these diseases) to controlling resistance. Efforts to reduce disease burden – through health interventions such as immunizations and improved socioeconomic conditions –

M.L. Bennish (✉)
Mpilonhle, Mtubatuba, South Africa; Department of Population,
Family and Reproductive Health, Johns Hopkins Bloomberg School of Public Health,
Baltimore, MD, USA
e-mail: michael@mpilonhle.org

A. de J. Sosa et al. (eds.), *Antimicrobial Resistance in Developing Countries*,
DOI 10.1007/978-0-387-89370-9_4, © Springer Science+Business Media, LLC 2010

have the potential to have profound effects on the burden of resistance. Ultimately, control of resistance will depend on an integrated, multidimensional effort, the components of which can be implemented if the commitment, political will, and resources are made available.

4.1 Introduction

Controlling antimicrobial resistance in developing countries in the future will require both diminishing the inappropriate use of antimicrobial agents that leads to the development of resistance and improving the social and environmental conditions that currently cause the high incidence of infectious diseases among the poor of developing countries.

Although there is no crystal ball that allows us to predict with certainty how these two determinants of resistance will play out in the coming years, there is reason for some optimism. Worldwide deaths of children under age 5 have dropped, for the first time, to below 10,000,000 (Unicef, 2007). This reflects improved socioeconomic standards in many countries, improved efforts in disease prevention, especially increased access to childhood immunization, and the development of new vaccines against common infectious diseases (Andre et al., 2008). Although it is unlikely that the millennium development goals for 2015 – including the target of decreasing child mortality by 67% and halting and beginning to reverse the spread of HIV/AIDS, malaria, and other major diseases – will be fully met (Mdg Africa Steering Group, 2008), the goals have provided an impetus for sustained attention and investment in the diseases and conditions of impoverished persons in developing countries. This can only have the ancillary benefit of reducing the need for antimicrobial therapy, and diminishing the likelihood that resistance will develop. Three of the millennium development goals – reduce child mortality; improve maternal health; and combat HIV/AIDS, malaria, and other diseases – are dependent in part on ensuring that effective antimicrobial agents are available (Mdg Africa Steering Group, 2008).

Efforts at improved health systems management may also result in diminished antibiotic resistance. The ability to monitor the performance of health professionals more readily with the use of health information systems, including computerized systems, will enhance the ability to control the use of antimicrobials (Wilkins et al., 2008).

There remain immense challenges, however. These include a drug market that is effectively unregulated in many developing countries, with antimicrobials widely sold by commercial establishments without trained personnel and without prescription. There is also the continued increase in patients with HIV and AIDS, with their risk for opportunistic infections, including tuberculosis, that require antimicrobial therapy. There is also the widespread use of prophylactic antimicrobial agents in such patients, which will inevitably result in an increase in resistance, an effect not often included in the cost-effectiveness analysis of such strategies (Ryan et al., 2008).

The dynamics of resistance in developing countries has to date been inexorably in one direction – a higher prevalence of resistance and development of resistance at an ever quickening pace. If we are to avert a public health catastrophe (or diminish one that has already occurred in many places), this dynamic has to change.

The increasing prevalence of resistance has been true for almost all bacterial pathogens of public health concern in developing countries – those causing enteric infections, those causing respiratory infections, including and especially tuberculosis, those causing sexually transmitted infections, and those causing bloodstream infections (Okeke et al., 2005a). Resistance to parasitic infections has been somewhat less relentless – but has been pronounced in the pathogen with the greatest health burden, such as Plasmodium (Greenwood et al., 2008). Viral resistance has until recently been less of a problem as treatment of viral infections has been exceptionally rare in the developing country setting. This of course has now changed with the advent of widespread treatment of HIV infections. Though resistance among primary HIV infections in developing countries remains rare, the possibility that resistant strains may cause primary infections is of concern (Shekelle et al., 2007; Bennett et al., 2008).

Control of resistance has now been identified as a public health priority in both poor and rich countries (Lord Soulsby of Swaffham Prior, 2008). How to control resistance, especially in impoverished countries, is less certain. Efforts to control resistance have centered on ensuring the appropriate use of antimicrobial agents (Okeke et al., 2005b; Simonsen et al., 2004). Means to this end include enhancing knowledge of caregivers, improving the knowledge base on which caregivers can make decisions, imposing regulatory controls, and improving management systems that can ensure that regulations are enforced. All of this, hopefully, would be done in a setting of greater prosperity that allows for the financing and development of these programs and systems within developing countries, rather than being dependent on resource transfers from industrialized countries.

This chapter examines the prospects for implementing systems and approaches that will allow for control of resistance in developing countries in the coming years.

4.2 Improving the Knowledge Base

4.2.1 Improving the Knowledge Base in the Care of Individual Patients

Health practitioners in developing countries face a number of dilemmas in attempting to ensure the appropriate use of antimicrobial agents. First, they for the most part lack the laboratory tools to identify a pathogen when they suspect infection (Girosi et al., 2006). In the poorest countries, it would probably be a generous estimate that 1% of patients are treated at clinical facilities that have diagnostic microbiologic laboratories. Even in those settings where diagnostic facilities are available, methods for determining susceptibility of pathogens are uncommon.

Second, patients in developing countries often present late in the course of their illness, and thus treatment often has to be administered urgently. Practitioners most often do not have the luxury of waiting for the availability of microbiologic test results to initiate treatment.

For these reasons, most infections in developing countries are treated empirically – without the benefit of diagnostic microbiologic studies (Chowdhury et al., 2008). If health practitioners are to use antimicrobials appropriately they require both good syndromic algorithms for identifying with a high degree of certainty the infection requiring treatment and a knowledge of the prevailing resistance pattern to ensure that the right antimicrobial agent is selected.

The treatment algorithms used by most health practitioners in developing countries usually derive from international organizations (such as the WHO) and are then promulgated through national departments of health (Victora et al., 2006). While these algorithms usually have reasonable sensitivity and specificity, the development of newer diagnostic technologies appropriate for developing country settings could greatly improve diagnostic sensitivity and specificity and hence reduce the overuse of antimicrobials, which is an inevitable consequence of relying on algorithmic diagnosis (Girosi et al., 2006; Lim, et al. 2006).

Identifying the pathogen is only the first step in ensuring that appropriate antimicrobial therapy is provided. The second, and more difficult, is identifying the susceptibility of the organism causing infection. For the near future, the choice of antimicrobial agent will also be based on knowledge on the prevailing pattern of susceptibility in the community, rather than identification of the resistance pattern of isolates from patients under care.

Because treatment will remain largely algorithmic in approach, there remains a need to develop (or refine) algorithms for treatment that have reasonable sensitivity and specificity and assuring that there is good surveillance data on resistance to allow for the use of antimicrobial agents with a high likelihood of being effective in the treatment of the suspected infection.

4.2.2 Improving the Public Health Knowledge Base

Designing algorithms for treatment of infections is complicated by the great diversity of countries included under the rubric of "developing countries". The latter is a broad term that is poorly defined. Within United Nations' taxonomy

"There is no established convention for the designation of 'developed' and 'developing' countries or areas". What is clear is that most – 33 of 50 – of the countries that the United Nations designates as "least developed" are within Africa (United Nations Statistics Division).

However, even in these least developed countries there is substantial diversity in the pattern of disease prevalence. There are also differences in the resistance patterns. What the iconic American politician Thomas "Tip" O'Neil said about politics – "all politics is local" – is also germane to resistance (Jacobs, 2003). Guidelines for treatment need to be based on patterns of infection and resistance that are derived from local conditions.

How local that information needs to be is uncertain – within the same country there can be widely divergent patterns of antimicrobial resistance – but it is clear that currently in the least developed countries diagnostic and treatment algorithms are often developed based on information that is external to the country where it is being applied. Increased efforts need to be made to collect local data that can be used to inform treatment guidelines. Unfortunately, those countries where the need is greatest – the most impoverished countries – are the ones that are least likely to collect such information on an ongoing basis.

The increased expenditure for new and emerging diseases – some of it under the umbrella effort to combat bioterrorism – has provided a major impetus to establish surveillance systems in developing countries.(Koplan, 2001) Surveillance efforts, once established, can be used for varied purposes. Thus, the infrastructure established to track new diseases – such as avian influenza (influenza H5N1) – can also be used to track patterns of other infections and resistance patterns (Greene et al., 2007). In addition, antimicrobial resistance is included as a Category C (emerging infectious disease threats) priority by the United States National Institute of Health under its effort to combat new and reemerging infectious diseases. A number of efforts to track resistance, including establishing common databases – such as INSPEAR (Richet et al., 2001) and WHONET (Stelling and O'brien, 1997) – already exist and can provide the framework to expand local efforts to track resistance and improve the quality of testing (Tenover et al., 2001).

4.3 Improving Information Flow

4.3.1 Improving Information Flow to Health Providers

Educational messages must reach both health-care providers and consumers (Finch et al., 2004). Most health-care workers in developing countries continue to work in isolated conditions, with little in the way of continuing education. This leaves them dependent on information that was most likely already out of date at the time they received their initial training – information that only becomes progressively more out of date as their initial training becomes more distant. Although comparatively well-endowed programs – such as those directed at HIV and AIDS – can afford to bring program staff to central sites for regular training, such continued in-service training remains the exception rather than the rule for most primary health-care workers. Updated journals or texts are rarely available at health facilities in developing countries: the cost remains prohibitive and the distribution networks uncertain. Thus, health-care workers are at best dependent on the occasional information leaflet that comes their way or the rare training that addresses issues of resistance.

The use of newer technologies may in the coming years become an effective means of updating health-care workers on changing resistance patterns and changes in drugs of choice for treating infection. Web-based education is one option (Geissbuhler et al., 2007). Access to computers and to the web remains inadequate in most developing countries, however. Web access, which would allow for real-time updates of resistance patterns and drugs of choice, is especially limited in rural areas and in Africa, where less than 5% of the population has access to the Internet. There are other options for electronic transmission of information that can be exploited. Perhaps the easiest and most effective is the use of mobile phones (Ybarra and Bull, 2007). In contrast to web access via fixed lines, mobile phone access in developing countries – including even rural Africa – is quite high – with over 120 million mobile phones being in use in Africa. Most health workers will have a mobile phone of their own, or access to a phone. Information updates can be sent at relatively low cost via text messaging, or phones can be used (at somewhat greater cost) to access Department of Health web sites dedicated to information provision on antimicrobial resistance. From the point of view of the health-care provider, updates via text messaging are probably the most appealing – as there is no cost to the health provider and they are not required to take the initiative to access the information.

If a text messaging solution is used, health departments will have the challenge of maintaining mailing lists for sending out text messages. Although this might sound trivial, the ability of Departments of Health to maintain updated list of health-care workers' telephone numbers may be a challenge that is beyond the current capacity of the many of the health departments in the poorest countries

4.3.2 Improving Information Flow to Consumers

Consumers are perceived to be in part the driver for inappropriate use of antimicrobials (Coenen et al., 2006). The concept that most illnesses are self-limited and do not require treatment – especially antimicrobial treatment – remains a difficult one to convey to patients or their family members. This problem is almost universal.

Some progress has been made in changing that perception – especially in European countries. In Europe, there is an increasing perception that the so-called "natural" remedies are better than pharmacologic ones (Ritchie, 2007). The trade off is that many remedies that are perceived or promoted as "natural" have little value – but they may be less likely to cause harm than the inappropriate use of antibiotics.

There are now efforts in developing countries to design public information campaigns aimed at consumers (Chetley et al., 2007). How to design culturally appropriate messages to address this issue remains a very substantial challenge – anywhere in the world. Nevertheless, such efforts are necessary if perceptions are to change and consumer demand for medicines is to be reduced. In addition, the changing health context in which the message is to be delivered presents challenges. The AIDS treatment program is the largest therapeutic program ever undertaken in developing countries and is focused on the use of anti-infectives, including the use of prophylactic antimicrobials. There thus is the likelihood of dissonance in the two approaches that are to be taken – increased use of anti-infective agents for treatment of AIDS and decreased use of anti-infective agents for most common conditions and symptoms.

There can be synergy between the AIDS efforts and the efforts to increase the prudent use of antimicrobials. One common theme is adherence to drug regimens – a vital strategy for both treating HIV and TB and reducing resistance to other pathogens (Thiam et al., 2007). In addition, efforts to make messages culturally appropriate are likely to transcend the specific disease that is targeted.

If an effective message can be designed, there are means of transmitting those messages – ranging from mass media to locally produced plays and songs (Finch et al., 2004, Chetley et al., 2007). Infrastructure that is being supported with funds specifically directed for prevention and treatment of HIV and AIDS can be used, if funding agencies allow, for purposes of addressing the issues of antimicrobial resistance. Large funders of HIV and AIDS efforts – such as the United States Presidents Emergency Plan for AIDS Relief (PEPFAR) – now recognize the need for their programs, which receive a large proportion of bilateral and multilateral aid funds directed for health (whether that share is disproportionate is in the eye of the beholder) to strengthen health-care systems, including health information systems (El-Sadr and Hoos, 2008).

4.4 Regulatory Environment

4.4.1 Regulating the Private Sector

Control of antimicrobial use, if it is to be effected, cannot be voluntary. From the 1980s onward, has been a period (until the recent economic collapse) of both anti-regulatory sentiment and attempts to bring the free market into the provision of health care (Reddy, 2008, Adeyi et al., 1997). The impetus for these initiatives were clear – the dysfunction of many government services in developing countries, including health services, and the effect of regulation on decreasing competitiveness and innovation. If antimicrobials are to be used appropriately, there need to be effective regulatory controls in addition to rhetorical support.

A primary challenge in developing countries remains the distribution and sale of antimicrobial agents without prescription by the private pharmacies that are a fixture in these countries (Gul et al., 2007; Volpato et al., 2005; Paphassarang et al., 2002). Even if the untrained owners and salespersons at these pharmacies had knowledge of what constitutes appropriate therapy, they have little incentive to reduce their inappropriate prescribing (Paphassarang et al., 2002; Stenson et al., 2001a). Their raison d'être is profit, and thus voluntary acceptance and implementation of appropriate prescribing would only undermine their bottom line. Similarly, many (if not most) of the physicians in private practice in many developing countries make a substantial portion of their income by selling drugs to their patients, a practice that has been prohibited in most industrialized countries for some decades now.

A fundamental challenge in to assuring the appropriate use of antimicrobials, and reducing the spread of resistance, will be the ability of governments in developing countries to control the sale and dispensing of drugs in private shops (Stenson et al., 2001b). The forces arrayed against this control are considerable. Pharmaceutical companies (many of them indigenous, rather than multinationals, and many of them well connected politically) would see a drop in revenue and profits, the shopkeepers themselves would protest, and patients who, because of the absence of alternative sources of care often rely on such local pharmacies, would likely object unless alternate ways to make medicines available were undertaken.

If governments desire, they probably could (with the exception of the poorest and least functional of governments) close such pharmacies. Most businesses need some sort of license to operate, and governments should be able to control the import and production of pharmaceuticals. That of course would require that corrupt practices be reduced – as the drugs from many public hospitals and clinics currently are found for sale in the private market.

Controlling the prescribing practices of private shops absent shutting them down would be difficult. Auditing their performance would be almost impossible, as there is rarely a recording of a diagnosis. Patient interactions are at best cursory – the recitation of a chief complaint – before medicines are dispensed (Paphassarang et al., 2002). Improving the quality of dispensing – the mixing of drugs and selling one or two tablets – is likely to be easier than actually enforcing adherence to treatment algorithms (Stenson et al., 2001b).

One glimmer of hope in controlling prescribing is the expansion of health insurance schemes as the middle class expands in size in many developing countries (Drechsler and Jutting, 2007). Many such insurance schemes reimburse for medications, and thus they can exercise control over prescribing practices by withholding reimbursement for prescribing that does not follow guidelines – just as private or government schemes have done in industrialized countries for some years now. This is one more way that economic expansion, by providing infrastructure to monitor and regulate the use of antimicrobials, becomes perhaps the most effective means of controlling resistance.

The paradox of course is that as economies grow the use of pharmaceuticals is likely to increase (Dong et al., 2008) – but the hope has to be that use will become more appropriate. Being able to visit a private practitioner, rather than visiting a government health clinic, is a sign of status and affluence for those newly arrived in the middle class. What is disheartening is the likelihood that the care provided, especially the drugs prescribed, will be, if anything, less rational than what they would have received in government health centers.

Nothing is more discouraging than to see the laundry list of medications that persons visiting private practitioners still are likely to receive. As we wrote this chapter, one of us was asked by a staff member to see their 18-monthold child who had a mild gastroenteritis. She had proudly taken the child to one of the private practitioners in town and proudly returned with six different medicines: ampicillin, an anti-emetic, an anti-peristaltic agent, a herbal remedy, a lactobacillus preparation, and one medicine without any name on it. This in a county – South Africa – where by developing country standards there is good regulatory control over drugs, and for a patient whose costs of drugs were covered by a medical insurance scheme.

That is not to tar all private practitioners with the same brush – but profiting from the sale of medicines makes for perverse incentives if the goal is to reduce their inappropriate use.

Counterfeit and adulterated drugs remain as great, or greater, a problem now than when we investigated 15 years ago fatal poisoning from paracetamol preparations that contained diethylene glycol (Hanif et al., 1995). Counterfeit antimicrobial preparations, or preparations that are reduced in potency, can elicit resistance by consistently exposing pathogens to sub-therapeutic concentrations of drug. Absent sustained international efforts, this problem is likely to increase. The hope for the future is that concerns about enforcing intellectual property rights for a variety of products – such as computer software, and concerns about adulterated foodstuffs and other products that are shipped internationally – will lead to greater regulation and enforcement of all types of products, including pharmaceuticals. The problem is that small local manufacturers for the domestic market in developing countries make many of the illicit pharmaceuticals. As such, they pose less of a threat to industrialized countries that have the capacity to regulate their production, and thus industrialized countries are less likely to or put pressure on developing countries to control the production of such companies.

4.4.2 Regulating the Public Sector

It should be easier to regulate dispensing of antimicrobials in the public sector than in the private sector. In the public sector, the government controls the supply of drugs, algorithms for treatment are usually in place, the government can audit actual prescribing practices, and corrective action can be taken directly with staff who are government employees.

There have been extensive and successful efforts to codify drugs lists and supplies for governmental primary health-care services. The challenge has been to assure that drugs are in place and that "leakage" of drugs into the private market does not occur (Moszynski, 2008). In many developing countries, the pharmacy cupboard in primary health-care centers is bare, forcing patients to go to private pharmacies to obtain drugs. There have also been limited attempts to institute regular and effective audits of prescribing practices in government clinics and hospitals. New efforts – such as the recently launched Medicines Transparency Alliance – attempt to address problems of pharmaceutical use in developing country hospitals (Moszynski, 2008).

Even though algorithms may be closely followed, what is uncertain is how much "diagnosis creep" occurs (Hsia et al., 1988; Carter et al., 1990). For instance, in response to patient pressure, or their own inclinations, is clinic staff more likely to diagnose a lower respiratory tract infection requiring antimicrobial therapy, rather than upper respiratory tract infection, which algorithmically does not require antimicrobials, even though the signs and symptoms might be characteristic of the latter? Or diagnose dysentery (also requiring antimicrobial therapy) rather than watery diarrhea, which does not require antimicrobial therapy?

Regular audits of diagnostic and prescribing practices are required to determine and control prescribing behavior. Routine audits are currently rarely, if ever, performed. One-off audits for research purposes are more common (Hadi et al., 2008a). Auditing practices that have been put in place by third-party providers (private or public insurers) in industrialized countries could be adapted for use in the developing country setting. However, this will also require a cultural change in how oversight is perceived. It cannot, at least initially, be seen as punitive – but rather supportive in achieving a common goal such as improved patient care and reduction in the risk of the development of resistance. Even then, absent adequate diagnostic facilities, it is often difficult to make a definitive statement about the appropriateness of the prescribing that was audited (Hadi et al., 2008b).

4.5 Management Capacity and Systems

Many if not most of the changes required to improve the appropriate prescribing of antimicrobials in developing countries, and thereby hopefully lessen the development of resistance, require improved management capacity within developing countries (Hartwig et al., 2008). Management capacity has been widely recognized, along with lack of resources, as the crucial constraint on improving health-care provision in developing countries. Absence of effective management systems has been the bane of attempts to implement and sustain new initiatives and programs – no matter how well intended, no matter how rational, and no matter how effective those new initiatives are in pilot studies (Sinha et al., 2007).

Many of the essential components of appropriate use of antimicrobials are dependent on effective management systems. Good management is crucial in ensuring that staff selection is done appropriately, that staff are recompensed according to terms of their contract, that nepotism is not the basis for employment, that performance oversight is in place, that auditing systems are actually implemented, that drug supply chains work, that surveillance systems are actually collecting the data that they are intended to collect, and that laboratory systems meet quality control standards. Most importantly, management needs to instill a common sense of purpose, with a focus on the outcome (appropriate use of antimicrobials) and with an attention to the processes that help achieve that goal, including appropriate delegation of authority, rather than centralizing control and decision making. Crucially, there needs to be identification of those elements of the process that obstruct achievement of the desired goal (nonsensical rules, rigid interpretation of rules) and putting in place systems that allow staff to achieve goals.

These management strategies might seem contradictory – to implement guidelines you want clinical staff to adhere to, while at the same time giving management staff some autonomy in decision making. The crucial challenge is to define where staff can take initiative in establishing and implementing systems and where they need to follow guidelines strictly.

A major challenge to implementing new programs, and new management approaches, is that neither programs nor individual staff work in a vacuum. If corruption is widespread, work ethic limited, salaries low, and not paid on time (or not paid at all), the effect can be demoralizing no matter how good the original intentions. Thus, improving management capacity has to be something that is done throughout an organization.

The success of many vertical programs has come from their clear focus, their relatively generous funding, and their ability to establish cultures – including management cultures – that are distinct from that of the larger organization in which they are imbedded. Even assuming the value of vertical (as opposed to integrated) programs, programs to control antimicrobial resistance cannot be "vertical" in nature. Programs for the appropriate use of antimicrobials are of necessity integrated within the larger health-care structure – whether it is primary care programs that provide services to children with respiratory infections or diarrhea, or targeted programs such as those for persons with HIV or tuberculosis. Because programs addressing antimicrobial resistance are inherently crosscutting, they can provide benefit to other programs that need improved system management. They can in turn benefit from the systems – including management systems – which other programs have developed (Ooms et al., 2008).

4.6 Reducing Infectious Disease Burden

The most fundamental approach to diminishing antimicrobial resistance is to diminish the incidence of infectious disease. The health transition that is occurring in many countries – from a high burden of infectious disease, especially childhood infections, to a higher burden of chronic diseases (heart disease, diabetes, cancer, such as occurs in industrialized countries) – can affect the patterns of use and need for antimicrobial agents. This health transition may only shift the burden of resistance and where it occurs – from children in primary care settings to adults in hospitals (Zhang et al., 2006). The latter pattern represents what we now see in industrialized countries. Nonetheless, reducing common infections would greatly reduce the need for antimicrobials – if not necessarily their use, given that need and use are not always linearly related.

Immunizations are clearly the most effective way to reduce infectious disease burden. Although there have been some notable disappointments in vaccine development (the lack of an effective HIV vaccine for instance), there remain vaccines effective against common bacterial infections that are underutilized in developing countries – including *Haemophilus influenzae* and pneumococcal vaccines. Increased use of these vaccines, and continued improvement of the current vaccines, would clearly diminish the need for antimicrobials and reduce (especially with *Streptococcus pneumoniae*) one of the most troubling organisms in terms of current resistance patterns (Cherian, 2007; Morris et al., 2008).

The lack of a vaccine for sexually transmitted diseases (with the exception of human papilloma virus) is also a concern, as *Neisseria gonorrhea* remains both an important public health problem and an organism for which resistance has been an ongoing problem (Rupp et al., 2005).

4.7 Improving Social and Economic Conditions

Fundamental changes in socioeconomic conditions would have a beneficial effect on the development of resistance in poor countries (Planta, 2007). Improvements in economic conditions are almost inevitably associated with improvements in hygiene and sanitation, and a diminishment in household crowding. These changes historically have been associated with a decrease in infectious diseases. Improvements in economic conditions are also likely to provide governments in currently poor countries the resources for implementing programs directed to surveillance of resistance, monitoring of antimicrobial usage patterns, control of the distribution of pharmaceuticals and monitoring their quality, and education of dispensers.

Improved economic conditions might, however, result in an increase in the use of antimicrobials in sectors – such as farming – that they had not been used previously (Holmström et al., 2003).

The conundrum is that changes in our approach to using antimicrobials cannot await fundamental socioeconomic changes. There is an urgent need to implement programs to deal with resistance.

Programs addressing antimicrobial resistance will need both initial and continued support from international donors – both bilateral and multilateral. Donor motivations include both humanitarian concerns (such as the largely bipartisan effort in the United States to support PEPFAR – the largest bilateral health assistance program ever undertaken) and perceived self-interest – such as the bio-defense initiatives. It is not clear how the economic crisis now affecting most of the industrialized world will affect how these underlying motivations translate into support. In an era of increased need within their own countries, it is not certain that there will be the political will in industrialized countries to support increased (or even constant) levels of foreign assistance.

4.8 Alternatives to Standard Antimicrobial Therapy

The pipeline of new antimicrobials has almost run dry – with fewer and fewer new agents being brought to market and no new widely used class of drugs suitable for use in developing countries having come on line since the fluoroquinolones (Wenzel, 2004).

Alternative means of treating infectious diseases – such as immunologic therapies or blocking of target sites – are being developed. Such therapies, if they are successful, are not likely to be affordable or available in developing countries in the near future. Cycling of antimicrobial therapy – periodically changing the drugs of choice for treating a particular infection – is another approach that has been suggested for maintaining the utility of currently effective antimicrobials (Okeke et al., 2005b; Smith and Coast, 2002). With the dearth of effective drugs for treating many of the infections common in developing countries, rotating drugs may be a policy with limited applicability for many infections.

4.9 Summary

Prospects for control of antimicrobial resistance are now modestly more promising than they have been in the past. There is a clear recognition by many governments, and international health organizations, that resistance is now a major public health problem and that urgent action is needed. The increased commitment of the industrialized countries to global health – motivated by either self-interest or altruistic concerns, or both – provides a base from which the problem can be tackled. The large-scale international commitment to specific diseases – HIV, malaria, and tuberculosis, for all of which resistance is a major concern – provides important tie-ins for building and sustaining a larger effort at reducing antimicrobial resistance.

The rapid development of biotechnology should provide innovations that allow for simplified, rapid, point-of-care diagnosis of infecting pathogens and determination of susceptibility patterns. These innovations are crucial for both treatment of individual patients and obtaining knowledge of local patterns of infections and susceptibility. The development of standardized and accessible databases of antimicrobial resistance, quality control procedures for susceptibility testing, and enhanced communication allow for much greater sharing of knowledge of resistance patterns, and early identification of the development and spread of resistance.

Despite these hopeful advances and initiatives, there remain substantial concerns. The largely unregulated market for antimicrobials in most developing countries is a substantial obstacle to controlling resistance. The lack of government resources to put controls in place, the generally hostile atmosphere to greater regulation (at least until the recent financial meltdown), and the diminishment of government health services at the expense of private health services also represent substantial challenges to implementing effective policies on antimicrobial use in developing countries. The lack of new antimicrobials makes the need for controlling resistance to current agents all the more urgent.

As with many health problems in developing countries, many of the solutions to controlling antimicrobial resistance are known (Table 4.1). The overriding challenge is to ensure that they are implemented. Implementation will take political will as reflected in a commitment of resources. Resistance is a concept that is more difficult for both the public and the politicians to grasp than are the ravages caused by HIV and AIDS, tuberculosis, or malaria – three diseases that have garnered widespread attention and increased funding in recent years. The public health impact of resistance is substantial, and hopefully in the coming years the commitment and will (and funding) to tackle this problem will be commensurate with the extent of the problem.

Table 4.1 Current challenges and possible future solutions to controlling resistance in developing countries

Challenge	Possible solutions	Challenges and prospects
Enhancing microbiologic diagnostic capabilities	Increase laboratory capacity in developing countries. Increase point-of-care diagnostics using rapid testing. Reduce cost of point-of-care diagnostics	Traditional laboratory diagnostic facilities are resource intensive and require substantial management capacity. Even with increasing urbanization in developing countries, it is unlikely that the majority of patients will be treated at facilities with access to traditional diagnostic microbiologic facilities.

Table 4.1 (continued)

Challenge	Possible solutions	Challenges and prospects
		Rapid point-of-care diagnostics are ideal if they can be simplified, and rendered less expensive. The greatest current effort on point-of-care diagnostics has been on identifying the pathogen – rather than the antimicrobial susceptibility of the pathogen. Public investment in developing simplified point-of-care diagnostics, especially for susceptibility, is crucial
Improving algorithmic management of infectious diseases	Refining algorithms to be location specific, as pattern of disease differs extensively by region	The development of guidelines for the diagnosis and treatment of adult and pediatric illnesses has received considerable attention in recent years. These have been region specific to reflect the major endemic illnesses of the area – i.e., malaria endemic, HIV hyperendemic. There can be considerable variation within regions and these needs to be explored in greater depth. The most difficult regions to develop guidelines for will be the areas with the least infrastructure – where the least is known about disease endemicity
Improving knowledge of changes in patterns of infections and resistance	Improve ongoing surveillance	Recent efforts addressed at new and emerging diseases and bioterrorism have provided an impetus for developing surveillance in developing countries. Challenges are to ensure that resistance surveillance is included as part of these initiatives; to sustain surveillance systems after the initial enthusiasm for their use

Table 4.1 (continued)

Challenge	Possible solutions	Challenges and prospects
		has subsided, and after high-profile diseases such as avian influenza are no longer in the headlines; to ensure that surveillance reaches the most difficult to reach areas; and to ensure that information is acted upon
Improving information flow to health providers	Make use of newer communication systems such as text messaging and easily accessible web sites, to assure that practitioners are informed of current policies, and any changes to policies.	Assuring that Departments of Health maintain up to date lists of contact numbers of health providers is a major challenge. Assuring Internet connectivity if terrestrial lines are used remains a challenge in both urban and rural areas of developing countries. Reducing cost of Internet access remains a hurdle in many developing countries. Disseminating standard paper-based policy instruments remains erratic. Agreeing upon new diagnostic and treatment guidelines, even if current information on which to develop guidelines is available, has proved an arduous and often insurmountable task for many national Departments of Health, and hence guidelines are often out of date
Improving information flow to consumers	Use of mass media and local community groups to ensure that messages about the need to use antimicrobials judiciously is provided to consumers.	Battling long-ingrained assumptions among both providers and consumers on the need to treat many common illnesses with antimicrobials. Pharmaceutical company promotional budgets can overwhelm most public efforts and messages of pharmaceutical companies in developing

Table 4.1 (continued)

Challenge	Possible solutions	Challenges and prospects
		countries (both national and multinational) are often poorly regulated
Controlling distribution and provision of antimicrobials by the private sector and assuring the quality of pharmaceuticals	Increased regulation and control over the distribution of antimicrobials by pharmaceutical companies, their sale by pharmacies without a prescription, and the prescribing practices of private physicians. Eliminate non-approved manufacturers and assure quality of products of licensed manufacturers	Governments of many developing countries lack the capacity to enforce regulations, there remains a generally anti-regulatory environment, there are perverse financial incentives for inappropriate prescribing by private physicians and private pharmacies that would need to be changed. Assuring quality of pharmaceuticals would take greater laboratory, enforcement, and regulatory capacity
Ensuring appropriate proscribing by the public sector	Establishing and auditing adherence to guidelines; ensuring effective distribution and availability of pharmaceuticals; minimizing corruption and loss of pharmaceuticals	All of these solutions require enhanced management capacity. Increased commitment to global health, even if on specific vertical programs (HIV, TB, malaria) may result in improvement of overall management and performance of the health sector
Improving management capacity	Develop clearly defined management systems that can be implemented and sustained in the developing country context	Improved management capacity is central to achieving many of the challenges to controlling resistance in the coming years. Management capacity needs most critically to be improved in the public sector. Need systems to focus on outcome rather than process, have clear lines of responsibility and accountability, assure competitive payment of civil servants, and improve motivation and morale

Table 4.1 (continued)

Challenge	Possible solutions	Challenges and prospects
Reducing infectious disease burden	Improve living conditions, increase use of current vaccines, and develop new vaccines	Substantial investments are being made in vaccine development for an array of infections common in developing countries. Assuring affordability has still proven problematic. Improvement in living conditions has accompanied the rising income levels in many countries, especially in Asia and Latin American, and has reduced infectious disease burden in the young in those regions. This increase in wealth may have the paradoxical effect of increasing demand (and purchasing power) for antimicrobials by a new middle class that associates the use of antimicrobials with affluent status despite the decrease in infectious disease burden, and increasing antimicrobial use in hospitals as chronic diseases become more common
Alternatives to current antimicrobial therapy	Develop and test new antimicrobial compounds or new uses of current compounds. Develop alternatives to traditional antimicrobial therapy	Public–private partnerships have been developed to develop new antimicrobials, or adapt current antimicrobials, for the treatment of resistant infections in developing countries. Funding is still sparse for these initiatives and major funding for development of new classes of antimicrobial agents will still come from the private industry. Alternatives to antibiotics – such as immune therapies – have not achieved widespread success for common infections

References

Adeyi O., Chellaraj G., Goldstein E., Preker A. and Ringold D. 1997. Health status during the transition in Central and Eastern Europe: development in reverse? Health Policy Plan 12:132–145

Andre F. E., Booy R., Bock H. L., Clemens J., Datta S. K., John T. J., Lee B. W., Lolekha S., Peltola H., Ruff T. A., Santosham M. and Schmitt H. J. 2008. Vaccination greatly reduces disease, disability, death and inequity worldwide. Bull World Health Organ 86:140–146

Bennett D. E., Bertagnolio S., Sutherland D. and Gilks C. F. 2008. The World Health Organization's global strategy for prevention and assessment of HIV drug resistance. Antivir Ther 13 Suppl 2:1–13

Carter G. M., Newhouse J. P. and Relles D. A. 1990. How much change in the case mix index is DRG creep? J Health Econ 9:411–428

Cherian T. 2007. WHO expert consultation on serotype composition of pneumococcal conjugate vaccines for use in resource-poor developing countries, 26–27 October 2006, Geneva. Vaccine 25:6557–6564

Chetley A., Hardon A., Hodgkin C. and Haaland A. 2007 How to improve the use of medicines by consumers World Health Organization Geneva http://www.who.int/medicines/publications/WHO_PSM_PAR_2007.2.pdf Accessed August 22 2008

Chowdhury E. K., El Arifeen S., Rahman M., Hoque D. E., Hossain M. A., Begum K., Siddik A., Begum N., Rahman Q. S., Akter T., Haque T. M., Al-Helal Z. M., Baqui A. H., Bryce J. and Black R. E. 2008. Care at first-level facilities for children with severe pneumonia in Bangladesh: a cohort study. Lancet 372:822–830

Coenen S., Michiels B., Renard D., Denekens J. and Van Royen P. 2006. Antibiotic prescribing for acute cough: the effect of perceived patient demand. Br J Gen Pract 56:183–190

Dong L., Yan H. and Wang D. 2008. Antibiotic prescribing patterns in village health clinics across 10 provinces of Western China. J Antimicrob Chemother 62:410–415

Drechsler D. and Jutting J. 2007. Different countries, different needs: the role of private health insurance in developing countries. J Health Polit Policy Law 32:497–534

El-Sadr W. M. and Hoos D. 2008. The president's emergency plan for AIDS relief – is the emergency over? N Engl J Med 359:553–555

Finch R. G., Metlay J. P., Davey P. G. and Baker L. J. 2004. Educational interventions to improve antibiotic use in the community: report from the International Forum on Antibiotic Resistance (IFAR) colloquium, 2002. Lancet Infect Dis 4:44–53

Geissbuhler A., Bagayoko C. O. and Ly O. 2007. The RAFT network: 5 years of distance continuing medical education and tele-consultations over the Internet in French-speaking Africa. Int J Med Inform 76:351–356

Girosi F., Olmsted S. S., Keeler E., Hay Burgess D. C., Lim Y. W., Aledort J. E., Rafael M. E., Ricci K. A., Boer R., Hilborne L., Derose K. P., Shea M. V., Beighley C. M., Dahl C. A. and Wasserman J. 2006. Developing and interpreting models to improve diagnostics in developing countries. Nature 444 Suppl 1:3–8

Greene J. M., Plunkett G., 3rd, Burland V., Glasner J., Cabot E., Anderson B., Neeno-Eckwall E., Qiu Y., Mau B., Rusch M., Liss P., Hampton T., Pot D., Shaker M., Shaull L., Shetty P., Shi C., Whitmore J., Wong M., Zaremba S., Blattner F. R. and Perna N. T. 2007. A new asset for pathogen informatics – the Enteropathogen Resource Integration Center (ERIC), an NIAID Bioinformatics Resource Center for Biodefense and Emerging/Re-emerging Infectious Disease. Adv Exp Med Biol 603:28–42

Greenwood B. M., Fidock D. A., Kyle D. E., Kappe S. H., Alonso P. L., Collins F. H. and Duffy P. E. 2008. Malaria: progress, perils, and prospects for eradication. J Clin Invest 118:1266–1276

Gul H., Omurtag G., Clark P. M., Tozan A. and Ozel S. 2007. Nonprescription medication purchases and the role of pharmacists as healthcare workers in self-medication in Istanbul. Med Sci Monit 13:PH9–PH14

Hadi U., Duerink D. O., Lestari E. S., Nagelkerke N. J., Keuter M., Huis In't Veld D., Suwandojo E., Rahardjo E., van den Broek P. and Gyssens I. C. 2008a. Audit of antibiotic prescribing in two governmental teaching hospitals in Indonesia. Clin Microbiol Infect 14:698–707

Hadi U., Keuter M., van Asten H. and van den Broek P. 2008b. Optimizing antibiotic usage in adults admitted with fever by a multifaceted intervention in an Indonesian governmental hospital. Trop Med Int Health 13:888–899

Hanif M., Mobarak M. R., Ronan A., Rahman D., Donovan J. J., Jr. and Bennish M. L. 1995. Fatal renal failure caused by diethylene glycol in paracetamol elixir: the Bangladesh epidemic. BMJ 311:88–91

Hartwig K., Pashman J., Cherlin E., Dale M., Callaway M., Czaplinski C., Wood W. E., Abebe Y., Dentry T. and Bradley E. H. 2008. Hospital management in the context of health sector reform: a planning model in Ethiopia. Int J Health Plann Manage 23:203–218

Holmström K., Gräslund S., Wahlström A., Poungshompo S., Bengtsson B.-E. and Kautsky N. 2003. Antibiotic use in shrimp farming and implications for environmental impacts and human health. Int J Food Sci Technol 38:255–266

Hsia D. C., Krushat W. M., Fagan A. B., Tebbutt J. A. and Kusserow R. P. 1988. Accuracy of diagnostic coding for Medicare patients under the prospective-payment system. N Engl J Med 318:352–355

Jacobs M. R. 2003. Worldwide trends in antimicrobial resistance among common respiratory tract pathogens in children. Pediatr Infect Dis J 22:S109–S119

Koplan J. 2001. CDC's strategic plan for bioterrorism preparedness and response. Public Health Rep 116 Suppl 2:9–16

Lim Y. W., Steinhoff M., Girosi F., Holtzman D., Campbell H., Boer R., Black R. and Mulholland K. 2006. Reducing the global burden of acute lower respiratory infections in children: the contribution of new diagnostics. Nature 444 Suppl 1:9–18

Lord Soulsby of Swaffham Prior 2008. The 2008 Garrod Lecture: antimicrobial resistance – animals and the environment. J Antimicrob Chemother 62:229–233

MDG Africa Steering Group 2008 Achieving the Millennium Development Goals in Africa. Recommendations of the MDG Africa Steering Group June 2008 United Nations New York City http://www.mdgafrica.org/pdf/MDG%20Africa%20Steering%20Group%20Recom mendations%20-%20English%20-%20HighRes.pdf Accessed August 25 2008

Morris S. K., Moss W. J. and Halsey N. 2008. Haemophilus influenzae type b conjugate vaccine use and effectiveness. Lancet Infect Dis 8:435–443

Moszynski P. 2008. Alliance aims to help health workers in poor countries challenge corruption, excessive pricing, and waste of drugs. Bmj 336:1155

Okeke I. N., Klugman K. P., Bhutta Z. A., Duse A. G., Jenkins P., O'Brien T. F., Pablos-Mendez A. and Laxminarayan R. 2005b. Antimicrobial resistance in developing countries. Part II: strategies for containment. Lancet Infect Dis 5:568–580

Okeke I. N., Laxminarayan R., Bhutta Z. A., Duse A. G., Jenkins P., O'Brien T. F., Pablos-Mendez A. and Klugman K. P. 2005a. Antimicrobial resistance in developing countries. Part I: recent trends and current status. Lancet Infect Dis 5:481–493

Ooms G., Van Damme W., Baker B. K., Zeitz P. and Schrecker T. 2008. The 'diagonal' approach to Global Fund financing: a cure for the broader malaise of health systems? Global Health 4:6

Paphassarang C., Philavong K., Boupha B. and Blas E. 2002. Equity, privatization and cost recovery in urban health care: the case of Lao PDR. Health Policy Plan 17:72–84

Planta M. B. 2007. The role of poverty in antimicrobial resistance. J Am Board Fam Med 20:533–539

Reddy S. G. 2008. Death in China: market reforms and health. Int J Health Serv 38:125–141

Richet H. M., Mohammed J., McDonald L. C. and Jarvis W. R. 2001. Building communication networks: international network for the study and prevention of emerging antimicrobial resistance. Emerg Infect Dis 7:319–322

Ritchie M. R. 2007. Use of herbal supplements and nutritional supplements in the UK: what do we know about their pattern of usage? Proc Nutr Soc 66:479–482

Rupp R. E., Stanberry L. R. and Rosenthal S. L. 2005. Vaccines for sexually transmitted infections. Pediatr Ann 34:818–820, 822–814

Ryan M., Griffin S., Chitah B., Walker A. S., Mulenga V., Kalolo D., Hawkins N., Merry C., Barry M. G., Chintu C., Sculpher M. J. and Gibb D. M. 2008. The cost-effectiveness of cotrimoxazole prophylaxis in HIV-infected children in Zambia. AIDS 22:749–757

Shekelle P., Maglione M., Geotz M. B., Wagner G., Wang Z., Hilton L., Carter J., Chen S., Tringle C., Mojica W. and Newberry S. 2007. Antiretroviral (ARV) drug resistance in the developing world. Evid Rep Technol Assess (Full Rep) 156:1–74

Simonsen G. S., Tapsall J. W., Allegranzi B., Talbot E. A. and Lazzari S. 2004. The antimicrobial resistance containment and surveillance approach – a public health tool. Bull World Health Organ 82:928–934

Sinha T., Ranson M. K., Chatterjee M. and Mills A. 2007. Management initiatives in a community-based health insurance scheme. Int J Health Plann Manage 22:289–300

Smith R. D. and Coast J. 2002. Antimicrobial resistance: a global response. Bull World Health Organ 80:126–133

Stelling J. M. and O'Brien T. F. 1997. Surveillance of antimicrobial resistance: the WHONET program. Clin Infect Dis 24 Suppl 1:S157–168

Stenson B., Syhakhang L., Eriksson B. and Tomson G. 2001a. Real world pharmacy: assessing the quality of private pharmacy practice in the Lao People's Democratic Republic. Soc Sci Med 52:393–404

Stenson B., Syhakhang L., Lundborg C. S., Eriksson B. and Tomson G. 2001b. Private pharmacy practice and regulation. A randomized trial in Lao P.D.R. Int J Technol Assess Health Care 17:579–589

Tenover F. C., Mohammed M. J., Stelling J., O'Brien T. and Williams R. 2001. Ability of laboratories to detect emerging antimicrobial resistance: proficiency testing and quality control results from the World Health Organization's external quality assurance system for antimicrobial susceptibility testing. J Clin Microbiol 39:241–250

Thiam S., LeFevre A. M., Hane F., Ndiaye A., Ba F., Fielding K. L., Ndir M. and Lienhardt C. 2007. Effectiveness of a strategy to improve adherence to tuberculosis treatment in a resource-poor setting: a cluster randomized controlled trial. JAMA 297:380–386

UNICEF 2007 The State of the World's Children 2008 http://www.unicef.org/publications/files/The_State_of_the_Worlds_Children_2008.pdf Dated Accessed August 18, 2008

United Nations Statistics Division Composition of macro geographical (continental) regions, geographical sub-regions, and selected economic and other groupings http://unstats.un.org/unsd/methods/m49/m49regin.htm#developed Dated Accessed August 18

Victora C. G., Huicho L., Amaral J. J., Armstrong-Schellenberg J., Manzi F., Mason E. and Scherpbier R. 2006. Are health interventions implemented where they are most needed? District uptake of the integrated management of childhood illness strategy in Brazil, Peru and the United Republic of Tanzania. Bull World Health Organ 84:792–801

Volpato D. E., de Souza B. V., Dalla Rosa L. G., Melo L. H., Daudt C. A. and Deboni L. 2005. Use of antibiotics without medical prescription. Braz J Infect Dis 9:288–291

Wenzel R. P. 2004. The antibiotic pipeline – challenges, costs, and values. N Engl J Med 351:523–526

Wilkins K., Nsubuga P., Mendlein J., Mercer D. and Pappaioanou M. 2008. The Data for Decision Making project: assessment of surveillance systems in developing countries to improve access to public health information. Public Health 122:914–922

Ybarra M. L. and Bull S. S. 2007. Current trends in Internet- and cell phone-based HIV prevention and intervention programs. Curr HIV/AIDS Rep 4:201–207

Zhang R., Eggleston K., Rotimi V. and Zeckhauser R. J. 2006. Antibiotic resistance as a global threat: evidence from China, Kuwait and the United States. Global Health 2:6

Chapter 5
The Introduction of Antimicrobial Agents in Resource-Constrained Countries: Impact on the Emergence of Resistance

Carlos Franco-Paredes and Jose Ignacio Santos-Preciado

Abstract A potential undesirable effect of the introduction of a new class or type of antimicrobial drug is the selection and spread of resistant strains of microbial agents. As a result of this biological adaptation, antimicrobial agents become ineffective, leading to poor clinical outcomes at the individual level. At the community level, the spread of these resistant strains may lead to significant morbidity and mortality with potentially devastating economic and social consequences. In this chapter, we describe two case studies that demonstrate the occurrence of antimicrobial resistance after the introduction of antimicrobial drugs into developing settings. The first case describes the emergence of antiretroviral resistance after their recent introduction to control HIV/AIDS in resource-poor settings, particularly in Sub-Saharan Africa. The second case illustrates the negative consequences of the overuse of fluoroquinolones to treat enteric gram-negative bacteria in Southeast Asia. These events have rendered many of these bacterial pathogens resistant to this class of drugs and have caused a significant setback to control programs for diseases such as dysentery, typhoid fever, and gastroenteritis. We therefore suggest that after the introduction of a new antimicrobial agent into developing-country settings, the emergence and subsequent spread of antimicrobial resistance can be ameliorated by the concomitant implementation of programs targeting prudent use of these drugs. This could be achieved through the development of clinical guidelines and prospective surveillance activities to detect the early occurrence of resistance that may provide the opportunity to implement preventive strategies to impede further the spread of resistant microbial strains at the community and hospital levels.

J.I. Santos-Preciado (✉)
Facultad de Medicina, Departmento de Medicina Experimental, Universidad
Nacional Autonoma de Mexico
e-mail: jisantosp@gmail.com

A. de J. Sosa et al. (eds.), *Antimicrobial Resistance in Developing Countries*,
DOI 10.1007/978-0-387-89370-9_5, © Springer Science+Business Media, LLC 2010

5.1 Antimicrobial Resistance in Resource-Constrained Settings

In an era of increasing global migration, infectious diseases represent major national and international public health threats affecting resource-rich and resource-poor countries. The main strategies to prevent and control infectious diseases include public health improvements in sanitation and hygiene, safe water initiatives, as well as vaccines and the use of antimicrobial agents (WHO 2001). Regrettably, a potential outcome of antimicrobial use to control the clinical impact of these infections is the emergence of antimicrobial resistance as an evolutionary response of microbial pathogens (Livermore 2007; Moellering 1998; Rhem and Weber 2007).

While it is recognized that the burden of antimicrobial resistance represents a significant threat to the health-care costs and clinical outcomes of infectious diseases in resource-rich countries, the impact on resource-poor countries has been shown to be devastating (WHO 2001; Hart and Kariuki 1998; Gallant 2007). The issue is that limited resources become scarcer and therefore control strategies to prevent and treat infections become ineffective. Indeed, many of the infectious diseases that are potentially treatable with newly introduced antimicrobials are themselves frequently considered poverty-promoting conditions because they hamper economic and social human development (Byarugaba 2004, 2005; WHO 2001; Laxminarayan and Weitzmann 2003; Franco-Paredes et al. 2007). Moreover, the association of poverty with poor sanitation and hygiene, unsafe water, overcrowding, poor housing, and lack of environmental sustainability programs converges to facilitate transmission and spread of resistant infectious organisms (Tupasi 1999). Therefore, in many of these developing-country settings, infectious diseases have become a double threat due to the already existent burden of the infectious disease concomitant with the synergistic emergence and spread of drug-resistant microbial strains (Byarugaba 2004, 2005).

The selection and subsequent spread of antimicrobial resistance in developing countries is thus considered a societal and economic issue in health-care systems that may be already fractured (WHO 2001; Okeke et al. 2005a, Sirinavin and Dowell 2002). A resultant cycle of destitution and further underdevelopment ensues with high morbidity and mortality of previously treatable infectious diseases such as tuberculosis, malaria, typhoid fever, bacterial meningitis, and other life-threatening infections (Parry et al. 2002; Nosten et al. 2000; Dromigny and Perrier-Gros-Claude 2003; Huebner et al. 2000; Okeke et al. 2005a, 2005b; Corea et al. 2003; Weller 2007; Shekelle et al. 2007).

5.2 Risk Factors for the Development of Antimicrobial Resistance After the Introduction of Antimicrobial Agents into Resource-Constrained Settings

The deployment of new classes of antimicrobial drugs has been relentlessly followed by the selection and spread of resistant microbes (WHO 2001, Moellering 1998). As a result, antimicrobial resistance directly prevents the success of many infectious disease control programs in many resource-constrained settings (Okeke et al. 2005b; WHO 2001). Equally importantly, antimicrobial resistance may lead to a higher number of treatment failures that result in increasing morbidity and mortality of infectious diseases. An illustrative example of these phenomena is the significant increase in the incidence and mortality of *Plasmodium falciparum* in Sub-Saharan Africa (Barnes and White 2005). Many experienced researchers in this field have observed that the trends seen over the last few years can be directly linked to the continuous use and misuse of antimalarial drugs such as chloroquine and sulfadoxine–pyrimethamine combinations despite evidence of widespread resistance to these drugs in this area of the world (Nosten et al. 2000; Barnes and White 2005).

By adapting the WHO framework to study the development of antimicrobial resistance (WHO 2001), we suggest reviewing various interrelated determinants conducive to the selection and spread of resistance at many levels after the introduction of new antimicrobial agents in developing settings: (a) social determinants, (b) provider-related determinants, (c) patient-related determinants, (d) institutional determinants, and (e) governmental determinants (Fig. 5.1).

5.2.1 Social Determinants

It has been observed that rapid population growth and urbanization without a parallel increase in resources and services lead to overcrowding, poor sanitation, and inadequate health-care services in rural areas and periurban slums of many of major cities of the developing world (Byarugaba 2004, 2005; Riley et al. 2007, Franco-Paredes et al. 2007). The combination of these factors facilitates the spread of infections, such as typhoid, tuberculosis, and gastrointestinal infections, and if resistance occurs, these social determinants favor the spread of resistance in these populations (Laxminarayan 2003, Franco-Paredes et al. 2007; Riley et al. 2007). The social ecology of extreme poverty in most of these countries (poor sanitation, hunger, malnutrition, poor access to drugs, and

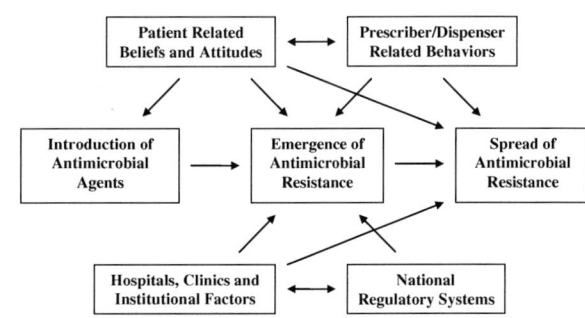

Fig. 5.1 Determinants of antimicrobial resistance after the introduction of antimicrobial agent into resource-poor settings

inadequate health care delivery) may precipitate the emergence and spread of antimicrobial resistance (Byarugaba 2004, 2005).

As an example, the spread of the extensively drug-resistant tuberculosis (XDR-TB) strain in Kwazulu Natal in South Africa in the last few years has been suggested to occur after the introduction and subsequent use of antimycobacterial agents in combination with detrimental social and economic factors (Dye et al. 2002; Pillay and Sturm 2007; Iscman 2007). Resistant strains of *Mycobacterium tuberculosis* that occur spontaneously are then selected and spread by the overuse or misuse of antimicrobial drugs increasing selective pressure. In this manner, the origin and subsequent spread of this antimicrobial resistance contributes to the increasing prevalence of resistance in any particular population (WHO 2001). The occurrence of multidrug-resistant tuberculosis (MDR-TB) or XDR-TB is associated with worse clinical outcomes in terms of morbidity and sequelae and mortality and a significant increase in health-care resources utilization and costs. For countries that have insufficient health-care services (resources for diagnosis, treatment, and adequate long-term follow-up), the spread of these resistant strains has made it impossible to provide appropriate care for many of these patients (Selgelid et al. 2008). It is not a coincidence to find XDR-TB in the most impoverished and forgotten populations of the world, where the confluence of lack of sufficient health-care services, low levels of sanitation, widespread poverty, and absence of health preventive services is at the beginning and at the end of this infinite vicious cycle. In this manner, social factors contribute to the selection and spread of *M. tuberculosis*-resistant strains after the introduction of antimicrobial chemotherapy.

5.2.2 Health-Care Provider Factors

In many resource-limited countries, health-care professionals are influenced by for-profit gains to inappropriately prescribe antimicrobial agents (WHO 2001). Some practitioners sell the antimicrobials to patients or receive a kick-back fee or other incentives by referring the patient to a particular pharmacy (Sirinavin and Dowell 2002). In this manner, many health-care practitioners presumptively prescribe antimicrobials even when there is no clinical indication (Byarugaba 2004, 2005). Furthermore, additional profit can be generated by the use of newer and more expensive antimicrobials in preference to previously used and effective therapies that are more affordable to the patient. In addition, in many developing settings, pharmaceutical companies have incentives for prescribers who use more of their newer and more expensive products (WHO 2001; Byarugaba 2004). Other factors related to health-care practitioners associated with the emergence of antimicrobial resistance in developing settings include lack of proper training with regard to the clinical diagnosis of infectious diseases and proper antimicrobial use (WHO 2001) (Fig. 5.1).

To illustrate the impact of prescriber practices in the emergence of antimicrobial resistance, a study conducted in Tamil Nadu State in India identified that the majority of health-care practitioners believed that antibiotics are over-prescribed (Vaypayee et al. 2007; Sirinavin and Dowell 2002). In this study, patient requests and expectations as well as patient satisfaction issues fostered health-care practitioners to inappropriately prescribe antibiotics (Vaypayee et al. 2007). Other studies conducted in other developing-country settings have shown similar patterns of behavior among health-care practitioners with regard to the misuse of antimicrobials (WHO 2001).

Many health-care practitioners operate on a daily basis without the support of ancillary diagnostic testing. There is evidence that the overuse of antimicrobials in these scenarios leads to the selection of resistance (Sirinavin and Dowell 2002). Indeed, the development of resistance is often directly linked to the absence of confirmatory testing for severe and often life-threatening infections such as typhoid fever, malaria, tuberculosis, and other highly prevalent infectious diseases (Okeke et al. 2005a, 2005b; Byarugaba 2004, 2005; Sirinavin and Dowell 2002).

5.2.3 Patient-Related Determinants

Patient's beliefs and perceptions regarding antimicrobials in developing settings are considered important determinants of the emergence of antimicrobial resistance after their introduction (WHO 2001, Sirinavin and Dowell 2002). Often a strong belief in resource-limited settings is that antibiotics are stronger and more powerful medications (WHO 2001, Sirinavin and Dowell 2002). Most patients believe that most infections respond to antimicrobials and expect to be prescribed antibiotics by their health provider for nonspecific symptoms that may not be even of any particular infection (WHO 2001, Sirinavin and Dowell 2002). Furthermore, in many developing countries, there is a widely dispersed belief that new and expensive medications are more efficacious than older and more affordable agents. A common belief, often shared and promoted by some prescribers and dispensers, is that the more expensive a drug is, the more effective it is (WHO 2001, Byarugaba 2004, 2005). Such perceptions often result in the unnecessary prescription dispensing or the use of newer agents (Hadi et al. 2006) with significant health-care consequences and with the potential outcome of selecting for resistant microbial strains (WHO 2001).

5.2.4 Health-Care Institutional Factors

Health-care centers, clinics, and hospitals are key components of the antimicrobial resistance problem in developing countries, similar to what occurs in more developed settings. However, the main difference in many developing settings is the consistent lack of sufficient staff, training of staff, and resources to enforce infection control practices and guidelines for rationale antibiotic use or drug shortages. In fact, these institutions are considered in many aspects the breeding grounds for the occurrence of antimicrobial resistance that subsequently may spread to other not previously uninfected patients and may also be a pool of resistant strains that may potentially spread to the community (Fig. 5.1). Transmission of highly pathogenic-resistant bacteria from patient to patient within the hospital environment amplifies the problem of antimicrobial resistance and may be responsible for spreading highly resistant bacteria to the community (WHO 2001; Duse 1998; Okeke et al. 2005a, 2005b). The absence or deficient use of guidelines in many clinical settings is conducive to a failure to implement and follow simple infection control practices such as hand washing in many resource-limited settings or to conduct infection control practices and supervision (Sirinavin and Dowell 2002; WHO 2001). Incorporating antibiotic use guidelines in clinical practice in combination with supportive interventions to oversee and modify inadequate behaviors for staff prescribing antimicrobials has provided the most successful interventions in some developing areas of the world (WHO 2001).

5.2.5 Governmental and Regulatory Issues

This group of determinants is a relevant contributing factor for the spread of antimicrobial resistance in developing countries, where a lack of regulatory agencies that oversee the production, distribution, and supervision of prescription practices by health-care practitioners is largely inexistent or insufficient (WHO 2001). Ideally, national drug procurement programs and policies should aim to contribute to the attainment of good standards of health for the population by ensuring the availability and adequate distribution of antimicrobial drugs and by promoting rational and prudent use (Hadi et al. 2006, WHO 2001).

Antibiotics can be obtained from pharmacies or other commercial locations without prescription in many developing countries (Hadi et al. 2006, WHO 2001, Sirinavin and Dowell 2002). As an example of these unregulated factors determining the establishment of antimicrobial resistance, a survey of rural medical practitioners in the Rajbari district of Bangladesh with regard to antibiotic use showed that they prescribed antibiotics to 60% of these patients. In this study, it was also described that 109 500 doses of antimicrobial drugs had been dispensed only by pharmacies, and a further 100 000 doses had been dispensed without a prescription (Hadi et al. 2006; Sirinavin and Dowell 2002).

In some developing countries, poor-quality antimicrobial preparations may lead to the presence of lower concentration of antibiotics, which may be associated with lower effective concentrations in human tissues, which may lead to therapeutic clinical failures and selection of drug-resistant strains of microbes (Hadi et al. 2006, WHO 2001). Similar problems may result from the distribution and use of counterfeit formulations in many developing settings (Newton et al. 2001). Counterfeit or substandard medications, particularly antimalarial drugs and other antimicrobials that provide significant financial profits, have been identified in many resource-limited regions, particularly in Sub-Saharan Africa and Southeast Asia (Newton et al. 2001; Lon et al. 2006). The use of counterfeit drugs, besides being clinically ineffective, provides a false sense of security and may result in community distrust of the health-care system. This may result in the long term in individuals attempting to reach pharmacies that will inadequately prescribe antibiotics without a prescription and a clinical indication that could potentially select for further antimicrobial resistance.

Finally, in many developing settings, hospital administrators and governments may not be readily familiar with the extent of the problem associated with antimicrobial resistance from an economic and societal perspective. Patients and the general population living in impoverished areas of the world may not have adequate access to health-related information in a health literacy-friendly format to assist in the prevention of the emergence of antimicrobial resistance (Hadi et al. 2006; Sirinavin and Dowell 2002; Byarugaba 2004; WHO 2001; Okeke et al. 2005a).

In summary, the introduction of antimicrobial drugs into developing countries is frequently followed by the emergence of resistance due to a combination of the above factors (Fig. 5.1). To illustrate the convergence of these multiple factors, we discuss two case scenarios. The first one is in regard to the introduction of antiretroviral drugs into Sub-Saharan Africa and Southeast Asia; although very successful at many levels, this activity has shown that antiviral resistance is becoming an important public health issue in many developing settings. In addition, the introduction of quinolones to control and prevent subsequent spread of gram-negative pathogens in many developing countries, particularly in Southeast Asia, has led to the emergence of highly resistant pathogens with significant economic and biomedical consequences.

5.3 Antimicrobial Resistance After the Introduction of Antiretroviral Therapy into Developing Countries

More recently, new antiretroviral combinations, the increased global support to treatment access programs, and measures to reduce the cost of antiretroviral agents have made HIV treatment accessible to many populations living in impoverished areas of the world. In fact, the World Health Organization/ Joint United Nations Programme on HIV/AIDS (WHO/UNAIDS) launched the initiative "3 by 5" in 2003. The objective of this program was to deliver antiretroviral therapy to 3 million people in low- and middle-income countries by the end of 2005 (Weller 2007). While this effort has fallen short of its original target, many resource-constrained settings are currently using antiretroviral drugs. As it occurs in developed countries, delivery of these highly effective drugs requires having an adequate infrastructure to assure adherence, surveillance, and other concomitant preventive strategies (Akileswaran et al. 2005). Indeed, close to 1 million people have been started on highly active antiretroviral treatment (HAART) in Sub-Saharan Africa, the most affected area of the world. Access to antiretroviral therapy in the developing world has had a dramatic impact on morbidity and mortality (Akileswaran et al. 2005). However, there are significant limitations of these life-saving drugs. Suboptimal laboratory monitoring and poor surveillance may lead to unrecognized spread of antiviral resistance that may impede future control efforts in the most HIV-stricken areas of the world. In fact, efforts to evaluate program and operational performance are needed to find the best models to integrate these programs where availability of resources is scarce (Weller 2007). This has prompted WHO and other international regulatory agencies to recommend CD4 T-lymphocyte count measurements along with routine clinical markers (Bagchi et al. 2007).

In the case of HIV, drug-resistant strains have been recognized as a serious threat to the efficacy of current antiretroviral treatment and could jeopardize efforts to increase access to treatment in countries most affected by the HIV pandemic (Gallant 2007). The potential for the transmission of HIV drug-resistant strains has been recognized as a serious threat to the efficacy of current antiretroviral treatments (Shekelle et al. 2007). The preferred HAART combination used in these settings includes the use of two nucleoside reverse transcriptase inhibitors (NRTIs), stavudine or zidovudine and lamivudine, and one non-nucleoside reverse transcriptase inhibitors (NNRTI), mostly nevirapine. This regimen is widely available, affordable for most countries particularly as the availability of generic antiviral drugs increases (Gallant 2007; Weller 2007). This regimen has major drawbacks, such as liver toxicity, mitochondrial toxicity, and other significant side effects, and concern about resistance and its effect on the availability of second-line therapy (Gallant 2007). In a recent report from Thailand, almost all patients that experienced treatment failure with a first-line regimen (stavudine, lamivudine, and nevirapine) developed resistance to NNRTs and lamivudine. Thus, in about 50% of them, alternative options for a second-line regimen were limited (Sungkanuparaph et al. 2007). This study demonstrates that strategies for prevention of HIV-associated resistance are important and the identification of early virologic failure may provide a better window of opportunity to provide better treatment outcomes in HIV-infected individuals living in resource-limited settings (Weller 2007; Gallant 2007).

A recent report has mapped antiretroviral drugs resistance in the developing world (Shekelle et al. 2007). This study has shown that patterns of antiretroviral resistance among treatment naïve populations worldwide appear to reflect geographic trends in the use of antiretroviral drugs. The rate of resistance was identified to be around 7.4% in Asia, 5.5% in Africa, and 6.4% in Latin America. Besides these identified geographic patters of resistance, certain subgroups of patients have been identified to be prone to develop resistance (Shekelle et al. 2007). In this regard, women taking intrapartum single-dose nevirapine to prevent mother-to-child transmission is the population with the highest identified risk for development of mutant viral strains (Gallant 2007; Kurle et al. 2007; Weller 2007). In some settings, the emergence of nevirapine resistance in this subgroup of patients has prompted significant concern to the point of challenging the simplicity, efficacy, and cost-effectiveness of nevirapine use in developing countries (Kurle et al. 2007; Weller 2007).

While there is compelling evidence that HAART can be feasibly administered in resource-limited settings, ongoing prospective evaluations to detect early virologic failure and determine antiviral resistance remain a public health and biomedical priority (Akileswaran et al. 2005; Shekelle et al. 2007). This significant benefit is demonstrated by an assessment of 28 studies analyzing antiretroviral programs in Sub-Saharan Africa showing evidence that in most populations in this region of the world, there are high rates of adherence and optimistic outcomes (Akileswaran et al. 2005). Avoiding further antiretroviral resistance is critical to preserve the efficacy of antiretroviral drugs when only first-line regimens might be available in some resource-limited settings (Lazzari et al. 2004). In addition, there is clear evidence that prospective studies to evaluate adherence to antiretroviral drugs in developing regions should also be conducted in association with resistance testing to guide the use of these highly valuable antiviral drugs.

5.4 Antimicrobial Resistance After the Introduction of Quinolones into Developing Countries

Increased incidence of fluoroquinolone resistance among gram-negative isolates has been reported from various studies in many resource-constrained countries. Indeed, studies from Pakistan, India, and Vietnam have demonstrated the emergence of resistance to quinolones among gram-negative bacteria such as *Neisseria gonorrhoeae, Salmonella* Typhi, *Salmonella* Paratyphi, *Shigella dysenteriae, Escherichia coli*, and other gram-negative enterobacteria (Sabir et al. 2004; Ray et al. 2000; Jorgensen et al. 2005; Kauser et al. 2006; Tupasi 1999; Okeke et al. 2005a, 2005b; Dromigny and Perrier-Gros-Claude 2003). Correlation between the use of fluoroquinolones and the emergence of resistance to these drugs has been established by evidence of their use in the community and in the hospital setting (Sabir et al. 2004).

In many of these settings, the opportunity and early enthusiasm for controlling dysentery, salmonellosis, and typhoid fever with the availability of fluoroquinolones has been replaced with pessimism brought in by the emergence of resistance to these valuable drugs. As mentioned above, the development of resistance to quinolones with their use in developing countries follows their overuse or misuse due to provider, patient, institutional, or governmental determinants (Fig. 5.1). In addition, sociocultural attitudes and economic policies are also responsible for the emergence of antimicrobial resistance to quinolones and other antibiotics (Tupasi 1999). To minimize selective pressure with newer fluoroquinolones, surveillance of antimicrobial resistance in sentinel organisms responsible for community-acquired infections, education policies to educate medical professional on rational antibiotic use through the development of guidelines, and other programs are urgently needed in many developing countries (Tupasi 1999; Sabir et al. 2004; Ray et al. 2000; Jorgensen et al. 2005; Jabeen et al. 2006; Kauser et al. 2006).

5.5 The Way Forward

As discussed above, antimicrobial agents, including antiretroviral agents, have the potential to select drug-resistant populations of microorganisms when their use becomes widespread after their introduction into developing countries. The natural process of microbial strains that are biologically resistant can be accelerated by inappropriate or irrational antimicrobial drug usage. Therefore, monitoring of antimicrobial resistance is essential for clinical or public health programs focused in the control of infectious diseases (WHO 2001). In addition, the implementation of containment strategies to ameliorate the impact of antimicrobial resistance in developing settings is also fundamental.

Several important challenges to contain the emergence and spread of antimicrobial resistance remain. Behavioral changes of providers and patients in combination with improvements in regulatory activities at a national and local levels offer an important opportunity to prevent and contain antimicrobial resistance in developing settings. Thus, paying attention to the emergence and subsequent spread of antimicrobial resistance in developing countries is a critical public health priority. In particular, establishing surveillance programs to detect resistance concomitantly to the introduction of a new class of drugs represents potentially the best possible strategy to reduce and contain the spread of resistance in these settings.

However, we must place social determinants as a key component of the strategies for the control and elimination strategies against the emergence and spread of infectious diseases in the developing world. We need to continue to directly address the larger, socially determined factors such as poverty, inadequate levels of sanitation and hygiene, and the lack of adequate health care services. Indeed, larger societal issues emerging from underdevelopment and poverty further contribute to the problem of antimicrobial resistance (Laxminarayan and Weitzman 2002; Byarugaba 2004, 2005). Unregulated construction and urbanization in slums and periurban slums characterized by a lack of clean water, sanitation, and medical infrastructure constitute in many developing settings the mixing vessel for antimicrobial-resistant strains of microbes to spread in the community and produce devastating health consequences (WHO 2001, Riley et al. 2007).

References

Akileswaran, C., Lurie, M. N., Flanigan, T. P., and Mayer, K. H. 2005. Lessons learned from use of highly active antiretroviral therapy in Africa. Clin. Infect. Dis. 41:376–385.

Bagchi, S., Kempf, M. C., Westfall, A. O., Maherya, A., Willig, J., and Saag, M. S. 2007. Can routine clinical markers be used longitudinally to monitor antiretroviral therapy success in resource-limited settings? Clin. Infect. Dis. 44:135–138.

Barnes, K. I., and White, N. J. 2005. Population biology and antimalarial resistance: The transmission of malaria drug resistance in *Plasmodium falciparum*. Acta Trop. 94: 230–240.

Byarugaba, D. K. 2004. A view on antimicrobial resistance in developing countries and responsible risk factors. Int. J. Antimicrob. Agents 24:105–110.

Byarugaba, D. K. 2005. Antimicrobial resistance and its containment in developing countries. In Antibiotic policies: Theory and practice, ed. I. M. Gould, and J. W. M. van der Meer, pp. 617–647. New York: Kluwer Academic/Plenum Publishers.

Corea, E., de Silva, T., and Perera, J. 2003. Methicillin-resistant *Staphylococcus aureus*: Prevalence, incidence and risk factors associated with colonization in Sri Lanka. J. Hosp. Infect. 55:145–148.

Duse, A. G. 1998. Antibiotic resistance in developing countries. In Infection control practices, ed. A. Emerson, and M. Arrowsmith, pp. 38–44. Borken, Germany: 3 M Medical Markets Laboratory.

Dromigny, J. A., and Perrier-Gros-Claude, J. D. 2003. Antimicrobial resistance of *Salmonella enterica* serotype Typhi in Dakar, Senegal. Clin. Infect. Dis. 37:465–466.

Dye, C., Williams, B. G., Espinal, M. A., and Raviglione, M. C. 2002. Erasing the world's slow stain: Strategies to beat multidrug-resistant tuberculosis. Science 295:2042–2046.

Franco-Paredes, C., Jones, D., Rodriguez-Morales, A. J., and Santos-Preciado, J. I. 2007. Improving the health of neglected populations in Latin America. BMC Public Health 7:11. DOI: 10.1186/1471-2458-7-11.

Gallant, J. E. 2007. Drug resistance after failure of initial antiretroviral therapy in resource-limited countries. Clin. Infect. Dis. 44:453–455.

Hadi, U., Kolopaking, E. P., Gardjito, W., Gyssens, I. C., and van den Broek, P. J. 2006. Antimicrobial resistance and antibiotic use in low-income and developing countries. Folia Medica Indonesiana, 42:183–195.

Hart, C. A., and Kariuki, S. 1998. Antimicrobial resistance in developing countries. Br. Med. J. 317:647–650.

Huebner, R. E., Wasas, A. D., and Klugman, K. P. 2000. Trends in antimicrobial resistance and serotype distribution of blood and cerebrospinal fluid isolates of *Streptococcus pneumoniae* in South Africa, 1991–1998. Int. J. Infect. Dis. 4:214–218.

Iseman, M. D. 2007. Extensive drug-resistant *Mycobacterium tuberculosis*: Charles Darwin would understand. Clin. Infect. Dis. 45:1415–1416.

Jabeen, K., Khan, E., and Hasan, R. 2006. Emergence of quinolone-resistant Neisseria gonorrhoeae in Pakistan. Int. J. STD AIDS. 17(1): 30–33.

Jorgensen, J. H., Crawford, S. A., and Fiebelkorn, K. R. 2005. Susceptibility of *Neisseria meningitidis* to 16 antimicrobial agents and characterization of resistance mechanisms affecting some agents. J. Clin. Microbiol. 43:3162–3171.

Kauser, J., Khan, E., and Hasan, R. 2006.Emergence of quinolone-resistant *Neisseria gonorrhoeae* in Pakistan. Int. J. STD AIDS 17: 30–34.

Kurle, S. N., Gangakhedkar, R. R., Sen, S., Hayatnagarkar, S. S., Tripathy, S. P., and Paranjape, R. S. 2007. Emergence of NNRT drug resistance mutations after single-dose nevirapine exposure in HIV type 1 subtype C-infected infants in India. AIDS Res. Hum. Retroviruses 23:682–685.

Lazzari, S., de Felici, A., Sobel, H., and Bertagnolio, S. 2004. HIV drug resistance surveillance: Summary of an April 2003 WHO Consultation. AIDS 18 (suppl 3):S49–S53.

Laxminarayan, R., and Weitzman, M. L. 2002. On the implications of endogenous resistance to medications. J. Health Econ. 21:709–718.

Laxminarayan, R., and Weitzman M. L. 2003. Value of treatment heterogeneity for infectious diseases. In Battling resistance to antibiotics and pesticides: An economic approach. ed. R. Laxminarayan, pp 63–75. Washington, DC: Resources for the Future.

Livermore, D. M. 2007. Introduction: The challenge of multiresistance. Int. J. Antimicrob. Agents 29 (suppl 3):S1–S7.

Lon, C. T., Tsuyuoka, R., Phanouvong, S., Nivanna, N., Socheat, D., Sokhan, C., Blum, N., Christophel, E. M., and Smine, A. 2006. Counterfeit and substandard antimalarial drugs in Cambodia. Trans. R. Soc. Trop. Med. Hyg. 100:1019–1024.

Moellering, R. C. Jr. 1998. Antibiotic resistance: Lessons for the future. Clin. Infect. Dis. 27 (suppl 1):S135–S140.

Newton, P., Proux, S., Green, M., Smithuis, F., Rozendaal, J., Prakongpan, S., Chotivanich, K., Mayxay, M., Looareesuwan, S., Farrar, J., Nosten, F., and White, N. J. 2001. Fake artesunate in Southeast Asia. Lancet 357:1948–1950.

Nosten, F., van Vugt, M., Price, R., Luxemberger, C., Thway, K. L., Brockman, A., McGready, A., ter Kuile, F., Looareesuwan, S., and White, N. J. 2000. Effects of artesunate–mefloquine combination on incidence of *Plasmodium falciparum* malaria and mefloquine resistance in Western Thailand: A prospective study. Lancet 356:297–302.

Okeke, I. N., Laxminarayan, R., Bhutta, Z. A., Duse, A. G., Jenkins, P., O'Brien, T. F., Pablos-Mendez, A., and Klugman, K. P. 2005a. Antimicrobial resistance in developing countries. Part 1: Recent trends and current status. Lancet 5:481–493.

Okeke, I. N., Klugman, K. P., Bhutta, Z. A., Duse, A. G., Jenkins, P., O'Brien, T. F., Pablos-Mendez, A., and Laxminarayan, R. 2005b. Antimicrobial resistance in developing countries. Part II: Strategies for containment. Lancet 5:568–580.

Parry, C. M., Duong, N. M., Zhou, J., Hoang Mai, N. T., Diep, T. S., Thinh, L. Q., Wain, J., Van Vinh Chau, N., Griffings, D., Day, N. P., White, N. J., Hien, T. T., Spratt, B. G., and Farrar, J. J. 2002. Emergence in Vietnam of *Streptococcus pneumoniae* resistant to multiple antimicrobial agents as a result of dissemination of the multiresistant Spain 23F^{-1} clone. Antimicrob. Agents Chemother. 46:3512–3517.

Pillay, M., and Sturm, A. W. 2007. Evolution of the extended drug-resistant F15/LAM4/KXZN strain of *Mycobacterium tuberculosis* in KwaZulu-Natal, South Africa. Clin. Infect. Dis. 45:1409–1414.

Ray, K., Bala, M., Kumar, J., and Misra, R. S. 2000. Trend of antimicrobial resistance in *Neisseria gonorrhoeae* at New Delhi, India. Int. J. STD AIDS 11:115–118.

Rhem, S. J., and Weber, T. 2007. The far-reaching impact of antimicrobial resistance. Clin. Infect. Dis. 45:S97–S98.

Riley, L. W., Ko, A. I., Unger, A., and Reis, M. G. 2007. Slum health: Diseases of neglected populations. BMC Int. Health Hum. Rights 7:2.

Sabir, N., Khan, E., Sheikh, L., and Hasan, R. 2004. Impact of antibiotic usage on resistance in microorganisms; urinary tract infections with *E. coli* as a case in point. J. Pak. Med. Assoc. 54:472–475.

Selgelid, M. J., Kelly, P. M., and Sleigh A. 2008. Ethical challenges in TB control in the era of XDR-TB. Int. J. Tuberc. Lung Dis. 12:231–235.

Shekelle, P., Maglione, M., Geotz, M. B., Wagner, G., Wang, Z., Hilton, L., Carter, J., Chen, S., Tringle, C., Mojica, W., and Newberry, S. 2007. Antiretroviral (ARV) drug resistance in the developing world. Evid. Rep. Technol. Assess. 156:1–74.

Sirinavin, S., and Dowell, S. F. 2002. Antimicrobial resistance in countries with limited resources: Unique challenges and limited alternatives. Semin. Pediatr. Infect. Dis. 15: 94–98.

Sungkanuparaph, S., Manosuthi, W., Kiertiburanakul, S. Piyavong, B., Chumpathat, N., and Chantratita, W. 2007. Options for a second-line antiretroviral regimen for HIV type-1 infected patients whose initial regimen of a fixed-dose combination of stavudine, lamivudine, and nevirapine falls. Clin. Infect. Dis. 44:447–452.

Tupasi, T. E. 1999. Quinolone use in the developing world. Drugs 58 (Supp.2):55–59.

Vaypayee, M., Kaushik, S., Mojumdar, K., and Sreenivas, V. 2007. Antiretroviral treatment in resource-poor settings: A View from India. Indian J. Med. Sci. 61:390–397.

WHO 2001. WHO Global strategy for containment of antimicrobial resistance. Geneva, Switzerland: World Health Organization.

Weller I (2007). Delivery of antiretroviral therapy in Sub-Saharan Africa. Clin. Infect Dis. 43: 777–778.

Part II
The Human Impact of Resistance

Chapter 6
Human Immunodeficiency Virus: Resistance to Antiretroviral Drugs in Developing Countries

Rebecca F. Baggaley, Maya L. Petersen, Marcelo A. Soares,
Marie-Claude Boily, and Francisco I. Bastos

Abstract This chapter reviews issues central to understanding the emergence and transmission of drug-resistant human immunodeficiency virus (HIV) and its impact on developing countries. We first give an overview of HIV, HIV treatment using antiretroviral drugs, and access to treatment in developing countries. Then we review current understanding of the impact of adherence and treatment interruption on the emergence of resistance among treated individuals (secondary resistance) and factors contributing to secondary resistance in resource-poor settings. Transmitted (or primary) resistance, which can threaten the effectiveness of antiretroviral regimens among treatment-naïve individuals, is also discussed. Furthermore, we address how antiretroviral delivery systems in developing countries may impact resistance.

Mathematical models of HIV transmission offer important insights into the course of HIV epidemics and how expanded access and policies for antiretroviral delivery in developing countries may impact resistance. We summarize the major findings from published modeling studies and discuss their predictions and limitations. We then review available empirical data on antiretroviral resistance in developing countries. Finally, we discuss the implications of these findings for policy and the monitoring of epidemic trends.

6.1 Introduction

Human immunodeficiency virus (HIV) is transmitted sexually, in blood or blood products, and perinatally. The World Health Organization (WHO) estimates that 33.2 million people worldwide are currently infected with HIV; of these, more than 95% live in low- and middle-income countries. Infection with HIV leads to immune dysfunction including depletion of $CD4^+$ T

R.F. Baggaley (✉)
Department of Infectious Disease Epidemiology, Faculty of Medicine,
Imperial College, London, UK; Département de pathologie et microbiologie, Faculté
de médecine vétérinaire, Université de Montréal, Montréal, Québec, Canada
e-mail: r.baggaley@imperial.ac.uk

A. de J. Sosa et al. (eds.), *Antimicrobial Resistance in Developing Countries*,
DOI 10.1007/978-0-387-89370-9_6, © Springer Science+Business Media, LLC 2010

lymphocytes, leaving the host vulnerable to opportunistic infections. As a result, if left untreated, HIV infection leads ultimately to acquired immune deficiency syndrome (AIDS) and death.

For the first decade of the epidemic, options for treating HIV infection were limited to therapies aimed at preventing and treating the range of opportunistic infections to which HIV-infected individuals were susceptible. In 1987, zidovudine (ZDV), also known as azidothymidine (AZT), was approved in the United States as the first antiretroviral drug directly targeting HIV. Other drugs of the same class, called nucleoside or nucleotide reverse transcriptase inhibitors (NRTIs), were later developed. While ZDV and other early NRTIs improved short-term survival among patients with AIDS, the virus-infected patients treated with one or two drugs from this class (mono or dual therapy) quickly developed resistance, limiting the ability of these early regimens to offer durable benefits.

The treatment of HIV was revolutionized with the introduction in 1996 of a second major antiretroviral drug class, the protease inhibitors (PIs), shortly followed by the introduction of a third class, the non-nucleoside reverse transcriptase inhibitors (NNRTIs). By combining drugs from classes that use distinct mechanisms and that target different points in the HIV lifecycle, regimens containing a PI or NNRTI drug together with two or more NRTIs were able to substantially delay the emergence of resistance. For the first time, the resulting triple-combination antiretroviral therapy (often known as highly active antiretroviral therapy or HAART, referred to here as ART) was able to successfully suppress viral replication for prolonged periods, particularly among subjects without pre-existing antiretroviral resistance (Palella et al. 1998). As a result, ART has dramatically decreased HIV-related morbidity and mortality in the developed world, as well as in developing countries, such as Brazil, where access to ART has been consistently provided.

There has been continued improvement in the effectiveness and tolerability of antiretroviral regimens. Low doses of ritonavir were added to PI-based combination regimens to create what is known as a "boosted" PI regimen. Ritonavir inhibits PI drug metabolism, helping to maintain elevated drugs levels in the blood, and thus improves viral suppression and reduces the risk of the emergence of resistant virus. Antiretroviral regimen development has also focused on the simplification of complicated dosing schemes (such as developing and using fixed-dose combinations (FDCs), which include two or more drugs in one tablet) and the reduction of adverse effects, targeting two common barriers to regimen adherence. Several new antiretroviral drug classes are also currently under development, or have been recently released (Dunne 2007).

When patients "fail" their first-line regimen, either due to intolerable side effects or treatment failure (which can be defined as rebound of HIV viral load, or immune decline detected through decline in CD4 count, or clinical failure through development of new opportunistic infections), patients are switched to a second-line regimen. While treating patients who have developed resistance to previous antiretroviral regimens is less effective than treating patients infected with drug-sensitive virus, ART still generally achieves durable viral suppression (Deeks et al. 1999).

6.1.1 Emergence and Transmission of Resistant HIV

While triple regimens are able to achieve durable viral suppression, resistance evolves when HIV replicates in the presence of suboptimal drug levels, which is commonly caused by intermittent therapy and poor adherence (Bangsberg et al. 2004). If a drug dose is delayed or missed, drug concentrations fall below the level required to suppress viral replication. As a result, the virus can evolve under the selective pressure of drug, where resistant strains have a competitive advantage over drug-sensitive (wild-type) virus.

The frequent evolution of resistance is attributable to certain characteristics of the virus. HIV replicates very quickly and has built-in mechanisms that predispose to a high rate of mutations: (1) the enzyme responsible for replicating HIV genetic information lacks a proofreading mechanism; and (2) the virus can recombine at a high rate. As a result, HIV exhibits remarkable genetic diversity, existing even within a single human host as a diverse population of viral quasispecies. While most mutations reduce viral fitness compared to wild-type, some confer resistance to one or more antiretrovirals, and thus can improve fitness in the presence of treatment, resulting in rapid selection of drug resistant strains from pre-existing variants upon drug selective pressure.

Antiretroviral-resistant HIV can be transmitted by any of the known HIV transmission routes. Any risk factors affecting transmission of HIV, such as risky behavior, presence of co-infections (which can increase infectivity), and ART (which decreases infectiousness due to viral suppression) may also influence transmission of resistance. In addition, factors favoring the emergence of resistance among infected individuals will also tend to contribute to primary resistance. Patients on suboptimal therapy in particular are both more likely to develop resistance and to have higher levels of viral replication, and thus higher infectiousness than patients on fully suppressive therapy.

Emergence and transmission of drug resistance have become major challenges to efforts to treat and prevent HIV in the developed world. A decade after the advent of triple ART, an estimated 76% of patients receiving ART in the United States carry HIV strains resistant to one or more antiretroviral drugs (Richman et al. 2004). In addition, resistant viruses continue to be transmitted via sexual and intravenous contact, infecting up to 20% of drug-naive individuals in the United States (Blower et al. 2005). Infection of a subject with a resistant strain may compromise their initial and/or subsequent therapies, leading to early treatment failure, as well as making the individual more likely to transmit resistant virus to newly infected individuals.

6.1.2 Antiretroviral Therapy in Developing Countries

The burden of HIV infection and disease falls overwhelmingly on the developing world. However, until recently, access to ART in these settings (with some notable exceptions among middle income countries such as Brazil) has been extremely limited, mainly due to associated costs. These include the cost of drugs, diagnosis, distribution, and monitoring requirements. The political will for scale-up of ART coverage in resource-poor settings has grown substantially in recent years and financial barriers have also been lowered as ART regimen costs have decreased dramatically. This has been driven by several factors, including the production of generic versions of several drugs and agreements by pharmaceutical companies to lower prices of key drugs for sale to resource-poor settings.

As a result of increased funding and political will and decreased drug costs, access to ART in developing countries has been vastly improved in recent years. According to UNAIDS, as of December 2007, approximately 3 million people in low- and middle-income countries were receiving treatment, 31% of those in need (UNAIDS 2008). While far short of the level of antiretroviral access needed, this represents a dramatic increase over the estimated 400,000 people in the developing world who were receiving ART in 2003.

The success of efforts to expand ART access has varied significantly between countries. ART rollout has been very successful in some middle-income countries, such as Botswana and Thailand, but has been slow or nonexistent in countries plagued by civil wars, such as Sudan. In other nations, such as Mozambique and Zimbabwe, dire poverty, former wars, and lack of trained personnel or adequate infrastructure have seriously compromised the achievement of even modest goals (WHO 2007).

While cost presented the greatest barrier to expanding access to life-saving ART, early opposition also stemmed in part from the concern that "widespread, unregulated access to antiretroviral drugs in sub-Saharan Africa could lead to the rapid emergence of drug-resistant viral strains, spelling doom for the individual, curtailing future treatment options, and leading to transmission of resistant virus" (Harries et al. 2001). It was debated whether access to ART should be expanded only in areas that were first able to achieve a level of healthcare infrastructure deemed sufficient to control the emergence and spread of resistant virus. The consensus has now shifted to favor the rapid expansion of access to ART in developing countries, with emphasis shifting to questions of how best to accomplish this goal.

Although experience with ART delivery has not resulted in the "antiretroviral anarchy" predicted by some (Harries et al. 2001), emergence and transmission of resistant HIV in developing countries (as in industrialized countries) remains a pressing concern. Continued expansion of antiretroviral access is clearly necessary to meet the need for access to these life-saving therapies. As the numbers on ART increase worldwide, so do the opportunities for emergence of resistance. While this should not be considered a barrier to increased rollout, understanding of ways in which resistance emerges and is transmitted and how this can be limited is needed to maximize the effectiveness of ART rollout programs.

6.2 Emergence and Transmission of Resistant HIV

6.2.1 Adherence and Emergence of Resistance

Adherence to prescribed drug regimens is crucially important in limiting treatment failure and the evolution of antiretroviral-resistant viral variants by preventing exposure of replicating virus to suboptimal drug levels. The relationship between poor adherence and the evolution of resistance is complex and can vary depending on the ART regimen used. Among patients treated with a nonboosted PI regimen, near-perfect adherence levels are required to achieve lasting health benefits and to prevent the emergence of drug resistance to that regimen in the short term (Bangsberg et al. 2004). In contrast, ritonavir-boosted, PI-based regimens reduce the risk of resistance at high to moderate adherence levels, while resistance to NNRTI-only-based regimens emerges most quickly at low levels of adherence (<40%) (Bangsberg et al. 2004). The vast majority of first-line regimens in developing countries are NNRTI-based, which may influence the rate at which resistance will emerge. Timing of missed doses also plays a role. Poor adherence immediately after initiating ART, when viral loads are still high, can accelerate the emergence of resistance by exposing a large pool of replicating virus to the selective pressure of drug exposure.

Adherence to ART must be maintained not only when the patient feels ill but for life, often in the absence of any symptoms. Regimens are often complicated, can involve multiple doses in a given day, and can entail serious side effects. While simpler and less-toxic regimens continue to be developed, maintaining high levels of adherence across a lifetime remains a significant challenge. Several factors are associated with adherence to ART, including complexity of regimen, side effects, depression, and other mental illness (Chesney 2000). However, no characteristics are fully predictive in individual patients, and all patients may be amenable to interventions to improve adherence.

Despite initial concerns about potentially low adherence levels to ART in resource-poor settings (Weiser et al. 2003; Stevens et al. 2004), emerging evidence from pilot ART programs suggests that reported adherence rates may be as high as or higher than those in industrialized countries (Mills et al. 2006). However, treated patients in developing countries are often sicker and have generally been on therapy for less time than those in rich countries. Both factors may contribute to increased motivation and higher adherence levels. Thus, adherence in developing countries is expected to decline over time, as a wider range of patients remain on therapy for longer. A study in Uganda showed that adherence declined over the first 6 months of therapy; however, in this and other African studies, adherence remained high (above 80%) after 6-months follow-up (Oyugi et al. 2007). Long-term retention of patients within ART programs is also crucial for success and appears to be a major challenge in sub-Saharan Africa, where a recent review found that an average of 78% patients were retained over 10 months among 33 independent cohorts, with the worst retaining only 46% (Rosen et al. 2007). Loss to follow-up was the major cause of this, followed by death.

While adherence observed in developing countries has been generally good, there are several reports of ART resistance emerging in Africa in programs where the ideal conditions for ART distribution are not met (Adje et al. 2001). For example, in Senegal, the main stated reasons for nonadherence were forgetfulness and financial difficulties (Laurent et al. 2002). Common reasons for nonadherence to ART in resource-poor settings vary by region but include financial constraints, forgetfulness, stigma, transportation and/or migration issues, and tolerability.

6.2.2 Treatment Interruption and Emergence of Resistance

In general, treatment interruptions (even structured, i.e., deliberate interruptions, sometimes referred to as "drug holidays") can contribute to the emergence of resistance by essentially creating a period of effective monotherapy. Because the different antiretroviral drugs of a given ART regimen are metabolized at different rates, the virus is able to replicate at a moderate rate in the presence of residual levels of whichever drug is metabolized most slowly, favoring the evolution of resistance to this drug. Phillips et al. (2007) predicted that the risk of resistant virus evolving was far higher for frequent short-term treatment cessations than for longer term interruptions because the "riskiest" time for resistance emergence, immediately after stopping ART (when many target CD4 cells have been produced due to suppression of viral replication and are ready to be rapidly infected), is repeated many times. Available data from both the developing and the developed world suggest that multiple treatment interruptions of at least 48 hours each are a good predictor of virologic failure and the emergence of resistance (Spacek et al. 2006; Oyugi et al. 2007). More generally, recent findings point to the high rates of immunologic virologic, and clinical failure associated with any ART discontinuation, even in the context of structured interruptions (Benson 2006).

In a treated cohort in Uganda, Oyugi et al. found that treatment interruptions were primarily attributed to financial constraints and pharmacy stockouts due to interruptions in the drug supply. They reported that 90% of all missed doses observed over 6 months occurred during treatment interruptions (or gaps) where patients stopped all drugs for over 48 hours (Oyugi et al. 2007).

In this cohort, treatment interruptions also played a more significant role in generating resistance than did poor adherence at the level of missed or delayed doses rather than gaps. Similarly, in Botswana, gaps in treatment (29%) were reported more frequently than day-to-day nonadherence (17%) most often due to inadequate insurance coverage for expensive medication regimens (69%), frequent travel and migration (16%), and lack of access to alternative regimens when side effects proved intolerable (22%) (Weiser et al. 2003). While these findings may not generalize to other resource-poor settings, they suggest the importance of ensuring a constant drug supply and of removing financial barriers contributing to intermittent use of treatment.

6.2.3 Measurement of Primary Resistance

Current prevalence estimates of transmitted HIV drug resistance (primary resistance) are generally based on standard genotyping assays, where HIV genes corresponding to current targets of ART are sequenced. These sequences are then compared to a laboratory wild-type strain that represents standard drug sensitivity. Although several automated algorithms are available to interpret results, expert opinion is required. More expensive and therefore less available are phenotypic assays, which measure in vitro susceptibility of patient-derived HIV genes to each antiretroviral compound. Unfortunately, these standard genotypic or phenotypic assays are imperfect tools to guide therapy choices and assess the degree of transmitted resistance. They generally detect only viral strains that represent at least 15–20% of the total viral population within an individual, yet the presence of any minor but undetected resistant strains may compromise future therapeutic schemes and impact on the selection of drug resistance. New, more sensitive technologies have been developed, but remain expensive.

The distinction between acutely and chronically HIV-infected, drug-naïve subjects is important when measuring primary drug resistance. In acutely infected individuals, there is a short time between infection and resistance testing, and therefore there is little time for the evolution of the virus that was transmitted. Conversely, in chronically infected individuals, a long time (sometimes years) may have elapsed between transmission and testing, allowing more time for viral evolution in the absence of antiretroviral selective pressure. Variants carrying mutations tend to be archived over time as minor subspecies within the viral population that are too small to be detected by currently used assays, but which can quickly reappear and become the dominant strain if the infected patient starts ART. This is because resistance mutations usually result in lower fitness and/or replicative capacity in the absence of drug. As a result, even if an individual was infected with a virus carrying resistance mutations, a genotypic test taken years later may fail to identify these mutations amongst the circulating virus sampled, despite resistance mutations persisting within viral subspecies present in low levels. Therefore testing performed in acutely infected subjects usually provides higher rates of resistance compared to testing chronic infections (Wensing et al. 2005). Furthermore, there are problems with identifying truly ART-naïve individuals, as people may have forgotten previous therapy or be reluctant to report it, sometimes perceiving that they will not qualify for further treatment. Incorrect identification of ART-naïve patients can result in overestimation of the prevalence of primary drug resistance due to misclassification of secondary resistance as primary resistance.

6.2.4 Transmissibility of Resistant Viral Strains

It is likely that resistant HIV is less fit and therefore less transmissible than wild-type virus, although the relative difference is likely to vary by the pattern of resistance mutations accumulated. For example, HIV strains with resistance to the NRTI drug lamivudine seem to be less transmissible than wild-type viruses or those carrying other resistance mutations (Turner et al. 2004). This decreased transmission efficiency may be due to two interdependent factors: lower plasma viral load of the transmitters and a potentially lower fitness of drug-resistant mutant viruses.

6.3 ART Delivery Systems in Developing Countries: Implications for Resistance

Several aspects of ART delivery in developing countries may pose a particular risk for resistance emergence. Potential problems with the supply chain for delivery of drugs may lead to widespread treatment interruptions, while inability to pay may result in interruptions for individuals in settings where therapy is not provided free of charge. Shortage of trained healthcare workers, limited availability of second-line regimens, and the possibility of a black-market ART trade are additional concerns. In addition, different HIV clades may preferentially develop different resistance mutations, which may have far-reaching implications, as our current knowledge of mutations is primarily from the treatment of subtype B infections, dominant in industrialized countries but accounting for only 14% of infections worldwide (Soares et al. 2007).

In the developed world, as well as in many middle-income settings, laboratory monitoring of CD4 count, viral load and in some cases, viral genotype is used to assess whether a given ART regimen is working, to detect when resistance occurs, and to guide choice of second-line regimen and timing of switching drugs. However, lack of infrastructure, funding, and trained personnel may preclude the use of some or all of these tests in some resource-limited settings. The degree of laboratory monitoring required to successfully implement an ART delivery program and the likely impact of such a decision on the generation and transmission of resistant virus are subjects of vigorous ongoing debate. Some clinicians believe that expensive viral load and CD4 count testing are unnecessary and that surrogate, clinical markers are sufficient (Farmer et al. 2001). A program in Haiti, for example, initiates ART based on simple clinical criteria designed to identify patients most likely to benefit (Mukherjee et al. 2006). At the same time, however, Brazil has managed to introduce sophisticated viral load assays costing only a small fraction of test costs in the United States (Teixeira et al. 2004), illustrating that such monitoring may be feasible in some resource-limited settings.

Current WHO recommendations suggest that viral load monitoring is not a high priority for resource-limited settings (WHO 2004). However, the majority of studies that have reported on ART programs in resource-poor settings so far have used viral load as well as CD4 count testing as part of their monitoring (Akileswaran et al. 2005). The absence of viral load and perhaps CD4 count testing from ART programs may accelerate the evolution and spread of drug resistance by reducing the ability to monitor treatment failure and to discontinue or replace failing regimens. Without such monitoring facilities, clinicians must rely on clinical relapse to detect treatment failure, by which point opportunities to limit cross-resistance may have been lost.

The optimum frequency for clinical follow-up of patients treated with ART involves a trade-off between the use of resources (healthcare workers' time, laboratory reagents, and laboratory workers) and the timeliness with which problems with treatment are identified and addressed. Where only one or at most two regimens are available, the importance of identifying early failure is minor because there are no alternative regimens accessible. Although the emergence of resistance reduces the effectiveness of ART, continuing a regimen that cannot achieve full viral suppression still provides considerable clinical benefits to patients (Ledergerber et al. 1999). Continuing the same ART regimen in the presence of detectable viral loads will promote the emergence of drug resistance and increase the risk of accumulating additional resistance mutations conferring high levels of resistance. However, discontinuing therapy leads to a rapid increase in viral load, decrease in CD4 count, and increase in risk of clinical progression (Miller et al. 2000) and is therefore not recommended.

Financial constraints can result in a limited range of second-line regimens being available. Given limited financial resources and a shortage of trained healthcare workers, the "public health approach" to ART delivery in developing countries, as advocated by WHO (WHO 2004), has recommended standardized guidelines for first- and second-line therapies. Such an approach can be contrasted with the individualized approach to ART adopted by most of the developed world, as well as by middle-income countries such as Brazil. Postponement of transfer to second- and third-line regimens until patients display clinical symptoms of treatment failure is expected and understandable in settings where second-line regimen options are limited, where the costs of these drugs are higher than first-line drugs, or where infrastructure is not available to detect early treatment failure. In some settings, the resources expended identifying failure early may indeed be better invested elsewhere. However, understanding the potential impact of such a strategy on the emergence and transmission of resistant virus is important.

ART use outside of official treatment programs is another potential contribution to resistance emergence and transmission. Currently, even subsidized prices for ART are unimaginable sums for the vast majority in developing countries. These people will be able to access ART only through public- or private-funded programs, or by haphazard and irregular drug supplies through the black market, when money is available. This kind of therapy will almost certainly result in treatment failure and resistance. In addition, increasing access to ART for only a minority of those in need could increase inappropriate use by others through theft, corruption, and distribution on the black market.

6.4 Insights from Mathematical Modeling

Predicting the short- and long-term impact of ART on the rate of emergence and transmission of drug resistance in resource-poor settings is far from intuitive; thus mathematical models can offer important insights. Different types of models have been used to address a variety of questions including the following: (1) operational questions on the benefits of different strategies with regard to patient management (Holmes et al. 2006; Walensky et al. 2007); (2) questions regarding the course of drug resistance evolution within individuals (McLean and Nowak 1992; Ribeiro et al. 1998); and (3) epidemiological questions such as resistance transmission between individuals (Blower et al. 2000; Nagelkerke et al. 2002).

6.4.1 Operational Questions

Operational research methods used to inform policy makers on issues such as resource allocation, patient management, and healthcare financing are typically based on models that can give short-term predictions. Examples of such studies of ART resistance include comparing the cost-effectiveness of initiating ART with PI- or NNRTI-based regimens in the context of various levels of NNRTI resistance in the population (resulting from single-dose administration of NNRTIs in an effort to prevent mother-to-child transmission) and examining the implications of excluding viral load and CD4 count monitoring facilities on the emergence of secondary resistance (Phillips et al. 2008). Although very useful, these types of models cannot be used to study the long-term impact of ART or the spread of resistance in the population and changes in HIV incidence because they do not explicitly model transmission of HIV infection from one person to another and they do not keep track of the changing prevalence of infection over time.

6.4.2 Emergence and Evolution of Resistance within Individuals

Dynamic within-host models that explicitly model viral and host cell dynamics and infection of cells over time have played an important role in understanding HIV pathogenesis. For investigation of drug resistance, they also model the pharmacokinetic properties of antiretroviral drugs and accumulation of drug resistance mutations and recombination events within the patient in order to investigate why and how secondary drug-resistant strains of HIV emerge during therapy. As with operational models, within-host models do not include the process of transmission of resistance between individuals.

6.4.3 Transmission of Drug Resistance

Most dynamic HIV transmission models incorporating resistance include key assumptions regarding treatment coverage, treatment failure, and emergence of resistance, among many others. These models can be used to study the short- and long-term impact of ART. Early studies concentrated on developed countries (Blower et al. 2000; Blower et al. 2001), while developing countries are a more recent focus, as treatment scale-up has become more feasible (Baggaley et al. 2006; Wilson et al. 2006; Vardavas and Blower 2007). The majority of models have assumed one triple regimen and one type of resistant viral strain. This resistant strain emerges among some treated patients, confers resistance to the single regimen, and can be transmitted to others.

Predictions of the extent of drug resistance transmission in resource-poor settings vary between studies based on the use of optimistic or pessimistic assumptions. Predictions also vary because the time horizon of predictions differs [e.g., resistance prevalence after 2 years (Wilson et al. 2006) versus 30 years (Nagelkerke et al. 2002)] and because some studies report the overall prevalence of all primary and secondary resistance combined, whereas other studies make the distinction between the two types of resistance. The fact that past models have been informed by relatively sparse data has led to variation in parameter values and structural assumptions, which in turn has resulted in diverse predictions. In general, however, more recent models have predicted relatively low levels of transmitted resistance in developing countries in the short term (Baggaley et al. 2006; Wilson et al. 2006; Vardavas and Blower 2007). The longer term outlook is more varied, and predictions are highly dependent on assumptions regarding the fitness and infectiousness of resistant strains and their ability to superinfect individuals already infected with wild-type virus (Baggaley et al. 2005).

Further elaboration of models based on new data from scaled-up ART programs and cohorts in developing countries is required to refine assumptions regarding sexual behavior change and resistance evolution and transmission and more adequately capture HIV transmission dynamics to predict transmission of drug-resistant HIV in developing country settings. More specific modeling of treatment decision algorithms is also required. So far, transmission models have assumed a single, standard triple-combination therapy regimen, with no second-line therapy available for those who experience treatment failure. All modeling studies are eventually dependent on the availability of setting-specific surveillance and behavioral data, and collection of such data is a clear priority for all regions where large-scale ART use is introduced.

6.5 Empirical Data on HIV Primary Resistance in Developing Countries

Most estimates of rates of HIV primary resistance in developing countries are derived from nonstandardized, small and local studies, limiting comparison of results, even within countries. Very few studies follow guidelines of structured surveillance methods. Comparability between countries is also limited because the extent of resistance transmission within a population depends on the duration and extent of availability of ART in that area. For example, in Brazil, there has been universal access to triple ART for all those eligible since 1996, whereas ART scale-up to any meaningful extent did not occur in sub-Saharan African countries until 2003.

The WHO recently launched the Global HIV Drug Resistance Surveillance Network (HIVResNet), an important initiative that aims to provide technical and analytical standards to measure resistance in developing countries Bennett et al. 2008; however, results are only now starting to emerge. Before such data become widely available, trends in primary resistance must be interpreted from more ad hoc surveillance studies, whose quality may be variable. A selection of such data is summarized in Table 6.1. As new prevalence estimates are continually published, readers are encouraged to search for the most recent studies and reviews through search engines such as PubMed (http://www.ncbi.nlm.nih.gov/pubmed/).

6.5.1 Asia

Prevalence of HIV drug resistance in northern and southeastern Asia is still low, although it is of increasing concern. Primary drug resistance patterns in India are probably of most concern for Asian public health, as the country was estimated to harbor 2.5 million infected individuals in 2006, the most of any country in the region. Estimates vary widely when comparing different studies and different regions of India. Studies conducted in Chennai and Pune failed to detect any primary drug resistance mutations among 50 and 12 patients, respectively, recruited from 1999 to 2003 (Balakrishnan et al. 2005; Eshleman et al. 2005),

Table 6.1 Prevalence of primary antiretroviral resistance by drug class and country for some low- and middle-income regions

Region, country	Study	NRTI (%)	PI (primary mutations) (%)	NNRTI (%)
Asia				
India	Deshpande et al. (2004)	1.6	NI	NI
India	Hira et al. (2004)	6.5	2.5	NI
India	Balakrishnan et al. (2005)	0	0	0
India	Eshleman et al. (2005)	0	0	0
Korea	Chin et al. (2006)	3.8	NI	3.8
China	Jiang et al. (2006)	4.2	0	0
China	Liao et al. (2007)	0	0	0
Thailand	Sukasem et al. (2007)	12.4	0	0
Africa				
Nigeria	Ojesina et al. (2006)	17	NI	NI
Mozambique	Parreira et al. (2006)	11.6	NI	NI
Madagascar	Razafindratsimandresy et al. (2006)	0	3.5	NI
Burkina Faso	Tebit et al. (2006)	2.4	0	4.8
Ethiopia	Kassu et al. (2007)	1.1	0	2.2
Latin America				
Brazil	Brindeiro et al. (2003)	2.4	2.2	2.1
Brazil	Pires et al. (2004)	14.0	0	0
Argentina	Dilernia et al. (2007)	1.4	1.4	2.1
Brazil	Sa-Ferreira et al. (2007)	1.4	0	0
Mexico	Valle-Bahena et al. (2006)	2.8	0	NI
Mexico	Viani et al. (2007)	2.5	0	2.5
Russia and Eastern Europe				
Slovenia	Babic et al. (2006)	3.9	0	0
Romania	Paraschiv et al. (2007)	0	0	0

NI – no information provided; NNRTI – non-nucleoside reverse transcriptase inhibitor; NRTI – nucleoside reverse transcriptase inhibitor; PI – protease inhibitor.

while rates as high as 6.5% for RTIs and 2.5% for PIs were observed among treatment-naïve patients from Bombay recruited from 1997 to 2003 (Hira et al. 2004). Although still relatively low, such rates of primary resistance suggest patchy dissemination of resistant virus within the region, which may contribute to increased transmission in India in the near future with increasing ART use, as observed in other settings (Maia Teixeira et al. 2006).

6.5.2 *Africa*

Although the African continent harbors approximately two-thirds of HIV-1 infections worldwide, few studies examining primary resistance have been conducted in this region. In West Africa, 5–8% of viruses from drug-naïve individuals in Burkina Faso, Cameroon, and Cote d'Ivoire were found to harbor primary resistance mutations (Toni et al. 2003; Tebit et al. 2006; Vergne et al. 2006). A very low prevalence of drug resistance was found in the island of Madagascar (Razafindratsimandresy et al. 2006), while an unexpectedly high prevalence of RTI-related mutations (17%) was found among drug-naïve subjects in Oyo State, Nigeria (Ojesina et al. 2006). In Mozambique, a recent survey found a rather high (11.6%) rate of strains harboring RTI-associated mutations (Parreira et al. 2006). A low rate of primary drug resistance was observed among 92 individuals in northern Ethiopia (Kassu et al. 2007).

6.5.3 Latin America

The prevalence of PI and NNRTI resistance among treatment-naïve individuals seems to have remained low in Brazil since ART became widely available in 1996, although discordant rates have been documented for NRTI resistance (Brindeiro et al. 2003; Pires et al. 2004; Sa-Ferreira et al. 2007). A recent study conducted with 74 samples from a Brazilian blood bank showed a low prevalence (1.4%) of NRTI primary resistance and no strains harboring NNRTI or PI resistance (Sa-Ferreira et al. 2007), yet a study published 3 years earlier reported NRTI primary resistance prevalence at 14.0% (Pires et al. 2004). In Argentina, Dilernia et al. (2007) have found rates of HIV primary resistance similar to those reported in Brazil (ranging from 1.4% to 2% for each drug class). In general Latin American rates of primary resistance can still be considered low when compared to that of developed nations, especially considering that wide-scale ART was introduced nearly a decade earlier than for other middle-income and developing countries.

6.5.4 Russia and Eastern Europe

Primary HIV drug resistance is low in Eastern European countries although currently there are few data available. A study in Slovenia identified only three strains (3.9%) carrying resistance mutations among 77 newly diagnosed individuals (Babic et al. 2006), while in Romania, a recent study conducted with 29 drug-naïve patients did not find any primary mutations in the protease or reverse transcriptase genes, reporting rates below 3.4% (Paraschiv et al. 2007).

6.6 Implications and Policy Recommendations

This chapter has briefly described the key determinants of the rates of emergence and transmission of HIV drug resistance, with a particular emphasis on resource-constrained settings. While emergence of resistance within a treated patient seems virtually inevitable at some point, it can be delayed, and its transmission should be prevented. Maximizing adherence and the effectiveness of safe sex counseling, minimizing interruption to drug supplies, and addressing financial barriers to drug procurement at the individual as well as the national level are all crucial. To date, adherence has generally been reported to be high in developing countries. However, many ART programs in the developing world are in their infancy and there is no room for complacency – counseling programs are crucial in order to keep adherence rates as high as possible, for as long as possible.

To aid the design of successful ART programs in resource-poor settings, more information on the impact of ART provision is crucial, including data on tolerability, treatment failure, and emergence of drug-resistant strains. Such data would help in the design of cost-effective programs that would maximize benefits to patients and increase the longevity of drug effectiveness. The most suitable drug regimens for the target population must be procured at the lowest possible cost (in generic versions where possible). Factors to consider include drug potency, efficacy in inhibiting viral replication, convenience, tolerability, and the implications of failure for future treatment options. Drugs with greater potency, which are metabolized slowly and which require many mutations in the viral genome to confer significant resistance, will generally be more robust to the emergence of resistance in the context of suboptimal adherence or treatment interruption. Such robustness is critical, given that choice of second-line regimens in much of the developing world will be limited. Antiretrovirals should be rationally sequenced in order to preserve future treatment options for as long as possible, although such sequencing may be complicated by cost constraints.

According to the WHO, successful application of ART in resource-poor settings depends on keeping drug regimens as simple as possible by limiting the choice of drugs for first-line treatment (WHO 2004). This will simplify the job of the clinician in prescribing and monitoring patients and may foster an increased role for nonspecialized doctors and nurses in ART administration (WHO 2006). Obviously, simpler programs as well as tailor-made ones will not fit the specific needs of individual patients. For example, the regimen most tolerable to treatment interruption may not be suitable for pregnant women (in whom efavirenz and combinations of didanosine (ddI) with stavudine (d4T) are contraindicated). A well-designed program should also include a carefully designed support infrastructure so that issues such as consistent drug supply are adequately addressed (WHO 2004).

The relative performances of ART programs with viral load or CD4 count testing, or neither of these, is a key research question when considering the design of treatment programs in resource-poor settings. The relative effects of the use of such monitoring systems, as well as the impact of second-line therapy availability, in terms of benefits to the patient and emergence of drug resistance should be addressed.

In summary, expansion of access to ART in the developing world has not had the catastrophic consequences on the emergence and transmission of resistant HIV predicted by some. However, since these programs are fairly recent, maintaining surveillance of resistance remains a priority. The success of nations such as Brazil in providing high-quality ART to all who need it, and the scale-up of access in many developing countries, suggests that the obstacles to ART program implementation where resources are scarce can be overcome. Furthermore, patients in resource-poor settings have not been subjected to the sequential mono and dual antiretroviral exposure that patients in industrialized countries experienced as technology and drug development advanced in the 1990s. Such treatment patterns favored the emergence of multiple drug resistance, whereas new, better tolerated triple regimens initiated in ART-naïve patients in developing nations result in a better prognosis. Nevertheless, there is no question that resistance presents a major challenge to successful treatment of HIV infection in developing countries. As the number of individuals treated with life-saving ART continues to grow, continuous surveillance to detect drug-resistant viruses must be organized to guide antiretroviral treatment strategies and policies.

References

Adje, C., R. Cheingsong, T. H. Roels, C. Maurice, G. Djomand, W. Verbiest, K. Hertogs, B. Larder, B. Monga, M. Peeters, S. Eholie, E. Bissagene, M. Coulibaly, R. Respess, S. Z. Wiktor, T. Chorba and J. N. Nkengasong (2001). "High prevalence of genotypic and phenotypic HIV-1 drug-resistant strains among patients receiving antiretroviral therapy in Abidjan, Cote d'Ivoire." J Acquir Immune Defic Syndr 26(5): 501–6.

Akileswaran, C., M. N. Lurie, T. P. Flanigan and K. H. Mayer (2005). "Lessons learned from use of highly active antiretroviral therapy in Africa." Clin Infect Dis 41(3): 376–85.

Babic, D. Z., M. Zelnikar, K. Seme, A. M. Vandamme, J. Snoeck, J. Tomazic, L. Vidmar, P. Karner and M. Poljak (2006). "Prevalence of antiretroviral drug resistance mutations and HIV-1 non-B subtypes in newly diagnosed drug-naive patients in Slovenia, 2000–2004." Virus Res 118(1–2): 156–63.

Baggaley, R. F., N. M. Ferguson and G. P. Garnett (2005). "The epidemiological impact of antiretroviral use predicted by mathematical models: a review." Emerg Themes Epidemiol 2: 9.

Baggaley, R. F., G. P. Garnett and N. M. Ferguson (2006). "Modelling the impact of antiretroviral use in resource-poor settings." PLoS Med 3(4): e124.

Balakrishnan, P., N. Kumarasamy, R. Kantor, S. Solomon, S. Vidya, K. H. Mayer, M. Newstein, S. P. Thyagarajan, D. Katzenstein and B. Ramratnam (2005). "HIV type 1 genotypic variation in an antiretroviral treatment-naive population in southern India." AIDS Res Hum Retroviruses 21(4): 301–5.

Bangsberg, D. R., A. R. Moss and S. G. Deeks (2004). "Paradoxes of adherence and drug resistance to HIV antiretroviral therapy." J Antimicrob Chemother 53(5): 696–9.

Bennett, D. E., S. Bertagnolio, D. Suthekland, C. F. Gilks (2008). "The World Health Organization's Global Strategy for prevention and assessment of HIV drug resistance." Antivir Ther 13(Suppl 2): 1–13.

Benson, C. A. (2006). "Structured treatment interruptions – new findings." Top HIV Med 14(3): 107–11.

Blower, S., E. Bodine, J. Kahn and W. McFarland (2005). "The antiretroviral rollout and drug-resistant HIV in Africa: insights from empirical data and theoretical models." AIDS 19(1): 1–14.

Blower, S. M., A. N. Aschenbach, H. B. Gershengorn and J. O. Kahn (2001). "Predicting the unpredictable: transmission of drug-resistant HIV." Nat Med 7(9): 1016–20.

Blower, S. M., H. B. Gershengorn and R. M. Grant (2000). "A tale of two futures: HIV and antiretroviral therapy in San Francisco." Science 287(5453): 650–4.

Brindeiro, R. M., R. S. Diaz, E. C. Sabino, M. G. Morgado, I. L. Pires, L. Brigido, M. C. Dantas, D. Barreira, P. R. Teixeira and A. Tanuri (2003). "Brazilian network for HIV drug resistance surveillance (HIV-BResNet): survey of chronically infected individuals." AIDS 17(7): 1063–9.

Chesney, M. A. (2000). "Factors affecting adherence to antiretroviral therapy." Clin Infect Dis 30 Suppl 2: S171–6.

Chin, B. S., J. Choi, J. G. Nam, M. K. Kee, S. D. Suh, J. Y. Choi, C. Chu and S. S. Kim (2006). "Inverse relationship between viral load and genotypic resistance mutations in Korean patients with primary HIV type 1 infections." AIDS Res Hum Retroviruses 22(11): 1142–7.

Deeks, S. G., N. S. Hellmann, R. M. Grant, N. T. Parkin, C. J. Petropoulos, M. Becker, W. Symonds, M. Chesney and P. A. Volberding (1999). "Novel four-drug salvage treatment regimens after failure of a human immunodeficiency virus type 1 protease inhibitor-containing regimen: antiviral activity and correlation of baseline phenotypic drug susceptibility with virologic outcome." J Infect Dis 179(6): 1375–81.

Deshpande, A., P. Recordon-Pinson, R. Deshmukh, M. Faure, V. Jauvin, I. Garrigue, M. E. Lafon and H. J. Fleury (2004). "Molecular characterization of HIV type 1 isolates from untreated patients of Mumbai (Bombay), India, and detection of rare resistance mutations." AIDS Res Hum Retroviruses 20(9): 1032–5.

Dilernia, D. A., L. Lourtau, A. M. Gomez, J. Ebenrstejin, J. J. Toibaro, C. T. Bautista, R. Marone, M. Carobene, S. Pampuro, M. Gomez-Carrillo, M. H. Losso and H. Salomón (2007). "Drug-resistance surveillance among newly HIV-1 diagnosed individuals in Buenos Aires, Argentina." AIDS 21(10): 1355–60.

Dunne, M. (2007). "Antiretroviral drug development: the challenge of cost and access." AIDS 21 Suppl 4: S73–9.

Eshleman, S. H., S. E. Hudelson, A. Gupta, R. Bollinger, A. D. Divekar, R. R. Gangakhedkar, S. S. Kulkarni, M. R. Thakar, R. S. Paranjape and S. Tripathy (2005). "Limited evolution in the HIV type 1 pol region among acute seroconverters in Pune, India." AIDS Res Hum Retroviruses 21(1): 93–7.

Farmer, P., F. Leandre, J. S. Mukherjee, M. Claude, P. Nevil, M. C. Smith-Fawzi, S. P. Koenig, A. Castro, M. C. Becerra, J. Sachs, A. Attaran and J. Y. Kim (2001). "Community-based approaches to HIV treatment in resource-poor settings." Lancet 358(9279): 404–9.

Harries, A. D., D. S. Nyangulu, N. J. Hargreaves, O. Kaluwa and F. M. Salaniponi (2001). "Preventing antiretroviral anarchy in sub-Saharan Africa." Lancet 358(9279): 410–4.

Hira, S. K., K. Panchal, P. A. Parmar and V. P. Bhatia (2004). "High resistance to antiretroviral drugs: the Indian experience." Int J STD AIDS 15(3): 173–7.

Holmes, C. B., H. Zheng, N. A. Martinson, K. A. Freedberg and R. P. Walensky (2006). "Optimizing treatment for HIV-infected South African women exposed to single-dose nevirapine: balancing efficacy and cost." Clin Infect Dis 42(12): 1772–80.

Jiang, S., H. Xing, X. Si, Y. Wang and Y. Shao (2006). "Polymorphism of the protease and reverse transcriptase and drug resistance mutation patterns of HIV-1 subtype B prevailing in China." J Acquir Immune Defic Syndr 42(4): 512–4.

Kassu, A., M. Fujino, M. Matsuda, M. Nishizawa, F. Ota and W. Sugiura (2007). "Molecular epidemiology of HIV type 1 in treatment-naive patients in north Ethiopia." AIDS Res Hum Retroviruses 23(4): 564–8.

Laurent, C., N. Diakhate, N. F. Gueye, M. A. Toure, P. S. Sow, M. A. Faye, M. Gueye, I. Laniece, C. Toure Kane, F. Liegeois, L. Vergne, S. Mboup, S. Badiane, I. Ndoye and E. Delaporte (2002). "The Senegalese government's highly active antiretroviral therapy initiative: an 18-month follow-up study." AIDS 16(10): 1363–70.

Ledergerber, B., M. Egger, M. Opravil, A. Telenti, B. Hirschel, M. Battegay, P. Vernazza, P. Sudre, M. Flepp, H. Furrer, P. Francioli and R. Weber (1999). "Clinical progression and virological failure on highly active antiretroviral therapy in HIV-1 patients: a prospective cohort study. Swiss HIV Cohort Study." Lancet 353(9156): 863–8.

Maia Teixeira, S. L., F. I. Bastos, M. A. Hacker, M. L. Guimaraes and M. G. Morgado (2006). "Trends in drug resistance mutations in antiretroviral-naive intravenous drug users of Rio de Janeiro." J Med Virol 78(6): 764–9.

McLean, A. R. and M. A. Nowak (1992). "Competition between zidovudine-sensitive and zidovudine-resistant strains of HIV." AIDS 6(1): 71–9.

Miller, V., C. Sabin, K. Hertogs, S. Bloor, J. Martinez-Picado, R. D'Aquila, B. Larder, T. Lutz, P. Gute, E. Weidmann, H. Rabenau, A. Phillips and S. Staszewski (2000). "Virological and immunological effects of treatment interruptions in HIV-1 infected patients with treatment failure." AIDS 14(18): 2857–67.

Mills, E. J., J. B. Nachega, I. Buchan, J. Orbinski, A. Attaran, S. Singh, B. Rachlis, P. Wu, C. Cooper, L. Thabane, K. Wilson, G. H. Guyatt and D. R. Bangsberg (2006). "Adherence to antiretroviral therapy in sub-Saharan Africa and North America: a meta-analysis." JAMA 296(6): 679–90.

Mukherjee, J. S., L. Ivers, F. Leandre, P. Farmer and H. Behforouz (2006). "Antiretroviral therapy in resource-poor settings. Decreasing barriers to access and promoting adherence." J Acquir Immune Defic Syndr 43 Suppl 1: S123–6.

Nagelkerke, N. J., P. Jha, S. J. de Vlas, E. L. Korenromp, S. Moses, J. F. Blanchard and F. A. Plummer (2002). "Modelling HIV/AIDS epidemics in Botswana and India: impact of interventions to prevent transmission." Bull World Health Organ 80(2): 89–96.

Ojesina, A. I., J. L. Sankale, G. Odaibo, S. Langevin, S. T. Meloni, A. D. Sarr, D. Olaleye and P. J. Kanki (2006). "Subtype-specific patterns in HIV Type 1 reverse transcriptase and protease in Oyo State, Nigeria: implications for drug resistance and host response." AIDS Res Hum Retroviruses 22(8): 770–9.

Oyugi, J. H., J. Byakika-Tusiime, K. Ragland, O. Laeyendecker, R. Mugerwa, C. Kityo, P. ugyenyi, T. C. Quinn and D. R. Bangsberg (2007). "Treatment interruptions predict resistance in HIV-positive individuals purchasing fixed-dose combination antiretroviral therapy in Kampala, Uganda." AIDS 21(8): 965–71.

Palella, F. J., Jr., K. M. Delaney, A. C. Moorman, M. O. Loveless, J. Fuhrer, G. A. Satten, D. J. Aschman and S. D. Holmberg (1998). "Declining morbidity and mortality among patients with advanced human immunodeficiency virus infection. HIV Outpatient Study Investigators." N Engl J Med 338(13): 853–60.

Paraschiv, S., D. Otelea, M. Dinu, D. Maxim and M. Tinischi (2007). "Polymorphisms and resistance mutations in the protease and reverse transcriptase genes of HIV-1 F subtype Romanian strains." Int J Infect Dis 11(2): 123–8.

Parreira, R., J. Piedade, A. Domingues, D. Lobao, M. Santos, T. Venenno, J. L. Baptista, S. A. Mussa, A. T. Barreto, A. J. Baptista and A. Esteves (2006). "Genetic characterization of human immunodeficiency virus type 1 from Beira, Mozambique." Microbes Infect 8(9–10): 2442–51.

Phillips, A. N., C. Sabin, D. Pillay and J. D. Lundgren (2007). "HIV in the UK 1980–2006: reconstruction using a model of HIV infection and the effect of antiretroviral therapy." HIV Med 8(8): 536–46.

Phillips, A. N., D. Pillay, A. H. Miners, D. E. Bennett, C. F. Gilks, J. D. Lundgren (2008). "Outcomes for monitoring of patients on Antiretroviral Therapy in resource-limited settings with viral, CD4 cell count, or clinical observation alone: A Computer Simulation Model." Lancet 371(9622): 1443–51.

Pires, I. L., M. A. Soares, F. A. Speranza, S. K. Ishii, M. C. Vieira, M. I. Gouvea, M. A. Guimaraes, F. E. de Oliveira, M. M. Magnanini, R. M. Brindeiro and A. Tanuri (2004). "Prevalence of human immunodeficiency virus drug resistance mutations and subtypes in drug-naive, infected individuals in the army health service of Rio de Janeiro, Brazil." J Clin Microbiol 42(1): 426–30.

Razafindratsimandresy, R., D. Rajaonatahina, J. L. Soares, D. Rousset and J. M. Reynes (2006). "High HIV type 1 subtype diversity and few drug resistance mutations among seropositive people detected during the 2005 second generation HIV surveillance in Madagascar." AIDS Res Hum Retroviruses 22(6): 595–7.

Ribeiro, R. M., S. Bonhoeffer and M. A. Nowak (1998). "The frequency of resistant mutant virus before antiviral therapy." AIDS 12(5): 461–5.

Richman, D. D., S. C. Morton, T. Wrin, N. Hellmann, S. Berry, M. F. Shapiro and S. A. Bozzette (2004). "The prevalence of antiretroviral drug resistance in the United States." AIDS 18(10): 1393–401.

Rosen, S., M. P. Fox and C. J. Gill (2007). "Patient retention in antiretroviral therapy programs in sub-Saharan Africa: a systematic review." PLoS Med 4(10): e298.

Sa-Ferreira, J. A., P. A. Brindeiro, S. Chequer-Fernandez, A. Tanuri and M. G. Morgado (2007). "Human immunodeficiency virus-1 subtypes and antiretroviral drug resistance profiles among drug-naive Brazilian blood donors." Transfusion 47(1): 97–102.

Soares, E. A., A. F. Santos, T. M. Sousa, E. Sprinz, A. M. Martinez, J. Silveira, A. Tanuri and M. A. Soares (2007). "Differential drug resistance acquisition in HIV-1 of subtypes B and C." PLoS ONE 2(1): e730.

Spacek, L. A., H. M. Shihab, M. R. Kamya, D. Mwesigire, A. Ronald, H. Mayanja, R. D. Moore, M. Bates and T. C. Quinn (2006). "Response to antiretroviral therapy in HIV-infected patients attending a public, urban clinic in Kampala, Uganda." Clin Infect Dis 42(2): 252–9.

Stevens, W., S. Kaye and T. Corrah (2004). "Antiretroviral therapy in Africa." BMJ 328(7434): 280–2.

Sukasem, C., V. Churdboonchart, K. Sirisidthi, S. Riengrojpitak, S. Chasombat, C. Watitpun, W. Piroj, M. Tiensuwan and W. Chantratita (2007). "Genotypic resistance mutations in treatment-naive and treatment-experienced patients under widespread use of antiretroviral drugs in Thailand: implications for further epidemiologic surveillance." Jpn J Infect Dis 60(5): 284–9.

Tebit, D. M., J. Ganame, K. Sathiandee, Y. Nagabila, B. Coulibaly and H. G. Krausslich (2006). "Diversity of HIV in rural Burkina Faso." J Acquir Immune Defic Syndr 43(2): 144–52.

Teixeira, P. R., M. A. Vitoria and J. Barcarolo (2004). "Antiretroviral treatment in resource-poor settings: the Brazilian experience." AIDS 18 Suppl 3: S5–7.

Toni, T. D., P. Recordon-Pinson, A. Minga, D. Ekouevi, D. Bonard, L. Bequet, C. Huet, H. Chenal, F. Rouet, F. Dabis, M. E. Lafon, R. Salamon, B. Masquelier and H. J. Fleury (2003). "Presence of key drug resistance mutations in isolates from untreated patients of Abidjan, Cote d'Ivoire: ANRS 1257 study." AIDS Res Hum Retroviruses 19(8): 713–7.

Turner, D., B. Brenner, J. P. Routy, D. Moisi, Z. Rosberger, M. Roger and M. A. Wainberg (2004). "Diminished representation of HIV-1 variants containing select drug resistance-conferring mutations in primary HIV-1 infection." J Acquir Immune Defic Syndr 37(5): 1627–31.

UNAIDS Report on the Global Aids Epidemic (2008). Available at www.unaids.org.

Valle-Bahena, O. M., J. Ramos-Jimenez, R. Ortiz-Lopez, A. Revol, A. Lugo-Trampe, H. A. Barrera-Saldana and A. Rojas-Martinez (2006). "Frequency of protease and reverse transcriptase drug resistance mutations in naive HIV-infected patients." Arch Med Res 37(8): 1022–7.

Vardavas, R. and S. Blower (2007). "The emergence of HIV transmitted resistance in Botswana: "when will the WHO detection threshold be exceeded?"" PLoS ONE 2: e152.

Vergne, L., S. Diagbouga, C. Kouanfack, A. Aghokeng, C. Butel, C. Laurent, N. Noumssi, M. Tardy, A. Sawadogo, J. Drabo, H. Hien, L. Zekeng, E. Delaporte and M. Peeters (2006). "HIV-1 drug-resistance mutations among newly diagnosed patients before scaling-up programmes in Burkina Faso and Cameroon." Antivir Ther 11(5): 575–9.

Viani, R. M., K. Hsia, P. Hubbard, J. Ruiz-Calderon, R. Lozada, J. Alvelais, M. Gallardo and S. A. Spector (2007). "Prevalence of primary HIV-1 drug resistance in pregnant women and in newly diagnosed adults at Tijuana General Hospital, Baja California, Mexico." Int J STD AIDS 18(4): 235–8.

Walensky, R. P., M. C. Weinstein, Y. Yazdanpanah, E. Losina, L. M. Mercincavage, S. Toure, N. Divi, X. Anglaret, S. J. Goldie and K. A. Freedberg (2007). "HIV drug resistance surveillance for prioritizing treatment in resource-limited settings." AIDS 21(8): 973–82.

Weiser, S., W. Wolfe, D. Bangsberg, I. Thior, P. Gilbert, J. Makhema, P. Kebaabetswe, D. Dickenson, K. Mompati, M. Essex and R. Marlink (2003). "Barriers to antiretroviral adherence for patients living with HIV infection and AIDS in Botswana." J Acquir Immune Defic Syndr 34(3): 281–8.

Wensing, A. M., D. A. van de Vijver, G. Angarano, B. Asjo, C. Balotta, E. Boeri, R. Camacho, M. L. Chaix, D. Costagliola, A. De Luca, I. Derdelinckx, Z. Grossman, O. Hamouda, A. Hatzakis, R. Hemmer, A. Hoepelman, A. Horban, K. Korn, C. Kucherer, T. Leitner, C. Loveday, E. MacRae, I. Maljkovic, C. de Mendoza, L. Meyer, C. Nielsen, E. L. Op de Coul, V. Ormaasen, D. Paraskevis, L. Perrin, E. Puchhammer-Stockl, L. Ruiz, M. Salminen, J. C. Schmit, F. Schneider, R. Schuurman, V. Soriano, G. Stanczak, M. Stanojevic, A. M. Vandamme, K. Van Laethem, M. Violin, K. Wilbe, S. Yerly, M. Zazzi and C. A. Boucher (2005). "Prevalence of drug-resistant HIV-1 variants in untreated individuals in Europe: implications for clinical management." J Infect Dis 192(6): 958–66.

WHO (2004). Scaling up antiretroviral therapy in resource-limited settings: treatment guidelines for a public health approach (2003 revision). Geneva, World Health Organization.

WHO (2006). Treat, train, retain: The AIDS and health workforce plan. Geneva, World Health Organization.

WHO (2007). Towards universal access. Scaling up priority HIV/AIDS interventions in the health sector Geneva, Switzerland, World Health Organization.

Wilson, D. P., J. Kahn and S. M. Blower (2006). "Predicting the Epidemiological impact of Antiretroviral allocation strategies in Kwazulu-natal: The effect of the urban-rural divide." Proc Natl Acad Sci USA 103(38): 14228–33.

Chapter 7
Drug Resistance in Malaria in Developing Countries

Quique Bassat and Pedro L. Alonso

Abstract A combination of complementary strategies is needed to control malaria, the most important parasitic infection causing an enormous burden of disease throughout the world. Antimalarial drugs play a crucial role among such strategies, as they may be used both for the treatment of cases and for their prevention. In the recent years, the treatment of malaria has been hampered by the emergence of widespread drug resistance to many of the available antimalarial drugs. This review attempts to give an insight of the history of malarial treatment and to describe the current status of drug-resistant malaria. The general determinants of drug resistance and the potential confounding factors for treatment failure will be assessed before a more thorough description of the specific mechanisms among host, vector, parasite and environment is summarized. Finally, the most important molecular pathways of drug resistance will be reviewed for each of the main drug families, together with an outline on the current methods for drug resistance diagnosis and surveillance.

7.1 Introduction

Malaria has consistently been, throughout the centuries, one of the major scourges for humankind. At the beginning of the 21st century, this parasitic disease remains unconquered, causing significant morbidity and mortality across the world. *Plasmodium falciparum,* one of the four species causing human malaria and the most virulent one, causes up to 600 million cases and more than 2 million deaths annually. While the remaining three species (*P. vivax, P. ovale* and *P. malariae*) are considered more benign, they still are responsible for an enormous morbidity burden, *P. vivax* being accountable for

Q. Bassat (✉)
Barcelona Center for International Health Research (CRESIB), Hospital Clínic/
Institut d'Investigacions Biomèdiques August Pi i Sunyer, Universitat de Barcelona,
Barcelona, Spain; Centro de Investigação em Saúde de Manhiça (CISM), Manhiça,
Maputo, Mozambique
e-mail: quique.bassat@cresib.cat

A. de J. Sosa et al. (eds.), *Antimicrobial Resistance in Developing Countries,*
DOI 10.1007/978-0-387-89370-9_7, © Springer Science+Business Media, LLC 2010

an estimated 400 million cases annually. Nowhere as in sub-Saharan Africa is the impact of malaria more blatant, with more than 90% of the world's malarial deaths occurring there (Snow et al. 2005; World Health Organization 2005b), predominantly among children under 5 years of age.

Strategies to tackle malaria rely on a multidisciplinary synergistic approach. While an effective vaccine is an exciting perspective (Aide et al. 2007), a hypothetical deployment is likely to occur only in a few years. Vector control strategies are currently being reconsidered, and the widespread mass distribution of insecticide-treated bednets (ITNs) is being promoted in malaria-endemic areas. However, since the World Health Organization's Global Malaria Eradication campaign was abandoned in 1969, focus has been placed on treatment rather than in prevention. Effective case management remains therefore the cornerstone of malaria control strategies, and in the past decades, this has relied essentially on the use of inexpensive and widely available antimalarial drugs, such as chloroquine or more recently sulfadoxine–pyrimethamine (SP). An alarming waning efficacy has been observed for these two drugs in several parts of the world and is now affecting Africa, where the major disease burden is observed and where these two drugs have managed to constrain the pandemic. While in recent years the death toll for other major child killers in developing countries (notably pneumonia, diarrheal diseases and measles) has fallen, deaths from malaria have increased. The main reason behind such an increase in malaria-related mortality, and for the global resurgence of malaria in the last three decades (Marsh 1998), lies in the continuous deployment of ineffective antimalarial drugs in the face of increasing resistance (Trape et al. 1998; Korenromp et al. 2003). Drug resistance has also been implicated in the reappearance of malaria in areas where the disease had previously been eradicated and in the occurrence and severity of epidemics in some parts of the world (Bloland 2001).

The inexorable spread of drug-resistant parasite strains has also threatened some of the existing limited alternative antimalarial drugs, particularly in Southeast Asia. Misuse and poor adherence to treatment regimens, inadequate absorption, intolerability, or the disturbing increase in counterfeit or inappropriate quality drugs have all contributed to the reduced effectiveness of antimalarial drugs. Parasites have managed to develop resistance against most of the existing drugs, but to date, there is no clear evidence of resistance to artemisinins, a family of highly effective antimalarial compounds. Combined therapies, including artemisinin-derived drugs, have emerged as the best current approach against resistance and more than 40 countries have adopted them in their National treatment policies (Mutabingwa 2005), despite their prohibitive cost. These same economic reasons have, however, constrained many other poor countries to maintain the older therapies, despite clear evidence of their weak efficacy.

Antimalarial drug resistance remains one of the greatest challenges facing malaria control today. This review aims to give a general overview of the geographical distribution, causes, mechanisms and consequences of antimalarial drug resistance in the first decade of the 21st century.

7.2 History of Malarial Treatment and the Emergence of Resistance

Malaria has infected humans for over 50 000 years, and many treatments have historically been used to counteract its main symptoms. One of the first effective treatments for malaria was the powder obtained from the bark of the cinchona tree, which contains quinine. This tree grows on the slopes of the Andes, mainly in Peru, and first reports of its use date back to 1632. Cinchona bark was subsequently introduced by the Jesuits into Europe as a treatment for what was known as "the ague", and it took almost two centuries (1820) to isolate the alkaloid quinine from the cinchona bark. Quinine remained for many centuries the only available treatment for the disease.

The development of other antimalarial drugs runs in parallel with the military history of the 20th century. World War I triggered research to find alternative treatments to quinine, the hitherto only available treatment. Quinacrine and several other compounds were developed, but the breakthrough came in 1934 with the synthesis of chloroquine, a new class of antimalarial of the 4-aminoquinoline family. This drug, first synthesized in Germany, was only recognized as a powerful antimalarial in the beginning of the 1940s, as part of the US World War II military effort. By the end of the war, in 1946, chloroquine was designated the drug of choice for the treatment of malaria, becoming the cornerstone treatment for this disease for the following four decades, as it was safe, highly effective and cheap. Other antimalarials such as proguanil (1946), amodiaquine and primaquine (1950's), the antifolate sulfadoxine–pyrimethamine (1967), halofantrine and mefloquine (developed by the US army as a result of the Vietnam War in the 1960's) appeared as alternatives to chloroquine. In China, piperaquine replaced chloroquine in 1978 as the first-line treatment and was used extensively as monotherapy for over 15 years (Myint et al. 2007). Lumefantrine was introduced in the 1980s as a response to the parasite's resistance to both chloroquine and the antifolates. During the last quarter of the 20th century, almost no new malarial drugs were developed, a proof of the little interest that drugs targeted at neglected diseases originated in the phar-maceutical industry. Among the approx. 1400 drugs licensed in the world during this period, only four corresponded to drugs specifically designed to cure malaria (Trouiller et al. 2002).

The most recent addition to the global antimalarial arsenal (artemisinin) is also one of the oldest. Infusions prepared from wormwood (*Artemisia annua*) were a component of Chinese herbal medicine for the treatment of fevers for more than 2000 years. However, the deployment of this drug on a large scale started only in the 1960s as a result of an antimalarial research programme undertaken by the Chinese army. The benefits of this highly effective reborn drug were kept secret by the Chinese communist government for many years, and widespread availability of the semisynthetic derivatives of this plant was upheld until the last decades of the century. Atovaquone (1996) and various other drugs have emerged at the turn of the century, a period that is witnessing an unprecedented boom in malarial drug research.

Emergence of resistance has been observed for all clinical antimalarials with the exception of artemisinin-derived drugs. Of the four human malarial species, only two (*P. falciparum* and *P. vivax*) have developed resistance to antimalarial drugs. Chloroquine resistance in *P. malariae* has also been reported from Indonesia (Maguire et al. 2002) but is possibly debatable (Collins and Jeffery 2002).

Chloroquine resistance from *P. falciparum* was first observed in Thailand and in the Colombian–Venezuelan border in the late 1950s (Harinasuta et al.1965). By the late 1970s, it had spread to New Guinea and eastern Sub-Saharan Africa (Bruce-Chwatt 1970). Resistance spread from the African coasts inland, and by 1989, chloroquine resistance was widespread in sub-Saharan Africa. Although the drug has lost its efficacy in most parts of the world, it remains effective in some areas of Central America and of South-western Asia. Despite the widespread chloroquine use, *P. vivax* chloroquine resistance remains very limited to certain areas in Asia (Indonesia, Papua New Guinea and Myanmar). *P. vivax* remains chloroquine-sensitive in Africa and the Americas. Except when stated, sensitivity patterns for the other drugs will refer to *P. falciparum*.

Resistance to SP and other antifolates, unlike that to chloroquine, seemed to appear independently, fast and simultaneously in different parts of the world. Resistance to proguanil, a type-2 antifolate, appeared in 1947, only 1 year after its introduction, and SP resistance was detected in 1967 in Thailand, the very same year of being launched. This did not prevent it from becoming the standard second-line therapy against chloroquine-resistant *P. falciparum* malaria. The shorter lag from introduction to appearance of resistance for SP is probably related to the smaller number of genetic mutations needed when compared to those involved in chloroquine resistance and to the longer half-life of the drugs. Currently, high level of resistance is seen in most Southeast Asia, southern China and the Amazon region. In Africa, SP sensitivity started declining in the late 1980s, and now high-level resistance is found in large areas of East Africa, and in a lesser degree in some western African countries, with a predicted rapid spread to the rest of the continent. Because of its pharmacokinetic properties, SP has also been used for intermittent preventive treatment (IPT) both in pregnant women and in infants. IPT is a malaria control strategy that consists in the administration of a full course of an antimalarial treatment to a population at risk at specified time points, regardless of whether or not they are known to be infected (Vallely et al. 2007). However, there is a need to develop alternative new regimens for IPT as increasing levels of parasite resistance may undermine this useful intervention (Greenwood 2006).

Quinine has been extensively used for the treatment of malaria since the 17th century. The first reported evidence of resistance to this drug came from Brazil in 1910. More recent evidence of quinine resistance has been regularly found in the Thai–Cambodian border, an area of intense drug pressure. Quinine efficacy has therefore only been jeopardized in some countries in Southeast Asia and Oceania, but seems uncompromised in Africa and South America. A poor compliance with the 7-day regimen, undesirable side effects and the risk of losing a precious treatment that may be effectively used for the treatment of severe malaria has discouraged its use as monotherapy for the treatment of *P. falciparum* infections.

The extensive use of piperaquine in China, as monotherapy, in mass treatment and mass prophylactic campaigns triggered the development of resistance to the drug. This drug has subsequently been combined with dihydroartemisinin, and the combination appears to be safe and highly effective (Myint et al. 2007).

Mefloquine resistance was first documented near the Thai–Cambodian border, 5 years after its introduction in 1977. Emergence of resistance was probably associated with the previous widespread use of quinine in the area, a compound to which mefloquine is structurally related (Wongsrichanalai et al. 2002). Resistance in that border area remains high, but decreases substantially in nearby regions, and the risk of therapeutic or prophylactic failure outside that particular region remains low in the rest of Asia. In Africa and South America, reports of mefloquine resistance have been published, but its overall efficacy remains high.

Primaquine is a unique antimalarial showing a broad action against most of the parasite's different stages. Moreover, it is the only licensed tissue-stage schizontocidal drug and therefore is widely used for the prevention of relapses after infection in the two plasmodia species (*vivax* and *ovale*) that have such a characteristic. Resistance to this drug has been observed for *P. vivax*, but is yet not fully understood, and appears to be of no clinical consequence (Baird 2005). Resistance to the liver stages may already have developed, and if proven, could be disastrous due to the lack of licensed therapeutic alternatives.

Plant-derived drugs such as quinine or the artemisinin derivatives, used during centuries, seem to have outlived most of the synthetic antimalarial drugs that were developed subsequently. Confirmed resistance to the artemisinin derivatives has not been reported yet (White 2004), although recrudescence among patients treated with short therapeutic courses has been observed (Magesa et al. 2001), and is probably a consequence of the pharmacodynamic properties of these agents (White 1999a). The development of resistance to these drugs has probably been delayed by its very short half-life (4 hours) and the drug's ability to reduce gametocytaemia, two important factors in the development of drug resistance. In this context, and to safeguard the effectiveness of these drugs, a consensus has emerged that they should never be used as monotherapy and should be combined with longer acting antimalarials. Similarly, atovaquone resistance appeared as soon as it was introduced as monotherapy in 1996, but its use in combination with proguanil (Malarone®) has guaranteed high efficacy.

7.3 Current Status of Drug-Resistant Malaria

Although drug resistance levels can change very fast, and that marked differences may coexist in neighbouring areas, it is useful to divide the world into broad zones, reflecting similar resistance patterns. Southeast Asia and Oceania are the paradigm and epicentre of multidrug-resistant malaria, defined as resistance to more than two operational compounds of different chemical classes (Wernsdorfer 1994). Chloroquine, SP, mefloquine and other drugs used as monotherapy have become ineffective there, and consequently these areas were the birthplace of artemisinin combination therapy (ACT) deployment. The rest of the Asian subcontinent, China and India and many parts of South America maintain adequate responses to mefloquine treatment, although resistance to both SP and chloroquine is high. Certain focal areas within the Amazon Basin may, however, show multidrug resistance, as mefloquine failures have been reported. In Africa, chloroquine resistance is virtually omnipresent in the continent, although reports from Malawi, the first country to abandon chloroquine treatment in 1993, have observed a resurgence of chloroquine-sensitive parasites (Laufer et al. 2006). SP resistance is markedly high in the eastern countries and increases rapidly threatening the rest of the continent, although some eastern areas remain moderately sensitive. Mefloquine has seldom been used in Africa, but it is believed to remain effective. Finally, in Central America (north of Panama) and the Caribbean, sensitivity to all antimalarials, including chloroquine, remains adequate. Figure 7.1 illustrates the geographical distribution of malarial resistance in the world.

Fig. 7.1 Geographical distribution of drug-resistant malaria in the world, according to WHO. Reproduced from World malaria report, 2005 (http://rbm.who.int/wmr2005/html/map5.htm)

7.4 Determinants of Drug Resistance

Several factors contribute to the development and spread of resistance to antimalarial drugs. Parasites causing malaria exhibit a range of susceptibility to the different antimalarial agents. Antimalarial drugs are not mutagenic per se, and it is believed that gene mutations conferring resistance occur in nature, independently of drug effect (Wongsrichanalai et al. 2002). These genetic events include mutations in or changes in the copy number of genes encoding or related to the drug's parasite target, or affecting the proteins that control intraparasitic concentrations of the drug (White 2004). As these mutations are rare, the proportion of mutant parasites within the parasite population is low. However, under drug pressure, an active selection of the "fitter" parasites, often corresponding to the mutant ones, occurs. Drug selection pressure seems therefore a critical and at the same time essential prerequisite for the development of resistance (Talisuna et al. 2004). A single genetic event may be enough to render a parasite resistant to a specific drug, although sometimes multiple unlinked events may be necessary, a phenomenon known as epistasis. This is, however, a rarer event, as the probability of multigenic resistance developing is the product of the individual component probabilities. Provided the mutations are not deleterious to the survival or reproduction of the parasite, drug pressure will remove the susceptible parasites allowing the resistant ones to survive. Over time, such resistant clones may become established in the population and can be very stable, persisting even after the removal of the specific drug pressure.

Aspects determining the emergence of resistance are generally better understood than those influencing the rate at which resistance spreads. Except for the importance attributable to drug selective pressure, the different theoretical models proposed (Curtis and Otoo 1986; Cross and Singer 1991; Mackinnon and Hastings 1998; Hastings and D'Alessandro 2000) offer conflicting and inconsistent explanations. Factors responsible for the stabilization and ulterior spread of resistance depend upon the interaction between a series of factors, including those related to the drug itself, the human host and the parasite–vector–environment complex.

7.4.1 Confounding Factors Responsible for Treatment Failure

Before starting to describe the different determinants of drug resistance, it is essential to establish the distinction between failure to clear malarial parasitemia or resolve clinical disease following a treatment with an antimalarial drug and true antimalarial resistance. Drug resistance per se is an important cause of treatment failure, but not all treatment failures are due to drug resistance. Several factors can be responsible for treatment failure, conditioning the effectiveness of malarial treatment, and it is possible that by causing treatment failure, such factors may all indirectly contribute to the development and intensification of true drug resistance by increasing the probability of exposing parasites to suboptimal drug levels.

The quality of drugs used to treat malarial episodes is clearly related to their efficacy. Poor manufacturing practices or deterioration due to inadequate handling or bad storage conditions can cause drugs to contain insufficient quantities of active ingredients. Prescription of the correct drugs at incorrect dosages may have the same consequences. The lucrative and qualm-less industry of counterfeit drugs poses a serious threat in regions where the trade in pharmaceuticals is not rigorously regulated (Newton et al. 2006) and surely should be held responsible for not only the emergence of resistance but also many avoidable deaths.

Incorrect adherence to the therapeutic regimens is another important determinant for treatment failure. Complex, inconvenient or poorly tolerated regimens carry a substantial risk of inadequate adherence. Often patients feel better after the first doses of treatment and abandon it, although they might have not yet cleared all parasites. Long regimens (such as quinine or primaquine) condition poor treatment compliance and put both the patient and the drug at risk. User-friendly packaging and education of the patients have become important to improve adherence and with it increase treatment success. Interactions with other drugs may modify absorption or metabolism of antimalarials, causing treatment failures, which are not related to the efficacy of the drugs. Home-based treatment strategies, in which patients may treat fevers at a village level with antimalarials, are increasingly considered in many developing settings but may enhance drug resistance if compliance is not adequate. Misdiagnosis is an extremely frequent problem, since symptoms of malaria are not specific and other infections may present with identical clinical pictures. In this context, guidelines proposed by the WHO in their Integrated Management of Childhood Illnesses (IMCI) programme in malaria-endemic countries suggest that all fevers should presumptively be treated with antimalarials. This strategy has surely saved many lives but is probably leading to excessive over-diagnosis and unnecessary treatments, increasing drug pressure.

7.4.2 Definitions of Antimalarial Drug Resistance

The standard definition for antimalarial drug resistance considers "the ability of a parasite strain to survive and/or multiply despite the administration and absorption of a drug given in doses equal to or higher than those usually recommended but within the tolerance of the subject", and its posterior addendum "the drug in question must gain access to the parasite or the infected red blood cell for the duration of the time necessary for its normal action" (Bloland 2001). This definition is generally interpreted referring only to persistence of parasites after the administration of an antimalarial drug and needs to be differentiated from a prophylaxis failure, which implies a reinfection. It also requires a demonstration of malarial parasitemia in the presence of adequate plasmatic drug and metabolite concentrations, established using any of the different available methodologies. In practice, this is rarely done, and in general, demonstration of the persistence of parasites in a patient receiving directly observed therapy is usually accepted as proof. Therefore, when serum drug concentrations are not measured, caution should guide the interpretation of in vivo therapeutic failure data.

7.4.3 Mechanisms of Antimalarial Resistance: The Host

In areas where malaria is endemic, partial immunity against the most severe forms of disease (death and severe disease) is progressively acquired (Gupta et al. 1999), followed by a protection against clinical episodes and finally a suppression of the parasitemia to low or undetectable levels. Such protection requires a continued booster effect, not conferring, however, a sterilising immunity, as individuals may get infected despite not developing clinical symptoms. Acquired immunity plays a central role in preventing the emergence and spread of resistance in high-transmission settings, as it reduces the parasite burden, which is a well-known determinant of antimalarial effectiveness. This phenomenon may partially explain the relative late increase in resistance in highly endemic areas like sub-Saharan Africa. In such settings, the immune system removes the parasites that have not been adequately tackled by the antimalarial drug. This explains why infections treated with inefficient drugs may still be cured. In low-transmission areas, the degree of acquired immunity is lower or inexistent, and the majority of the infections are symptomatic and associated to a higher parasite biomass. In addition, such infections are generally treated far more often than asymptomatic infections in areas of higher endemicity, increasing the drug pressure as parasites are confronted to antimalarial drugs. Factors that affect the immune system, either physiologically (pregnancy) or pathologically (drugs, diseases), could therefore have a critical role in the development of antimalarial drug resistance. There have been suggestions that immunosuppression states secondary to malnutrition (Wolday et al. 1995) or to HIV infection (Ayisi et al. 2003) may compromise acquired immunity to malaria and possibly enhance drug resistance. This needs to be confirmed (Laufer et al. 2007), but if true, the high prevalence of these illnesses in malaria-endemic areas could pose a tremendous threat to existing and future antimalarial drugs.

Variant alleles of the human genes codifying enzymes, which are important in certain metabolic pathways, may also have a role in the development of resistances, by affecting the pharmacokinetics of antimalarial drugs. Variants of the gene for cytochrome CYP2C19 correlate with slow or rapid metabolism of some antimalarial agents (Baird 2005). Interactions observed when using certain drugs (oral contraceptives) could be explained by the effect of such drugs in this same enzyme (McGready et al. 2003).

7.4.4 Mechanisms of Antimalarial Resistance: The Parasite, the Vector and the Environment

Some evidence suggests that certain combinations of drug-resistant parasites and vector species could enhance transmission of drug resistance. Two important malarial vectors in Southeast Asia (*Anopheles stephensi* and *Anopheles dirus*) appear to be more susceptible to drug-resistant parasites than to drug-sensitive ones (Sucharit et al. 1977). The opposite may also be true, and partially explain the pockets of chloroquine sensitivity that remain in the world, in spite of very similar human populations and drug pressure conditions (Bloland 2001).

The initial burden of parasitemia also plays a role in the risk for resistance. High levels of parasitemia, as compared with lower ones, require longer exposure to effective drug levels and have a relatively higher risk of treatment failure.

Although it is generally recognized that the level of transmission influences the rate of development and spread of drug resistance, whether malarial transmission intensity plays an independent role in this spread is still a matter of debate. It has been suggested that intensity of transmission is an important determinant of drug resistance as a result of its relationship to clone multiplicity (Babiker and Walliker 1997). A higher transmission would increase the number of infected clones per infected individual and therefore multiply the chances of developing resistance. The exact nature of this relationship is not yet well understood and has given rise to contrasting theories (Talisuna et al. 2004), postulating that the development of drug resistance was increased in both low (Hastings and D'Alessandro 2000) and high (Molyneux et al. 1999) transmission settings. The contradictory implications of such findings jeopardize their use for malaria control strategies, but a recent review (Talisuna et al. 2007) in East Africa suggested that *P. falciparum* resistance to chloroquine and SP was highest where malarial transmission was most intense and that vector control was associated with an increase in the efficacy of these drugs, presumably by decreasing transmission intensity.

7.4.5 Mechanisms of Antimalarial Resistance: The Drugs

Drug abuse has certainly contributed to the development of resistances, and circumstantial evidence for its particular role are highlighted by the observation that resistance to chloroquine developed from different areas whose common denominator was the long-term use of this drug for either prophylaxis or treatment (Payne 1988). However, the appearance of resistance to drugs such as SP in areas with relatively low SP use suggests that other factors may also have a role in the spread of resistance in addition to drug pressure.

Drug elimination half-life plays an important role in the evolution of parasite resistance (Hastings et al. 2002). Drugs with a long elimination half-life (SP, mefloquine and so on) have multiple therapeutic advantages. Patient's compliance is improved, as treatments are normally taken as a single dose or a short regimen that can be supervised by the clinician. The prolonged elimination periods maintain plasmatic therapeutic drug levels, which offer certain protection against the re-emergence of parasitemia for several weeks and give time to patients to recover from anaemia, a major cause of malaria-related morbidity in areas of intense malarial transmission. Subtherapeutic drug concentrations eliminate the most susceptible parasites and leave those that may be more prepared to recover and reproduce. Therefore, during this long elimination period, parasites from new infections or recrudescent ones originating in infections that did not fully clear will be exposed to drug blood levels insufficient to provide protection but high enough to exert selective pressure (Watkins and Mosobo 1993). The initial individual benefit conferred by drugs with long half-lives may be therefore counterproductive for the population, as they can create a potent selective pressure capable of accelerating the evolution of resistance.

The development of resistance to one drug can facilitate the development of resistance to others, provided they are closely related chemically. This phenomenon, cross-resistance, has been well described particularly between chloroquine and amodiaquine, two drugs structurally related within the 4-aminoquinoline family (Bray et al. 1996; O'Neill et al. 1998). Similarly, development of resistance to mefloquine may also lead to resistance to quinine or halofantrine. Due to their structural similarities to other antifolates, the promising combination chlorproguanil–dapsone (LapDap) may be at risk even before it is deployed. Antimalarials with new modes of action need to be developed to avoid this problem.

We have seen that certain genotypes may influence the bioavailability of antimalarial drugs. Some treatments may also interact with antimalarial drugs by competing with them throughout metabolic pathways. Similarly, certain foods may further determine bioavailability by increasing or reducing gastrointestinal absorption, thus modulating drug effectiveness. An example of this is the enhancement of artemether–lumefantrine's absorption by fatty foods (Piola et al. 2005).

7.5 Molecular Markers of Antimalarial Resistance

As we have seen, single-point mutations are sufficient to render a parasite resistant to an antimalarial drug. However, for certain drugs, more complicated genetic events involving sequential mutations may be necessary. Table 7.1 describes the major molecular markers of antimalarial resistance for the most important drugs.

Table 7.1 Major molecular markers of antimalarial resistance for the most important drugs

Organism	Drug resistant to	Gene involved	Product	Genetic determinant of resistance*
P. falciparum	Chloroquine	Pfcrt	Transporter	Thr76
		Pfmdr1	Transporter	Tyr86
	Mefloquine	Pfmdr1	Transporter	Tyr86
		Pfmdr1	Transporter	Copy number**
	Sulfadoxine	dhps	Dihydropteroate synthetase	Gly437,Glu540, Gly581
	Pyrimethamine	dhfr	Dihydrofolate reductase	Asn108, Arg59, Ile51,Leu164
	Atovaquone	cyb	Cytochrome b	Ser 268
P. vivax	Pyrimethamine	dhfr	Dihydrofolate reductase	Ser58, Ser117
	Chloroquine	Unknown	Unknown	Unknown

*Only the predominant determinants of resistance are shown.
**This mechanism also described for halofantrine and lumefantrine.

7.5.1 Chloroquine

Malarial parasites digest haemoglobin producing large quantities of a by-product known as malaria pigment. Some drugs, like chloroquine, base their antimalarial activity on their ability to bind to and interrupt haematin detox-ification, resulting in the accumulation of large quantities of toxic haematin, which eventually kills the parasite. Parasite resistance seems to depend on the ability of that resistant parasite to limit this accumulation. Polymorphisms in two genes of the *P. falciparum* genome seem to be linked to chloroquine resistance. The *P. falciparum* multidrug resistance (*pfmdr1*) gene, located on chromosome 5, encodes the P-glycoprotein1 (Pgh-1). The aspartic acid to tyrosine point mutation in codon 86 has been associated with chloroquine resistance in some clinical and in vitro studies (Wongsrichanalai et al. 2002). The *pfcrt* gene is located on chromosome 7 and codes for PfCRT, a vacuolar membrane transporter protein. Many polymorphisms have been identified, but the substitution of threonine for lysine in codon 76 has consistently been associated with chloroquine resistance. Chloroquine resistance seems to involve progressive accumulation of mutations in the *pfcrt* gene, and the one at position 76 seems to be the last in the long process leading to chloroquine resistance (Djimde et al. 2001). The molecular mechanisms for chloroquine resistance in *P. vivax* have not yet been ascertained (Baird 2004).

7.5.2 Sulfadoxine–Pyrimethamine (SP) and Antifolates

The molecular basis for resistance to antifolate combination drugs, such as sulfadoxine–pyrimethamine, has been extensively studied and seems more straightforward than that of chloroquine. Antifolate drugs interrupt the parasite's DNA replication by competitively inhibiting folate synthesis, which is essential for the synthesis of pyrimidines (Foote and Cowman 1994). Sulfadoxine and pyrimethamine act synergistically. While pyrimethamine and related compounds inhibit the step mediated by the enzyme dihydrofolate reductase (*dhfr*), sulfones and sulphonamides inhibit the step mediated by dihydropteroate synthase (*dhps*). Specific gene mutations encoding for resistance to both *dhps* and *dhfr* have been identified (Wernsdorfer and Noedl 2003), and specific combinations of these mutations are related with varying degrees of resistance to antifolate combination drugs. It has been suggested that the sequence of mutations occurs in a stepwise fashion, with selection for mutations in *dhfr* gene probably occurring first and the *dhps* gene mutations following later (Sibley et al. 2001). Overall, a change from serine at codon 108 to asparagine is thought to be the key mutation for pyrimethamine resistance. Additional successive point mutations (Ile51, Arg59 and Leu164) are known to increase progressively the degree of resistance. Thus, quadruple mutants, including the Leu164 change, confer the most severe resistance. Even if each mutation seems to confer a stepwise reduction in sensitivity, some data suggest that the presence of a sensitive *dhfr* allele is highly predictive of SP treatment success, irrespective of *dhps* allele.

7.5.3 Mefloquine, Quinine and Atovaquone

Resistance to mefloquine in *P. falciparum* results from amplifications (i.e. duplications rather than mutations) in *pfmdr*, the same gene involved in chloroquine resistance. As a result of these amplifications, there seems to be a decreased concentration of the drug at a cellular level. However, some studies have suggested a possible inverse relation between sensitivity to chloroquine and to mefloquine, as increased sensitivity (rather than resistance) to the latter has been associated with the *pfmdr1* Tyr86 mutation. Copy number and polymorphisms of the involved *pfmdr1* gene have been proposed as molecular markers for mefloquine resistance, but their applicability remains uncertain.

Resistance to quinine appears sporadic, and a moderate risk of associated treatment failure has remained limited to some regions of Southeast Asia and New Guinea. *Pfmdr1* polymorphism may modulate quinine susceptibility, but the exact molecular mechanisms of resistance are not yet clearly understood.

Atovaquone, one of the most recent acquisitions of malarial therapy, acts through the inhibition of electron transport at the cytochrome *bc*1 complex (Korsinczky et al. 2000). The emergence and spread of resistance is believed to occur rapidly, as it arises from single-point mutations in the cytochrome *b* gene.

Atovaquone monotherapy led to rapid emergence of resistance, and it is now deployed only in a fixed combination with proguanil.

7.5.4 Artemisinin

Except in animal models and in laboratory-selected parasite lines, there is no solid evidence of artemisinin resistance yet (Nosten and White 2007). A recent report (Jambou et al. 2005) that some parasites from French Guyana with mutations in the gene encoding the putative target PfATPase6 were highly resistant to artemether has raised concerns, but these findings must be confirmed as these parasites have not been cultured. Molecular studies have also associated the *pfmdr1* Tyr86 mutant with increased sensitivity to artemisinins.

7.6 Detection of Drug Resistance

As chloroquine and SP are replaced by the more effective, artemisinin-based combination therapies, strategies for monitoring and preventing drug-resistant malaria must be updated and optimized (Laufer et al. 2007). The timing of drug policy changes can be guided by data obtained from any of the methods routinely used to study or measure antimalarial drug resistance. The three principal methodologies involved in detecting drug resistance are in vitro and in vivo tests and molecular characterization. Other methods such as animal studies and passive surveillance detection are also useful. All of them have advantages and limitations and should be considered complementary rather than competing sources of data about resistance.

7.6.1 In Vitro Tests

In vitro assays are based on the inhibition of the growth and development of malarial parasites by different concentrations of a given drug, when compared to growth in drug-free controls. By removing the parasites from the host and placing them into a controlled experimental environment, these tests avoid many of the confounding factors that influence in vivo tests. They provide a direct and quantitative assessment of drug resistance and identify the phenotype of the parasite independently of the immune and physiopathological conditions in the host (Talisuna et al. 2004). The WHO in vitro microtest and the isotopic microtest are the two principal methods. Compared to in vivo assays, they present several drawbacks. They can be performed only by highly trained personnel and require expensive laboratory equipment, making them more difficult to apply in the field. Results obtained are difficult to interpret, as the correlation of in vitro response with clinical response in patients is not clear-cut. This correlation depends on the level of acquired immunity within the population being tested. Moreover, previous antimalarial treatments taken by the patient in the days preceding the sample may delay or impede parasite's growth. Some prodrugs, like proguanil, which require host conversion into active metabolites, cannot be tested, and neither can drugs that require some level of synergism with the patient's immune system. Despite these limitations, the in vitro tests remain useful to test drugs that are new and have not been used previously, and they should be considered in the front line of resistance monitoring for artemisinins and ACTs (Laufer et al. 2007).

7.6.2 *In Vivo Tests*

In vivo tests are considered the "gold standard" method for detecting drug resistance (World Health Organization 1996) and rely on the measurement of the parasitological and/or clinical response in a group of symptomatic and parasitemic individuals treated with a controlled dosage of the drug. Originally these tests required prolonged periods of follow-up (minimum of 4 weeks) and were conducted under seclusion in screened rooms so as to prevent reinfections. Nowadays, seclusion has become unnecessary, and these tests can easily be conducted in the field with little equipment and personnel, the results being easy to interpret. In contrast to in vitro tests, they closely reflect the biological nature of treatment response in the true clinical and epidemiological context, which involves a complex interaction between the drugs, the parasites and the host response. Because of the influence of these external factors (immunity, pharmacokinetics and so on), these tests offer valuable data on the effectiveness of antimalarial treatment under close to real-life conditions but may not necessarily reflect the true level of drug resistance.

For ongoing monitoring of the therapeutic efficacy of current antimalarial treatment policies, most malaria control programmes and malarial researchers have relied on protocols published by the World Health Organization. In one of the first standard protocols, response to treatment was categorized purely on parasitological grounds as sensitive or resistant (with three levels: RI, RII and RIII) (World Health Organization 1973). Different classifications have subsequently been proposed (Talisuna et al.2004), and Table 7.2 summarizes WHO's 2003 guidelines for the assessment of response to treatment (World Health Organization 2003). This latest classification involves a follow-up period of either 14 or 28 days. The timing of recurrent parasitemia or clinical disease reflects the degree of resistance. In areas of high transmission, later recurrence may be confounded by reinfection, and comparison of genotypes of the strains causing the original and recurrent parasitemias can be used to

Table 7.2 Revised WHO guidelines (2003) assessing in vivo response to treatment

Early treatment failure (ETF): Days 0, 1, 2 and 3

Development of danger signs of severe malaria on days 0–3 in the presence of parasitemia

Parasitemia on day 2 > day 0, irrespective of temperature

Parasitemia on day 3 with axillary temperature > 37.5 °C

Parasitemia on day 3 > 25% of day 0 count

Late clinical failure (LCF): Days 4–14 or 28

Development of danger signs of severe malaria in the presence of parasitemia, without previously meeting any of the ETF criteria

Axillary temperature > 37.5 °C or history of fever in previous 24 h on days 4–14 or 28 in the presence of parasitemia, without meeting any of the ETF criteria

Late parasitological failure (LPF): Days 7–14 or 28

Presence of parasitemia on any day from day 7 to day 14 or 28, and axillary temperature < 37.5 °C, without previously meeting any of the criteria of early or late treatment failure

Adequate clinical and parasitological response (ACPR)

Absence of parasitemia on day 14 or 28 irrespective of temperature, without previously meeting any of the criteria for early or late failure

address this problem. However, few clinics or laboratories have the capacity to conduct such testing.

7.6.3 Molecular Markers

Molecular tests use polymerase chain reaction (PCR) to indicate the presence of mutations encoding biological resistance to antimalarial drugs (Plowe et al. 1995). Genetic assessment of molecular markers of resistance can easily be determined using fingerprick samples from infected patients. Genetic markers have a demonstrable clear association with clinical failure rates (World Health Organization 2005a), but policy makers usually regard them as less compelling than clinical data (Hastings et al. 2007). Advantages include the need for only small amounts of genetic material as opposed to live parasites, the possibility of conducting large number of tests in a relatively short period of time, and the independence from host and environmental factors. However, they are expensive, technologically demanding and restricted to a limited number of known gene mutations. Moreover, confirmation of the association between given gene mutations and real-life drug resistance is difficult and not clear-cut, especially when resistance involves more than one gene locus and multiple mutations. The role of host factors such as immunity may undermine this association too, as some patients with "resistant" parasites still manage to clear the infection. These correlations, albeit not conclusive, are nevertheless sufficiently robust for genetic mutations to serve as good indicators for impending drug resistance. The logistical and ethical ease of collecting these data among ample communities has placed genetic markers at the mainstay for early monitoring of antimalarial efficacy in the era of ACTs (Hastings et al. 2007; Laufer et al. 2007).

7.7 Future Prospects

Three major issues are particularly relevant to the future scenario of drug-resistant malaria. The first one considers the future use of molecular information for a better control of parasite drug resistance. The second one regards evidence-based recommendations for antimalarial regimens in malaria-endemic areas. The last one is related to the second and considers the economic hindrances that such recommendations convey.

The genetic sequencing of the triangle that defines malaria (the parasite *P. falciparum*, the principal vector *Anopheles gambiae* and the human host) has offered a unique opportunity for a more comprehensive understanding of the genetic susceptibility to both infection and clinical disease (Ekland and Fidock 2007) and will certainly generate unparalleled data for the development of new and improved diagnostic, prophylactic and therapeutic tools. Moreover, this valuable genetic information must be efficiently used to identify new drug targets or design tools for rapid diagnosis and confirmation of antimalarial resistance, as well as for distinguishing recrudescence from new infection. Assuming that the current generation of ACTs will not maintain their efficacy indefinitely, comprehensive monitoring methods for early detection and evaluation of drug resistance will become critical to extend their useful therapeutic lives (Laufer et al. 2007).

As for other infectious diseases such as leprosy, tuberculosis and HIV, the increasing resistance to the available drugs has pushed for the adoption of combination treatment strategies. For such diseases, it is no longer considered ethical to attempt to treat them with a single drug, and this principle should be applied to malaria also. Theoretically, combining different treatments with diverse modes of action should delay the emergence of resistance, as the parasite would need to mutate in several sites simultaneously to become resistant to the combination treatment, a very unlikely event in comparison to the probability of developing a single mutation conferring resistance to the individual drugs. Mathematically, it can be expressed as a simple calculation. The chance of a mutant arising that is simultaneously resistant to two different antimalarial drugs is the product of the mutation rates per parasite for the individual drugs and the number of parasites exposed to the drugs in that infection. If one in 10^9 parasites is resistant to drug A and one in 10^{13} is resistant to drug B, the probability of that parasite being simultaneously resistant to both drugs is an almost negligible one in 10^{22}, provided the genetic mutations that confer resistance are not linked. Most clinical malarial infections yield between 10^8 and 10^{12} parasites and a biomass larger than 10^{13} parasites in a single person is physically impossible. Therefore, the chances of simultaneous double resistance are about once in every 10^{12} treatments, or less than once in a century. Compared with the sequential use of single drugs, combinations should substantially hold back the development of resistance. The options for future malarial treatment rely therefore on combining an effective drug with either one of the traditional treatments (such as chloroquine, SP and amodiaquine) or a new one (lumefantrine or others). Using drugs with existing resistance-conferring mutations as partner drugs may be a dangerous short-term solution, but the synergistic effect of the two drugs together may overcome this problem. As an example, combining atovaquone with proguanil delays the emergence of atovaquone resistance, which was fast when the drug was used alone.

It is now widely recognized that ACTs are probably the best option available for the treatment of malaria (White 1999b). Artemisinins are very powerful drugs capable of rapidly reducing the parasite biomass. Their short lifespan guarantees that they are never presented to infecting malarial parasites at intermediate selective drug concentrations, because they are eliminated completely within the two-day life cycle of the asexual parasite. However, it has been argued that in areas of high transmission, where infective bites are extremely frequent, combinations including an artemisinin derivative (with a short half-life) may increase the risk of resistance development of the partner drug that is eliminated more slowly, as the latter would be effectively acting as monotherapy during the elimination "tail" of declining blood concentrations. As artemisinins are capable of reducing the parasite load by 10^4-fold per cycle, partner drugs should be exposed only to a tiny fraction of the parasite number present at the infection peak and should therefore be able to eradicate them easily (White 2004). Furthermore, due to their gametocidal properties, artemisinins reduce gametocyte carriage and transmissibility, and by doing so, they further decrease the possibilities of parasite resistance emerging from mutant gametocytes.

The widespread but judicious use of effective antimalarials should provide an excellent opportunity to control antimalarial resistances and to alleviate the enormous burden that this disease causes. However, the main factor affecting availability is economy, as countries that most need these drugs certainly cannot afford them. While a full adult treatment course with chloroquine or SP costs less than $0.15, prices rise for the other antimalarials. Quinine ($0.97) is expensive but is generally saved for the parenteral treatment of severe malaria. Negotiations between the WHO and the manufacturer have reduced the cost of artemether–lumefantrine combination to around $0.9–1.4 for a complete child's treatment and to $2.4 for an adult treatment course, and this price is similar to the costs of mefloquine ($2.55). A new ACT, dihydroartemisinin–piperaquine, with a simpler dosing schedule and a much cheaper price ($1 per adult treatment) should be licensed before the end of 2009 (Myint et al. 2007). In Africa, ACTs remain prohibitive in many countries, as the health budget is often restricted to less than $10 per capita annually to spend on all aspects of health (White 2004). In spite of this, by February 2005, more than 43 countries (mostly in Africa) had adopted ACTs in their treatment policies (Mutabingwa 2005), even if treatments may not reach all who need them. Product competition will surely enhance further price reductions, but there is a need for political commitment to guarantee an efficient and widespread distribution of these drugs. Only then African countries may be able to replace the cheap but ineffective drugs still widely used.

7.8 Conclusions

The last decade has witnessed unprecedented advances in malarial research, coupled with scientific achievements that have brought malaria on top of the policy makers' and scientific community's agenda. Recommendations for widespread use of ACTs as the best available weapon to challenge drug-resistant malaria have already been translated into national policy changes. Research coalitions, supported by public and private partnerships, have been injecting the malarial community with significant money and expertise contributions and have become fundamental in converting enthusiasm into action. For the first time in many decades, some countries are controlling malaria (UNICEF 2007). It is precisely in this period of contained optimism that mistakes done in the past have to guide the path to follow. The development of drug resistance to antimalarial drugs in some parts of the world has taken much less time than the estimated 12–17 years it takes to develop and market a new antimalarial compound (Bloland 2001). A sound and cautious utilization of the best affordable treatment options by the most severely affected countries needs to go coupled with a proactive monitoring of drug resistance, and these drugs must be accessible to the people who need them most. Only then, many malarial deaths will be avoided, and the remaining effective treatments will not be wasted.

References

Aide, P., Q. Bassat and P.L. Alonso (2007). "Towards an effective malaria vaccine." *Archives Disease in Childhood* **92**, 476–479.

Ayisi, J.G., et al. (2003). "The effect of dual infection with HIV and malaria on pregnancy outcome in western Kenya." *AIDS* **17**, 585–594.

Babiker, H.A. and D. Walliker (1997). "Current views on the population structure of *Plasmodium falciparum*: Implications for control." *Parasitology Today* **13**, 262–267.

Baird, J.K. (2004). "Chloroquine resistance in *Plasmodium vivax*." *Antimicrobial Agents and Chemotherapy* **48**, 4075–4083.

Baird, J.K. (2005). "Effectiveness of antimalarial drugs." *New England Journal of Medicine* **352**, 1565–1577.

Bloland, P.B. (2001). "Drug resistance in malaria." World Health Organization, Geneva.

Bray, P.G., S.R. Hawley and S.A. Ward (1996). "4-Aminoquinoline resistance of Plasmodium falciparum: insights from the study of amodiaquine uptake." *Molecular Pharmacology* **50**, 1551–1558.

Bruce-Chwatt, L.J. (1970). "Resistance of *P. falciparum* to chloroquine in Africa: true or false?" *Transactions of the Royal Society of Tropical Medicine and Hygiene* **64**, 776–784.

Collins, W.E. and G.M. Jeffery (2002). "Extended clearance time after treatment of infections with *Plasmodium malariae* may not be indicative of resistance to chloroquine." *American Journal of Tropical Medicine and Hygiene* **67**, 406–410.

Cross, A.P. and B. Singer (1991). "Modelling the development of resistance of *Plasmodium falciparum* to anti-malarial drugs." *Transactions of the Royal Society of Tropical Medicine and Hygiene* **85**, 349–355.

Curtis, C.F. and L.N. Otoo (1986). "A simple model of the build-up of resistance to mixtures of anti-malarial drugs." *Transactions of the Royal Society of Tropical Medicine and Hygiene* **80**, 889–892.

Djimde, A., et al. (2001). "A molecular marker for chloroquine-resistant falciparum malaria." *New England Journal of Medicine* **344**, 257–263.

Ekland, E.H. and D.A. Fidock (2007). Advances in understanding the genetic basis of antimalarial drug resistance. *Current Opinion in Microbiology* **10**, 363–370.

Foote, S.J. and A.F. Cowman (1994). "The mode of action and the mechanism of resistance to antimalarial drugs." *Acta Tropica* **56**, 157–171.

Greenwood, B. (2006). "Review: intermittent preventive treatment – a new approach to the prevention of malaria in children in areas with seasonal malaria transmission." *Tropical Medicine and International Health* **11**, 983–991.

Gupta, S., et al. (1999). "Immunity to non-cerebral severe malaria is acquired after one or two infections." *Nature Medicine* **5**, 340–343.

Harinasuta, T., P. Suntharasamai and C. Viravan (1965). "Chloroquine-resistant falciparum malaria in Thailand." *Lancet* **2**, 657–660.

Hastings, I.M. and U. D'Alessandro (2000). "Modelling a predictable disaster: the rise and spread of drug-resistant malaria." *Parasitology Today* **16**, 340–347.

Hastings, I.M., E.L. Korenromp and P.B. Bloland (2007). "The anatomy of a malaria disaster: drug policy choice and mortality in African children." *Lancet Infectious Diseases* **7**, 739–748.

Hastings, I.M., W.M. Watkins and N.J. White (2002). "The evolution of drug-resistant malaria: the role of drug elimination half-life." *Philosophical Transactions of the Royal Society London B (Biological Sciences)* **357**, 505–519.

Jambou, R., et al. (2005). "Resistance of Plasmodium falciparum field isolates to in-vitro artemether and point mutations of the SERCA-type PfATPase6." *Lancet* **366**, 1960–1963.

Korenromp, E.L., et al. (2003). "Measurement of trends in childhood malaria mortality in Africa: an assessment of progress toward targets based on verbal autopsy." *Lancet Infectious Diseases* **3**, 349–358.

Korsinczky, M., et al. (2000). "Mutations in *Plasmodium falciparum* cytochrome b that are associated with atovaquone resistance are located at a putative drug-binding site." *Antimicrobial Agents and Chemotherapy* **44**, 2100–2108.

Laufer, M.K., A.A. Djimde and C.V. Plowe (2007). "Monitoring and deterring drug-resistant malaria in the era of combination therapy." *American Journal of Tropical Medicine and Hygiene* **77**, 160–169.

Laufer, M.K., et al. (2006). "Return of chloroquine antimalarial efficacy in Malawi." *New England Journal of Medicine* **355**, 1959–1966.

Laufer, M.K., et al. (2007). "Malaria treatment efficacy among people living with HIV: the role of host and parasite factors." *American Journal of Tropical Medicine and Hygiene* **77**, 627–632.

Mackinnon, M.J. and I.M. Hastings (1998). "The evolution of multiple drug resistance in malaria parasites." *Transactions of the Royal Society of Tropical Medicine and Hygiene*, **92**, 188–195.

Magesa, S.M., et al. (2001). "Distinguishing Plasmodium falciparum treatment failures from re-infections by using polymerase chain reaction genotyping in a holoendemic area in northeastern Tanzania." *American Journal of Tropical Medicine and Hygiene* **65**, 477–483.

Maguire, J.D., et al. (2002). "Chloroquine-resistant *Plasmodium malariae* in south Sumatra, Indonesia." *Lancet* **360**, 58–60.

Marsh, K. (1998). "Malaria disaster in Africa." *Lancet* **352**, 924.

McGready, R., et al. (2003). "Pregnancy and use of oral contraceptives reduces the biotransformation of proguanil to cycloguanil." *European Journal of Clinical Pharmacology* **59**, 553–557.

Molyneux, D.H., et al. (1999). "Transmission control and drug resistance in malaria: a crucial interaction." *Parasitology Today* **15**, 238–240.

Mutabingwa, T.K. (2005). "Artemisinin-based combination therapies (ACTs): best hope for malaria treatment but inaccessible to the needy!" *Acta Tropica* **95**, 305–315.

Myint, H.Y., et al. (2007). "Efficacy and safety of dihydroartemisinin-piperaquine." *Transactions of the Royal Society of Tropical Medicine and Hygiene* **101**, 858–866.

Newton, P.N., et al. (2006). "Manslaughter by fake artesunate in Asia – will Africa be next?" *PLoS Medicine* **3**, e197.

Nosten, F. and N.J. White (2007). "Artemisinin-based combination treatment of falciparum malaria." *American Journal of Tropical Medicine and Hygiene* **77**, 181–192.

O'Neill, P.M., et al. (1998). "4-Aminoquinolines – past, present, and future: a chemical perspective." *Pharmacology and Therapeutics* **77**, 29–58.

Payne, D. (1988). "Did medicated salt hasten the spread of chloroquine resistance in *Plasmodium falciparum*?" *Parasitology Today* **4**, 112–115.

Piola, P., et al. (2005). "Supervised versus unsupervised intake of six-dose artemether-lumefantrine for treatment of acute, uncomplicated *Plasmodium falciparum* malaria in Mbarara, Uganda: a randomised trial." *Lancet* **365**, 1467–1473.

Plowe, C.V., et al. (1995). "Pyrimethamine and proguanil resistance-conferring mutations in *Plasmodium falciparum* dihydrofolate reductase: polymerase chain reaction methods for surveillance in Africa." *American Journal of Tropical Medicine and Hygiene* **52**, 565–568.

Sibley, C.H., et al. (2001). "Pyrimethamine–sulfadoxine resistance in *Plasmodium falciparum*: what next?" *Trends in Parasitology* **17**, 582–588.

Snow, R.W., et al. (2005). "The global distribution of clinical episodes of *Plasmodium falciparum* malaria." *Nature* **434**, 214–217.

Sucharit, S., et al. (1977). "Chloroquine resistant *Plasmodium falciparum* in Thailand: susceptibility of Anopheles." *Journal of the Medical Association of Thailand* **60**, 648–654.

Talisuna, A.O., P. Bloland and U. D'Alessandro (2004). "History, dynamics, and public health importance of malaria parasite resistance." *Clinical Microbiology Reviews* **17**, 235–254.

Talisuna, A.O., et al. (2007). "Intensity of malaria transmission and the spread of *Plasmodium falciparum* resistant malaria: a review of epidemiologic field evidence." *American Journal of Tropical Medicine and Hygiene* **77**, 170–180.

Trape, J.F., et al. (1998). "Impact of chloroquine resistance on malaria mortality." *C R Acad Sci III* **321**, 689–697.

Trouiller, P., et al. (2002). "Drug development for neglected diseases: a deficient market and a public-health policy failure." *Lancet* **359**, 2188–2194.

UNICEF (2007). Malaria and Children, Progress in Intervention Coverage. United Nations Children's Fund, New York.

Vallely, A., et al. (2007). "Intermittent preventive treatment for malaria in pregnancy in Africa: what's new, what's needed?" *Malaria Journal* **6**, 16.

Watkins, W.M. and M. Mosobo (1993). "Treatment of Plasmodium falciparum malaria with pyrimethamine–sulfadoxine: selective pressure for resistance is a function of long elimination half-life." *Transactions of the Royal Society of Tropical Medicine and Hygiene* **87**, 75–78.

Wernsdorfer, W.H. (1994). "Epidemiology of drug resistance in malaria." Acta Tropica **56**, 143–156.

Wernsdorfer, W.H. and H. Noedl (2003). "Molecular markers for drug resistance in malaria: use in treatment, diagnosis and epidemiology." *Current Opinion in Infectious Diseases* **16**, 553–558.

White, N. (1999a). "Antimalarial drug resistance and combination chemotherapy." *Philosophical Transactions of the Royal Society London B (Biological Sciences)* **354**, 739–749.

White, N.J. (1999b). "Delaying antimalarial drug resistance with combination chemotherapy." *Parassitologia* **41**, 301–308.

White, N.J. (2004). "Antimalarial drug resistance." *Journal of Clinical Investigation* **113**, 1084–1092.

Wolday, D., et al. (1995). "Sensitivity of Plasmodium falciparum in vivo to chloroquine and pyrimethamine–sulfadoxine in Rwandan patients in a refugee camp in Zaire." *Transactions of the Royal Society of Tropical Medicine and Hygiene* **89**, 654–656.

Wongsrichanalai, C., et al. (2002). "Epidemiology of drug-resistant malaria." *Lancet Infectious Diseases* **2**, 209–218.

World Health Organization (1973). "Chemotherapy of malaria and resistance to antimalarials." (ed. World Health Organization), World Health Organization, Geneva, Switzerland.

World Health Organization (1996). "Assessment of therapeutic efficacy of antimalarial drugs for uncomplicated malaria in areas with intense transmission." WHO/MAL/96.1077. (ed. World Health Organization), World Health Organization, Geneva, Switzerland.

World Health Organization (2003). "Assessment and monitoring of antimalarial efficacy for the treatment of uncomplicated falciparum malaria." (ed. World Health Organization), World Health Organization, Geneva, Switzerland.

World Health Organization (2005a). "Susceptibility of *Plasmodium falciparum* to antimalarial drugs: report on global monitoring 1996–2004." World Health Organization, Geneva.

World Health Organization (2005b). "World malaria report." World Health Organization, Geneva, Switzerland.

Chapter 8
Drug Resistance in *Mycobacterium tuberculosis*

Moses Joloba and Freddie Bwanga

Abstract Over 95% of tuberculosis (TB) cases and deaths among adults occur in developing countries. The emergence, management of cases, and spread of drug-resistant strains of *Mycobacterium tuberculosis* is one of the biggest challenges faced by national tuberculosis control programs. Multidrug-resistant TB (MDR-TB), defined as resistance to isoniazid and rifampicin, the two most potent anti-TB drugs, is increasing. MDR-TB is difficult and expensive to treat. Extensively drug-resistant TB (XDR-TB), defined as MDR-TB with additional resistance to a fluoroquinolone and one or more of the injectable anti-TB drugs, has been reported in many countries. Due to lack of diagnostic capacity, particularly in developing countries, the burden of MDR-TB and XDR-TB is not well known. Detection of drug-resistant TB is hindered by the difficulty and length of time required for its diagnosis while using conventional indirect drug susceptibility testing. Development and implementation of rapid methods for the diagnosis of drug-resistant TB are essential to long-term effective control of TB. Presented in this chapter are the terminology used in TB drug resistance, anti-TB drug groups, development and mechanism of drug resistance, and the conventional and new rapid susceptibility testing techniques. Finally, the treatment, control, and surveillance for TB drug resistance are introduced.

8.1 Introduction and Epidemiology

Tuberculosis (TB) is caused by *Mycobacterium tuberculosis (MTB)*. With an estimated 9 million new infections and 2 million deaths per year, TB is the world's number one cause of human suffering attributed to a single infectious agent. Eighty percent of all infections occur in sub-Saharan Africa and Asia where HIV is fueling the epidemic (WHO, 2006).

M. Joloba (✉)
Department of Medical Microbiology, Faculty of Medicine, Makerere University Medical School, Kampala, Uganda
e-mail: mljl0@cwru.edu

A. de J. Sosa et al. (eds.), *Antimicrobial Resistance in Developing Countries*, 117
DOI 10.1007/978-0-387-89370-9_8, © Springer Science+Business Media, LLC 2010

The discovery of effective anti-tuberculosis drugs in the 1940s made TB a curable disease. Before the current HIV-associated TB epidemic, it was assumed that most strains were susceptible to the first-line drugs. However, a worldwide survey from 1994 to 1999 documented resistant TB as a rapidly increasing health problem. The median prevalence of multidrug-resistant tuberculosis (MDR-TB) defined as resistant to at least rifampicin and isoniazid (two key drugs) was 1% (range 0–14.1%) among new cases and 9.3% (range 0–48%) among previously treated cases, respectively (Espinal et al. 2001). The highest prevalence of MDR-TB was in countries of the former Soviet Union (Russian federation, Estonia), parts of China, India, and Latvia, as shown in Table 8.1. Data in Table 8.1 also show that MDR-TB is much more common in previously treated than in new cases of TB.

Although HIV infection is the driving force of the current TB epidemic, surveillance data suggested that HIV is not an independent risk factor for multi-drug resistance and does not influence the choice of empirical TB treatment. MDR-TB requires treatment with second-line anti-TB drugs but cure may occur in only around 50–60% of cases. Treatment is also lengthy requiring 18–24 months, and it is expensive. The cost per case may be as high as

Table 8.1 Prevalence of MDR-TB in selected countries presented by region

Country/territory	MDR-TB burden	
	New cases (%)	Previously treated cases (%)
Eastern Europe		
Estonia	14.1	37.8
Latvia	9.0	23.7
Russia (Tomsk Oblast)	6.5	26.7
Russia (Ivanovo Oblast)	9.0	25.9
Western Europe		
Sweden	4.8	8.3
Switzerland	0.3	12.5
Asia		
China (Henan Province)	10.8	34.4
India (Tamil Nadu State)	3.4	25.0
Republic of Korea	2.2	7.1
Africa		
Botswana	0.5	9.0
Central African Republic	1.1	18.2
Uganda	0.5	4.4
South America		
Cuba	0.0	7.1
Peru	3.0	12.3
Puerto Rico	2.5	16.7
Mexico	2.4	22.4
Uruguay	0.0	6.3
North America		
USA	1.2	5.6

US$85,000–120,000 (Ormerod, 2005). In developing countries, second-line drugs are largely unavailable, and a standard WHO drug regimen is given to all patients at risk of MDR-TB (previously treated cases). In this kind of setting, cure may even be lower.

The recent appearance of extensively drug-resistant tuberculosis (XDR-TB), defined as MDR-TB that is also resistant to second-line injectable drugs (kanamycin, amikacin or capreomycin) plus a fluoroquinolone and that in many cases is untreatable, has the potential to make TB an incurable disease once again. Due to the limited capacity to perform drug susceptibility testing (DST) on these drugs, the burden of XDR-TB is not well known in developing countries. However, it is likely to be low since second-line drugs, which select for XDR-TB, are not yet fully accessible. In this chapter, drug resistance in *M. tuberculosis* is discussed. First, we define the terminology, then discuss how drug resistance develops, mechanism of resistance, methods for drug susceptibility testing, management and control of drug-resistant TB, and finally, an overview of drug-resistant TB surveillance is given.

8.2   Terminology – Definitions are based on Aziz MA et al. 2003

8.2.1   New Case of TB

Is a patient with tuberculosis who, in response to direct questioning, denies having had any prior anti-TB treatment for more than 1 month or for whom there is no record of such history.

8.2.2   Previously Treated TB Patient

Is a TB patient who, in response to direct questioning, admits having been treated for tuberculosis for 1 month or more or for whom there is a record of such history.

8.2.3   Primary Drug Resistance

Presence of resistant strains of *Mycobacterium tuberculosis* (MTB) in a new case of TB.

8.2.4   Acquired Drug Resistance

Presence of resistant strains of MTB in "previously treated patient." To ascertain acquired drug resistance, drug susceptibility pattern must be determined before the start of treatment, as well as at a later point in treatment or at the end of treatment. Such an approach is possible only in countries with the resources to perform serial susceptibility testing. In developing countries, routine drug susceptibility testing is not usually feasible. Therefore, drug resistance in "previously treated patients" is regarded as proxy for acquired drug resistance. Previously treated patients include the following.

8.2.4.1 Treatment Failure

New cases of TB who begin treatment for smear-positive pulmonary TB and who remain smear-positive, or become smear-positive again, at 5 months or later during the course of treatment.

8.2.4.2 Relapse

Patients who have completed their initial treatment, declared cured but become smear-positive again. The term relapse in the strict sense would require genetic comparison of the strains before and after. This is not always possible in developing countries. Thus, the term *recurrence* is now preferred to describe such patients.

8.2.4.3 Return After Default

Patients who interrupt their treatment for more than 2 months after having received a total of at least 1 month of anti-TB treatment and who then return with bacteriologically confirmed tuberculosis.

8.2.4.4 Chronic Case

Patients who continue to be smear-positive after a retreatment regimen. Classification of resistance as primary or acquired has recently been questioned because DNA fingerprinting studies have shown that re- or superinfection with a second strain that is drug resistant can occur during the period of therapy (Van Rie et al. 2000). It is therefore difficult to verify the exact time when drug resistance developed. Thus, the WHO now recommends the use of the term "drug resistance among new cases" instead of primary resistance and "drug resistance among previously treated patients" instead of acquired resistance.

8.2.5 Mono Resistance

In vitro resistance to one anti-TB drug.

8.2.6 *Multidrug-Resistant Tuberculosis*

In vitro resistance to isoniazid and rifampicin, with or without resistance to any other drugs.

8.2.7 *Extensively Drug-Resistant Tuberculosis*

In vitro resistance to kanamycin, amikacin, or capreomycin plus resistance to a fluoroquinolone in an MDR isolate.

8.3 Drugs for Tuberculosis

Anti-tuberculosis drugs are categorized into five groups (Drobniewski 1998, Rich et al. 2006). First-line drugs are used in treatment of new TB cases in whom the risk of MDR-TB is low. Second-line drugs are used for resistant TB such as MDR-TB. In Table 8.2, these groups are shown.

8.4 Development of Drug Resistance in *M. tuberculosis*

Drug resistance among previously treated patients usually results from exposure to a single drug that suppresses the growth of bacilli susceptible to that drug but permits the multiplication of pre-existing drug-resistant mutants. It is the most common type of resistance to the first-line drugs and can emerge against any anti-tuberculosis agent during chemotherapy (Cohn et al. 1997). Subsequent transmission of resistant bacilli to other persons leads to disease that is drug resistant from the outset. This is what is referred to as drug resistance among new cases, formerly termed as primary resistance.

Within a population of *M. tuberculosis* are naturally occurring mutants that arise due to spontaneous point mutations and/or deletions (Zhang and Telenti 2000). Mutations in genes encoding drug targets or drug-activating enzymes are responsible for resistance and such mutations have been found for all first-line drugs and some second-line drugs (Zhang and Young 1994). For a given drug, resistant mutants occur approximately once in every 10^7–10^{10} cells (Parsons 2004). Therapy with one drug results in rapid selection of drug-resistant mutants, which now dominate the lesions. This was first identified in the 1940s when therapy with streptomycin (SM) or *p*-aminosalicylic acid (Pai et al. 2005) as the sole drug resulted in the emergence of drug-resistant strains in the majority of treated TB patients (Pyle 1947). The occurrence of one mutant strain resistant to two drugs simultaneously requires a theoretical population of 10^{16} mycobacterial cells. Thus, combining two or more anti-TB drugs reduces effectively the chance of selecting for resistant mutants. Indeed by combining

Table 8.2 Anti-tuberculosis drugs, doses, mechanism of action, major side effects, and genes involved in resistance mutations

Drug	Dose per day	Mechanism of action	Major side effects	Genes affected by mutations
Group 1: First-line drugs (oral)				
Isoniazid (H)	300 mg (5 mg/kg)	Inhibits mycolic acid synthesis	Peripheral neuropathy, hepatitis	katG, inhA, oxyR
Rifampicin (R)	600 mg (10 mg/day)	Binds to RNA polymerase inhibiting RNA synthesis	GI upset, hepatitis	rpoB
Pyrazinamide (Z)	1.5–2.5 g (15–30 mg/kg)	Activated to pyrazinoic acid, which is bacteriocidal	GI upset, rash, joint pain	pncA
Ethambutol (E)	2.5 g max. (15–5 mg/kg)	Inhibits cell wall synthesis	Optic neuritis, decreased visual acuity	embA,B,C
Group 2: Injectable second-line drugs				
Streptomycin (S)	1 g (15 mg/kg)	Binds to ribosomal proteins and inhibits protein synthesis	Nephrotoxicity (usually reversible, occurs in 20–25% with capreomycin). Occasional irreversible vestibular damage with kanamycin	Rrs, rpsl
Amikacin (Am)	750 mg (15–20 mg/kg)	Disrupts ribosomal function		?rrs, ?rpsl
Kanamycin (Km)	750 mg (15 mg/kg)	Binds to 30S ribosomal subunit inhibiting protein synthesis		?rrs, ?rpsl
Capreomycin (Cm)	1 g (15–20 mg/kg)	Similar to aminoglycosides		?rrs, ?rpsl
Group 3: Fluoroquinolones (second line)				
Ciprofloxacin (Cfx)	20–30 mg/kg	Disrupts the DNA–DNA gyrase complex inducing double-strand cleavage of DNA, blocking DNA synthesis	GI upsets	gyrA
Ofloxacin (Ofx)	800 mg			
Levofloxacin (Lfx)	750 mg			
Moxifloxacin (Mfx)	400 mg			
Gatifloxacin (Gfx)	400 mg			

Table 8.2 (continued)

Drug	Dose per day	Mechanism of action	Major side effects	Genes affected by mutations
Group 4: Oral bacteriostatic anti-tuberculosis agents (second-line)				
Cycloserine (Cs)	750 mg (15–20 mg/kg)	Inhibits cell wall synthesis	Peripheral neuropathy, nervousness	–
Ethionamide (Eto)	750 mg (15–20 mg/kg)	Inhibits oxygen-dependent mycolic acid synthesis	GI upsets	–
P-Aminosalicylic acid (PAS)	8 g (150 mg/kg)	Disrupts folic acid metabolism	GI upsets	–
Rifabutin	750 mg (15–20 mg/kg)	Binds to RNA polymerase inhibiting RNA synthesis	GI upset, hepatitis	rpoB
Thioacetazone (Th†)				
Group 5: Anti-TB drugs§ of unclear efficacy				
Clofazimine (Cfz)	–		–	–
Amoxicillin / clavulanate (Amx/Clv)	–		–	–
Clarithromycin (Clr)	–		–	–
Linezolid (Lzd)	–		–	–

†Thioacetazone should be used only in patients documented to be HIV-negative to avoid the risk of Stevenson–Johnson Syndrome and usually should not be chosen over other drugs listed in Group 4.
§Drugs not recommended by WHO for routine use in MDR-TB patients.

SM and PAS, the resistance rate was reduced to 9% (MRC 1949). The above findings led to a decision on combining drugs in the treatment of TB, which has remained a fundamental principle in today's TB management. Multidrug-resistant and extensively drug-resistant strains arise by sequential accumulation of resistance mutations for individual drugs.

There are some lessons to learn. First, cavitary TB lesions, with abundant tubercle bacilli (usually 10^9/lesion), are likely to contain more mutants resistant to a given drug. Development of resistance is thus likely to be more rapid than in cases with non-cavitary lesions, which contain relatively few cells (about 10^{3-4}/lesion) (Canetti 1995). Second, drug resistance can be taken to be a man-made problem contributed to by the TB control program, health providers, and the patient (Aziz et al. 2003). The TB control program contribution to drug resistance may be through failure to avail a standardized, appropriate thera-peutic regimen; frequent or prolonged shortages of drugs; procurement of anti-TB drugs of unproven quality; and poorly regulated sale of anti-TB drugs. Health workers may incorrectly manage individual patients through prescrip-tion of inappropriate regimens or wrong dosage. Patients through non-adherence to prescribed treatment or through comorbid conditions that reduce absorption of oral drugs significantly contribute to the development of drug resistance. Where directly observed therapy short course (DOTs) treatment is excellent, non-adherence is likely to be insignificant in drug resistance development.

The contribution of each of these factors may differ between countries. However, TB program management issues, poor adherence, and health worker practices are all common in developing countries where DOTs implementation is still suboptimal.

8.5 Mechanism of Resistance to Anti-tuberculosis Drugs

Knowledge of the molecular genetic basis of resistance to anti-tuberculosis agents has advanced rapidly in the last 10 years. Genetic analysis shows that resistance is a result of alteration in the drug targets due to spontaneous mutations in the genome of *M. tuberculosis* (David 1970). The mechanisms of resistance to each drug have been extensively reviewed in an update by Ramaswamy and Musser (1998). A brief review of the mechanism for each drug is given here.

Resistance to rifampicin and related rifamycins is due to mutations that alter the sequence of a 27-amino-acid region of the β-subunit of ribonucleic acid (RNA) polymerase. Resistance to isoniazid (INH) is more complex. Many resistant organisms have mutations in the *katG* gene encoding catalase–peroxidase, which result in altered enzyme structure. These structural changes apparently result in decreased conversion of INH to its biologically active form. Some INH-resistant organisms also have mutations in the *inhA* locus or a recently characterized gene (*kasA*) encoding a β-ketoacyl-acyl carrier protein synthase. Resistance to streptomycin is mainly due to mutations in the 16S rRNA gene or the *rpsL* gene encoding ribosomal protein S12. Resistance to pyrazinamide (PZA) in the majority of organisms is caused by mutations in the gene (*pncA*) encoding pyrazinamidase that result in diminished enzyme activity. Pyrazinamidase is an enzyme required to catalyze the conversion of pyrazinamide to its active form pyrazinoic acid in the phagosome. Ethambutol resistance in approximately 60% of organisms is due to amino acid replacements at position 306 of an arabinosyltransferase encoded by the *embB* gene. Amino acid changes in the A subunit of deoxyribonucleic acid gyrase cause fluoroquinolone resistance in most organisms. Kanamycin resistance is due to nucleotide substitutions in the *rrs* gene encoding 16S rRNA.

Cross-resistance has also been reported. Cross-resistance can occur between drugs that are chemically related and/or have the same or similar target within the mycobacterial cell. For example, strains resistant to rifampicin are also resistant to rifabutin in 70–90% of cases (Yang B et al. 1998). It is therefore important that strains that are rifampicin resistant are screened for rifabutin resistance if the latter drug is being considered for MDR treatment. Some strains of *M. tuberculosis* with low-level resistance to INH are cross-resistant to ethionamide (Zhang and Young 1994). It is uncertain why no cross-resistance has been seen between INH and PZA, even though both drugs are structurally similar to nicotinamide.

Although remarkable advances have been made, a complete picture on the molecular genetic basis of drug resistance in *M. tuberculosis* remains to be learned. Thus, development of new diagnostic and therapeutic approaches based on knowledge obtained from the study of the molecular mechanisms of resistance will continue to occur.

8.6 Drug Susceptibility Testing of *M. tuberculosis*

In developing countries, tests to detect drug-resistant TB are rarely performed on a routine basis. Where this is occasionally done, the so-called conventional methods that involve initial bacterial culture isolation and drug susceptibility testing on solid media is common. Due to the slow growth of mycobacteria on solid media, the susceptibility testing process may require at least 2 months in ideal conditions to yield results. Moreover, only previously treated patients who go to hospitals near the location of the culture laboratory benefit. Recent advances in molecular biology and elucidation on the molecular mechanisms of drug resistance in TB have led to the development of new tools for rapid diagnosis and detection of drug resistance (Shamputa et al. 2004). In addition, the non-molecular approaches including detection of growth by measuring consumption of oxygen, observation of cord-like growth, and colorimetric and phage-based assays are being studied or introduced in some developing countries. In this section, the conventional and newer testing techniques for the detection of resistance will be reviewed.

8.6.1 Conventional Susceptibility Methods

Three solid-media-based DST methods (the proportion method, the absolute concentration and resistance ratio) have been standardized and are widely used in developing countries for first-line drugs.

8.6.1.1 Proportion Method

The proportion method depends on the inoculation of drug-free and drug-containing solid media with equal quantities of two dilutions of a standardized *M. tuberculosis* inoculum. Distinct, countable colony-forming units (CFU) should be present on the drug-free media. The number of CFU, if any, on drug-containing is compared with those on drug-free media to obtain the proportion of resistant mutants. A proportion greater than 1% indicates resistance to that agent (Canetti et al. 1963, 1969). That is if the CFU's on drug-containing medium are more than those on drug-free medium, resistant mutants exceed 1% of the original inoculation on drug medium and the strain is therefore resistant to the tested drug. If CFU's on the drug-containing medium are less than those on drug free medium, the resistant mutants in the original inoculation on the drug-medium were less than 1%, and the strain is susceptible to the tested drug. If CFU's on the drug-containing and drug-free media are equal, susceptibility results are uninterpretables and testing must be repeated. The advantage of this technique is the ease of estimation of the inoculum size from the CFU counts. However, a single CFU could have arisen from a clump of bacilli rather than from an individual cell, resulting in an inaccurate calculation of the proportion of resistant mutants in the population. The proportion method can be performed on Lowenstein–Jensen (Makinen et al. 2006) or Middlebrook agar medium. The LJ medium is recommended by the WHO and the IUATLD for developing countries as it is cheap, easy to set and read and has low contamination rates (WHO/IUATLD 1997). One disadvantage is that the composition of LJ is complex and media preparation is time consuming. Middlebrook 7H10 agar is the preferred medium in the United States [24]. The latter medium has a simpler composition, it is easy to prepare, and the transparency of the medium allows for early detection and quantification of colonies microscopically (NCCLS 2003). The cost of routine use of the 7H10 agar discourages its use in resource-limited settings as it requires supplementation with the expensive oleate-albumin-dextrose-catalase (OADC) and incubation of agar plates in a CO_2 environment.

Critical proportion of resistant mutants: We earlier stated that any given population of MTB has a certain proportion of naturally occurring mutants that are resistant to a particular drug. Although the proportion of pre-existing mutants based on a mutation rate of 10^{-7} would be much lower, it is theoretically assumed to be 1%, and this has been determined to predict therapeutic outcome. When performing DST with the proportional method, the drug-free medium is seeded with an inoculum that is diluted 100 times compared to that seeded on the drug containing medium. In other words, the drug-free medium will grow 1% of the number of colony-forming units (CFU) seeded on the drug medium. Assuming that 1% of the inoculum on the drug medium is resistant mutants, only these mutants will grow. By comparing the CFU's on drug medium with those on drug-free medium, it is possible to deduce that the isolate is susceptible ($\leq$1%) or resistant (>1%). Thus to interpret as susceptible, the number of CFU on the drug medium must not exceed those on drug-free medium. This is the fundamental principle underlying the proportional method of DST in MTB.

Critical concentrations of drugs: The critical concentration (CC) is defined as the concentration that inhibits *in vitro* growth of most MTB cells within the population of wild-type strains without appreciably affecting the growth of pre-existing resistant mutants (Canetti et al. 1969). Table 8.3 shows the CC for different anti-TB drugs. If resistant mutants exceed 1%, that critical concentration may not inhibit growth, and this predicts possible therapeutic failure. The critical concentration of the drug in the test medium used for DST may be different from the serum peak levels or the MIC of that drug. It may be higher (rifampicin, isoniazid), lower (ethambutol), or same as the MIC (streptomycin) (Heifets 2000).

Table 8.3 Critical concentration of anti-tuberculosis drugs

Drug	MIC (µg/ml)	Critical concentration (µg/ml)			
		BACTEC 460	MGIT 960	Agar (7H10 or 7H11)	LJ
First-line drugs					
Rifampicin	0.25–0.5	2	1.0	1	40
Isoniazid	0.05–0.2	0.1 and 0.4	0.1 and 0.4	0.2 and 1	0.2 and 1
Pyrazinamide	20–50	100	–	–	400
Ethambutol	1–5	2.5 and 7.5	5.0	5 and 10	2.5
Streptomycin	2–8	2 and 6	1 and 4.0	2 and 10	4
Second-line drugs					
Amikacin	1	1	1	4	–
Kanamycin	5	5	–	5	30
Capreomycin	1–50	1.25	–	10	40
Cycloserine	5–20	–	–	–	40
Ethionamide	0.6–2.5	1.25	–	5	40
Ofloxacin	0.5–2.5	2	–	2	2
PAS	1	4	–	2	0.5
Rifabutin	0.06–8	0.5	–	0.5 and 1	–

MIC = minimum inhibitory concentration; MGIT = mycobacterium growth indicator tube; LJ = Lowenstein–Jensen; PAS = *p*-aminosalicylic acid

8.6.1.2 The Absolute Concentration Method

An inoculum of *M. tuberculosis* is added to LJ or 7H10 agar containing several sequential dilutions of each drug. Resistance is indicated by the lowest concentration of the drug that inhibits growth, i.e., less than 20 colonies by the end of 4 weeks (IUATLD 1998).

8.6.1.3 The Resistance Ratio Method

The resistance ratio (RR) is the ratio of the minimum inhibitory concentration (MIC) for the patients' strain to the MIC of the drug-susceptible reference strain, H37Rv, both tested in the same experiment (Heifets 2000). After 4 weeks of incubation, growth on any slope is defined as the presence of 20 or more colonies, and MIC is defined as the lowest drug concentration where the number of colonies is less than 20. A resistance ratio of 2 or less indicates sensitive strain, and a resistance ratio of 8 or more indicates resistant strains (Canetti et al. 1969). The RR method is the most expensive of the three conventional methods that use solid media. Conventional methods are not rapid enough to cope with the high burden of TB in developing countries. New rapid susceptibility techniques (phenotypic and genotypic) preceded with or without culture are being studied or introduced in some developing countries.

8.6.2 New Phenotypic Methods

The new phenotypic tests include the mycobacteria growth indicator tube (MGIT), microscopic-observation drug susceptibility (MODS) assay, colorimetric assays, phage assays, and others such as the E-test.

8.6.2.1 The Mycobacterial Growth Indicator Tube

The mycobacterial growth indicator tube (Becton Dickinson, Sparks, Maryland, USA) is based on fluorescence detection of mycobacterial growth in a tube containing a modified Middlebrook 7H9 medium together with a fluorescence quenching-based oxygen sensor embedded at the bottom of the tube. Consumption of oxygen in the medium results in fluorescence under ultraviolet (UV) light. The MGIT system was introduced about 10 years ago as manual test but the automated version is now available. Studies on the MGIT system for first- and second-line anti-TB drugs have shown sensitivity, specificity, and agreement ranging from 90 to 100% compared to conventional tests (Johansen et al. 2004). Operational, cost-effectiveness, and validation studies assessing the MGIT system in low- and middle-income countries are limited. Operational issues might be major limitations to the wide use of this new technique in developing countries. For example, the cost of the MGIT instruments, their maintenance, and reagents can be too high for these settings.

8.6.2.2 Microscopic Observation Drug Susceptibility Assay

Microscopic-observation drug susceptibility assay is a broth-based technique for the drug-resistant MTB, directly from sputum. The test relies on the ability of MTB to form cord-like growth in liquid medium. Incorporation of drugs permits rapid and direct DST. Under an inverted microscope, observation of cords in a drug-containing tissue culture well indicates resistance to the drug. Studies on MODS have shown sensitivities and specificities above 90% for rifampicin and isoniazid resistance (Moore et al. 2006). On the basis of rapidity (5–14 days), technical ease, and low cost, this is a promising DST method for developing countries. Biosafety concerns and the risk of cross contamination may however limit its wide use.

8.6.2.3 Colorimetric Methods

Colorimetric methods include the nitrate reductase assay (NRA), the MTT (3-[4, 5-dimethylthiazol-2-yl]-2, 5-diphenyltetrazolium bromide), and alamar blue assays. For all these tests, resistance is detected by a change in the color of the indicator, when detection reagents are added in the medium containing viable mycobacteria. These tests have been reported to have sensitivities above 95% (Coban et al. 2004, Sethi et al. 2004, Montoro et al. 2005). Multicentre evaluations of these tests for first-line drugs showed accuracy for isoniazid, rifampicin, and ethambutol to be between 96 and 100%. Lower values were, however, obtained for streptomycin (Martin et al. 2005). Reproducibility was good and results were ready in 5–14 days.

The NRA can easily be adopted where the conventional LJ proportion method is used, does not require additional equipment, it is simple and could easily be implemented in countries with limited laboratory facilities (Angeby et al. 2002). However, most published studies on these tests had various biases. It is not clear whether the high diagnostic accuracy values indicated in the reports are reproducible if the tests are put to routine clinical use. The MTT and alamar blue are technically not simple tests in the perspective of developing countries. These tests use microtiter plates making them difficult to work with and prone to cross contamination. Reagent preparation for MTT is quite delicate as the pH must be strictly standardized and reagents refrigerated for specific periods. Biosafety is also of concern since the plates have to be reopened once or twice after incubation to add the detection reagents. For these reasons, the NRA seems to be a more promising alternative in developing countries than the MTT or alamar blue assays.

8.6.2.4 Other New Phenotypic Tests

Other new phenotypic tests include the E-test (Joloba et al. 2000) and phage-based assays such as *FastPlaque* TB assay (Biotec Laboratories Ltd, Ipswich, UK) and Luciferase reporter assay (Simboli et al. 2005)

8.6.3 New Genotypic Methods

Genotypic methods aim at the detection of genetic mutations that lead to resistance rather than the resistance phenotype. Genotypic tests involve nucleic acid amplification of sections of the *M. tuberculosis* genome known to be altered in resistant strains and a second step of assessing the amplified products for specific mutations correlating with resistance. Genotypic methods were initially directed to detect rifampicin resistance, which predicts MDR-TB in 90% of cases (Rossau et al. 1997). However, recent findings suggest that such prediction may not be true for all settings (WHO 2005). Examples of genotypic tests include the following.

8.6.3.1 Solid-Phase Hybridization Techniques

Two commercial solid-phase hybridization techniques are available: the Line Probe Assay (INNO-LiPA Rif TB Assay; Innogenetics, Ghent, Belgium) for the detection of mutations in the *rpoB* gene for rifampicin resistance and the Genotype MTBDR (Hain Lifesciences, Nehren, Germany) for the simultaneous detection of mutations in the *rpoB* and *katG* genes for rifampicin and isoniazid resistance, respectively. Cultured strains or clinical samples with MTB DNA may be amplified and hybridized on to probes covering the core region of the *rpoB* gene immobilized on a nitrocellulose strip (Rossau et al. 1997). An evaluation study of the INNO-LiPA Rif TB assay showed high sensitivity and specificity (Jureen et al. 2004). In one such study of the GenoType MTBDR assay, 99% of MDR strains with mutations in the *rpoB* gene and 88.4% of strains with mutations in the codon 315 of the *katG* gene were correctly identified (Hillemann et al. 2005). Other studies have reported sensitivity and specificity for rifampicin resistance to be 99–100%, while for isoniazid, it ranges from 60 to 70%. The lower values for INH are due to the design of the test, which detects only *katG* mutations, yet other mutations may be involved in resistance (Hillemann et al. 2005, 2006). A newer version of the genotype MTB test called the GenoType MTBDR*plus*, which detects additional mutations involved in INH resistance, is now available (Hillemann et al. 2007).

Although genotypic methods require some expertise in molecular biology, the above tests require very minimal skills that could be learnt in a matter of weeks (Quezada et al. 2007). The disadvantages of the molecular tests include cross contamination causing false results, failure to obtain results due to low amount of DNA present in the sample (if direct tests are performed), and presence of polymerase chain reaction (PCR) inhibitors.

8.6.3.2 DNA Sequencing

Sequencing DNA of PCR-amplified products has become the gold standard for mutation detection. Both manual and automated procedures may be used but the latter is now favored (Victor 2001). These techniques have been widely used to characterize mutations in the *rpoB* gene in rifampicin-resistant strains and to other anti-tuberculosis drugs (Sekiguchi et al. 2007). Cost may be prohibitive for routine use of sequencing tests in developing countries.

8.6.3.3 Other Molecular Tests

Other molecular tests include real-time PCR and microarrays. These new tests have been evaluated in some settings with promising results (Gryadunov et al. 2005). Real-time PCR techniques are ultra rapid (average of 1.5–2.0 hours after DNA extraction) and have a lower risk of contamination. However, the technical capacity and the cost are currently still high for developing countries.

All the genotypic methods have the advantage of a shorter turnaround time, possibility for direct application in clinical samples, less biohazard risks, and feasibility for automation.

8.6.4 Direct Detection of Drug Resistance

Early detection of drug resistance in TB allows timely initiation of the appropriate treatment. This is important in patient well being and in control of drug-resistant TB transmission. Some of the tests reviewed above are now being studied for possible direct application on smear positive sputum. The molecular tests have been more promising in this regard compared to phenotypic tests.

8.6.5 Implementation Issues in Drug Susceptibility Testing

Developing countries are constrained by monetary resources, highly trained personnel, and laboratory infrastructure for routine TB culture and DST. Fortunately, primary resistance, particularly MDR, is low in some developing countries. In view of these facts, the WHO recommends routine DST in only previously treated TB patients who are at higher risk of MDR-TB. Susceptibility testing of at least rifampicin is a prerequisite for drug resistance control program. Testing of second-line drugs is not mandatory for such control programs and is not recommended unless rigorous quality control is in place, including external proficiency testing by one of the supranational TB reference laboratories (Rich et al. 2006). Where resources are limited, hierarchy of DST capacity should include the following priority ranking: rifampicin and isoniazid; ethambutol and streptomycin; pyrazinamide (if broth-based DST systems are available); kanamycin (or amikacin) and capreomycin; fluoroquinolone (generally, the same fluoroquinolone that is used in the setting) and ethionamide/prothionamide, PAS, and cycloserine.

The conventional tests discussed earlier are often the only DST methods in developing countries. It is now generally felt that new rapid tests be used alongside the conventional tests to quicken the process of drug-resistant TB detection. The algorithm below is proposed for the drug-resistant TB detection and management, but it could be customized to particular settings (Fig. 8.1).

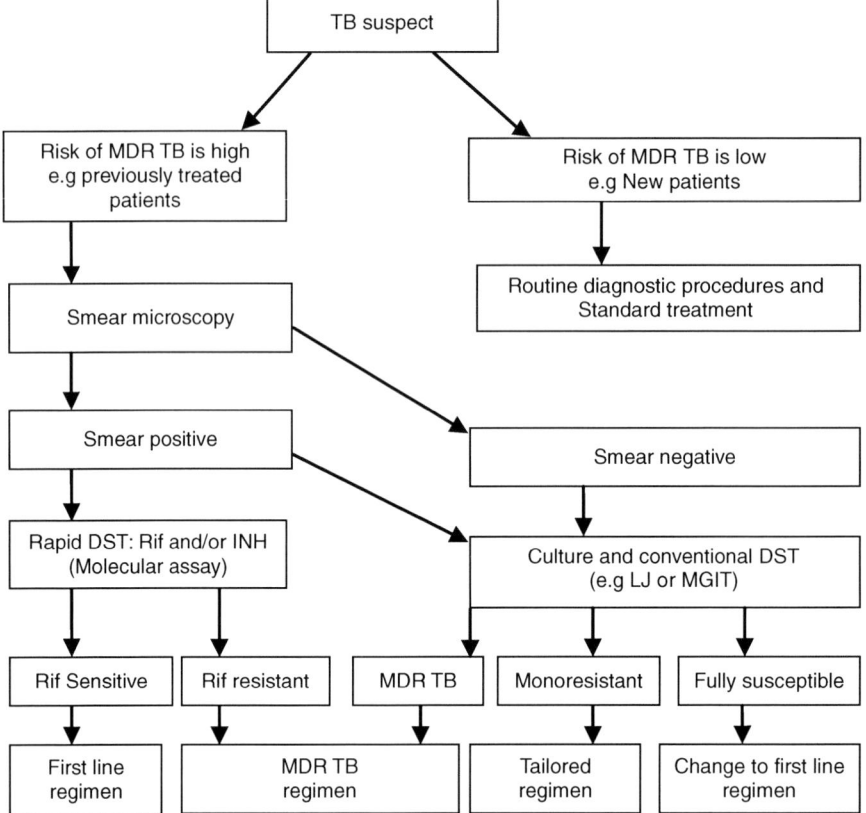

Fig. 8.1 Proposed algorithm for drug-resistant TB detection and management in developing countries

8.7 Treatment of Drug-Resistant Tuberculosis

Treatment of 'new cases' of tuberculosis in developing countries is based on a standard WHO regimen that consist of 8 months with first-line drugs, i.e., 2 months of rifampicin (R), isoniazid (H), ethambutol (E), and pyrazinamide (Z) and 6 months of EH. It is common practice to abbreviate the regimen as 2EHRZ–6EH. In other countries, 2EHRZ–4RH is used. In resource-limited settings, due to the lack of rapid DST, all "previously treated" cases are given one regimen for which only streptomycin (S), an injectable agent, is added to the above regimen, i.e., 2SEHRZ–1EHRZ–5EHR. Since one injectable drug is added, the regimen may not be optimal for MDR-TB. Ideal regimens for all forms of drug-resistant TB have been published in comprehensive guidelines (Rich et al. 2006, Iseman 1993). The reader is strongly urged to read these reviews, which describe in detail the principles of regimen selection for both susceptible and resistant TB. Briefly, early MDR-TB detection and prompt initiation of treatment constitute the cornerstone in achieving successful outcomes. To treat MDR-TB, regimens should include at least four drugs that are highly likely to be susceptible, based on DST and/or the drug history of the patient. Injectable agents (Group 2) must be used for a minimum of 6 months. Pyrazinamide can be used for the entire treatment if it is judged to be effective because MDR-TB patients have chronically inflamed lungs, which theoretically produce the acidic environment in which pyrazinamide is active. Treatment is for a minimum duration of 18 months beyond conversion. Each dose is given as DOTs Plus throughout the treatment. When DST is available from a reliable laboratory, it should be used to guide therapy.

8.8 Control of Drug-Resistant Tuberculosis

Patients with MDR-TB may respond to treatment slowly and remain infectious longer, so they may infect more contacts. Following outbreaks of MDR-TB in New York and Florida, it became apparent that MDR-TB is as transmissible as drug-susceptible TB, and transmission may occur unnoticed especially among highly vulnerable populations and in institutional settings (Frieden et al. 1993). According to guidelines by the WHO/CDC/IUATLD, infection control in MDR-TB does not significantly differ from the basic TB infection control strategies. However, because of the seriousness of MDR-TB, the highest level of compliance with guidelines (early case detection, complete treatment, administrative, environmental, or engineering controls and personal respiratory protection) is recommended for any institution attempting to treat MDR-TB.

8.9 Surveillance for Drug-Resistant Tuberculosis

The WHO recommends surveillance mechanisms to help in monitoring the efficacy of drug regimens. Comprehensive guidelines for drug resistance surveillance have been published by the WHO (Aziz et al. 2003).

Acknowledgment Thanks to Jessica Restucci, APUA executive assistant and Irma Stefani, APUA administrative assistant for their valuable assistance and support throughout the editing and proof-reading of selected chapters.

References

WHO., Global tuberculosis control: surveillance, planning and financing. WHO/HTM/TB/ 2006.362 2006, WHO.

Espinal, M.A., et al., Global trends in resistance to antituberculosis drugs. World Health Organization-International Union against Tuberculosis and Lung Disease Working Group on Anti-Tuberculosis Drug Resistance Surveillance. N Engl J Med, 2001. **344**(17): 1294–303.

Ormerod, L.P., Multidrug-resistant tuberculosis (MDR-TB): epidemiology, prevention and treatment. Br Med Bull, 2005. **73–74**: 17–24.

Aziz, M.A., et al., *Guidelines for surveillance of drug resistance in tuberculosis/Document WHO/CDS/CSR/RMD/2003.3* 2nd ed World Health Organization. 2003.

Van Rie, A., et al., *Classification of drug-resistant tuberculosis in an epidemic area.* Lancet, 2000. **356**(9223):. 22–5.

Drobniewski, F., *Drug resistant tuberculosis in adults and its treatment.* J R Coll Physicians Lond, 1998. **32**(4): 314–8.

Rich, M., et al., Guidelines for the programmatic management of drug-resistant tuberculosis. World Health Organisation. 2006.

Cohn, D.L., F. Bustreo, and M.C. Raviglione, Drug-resistant tuberculosis: review of the worldwide situation and the WHO/IUATLD Global Surveillance Project. International Union Against Tuberculosis and Lung Disease. Clin Infect Dis, 1997. **24 Suppl 1**: S121–30.

Zhang, Y. and A. Telenti, *Genetics of drug resistance in Mycobacterium tuberculosis.* In: *Molecular genetics of mycobacteria*, G.F. Hatfull, Editor. 2000, ASM Press, Washington, D.C. 2000, pp. 235–254.

Zhang, Y. and D. Young, *Molecular genetics of drug resistance in Mycobacterium tuberculosis.* J Antimicrob Chemother, 1994. **34**(3):. 313–9.

Parsons, L., *Laboratory diagnostic aspects of drug resistant tuberculosis.* Front Biosci 2004. **9**: 2086–2105.

Pai, M., et al., Bacteriophage-based assays for the rapid detection of rifampicin resistance in Mycobacterium tuberculosis: a meta-analysis. J Infect, 2005. **51**(3): 175–87.

Pyle, M.M., Relative numbers of resistant tubercule bacilli in sputa of patients before and during treatment with streptomycin. Proc Mayo Clin, 1947. **22**: 465–73

Medical, Research, and Council., Treatment of pulmonary tuberculosis with streptomycin and *para*-aminosalicylic acid. Br Med J, 1949. **2**: 1521–1525

Canetti, G., *Chapter two: the tubercle bacillus in the pulmonary tuberculosis lesion.* in *The tubercle bacillus*, G. Canetti, Editor. Springer Publishing Company, NY, 1955. 29–85.

David, H.L., Probability distribution of drug-resistant mutants in unselected populations of Mycobacterium tuberculosis. Appl Microbiol, 1970. **20**(5): 810–4.

Ramaswamy, S. and J.M. Musser, Molecular genetic basis of antimicrobial agent resistance in Mycobacterium tuberculosis: 1998 update. Tuber Lung Dis, 1998. **79**(1):3–29.

Yang, B., et al., Relationship between antimycobacterial activities of rifampicin, rifabutin and KRM-1648 and rpoB mutations of Mycobacterium tuberculosis. J Antimicrob Chemother, 1998. **42**(5): 621–8.

Shamputa, I.C., L. Rigouts, and F. Portaels, Molecular genetic methods for diagnosis and antibiotic resistance detection of mycobacteria from clinical specimens. APMIS, 2004. **112**(11–12): 728–52.

Canetti, G., et al., Mycobacteria: laboratory methods for testing drug sensitivity and resistance. Bull World Health Organ, 1963. **29**: 565–78.

Canetti, G., et al., Advances in techniques of testing mycobacterial drug sensitivity, and the use of sensitivity tests in tuberculosis control programmes. Bull World Health Organ, 1969. **41**(1): 21–43.

Makinen, J., et al., Comparison of two commercially available DNA line probe assays for detection of multidrug-resistant Mycobacterium tuberculosis. J Clin Microbiol, 2006. **44**(2): 350–2.

WHO/IUATLD., *Guidelines for surveillance of drug resistance in tuberculosis. WHO/TB/96. 216.* World Health Organization, Geneva. 1997.

NCCLS., Susceptibility testing of mycobacteria, nocardia and other aerobic actinomycetes, approved standard. NCCLS document M24-A (ISBN 1-56238-500-3). 2003, NCCLS, 940 West Valley Road, suite 1400, Wayne, PA 19087-1898. 2003.

Heifets, L., Conventional methods for antimicrobial susceptibility testing of Mycobacterium tuberculosis, in Resurgent and emerging infectious diseases. Multi-drug resistant tuberculosis, I. Bastian, Editor. Dordrecht, Kluwer Academic, 2000:135–141.

IUATLD., The Public Health Service National Tuberculosis Reference Laboratory and the National Laboratory Network: minimum requirements, role and operation in a low-income country. 1998, International Union Against Tuberculosis and Lung Disease, 1998. Paris.

Johansen, I.S., et al., Rapid, automated, nonradiometric susceptibility testing of Mycobacterium tuberculosis complex to four first-line antituberculosis drugs used in standard short-course chemotherapy. Diagn Microbiol Infect Dis, 2004. **50**(2): 103–7.

Moore, D.A., et al., *Microscopic-observation drug-susceptibility assay for the diagnosis of TB.* N Engl J Med, 2006. **355**(15): 1539–50.

Coban, A.Y., et al., Drug susceptibility testing of Mycobacterium tuberculosis with nitrate reductase assay. Int J Antimicrob Agents, 2004. **24**(3): 304–6.

Sethi, S., et al., Drug susceptibility of Mycobacterium tuberculosis to primary antitubercular drugs by nitrate reductase assay. Indian J Med Res, 2004. **120**(5): 468–71.

Montoro, E., et al., Comparative evaluation of the nitrate reduction assay, the MTT test, and the resazurin microtitre assay for drug susceptibility testing of clinical isolates of Mycobacterium tuberculosis. J Antimicrob Chemother, 2005. **55**(4): 500–5.

Martin, A., et al., Multicenter evaluation of the nitrate reductase assay for drug resistance detection of Mycobacterium tuberculosis. J Microbiol Methods, 2005. **63**(2): 145–50.

Angeby, K.A., L. Klintz, and S.E. Hoffner, Rapid and inexpensive drug susceptibility testing of Mycobacterium tuberculosis with a nitrate reductase assay. J Clin Microbiol, 2002. **40**(2): 553–5.

Joloba, M.L., S. Bajaksouzian, and M.R. Jacobs, *Evaluation of Etest for susceptibility testing of Mycobacterium tuberculosis.* J Clin Microbiol, 2000. **38**(10): 3834–6.

Simboli, N., et al., In-house phage amplification assay is a sound alternative for detecting rifampicin-resistant Mycobacterium tuberculosis in low-resource settings. Antimicrob Agents Chemother, 2005. **49**(1): 425–7.

Rossau, R., et al., Evaluation of the INNO-LiPA Rif. TB assay, a reverse hybridization assay for the simultaneous detection of Mycobacterium tuberculosis complex and its resistance to rifampicin. Antimicrob Agents Chemother, 1997. **41**(10): 2093–8.

WHO., Anti-tuberculosis drug resistance in the world. Report No. 3. WHO/HTM/TB/2004. 343. 2005, World Health Organisation: Geneva.

Jureen, P., J. Werngren, and S.E. Hoffner, Evaluation of the line probe assay (LiPA) for rapid detection of rifampicin resistance in Mycobacterium tuberculosis. Tuberculosis (Edinb) 2004. **84**(5): 311–316.

Hillemann, D., et al., Use of the genotype MTBDR assay for rapid detection of rifampicin and isoniazid resistance in Mycobacterium tuberculosis complex isolates. J Clin Microbiol, 2005. **43**(8): 3699–703.

Hillemann, D., S. Rusch-Gerdes, and E. Richter, *Application of the Genotype MTBDR assay directly on sputum specimens.* Int J Tuberc Lung Dis, 2006. **10**(9): 1057–9.

Hillemann, D., S. Rusch-Gerdes, and E. Richter, Evaluation of the GenoType MTBDRplus assay for rifampicin and isoniazid susceptibility testing of Mycobacterium tuberculosis strains and clinical specimens. J Clin Microbiol, 2007. **45**(8): p. 2635–40.

Quezada, C.M., et al., Implementation validation performed in Rwanda to determine whether the INNO-LiPA Rif.TB line probe assay can be used for detection of multidrug-resistant Mycobacterium tuberculosis in low-resource countries. J Clin Microbiol, 2007. **45**(9): 3111–4.

Victor, T., Detection of mutations in Mycobacterium tuberculosis by a dot blot hybridization strategy, in Mycobacterium tuberculosis protocols. Methods in molecular medicine., T. Parish, Editor. Humana Press, New Jersey, 2001, p. 155–64.

Sekiguchi, J., et al., *Detection of multidrug resistance in Mycobacterium tuberculosis.* J Clin Microbiol, 2007. **45**(1): 179–92.

Gryadunov, D., et al., Evaluation of hybridisation on oligonucleotide microarrays for analysis of drug-resistant Mycobacterium tuberculosis. Clin Microbiol Infect, 2005. **11**(7): 531–9.

Iseman, M.D., *Treatment of multidrug-resistant tuberculosis.* N Engl J Med, 1993. **329**(11): 784–91.

Frieden, T.R., et al., *The emergence of drug-resistant tuberculosis in New York City.* N Engl J Med, 1993. **328**(8): 521–6.

Chapter 9
Antifungal Drug Resistance in Developing Countries

David S. Perlin

Abstract Opportunistic fungal infections are a major cause of morbidity and mortality in immunosuppressed patients; and in HIV-positive individuals, infections due to *Candida*, *Cryptococcus*, and *Pneumocystis* are AIDS-defining illnesses. The widespread use of antifungal drugs, particularly triazole drugs, has led to the emergence of primary resistance, which largely reflects infection with inherently less susceptible strains. Secondary resistance in normally susceptible strains also occurs and involves a variety of mechanisms including target site modification and drug efflux transporters. Resistance is a clinical management issue, but it has remained relatively constant in most developed countries. In developing countries, resistance is minimal due to limited antifungal therapy. However, as access to these drugs increases, it is particularly important to evaluate trends that reflect evolving resistance issues observed elsewhere, especially among individuals with HIV/AIDS.

9.1 Introduction

Opportunistic fungal infections are a mounting threat to human health, especially in individuals with compromised immune systems resulting from AIDS, cancer, bone marrow or organ transplantation, acute leukemia, burns, gastrointestinal disease, and premature birth. Fungal infections are a consequence of immune suppression and in HIV patients, they are an AIDS-defining condition. Invasive disease due to these organisms has become a major cause of morbidity and mortality in immunocompromised individuals (Wenzel 1995), and they are the third/fourth leading cause of bloodstream infections in US hospitals (Wisplinghoff et al. 2004). Most fungi causing human disease have a low inherent virulence and are contained through the action of host defense

D.S. Perlin (✉)

Public Health Research Institute, New Jersey Medical School/UMDNJ at the International Center for Public Health (ICPH), Newark, NJ 07103-3535, USA
e-mail: perlinds@umdnj.edu

A. de J. Sosa et al. (eds.), *Antimicrobial Resistance in Developing Countries*,
DOI 10.1007/978-0-387-89370-9_9, © Springer Science+Business Media, LLC 2010

systems. Yet, suppression of these systems allows fungi to flourish, leading to superficial, subcutaneous, and/or invasive infections.

For example, the commensal organism *Candida albicans* is normally found on the mucosal surfaces of the mouth, gastrointestinal tract, and female reproductive system. It is usually contained through competition with other microorganisms and the action of host defense systems, such as an intact epithelium, salivary secretions, and antibody- and cell-mediated immunity. Yet, suppression of these systems by chemical (drug) or disease-related immune modulation gives this organism a sufficient competitive advantage such that it dominates the topical flora. This often results in superficial mucosal disease, but it may also result in invasive fungal disease. Such invasive fungal disease is associated with attributable mortality rates of nearly 40% (Tumbarello et al. 1999; Gudlaugsson et al. 2003), which surpass those of bacterial infections (Wisplinghoff et al. 2003).

The *Candida* spp. are the most common bloodstream infection (Pfaller et al. 2001; Wisplinghoff et al. 2004; Pfaller and Diekema 2007). AIDS-defining invasive mycoses also include *Cryptococcus neoformans* and *Pneumocystis jirovecii*. But increasingly, *Aspergillus* spp., Zygomycetes, other filamentous fungi, endemic fungi, and drug-resistant variants of common yeasts are important causes of invasive disease in high-risk patients (Goldman et al. 1999; Walsh and Groll 1999; Marr et al. 2002; Wingard 2005). This shift is a consequence of immunosuppressive conditions and procedures, environmental exposure, and the use of a new generation of highly active antifungal agents for prophylaxis and empiric therapy (Wingard 2005). More than 100 fungal species have been identified as human pathogens (Kwon-Chung 1994).

In the developing world, HIV is a predominant cause of immunosuppression leading to invasive fungal infections. These individuals are subject to a variety of opportunistic infections (OI) during late-stage disease, and timely and effective management of these infections is critical to improved survival rates. Every episode of diseases is likely to increase the probability that resistance will emerge. In the developed world, the widespread availability of highly active antiretroviral therapy (HAART) for AIDS patients has minimized the frequency of OI. However, in the developing world, due to limitations on the accessibility of HAART, the therapeutic management of OI is more important and has greater consequences for patient survival.

Antifungal drug resistance is an emerging issue in the developing world, but it remains low largely due to the limited availability of drugs. Yet, the consequences of broader access of patients to antifungal agents in the developing world to treat opportunistic *Candida*, *Cryptococcus*, and *Pneumocystis* infections suggest that the incidence of resistance will rise. To understand better emerging resistance, it is valuable to review the current state of therapy, mechanisms, and trends of antifungal resistance in HIV and other immunosuppressed patients around the world.

9.2 Antifungal Options

The treatment options for opportunistic fungal infections are limited. Unlike antibiotics for bacterial infections, there are relatively few chemical classes and targets represented by existing antifungal drugs. These classes include the azoles and triazoles (miconazole, fluconazole, itraconazole, voriconazole, and posaconazole), polyene macrolides (nystatin and amphotericin B), flucytosine (5-fluorocytosine), and echinocandins (caspofungin, micafungin, and anidulafungin). The azole and triazole antifungal drugs comprise the largest and most widely used class of compounds. They interfere with the cytochromeP-450-dependent enzyme, lanosterol 14α-demethylase (encoded by yeast *ERG11*), which is necessary for the biosynthesis of ergosterol, a fungal-specific membrane sterol. These drugs are fungistatic, requiring an active immune system to fully clear an infection. Resistant organisms commonly show mechanism-specific cross-resistance between drugs within a class (Sanglard and Odds 2002). Fluconazole, a first-generation triazole, is widely used against a wide variety of *Candida* and cryptococcal infections, although it has limited spectrum. The second-generation triazole drugs, like voriconazole and posaconazole, show better species reactivity and an ability to overcome some primary azole resistance in yeasts and molds (Diekema et al. 2003; Ostrosky-Zeichner et al. 2003). The polyene antibiotics are pore-forming molecules that disrupt nutrient uptake and induce metabolite leakage. Amphotericin B, the first commercially significant antifungal drug, has been used for more than 45 years (Gallis et al. 1990). It is a highly effective fungicidal drug against yeasts and molds, but it has serious drug-associated toxicities including infusion and hematologic effects and, most importantly, renal insufficiency (Sabra and Branch 1990). Newer lipid formulations can reduce toxicity but do not eliminate it (Walsh et al. 1998; Wong-Beringer et al. 1998). Flucytosine interferes with RNA synthesis and is a potent inhibitor of thymidylate synthesis. It acts synergistically with fluconazole and amphotericin, which avoids the problem of rapid resistance (Vermes et al. 2000), and is used against cryptococcosis. The echinocandins are the newest class of antifungal drugs. They are lipopeptides that target biosynthesis of β-1,3-D-glucan, a key fungal cell wall component. They are active against clinically relevant yeasts and molds and triazole-resistant *Candida* strains (Perlin 2007). Finally, pneumocystis pneumonia (PCP) in AIDS patients is treated with trimethoprim–sulfamethoxazole, inhibitors of folic acid biosynthesis, and atovaquone, a second-line drug that resembles ubiquinone and inhibits the mitochondrial electron transport cytochrome bc_1 complex.

9.3 Mechanisms of Resistance

The most widely used antifungal drugs to treat yeast and mold infections in the developing world are fluconazole and amphotericin B. The newer echinocandin class drugs are just starting to be available in parts of Latin America. Our understanding of resistance mechanisms has come from model studies and from analyses of clinical isolates in the United States and Europe.

9.3.1 Azoles

The development of azole resistance in *C. albicans* is most problematic in patients who receive extended or repeated courses of azole drugs for therapy or prevention of oral candidiasis (Marr et al. 2001; Brion et al. 2007). The molecular mechanisms responsible for triazole resistance in *C. albicans* are common to other *Candida* spp., as well as to less related fungi (Vanden Bossche et al. 1998; Perea et al. 2001; Sanglard 2002; Sanglard and Odds 2002; White et al. 2002). Three major mechanisms have been elucidated in recent years. First, mutations in the drug target, cytochrome P450 14α-demethylase, encoded by Erg11 (yeasts) and Cyp51A (molds), alter the apparent drug-binding domain. Second, overexpression of drug efflux transporters reduces the steady-state intracellular level of drug (Fig. 9.1). As in other eukaryotic systems, the two principal families of efflux pumps are the ATP-binding cassette (ABC) (e.g., *CDR1* and *CDR2*) and the major facilitator superfamily (MFS) (e.g., MDR1) (Perea et al. 2001; Sanglard

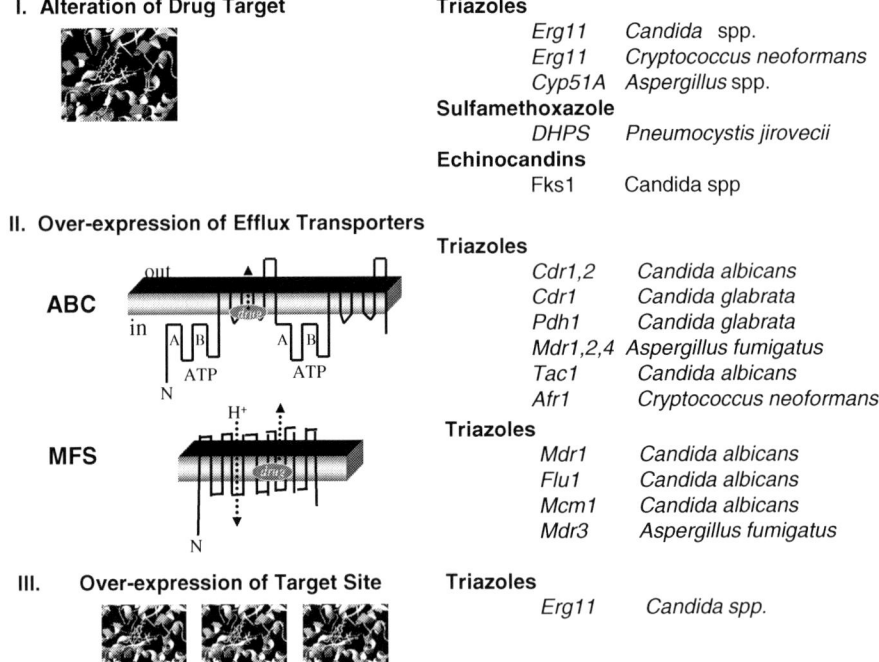

I. Alteration of Drug Target

Triazoles

Erg11	Candida spp.
Erg11	Cryptococcus neoformans
Cyp51A	Aspergillus spp.

Sulfamethoxazole

DHPS	Pneumocystis jirovecii

Echinocandins

Fks1	Candida spp

II. Over-expression of Efflux Transporters

ABC

Triazoles

Cdr1,2	Candida albicans
Cdr1	Candida glabrata
Pdh1	Candida glabrata
Mdr1,2,4	Aspergillus fumigatus
Tac1	Candida albicans
Afr1	Cryptococcus neoformans

MFS

Triazoles

Mdr1	Candida albicans
Flu1	Candida albicans
Mcm1	Candida albicans
Mdr3	Aspergillus fumigatus

III. Over-expression of Target Site

Triazoles

Erg11	Candida spp.

Fig. 9.1 Summary of major antifungal resistance mechanisms showing drug class, genetic loci, and fungal type

2002; Sanglard and Odds 2002; White et al. 2002). The genome of fungi contains hundreds of such transport systems, many of which have the potential to confer resistance (Rogers et al. 2001; De Hertogh et al. 2006). Upregulation of ABC or MFS transporters has been shown to arise from mutations in transcription factors (Harry et al. 2002; Coste et al. 2004; Vermitsky and Edlind 2004; Riggle and Kumamoto 2006) or from specific segmental aneuploidy (Selmecki et al. 2006). Azole resistance is mostly correlated with upregulation of genes of a single family (e.g., ABC), although there are examples of genes of both families being coregulated in the same azole-resistant isolate. These mechanisms confer cross-resistance among the class of triazole drugs (Pfaller et al. 2004a b, Wakiec et al. 2007). Fluconazole resistance can arise from a less robust mechanistic response, which in some strains may allow more active triazole drugs like voriconazole to be effective (Ostrosky-Zeichner et al. 2003). Finally, azole resistance has also been linked to upregulation of the demethylation target, Eg11p (Perea et al. 2001). The evolution of drug resistance in a susceptible strain like *C. albicans* involves either a single-step mechanism or a progressive accumulation of mutations, resulting in target site affinity changes and/or induction of various types of drug efflux transporters. This multifactorial basis of azole resistance in clinical isolates is widely observed (White et al. 1998; Perea et al. 2001). In contrast, a single resistance mechanism involving upregulation of an ABC-type transporter, or other related pumps, is sufficient to confer triazole resistance in *C. glabrata* (Sanguinetti et al. 2005). Although less prevalent, many of these systems have been observed with fluconazole resistance in *Cryptococcus* (Sanglard and Odds 2002). For molds like *Aspergillus fumigatus*, both target site mutations in Cyp51A and overexpression of drug pumps help confer resistance to broad-spectrum triazole drugs like itraconazole, voriconazole, and posaconazole (Nascimento et al. 2003; Mellado et al. 2004; Garcia-Effron et al. 2005; Howard et al. 2006; Mellado et al. 2007).

9.3.2 Polyenes

Amphotericin B resistance is uncommon even among to severely immuno-compromised patients. In most cases, poor clinical response involves colonization with fungi that shows inherently reduced susceptibility (*C. lusitaniae*, *C. guilliermondii* or *Trichosporon beigelii, Pseudallescheria boydii, Fusarium, Scopulariopsis*, and *A. terreus*) (Pfaller et al. 1994; 2006a; Walsh and Groll 1999; Dannaoui et al. 2000; Marr et al. 2002; Steinbach et al. 2004). Most recently, the mold *A. lentulus* was shown to have decreased susceptibility to a broad range of antifungal agents, including polyene, azole, and echinocandin drugs (Balajee et al. 2005). Resistance to amphotericin B rarely emerges in previously susceptible strains, although, in rare circumstances, it can emerge in a highly susceptible organism like *C. albicans* or *C. neoformans* following prior exposure to fluconazole, which reduces the level of membrane-bound ergosterol, the target for polyene antifungal drugs (Broughton et al. 1991; Kelly et al. 1997; Moosa et al. 2002). In *Cryptococcus*, the production of melanin may also play a role in decreasing the sensitivity to drug (Ikeda et al. 2003).

9.3.3 Echinocandins

Echinocandins are the newest approved class of drugs and resistance is relatively rare, although *Candida* strains with elevated MIC values and occasional treatment failure have been reported (Hakki et al. 2006; Laverdiere et al. 2006; Miller et al. 2006). Fungi display several adaptive physiological mechanisms that result in elevated MIC values, but these are not generally correlated with clinical resistance (Perlin 2007). Clinical resistance to echinocandin drugs is associated with amino acid substitutions in the major subunit of glucan synthase, Fks1p (Park et al. 2005; Kahn et al. 2007; Perlin 2007). These mutations confer cross-resistance to all echinocandin drugs and decrease the sensitivity of glucan synthase to drug by 1000-fold or more. The Fks1-mediated resistance mechanism is conserved in a wide variety of yeasts and molds and helps account for naturally occurring reduced susceptibility in *C. parapsilosis* and *C. guilliermondii* (Perlin 2007). Echinocandin resistance has been noted in the mold *A. lentilus* (Balajee et al. 2005), but resistance in other *Aspergillus* spp. has not been reported, except in laboratory strains (Gardiner et al. 2005; Rocha et al. 2007). As echinocandin usage broadens, it remains to be seen whether resistance will become a more significant factor influencing therapy with echinocandin drugs.

9.3.4 Flucytosine

Resistance to 5-FC is common and therefore it is rarely used in monotherapy. Resistance in yeasts is associated with changes in pyrimidine salvage, drug metabolism, and transport genes involving *FCA1*, *FUR1*, *FCY21*, and *FCY22*, which encode cytosine deaminase, uracil phosphoribosyltransferase (UPRT), and purine–cytosine permeases, respectively (Hope et al. 2004). *C. albicans* resistance is largely clustered in a single clade (Pujol et al. 2004), in which a single nucleotide change is observed in the uracil phosphoribosyltransferase gene (*FUR1*) (Dodgson et al. 2004). Due to the resistance problem, most clinicians in the developing world avoid 5-FC therapy.

9.3.5 Sulfamethoxazole

The combination of trimethoprim and sulfamethoxazole is used to treat PCP. They are broad-spectrum antimicrobial agents that target a variety of aerobic

Gram-positive and Gram-negative organisms, as well as *Pneumocystis*. Prophylaxis failure results from mutations in the folic acid biosynthetic enzyme dihydropteroate synthase (DHPS) (Helweg-Larsen et al. 1999; Meshnick 1999; Visconti et al. 2001; Nahimana et al. 2003; Kazanjian et al. 2004; Nahimana et al. 2004; Stein et al. 2004; Iliades et al. 2005a). Prominent mutations (T517A, P519S, 597A/T, and P599S) confer cross-resistance to a range of sulfa drugs (Iliades et al. 2005a; Iliades et al. 2005b).

9.4 Antifungal Resistance

Resistance is best defined as clinical failure following drug therapy due to organisms that are tolerant of drugs at a therapeutic dosage. The level of resistance within a community can be approximated by the prevalence of infecting strains that show reduced susceptibility to antifungal drugs. Standardized antifungal testing utilizes either broth microdilution or disk diffusion assays performed in accordance with guidelines of the Clinical Laboratory Standards Institute (CLSI) (Sandven 1999; Standards 2002). Drug threshold levels producing no growth yield minimum inhibitory concentrations (MICs), which in combination with clinical outcome data can be used to define break points for resistance, intermediate, and full susceptibility (Rex et al. 1997; Pfaller 2000; Pfaller et al. 2006b). Overall, the in vitro susceptibility (MIC) of an infecting organism is a valuable measure for predicting clinical outcome (Rex et al. 1997; Pfaller 2000).

The widespread use of antifungal agents in the past 15 years has led to the development of antifungal resistance. Like resistance to other antimicrobial agents, the development of antifungal resistance is complex and relies on multiple host and microbial risk factors (White et al. 2002). Antifungal resistance is a clinical management challenge for patients with serious mycoses. It typically involves either primary resistance, observed as a shift toward colonization with inherently less susceptible organisms, or secondary resistance in susceptible strains involving the emergence of cell-specific resistance mechanisms (target site mutations, drug pumps, etc.). Primary resistance comprises the majority of antifungal resistance, while secondary resistance occurs less frequently, but it is observed in patients with recurrent episodes of fungal infections requiring therapy, such as patients with HIV/AIDS. As *Candida* spp. represent the most abundant source of fungal mycoses, resistance to widely used triazole drugs is the most significant source of resistance. In particular, the widespread application of fluconazole and related triazole antifungals can induce resistant subpopulations of normally susceptible organisms like *C. albicans* but more commonly, they promote colonization with more intrinsically resistant species. In other AIDS-defining opportunistic infections like cryptococcosis and pneumocystis pneumonia, resistance to fluconazole and sulfamethoxazole, respectively, involves modification of target enzymes. Finally, triazole and polyene resistance observed in mold infections displays both primary and secondary resistance mechanisms.

The expanding use of triazole antifungal drugs for prophylaxis, empiric, and preemptive therapy contributes to the overall prevalence of resistance as selection pressure abounds for colonization with primary resistant organisms. Overall, the selection of antifungals in patients with invasive *Candida* or mold infections depends upon the identification of species involved, as some may show inherent resistance or reduced susceptibility. In the developing world, AIDS-related mycoses are paramount and the expanding availability of antifungal drugs is expected to drive the appearance of azole-resistant *Candida* spp. and *Cryptococcus* variants, as well as sulfa resistance in *Pneumocystis*.

9.4.1 Candidemia

Infections due to *Candida* spp. remain the most common source of invasive fungal infections and occur at a rate of 0.28–9.8 infections per 1000 hospital admissions in the United States (Rangel-Frausto et al. 1999; Blumberg et al. 2001; Eggimann et al. 2003) and 1.66 in Brazil (Colombo et al. 2007). They represent 8–10% of all nosocomial bloodstream infections (BSIs), making them the third leading cause of BSIs in intensive care units (Wisplinghoff et al. 2004). *Candida*-associated mortality is 0.4 deaths per 100,000 with a high attributable mortality rate of nearly 40% (Gudlaugsson et al. 2003). Factors associated with an increased risk of *Candida* bloodstream infections include prior surgery, acute renal failure, and parenteral nutrition (Blumberg et al. 2001). As *Candida* spp. colonize the gastrointestinal tract, chemotherapy or procedures that alter the integrity or ecology of the gastrointestinal tract promote such infections (Saiman et al. 2001; Nucci and Colombo 2002; Stiefel and Donskey 2004). *C. albicans* is the most significant species accounting for 56% of all such isolates, although other *Candida* spp. are increasingly important. Its frequency in global regions ranges from 47% in Latin America to 74% in the Asia-Pacific, and overall, its frequency increased from 44 to 60% from 1992 to 2001 (Pfaller and Diekema 2004). *C. albicans* occurs in five major clades: I, II, III, SA, and E, with the latter two unique to South Africa and Europe (Pujol et al. 2004). *C. glabrata*, an important source of azole resistance, accounts for 18% of *Candida* isolates in the United States and even higher in several countries in Western Europe (Pfaller et al. 1999a,b, 2004a, 2005a,b, 2007; Pfaller and Diekema 2004). Its prevalence has been correlated with the widespread use of azole antifungal drugs (Martins and Rex 1996; Nguyen et al. 1996; Pfaller et al. 1998a,b, Vanden Bossche et al. 1998). In contrast, *C. parapsilosis*, an azole-susceptible strain, is more important in Latin American and South American countries where it represents >18% of isolates (Pfaller and Diekema 2004; Colombo et al. 2007); it is also found as high as 49% in Malaysia (Pfaller and Diekema 2004). This is significant because of the inherent reduced susceptibility of *C. parapsilosis* to the new echinocandin drugs, which are just appearing in these regions. In countries like India, non-albicans *Candida* species are more important. In a recent 5-year study in North India, *C. tropicalis* represented 35% of isolates, followed by *C. albicans* (22%), *C. parapsilosis* (20%), and *C. glabrata* (18%) (Xess et al. 2007). As azole exposure broadens in the developing world, ongoing surveillance is required to assess shifts toward less susceptible strains like *C. glabrata* and *C. krusei*, especially in regions such as Latin America.

The expanding use of triazole drugs around the world, especially fluconazole, for prophylaxis, preemptive, and empirical therapy in high-risk patients has raised concern about resistant infections. The prevalence of azole resistance has been estimated to be 21–32% in symptomatic patients and up to 14% in asymptomatic patients (White et al. 1998). A prominent association has evolved between prior exposure to triazole antifungal drugs and emergence of resistant strains (Johnson et al. 1995; Ampel 1996; Nguyen et al. 1996; Cartledge et al. 1997; Brion et al. 2007). In Brazil, in a study of 270 cases of candidemia, breakthrough infections occurred in patients receiving at least 3 days of antifungal therapy (Nucci and Colombo 2002). Resistance in *C. albicans* has also been observed in HIV patients in Southern China following recurrent episodes with antifungal therapy (Samaranayake et al. 2001).

Global surveillance programs provide snapshots of antifungal resistance levels and trends around the world (Pfaller et al. 1998a, 1999a,b, 2000, 2001, 2005a,b,c, 2007; Meis et al. 2000; Liebowitz et al. 2001). The ARTEMIS Global Antifungal Surveillance Program, a long-standing comprehensive fungal surveillance program, involves a broad international network of 134 clinical sites in 40 different countries (Pfaller et al. 2003, 2004a,b, 2005b, 2006a, 2007). The sites perform CLSI-approved disk diffusion testing for fluconazole and other triazole drugs (e.g., voriconazole) against yeast isolates from a variety of clinical sources. Over an 8.5-year period ending in December 2005, fluconazole susceptibility was evaluated for more than 140,000 *Candida* spp., which showed an overall resistance of 6.2%. Voriconazole was evaluated on more than 137,000 isolates, which showed 3.1% resistance (Pfaller et al. 2005a,b, 2007). Overall, the level of resistance for all *Candida* spp. has remained relatively constant over a decade (Chen et al. 2003; Dodgson et al. 2004; Pfaller and Diekema 2004). *C. albicans* isolates were 98% susceptible with only a small increase in the overall percentage of resistant isolates from 0.9 to 1.6% in 2005. In North America, fluconazole resistance for *C. albicans* was 5.1%, while it was 2.4% in Latin America and 0.9% in Asia (Pfaller et al. 2007). In Brazil, resistance is exceedingly rare at only 0.2% (da Matta et al. 2007). A resistance level of 3–7% has been noted for *C. albicans* isolates from Southern India (Girish Kumar et al. 2006), although resistance frequencies exceeding 20% for fluconazole and voriconazole have been noted from India (Pfaller et al. 2007).

In contrast, *C. glabrata* and *C. krusei* were less susceptible to fluconazole with 17 and 77% resistance, respectively (Pfaller et al. 2006b). Fluconazole resistance varied somewhat with geographic region with 11% in Asia-Pacific region, 13% in Latin America, 17% in Europe, and 18% in North America (Pfaller et al. 2005b).

Paradoxically, despite the shift in abundance of *C. glabrata*, the percentage of fluconazole-resistant isolates actually fell from 18 to 5% from 1992 to 2000 (Pfaller and Diekema 2004), which may reflect the widening use of HAART in AIDS patients. Other non-albicans *Candida* species showed a range of resistance from 3.6% for *C. parapsilosis* to 47.8% for *C. rugosa* (Pfaller et al. 2005b, 2006b). Among the species (e.g., *C. glabrata*, *C. krusei*, and *C. guilliermondii*) with decreased susceptibility to fluconazole, more than 80% of isolates were susceptible to voriconazole (Pfaller et al. 2005b).

By and large, triazole resistance has remained remarkably constant. *C. albicans* has remained highly susceptible to azole antifungal drugs with only limited secondary resistance to fluconazole (Pfaller and Diekema 2004). The amount of resistance is relatively low in the developing world due to thin drug exposure. However, as exposure broadens, it is expected that resistance will approach levels observed in other parts of the world. In fact, such resistance was observed in a cohort of 751 South African HIV/AIDS patients, in which 3 of 72 patients failed oral miconazole therapy after prior antifungal exposure (Blignaut et al. 1999).

It is important to note that candidiasis is significantly higher among HIV-seropositive women (Sobel et al. 2000; Ohmit et al. 2003). Despite increased vaginal and oral colonization, azole resistance among *C. albicans* isolates is rare, even after repeated exposures (Mathema et al. 2001; Sobel et al. 2001). However, antifungal therapy does yield predictable shifts in the prevalence of less susceptible species like *C. glabrata*, and HIV-seropositive women were more likely to be colonized with such strains (Fidel et al. 1999; Sobel et al. 2000; Sobel et al. 2001; Ohmit et al. 2003). As azole drugs are used to treat yeast vaginitis, the widespread availability of topical over-the-counter antifungal drugs (e.g., miconazole) has raised the specter that they may promote resistance. However, this does not appear to be the case and resistance is generally less than 2–4% (Mathema et al. 2001; Richter et al. 2005). There is no association of resistant isolates contributing to recurrent vulvovaginal candidiasis, even after long-term azole exposure (Sobel 2002). Antifungal therapy for yeast vaginitis is not common in the developing world, where it is confined largely to homeopathic remedies (Dallabetta et al. 1995), except in cases of extreme immunosuppression due to HIV/AIDS (Pepin et al. 2006). Presently, there is no meaningful resistance reported and, given experience in the United States, it is likely that topical agents for vaginitis can be used without fear of significant resistance.

9.4.2 Cryptococcosis

Cryptococcal meningitis is a common opportunistic infection in AIDS patients, particularly in Southeast Asia and Africa, although other immunosuppressed individuals are also at risk. The fungus *C. neoformans* is commonly found in soil and bird droppings, and infections occur in HIV individuals with CD4-positive T lymphocyte counts <400. Serotypes include *C. neoformans* var. *grubii*, *C. neoformans* var. *neoformans*, and *C. neoformans* var. *gattii*. HIV-associated cryptococcal meningitis has a high mortality (10–30%), even in the developed world. In HIV-infected patients from sub-Saharan Africa, cryptococcosis accounts for 13–44% of all deaths (Bicanic and Harrison 2004). The treatment of cryptococcal meningitis requires lifelong suppressive antifungal therapy. Multidrug regimens include initial therapy with amphotericin B plus 5-FC followed by long-term fluconazole therapy. Resistance to amphotericin B is unusual, although flucytosine and fluconazole resistance is more common. After initial treatment with fluconazole, relapses are often associated with fluconazole resistance (Bicanic et al. 2006). For example, in Cambodia, where cryptococcal meningitis is the third most common opportunistic infection in HIV patients, a 2-year study of more than 400 patients showed that 2.5% of strains were resistant to fluconazole in the first year, while the prevalence increased to 14% in the second year (Sar et al. 2004). In a study of 80 patients from Kenya, 21% of isolates were resistant to flucytosine, while 11% were resistant to fluconazole (Bii et al. 2007). A similar profile was observed in a limited study from Egypt (Abdel-Salam 2005). A comprehensive global surveillance study involving 1811 clinical isolates from five broad geographic regions between 1990 and 2004 confirmed that isolates remained highly susceptible to amphotericin B. Flucytosine susceptibilities varied widely from 35% in North America to 68% in Latin America. North American isolates were 75% susceptible to fluconazole compared to 94–100% in the other regions; resistance to fluconazole was 14% (Pfaller et al. 2007). Nearly all isolates were fully susceptible to newer triazole drugs: voriconazole, posaconazole, and ravuconazole. Overall, the data indicate that antifungal resistance is significant, but it has not increased over a 15-year period (Pfaller et al. 2005c). Newer antifungal agents with expanding access enable patients to manage cryptococcal infections more readily. Given the difficulty of prolonged antifungal therapy for patients with cryptococcal meningitis, it is the expansion of antiretroviral therapy in the developing world that is likely to improve the prognosis for these patients (Bicanic and Harrison 2004).

9.4.3 Pneumocystis Pneumonia

Pneumocystis pneumonia (PCP) is an AIDS-defining illness and continues to be a significant cause of AIDS-related mortality in high-burden regions of the developing world (Chakaya et al. 2003). PCP accounts for up to 55% of admissions for community-acquired pneumonia in HIV patients, especially children (Zar 2001; Ruffini and Madhi 2002; Morris et al. 2004a, b). In South Africa, PCP in HIV+ young children is a major cause of hospital admissions with associated high crude mortality (47%) (Zar 2001). PCP is caused by

P. jirovecii (formerly *P. carinii*). The organism is difficult to culture, which precludes studies of its biology and pathogenicity. Molecular and immunological studies have revealed a great deal about its host range and transmission, which usually involve acquisition and not reactivation of disease. More than 50 different strains have been identified (Kovacs et al. 2001), and prior antifungal prophylaxis is an important risk factor for the disease (Stein et al. 2004). The incidence of PCP has decreased in AIDS patients in developed countries over the past decade largely due to the introduction of HAART (Morris et al. 2004a,b). However, an increasing number of PCP cases have been reported in HIV-negative immunosuppressed patients, such as those with malignancy, transplants, or connective tissue diseases (Miller 1999; Takahashi et al. 2002; Narasimhan et al. 2004). Therapy and long-term prophylaxis involve the use of the sulfa drug sulfamethoxazole in conjunction with trimethoprim (cotrimoxazole). Prophylaxis is an important management strategy for high-risk individuals. There is now compelling evidence from randomized clinical trials and observational cohort studies from several African countries, India, and Thailand that cotrimoxazole is effective in reducing morbidity and mortality among HIV-infected patients in resource-limited countries (WHO 2006). Resistance to sulfa drugs involving target site mutations in dihydropteroate synthase (DHPS) is common in developed countries, but it is much less prevalent in the developing world (Kazanjian et al. 2004; Wissmann et al. 2006). An analysis of isolates from 145 patients from 1983 to 2001 revealed that 40% of isolates had DHPS mutations and this prevalence increased to 70% in 2001. In contrast, isolates from China showed only 7% of such resistance markers from 1998 to 2001 (Kazanjian et al. 2004). While the prevalence of DHPS mutations is less frequent in the developing world, breakthrough infections are common. In a pediatric study of 105 HIV + children with severe pneumonia in South Africa, 49% were identified with *P. jirovecii* cysts and 28% of the children developed breakthrough infections despite prophylaxis with cotrimoxazole (Ruffini and Madhi 2002). The use of atovaquone, a second-line drug, has also led to resistance (Kazanjian et al. 2001; Kovacs et al. 2001), which is due to modification of the cytochrome *bc1* complex (Kessl et al. 2004).

9.4.4 Invasive Mold Infections

Mold infections are less of a problem in developing countries, since they are more closely associated with severely ill patients with cancer (Marr et al. 2002). The risk for mold infections is closely linked to the duration and severity of immunosuppression (Marr et al. 2002; Bow 2005). *A. fumigatus* is the most prevalent cause of invasive aspergillosis (Messer et al. 2006), and it remains highly susceptible to broad-spectrum triazole drugs like itraconazole and voriconazole (Messer et al. 2006: Cuenca-Estrella, 2006 #219). Secondary resistance is observed, but it is unusual and occurs with low frequency (Pfaller et al. 2005). Aspergillus species with reduced susceptibility to polyene and triazole drugs are encountered more frequently (Steinbach et al. 2004; Lionakis et al. 2005). Aggressive therapy with highly active antifungal agents like voriconazole can counter-select for rare fungi (e.g., Zygomycetes) with reduced susceptibility (Walsh and Groll 1999; Imhof et al. 2004; Lionakis et al. 2005). Thus, selective pressure during therapy can result in rare infections such as zygomycosis, which are largely refractory to conventional therapy (Imhof et al. 2004; Marty et al. 2004). In a review of more than 5500 patients who underwent hematopoietic stem cell transplantation over a 14-year period, *Fusarium* species and Zygomycetes increased in patients who received multiple transplants. Scedosporium infections were common in patients who had neutropenia, while infection caused by Zygomycetes generally occurred later after transplantation as patients developed graft-versus-host disease (Marr et al. 2002). Other azole-susceptible hyaline molds such as *Penicillium marneffei* are a growing problem among HIV-infected patients living in Southeast Asia (Ampel 1996), although these patients can be effectively managed with newer triazole drugs with limited breakthrough resistance noted (Chaiwarith et al. 2007).

9.5 Conclusion

Antifungal drug resistance is a consequence of increasing antifungal usage for empiric, preemptive, and prophylactic strategies to avoid the high morbidity and mortality associated with fungal infections. In developed countries, resistance to triazole drugs in *Candida* spp. and *Cryptococcus* is significant, but it has remained largely constant in the past decade. The use of highly active antiretroviral therapy in AIDS patients is changing the epidemiology of fungal infections and, as a consequence, resistance trends. In contrast, resistance is not a significant clinical management issue in developing countries, as antifungal exposure is limited. However, as triazole drugs such as fluconazole become more readily available, it will be important to monitor trends for the emergence of resistant strains, which should parallel those observed in developed countries in the pre-HAART era, in particular, a shift toward resistant non-albicans *Candida* spp. like *C. glabrata* in countries where such strains currently are not prevalent infecting strains.

References

Abdel-Salam, H. A. 2005. In vitro susceptibility of Cryptococcus neoformans clinical isolates from Egypt to seven antifungal drugs. Mycoses 48: 327–32.

Ampel, N. M. 1996. Emerging disease issues and fungal pathogens associated with HIV infection. Emerg Infect Dis 2: 109–16.

Balajee, S. A., J. L. Gribskov, E. Hanley et al. 2005. Aspergillus lentulus sp. nov., a new sibling species of A. fumigatus. Eukaryot Cell 4: 625–32.

Bicanic, T., T. Harrison, A. Niepieklo et al. 2006. Symptomatic relapse of HIV-associated cryptococcal meningitis after initial fluconazole monotherapy: the role of fluconazole resistance and immune reconstitution. Clin Infect Dis 43: 1069–73.

Bicanic, T., T. S. Harrison 2004. Cryptococcal meningitis. Br Med Bull 72: 99–118.

Bii, C. C., K. Makimura, S. Abe et al. 2007. Antifungal drug susceptibility of Cryptococcus neoformans from clinical sources in Nairobi, Kenya. Mycoses 50(1): 25–30.

Blignaut, E., M. E. Botes, H. L. Nieman 1999. The treatment of oral candidiasis in a cohort of South African HIV/AIDS patients. SADJ 54: 605–8.

Blumberg, H. M., W. R. Jarvis, J. M. Soucie et al. 2001. Risk factors for candidal bloodstream infections in surgical intensive care unit patients: the NEMIS prospective multicenter study. The National Epidemiology of Mycosis Survey. Clin Infect Dis 33: 177–86.

Bow, E. J. 2005. Long-term antifungal prophylaxis in high-risk hematopoietic stem cell transplant recipients. Med Mycol 43 Suppl 1: S277–87.

Brion, L. P., S. E. Uko, D. L. Goldman 2007. Risk of resistance associated with fluconazole prophylaxis: systematic review. J Infect 54: 521–9.

Broughton, M. C., M. Bard, N. D. Lees 1991. Polyene resistance in ergosterol producing strains of Candida albicans. Mycoses 34: 75–83.

Cartledge, J. D., J. Midgley, B. G. Gazzard 1997. Prior fluconazole exposure as an independent risk factor for fluconazole resistant candidosis in HIV positive patients: a case-control study. Genitourin Med 73: 471–4.

Chaiwarith, R., N. Charoenyos, T. Sirisanthana, K. Supparatpinyo 2007. Discontinuation of secondary prophylaxis against Penicilliosis marneffei in AIDS patients after HAART. AIDS 21: 365–7.

Chakaya, J. M., C. Bii, L. Ng'ang'a et al. 2003. Pneumocystis carinii pneumonia in HIV/AIDS patients at an urban district hospital in Kenya. East Afr Med J 80: 30–5.

Chen, Y. C., S. C. Chang, K. T. Luh, W. C. Hsieh 2003. Stable susceptibility of Candida blood isolates to fluconazole despite increasing use during the past 10 years. J Antimicrob Chemother 52: 71–7.

Colombo, A. L., T. Guimaraes, L. R. Silva et al. 2007. Prospective observational study of candidemia in Sao Paulo, Brazil: incidence rate, epidemiology, and predictors of mortality. Infect Control Hosp Epidemiol 28: 570–6.

Coste, A. T., M. Karababa, F. Ischer et al. 2004. TAC1, transcriptional activator of CDR genes, is a new transcription factor involved in the regulation of Candida albicans ABC transporters CDR1 and CDR2. Eukaryot Cell 3: 1639–52.

da Matta, D. A., L. P. de Almeida, A. M. Machado et al. 2007. Antifungal susceptibility of 1000 Candida bloodstream isolates to 5 antifungal drugs: results of a multicenter study conducted in Sao Paulo, Brazil, 1995–2003. Diagn Microbiol Infect Dis 57: 399–404.

Dallabetta, G. A., P. G. Miotti, J. D. Chiphangwi et al. 1995. Traditional vaginal agents: use and association with HIV infection in Malawian women. AIDS 9: 293–7.

Dannaoui, E., E. Borel, F. Persat et al. 2000. Amphotericin B resistance of Aspergillus terreus in a murine model of disseminated aspergillosis. J Med Microbiol 49: 601–6.

De Hertogh, B., F. Hancy, A. Goffeau, P. V. Baret 2006. Emergence of species-specific transporters during evolution of the hemiascomycete phylum. Genetics 172: 771–81.

Diekema, D. J., S. A. Messer, R. J. Hollis et al. 2003. Activities of caspofungin, itraconazole, posaconazole, ravuconazole, voriconazole, and amphotericin B against 448 recent clinical isolates of filamentous fungi. J Clin Microbiol 41: 3623–6.

Dodgson, A. R., K. J. Dodgson, C. Pujol et al. 2004. Clade-specific flucytosine resistance is due to a single nucleotide change in the FUR1 gene of Candida albicans. Antimicrob Agents Chemother 48: 2223–7.

Eggimann, P., J. Garbino, D. Pittet 2003. Epidemiology of Candida species infections in critically ill non-immunosuppressed patients. Lancet Infect Dis 3: 685–702.

Fidel, P. L., Jr., J. A. Vazquez, J. D. Sobel 1999. Candida glabrata: review of epidemiology, pathogenesis, and clinical disease with comparison to C. albicans. Clin Microbiol Rev 12: 80–96.

Gallis, H. A., R. H. Drew, W. W. Pickard 1990. Amphotericin B: 30 years of clinical experience. Rev Infect Dis 12: 308–29.

Garcia-Effron, G., E. Mellado, A. Gomez-Lopez et al. 2005. Differences in interactions between azole drugs related to modifications in the 14-alpha sterol demethylase gene (cyp51A) of Aspergillus fumigatus. Antimicrob Agents Chemother 49: 2119–21.

Gardiner, R. E., P. Souteropoulos, S. Park, D. S. Perlin 2005. Characterization of Aspergillus fumigatus mutants with reduced susceptibility to caspofungin. Med Mycol 43 Suppl 1: S299–305.

Girish Kumar, C. P., A. M. Hanafy, M. Katsu et al. 2006. Molecular analysis and suscept-ibility profiling of Candida albicans isolates from immunocompromised patients in South India. Mycopathologia 161: 153–9.

Goldman, M., P. C. Johnson, G. A. Sarosi 1999. Fungal pneumonias. The endemic mycoses. Clin Chest Med 20(3): 507–19.

Gudlaugsson, O., S. Gillespie, K. Lee et al. 2003. Attributable mortality of nosocomial candidemia, revisited. Clin Infect Dis 37: 1172–7.

Hakki, M., J. F. Staab, K. A. Marr 2006. Emergence of a Candida krusei isolate with reduced susceptibility to caspofungin during therapy. Antimicrob Agents Chemother 50: 2522–4.

Harry, J. B., J. L. Song, C. N. Lyons, T. C. White 2002. Transcription initiation of genes associated with azole resistance in Candida albicans. Med Mycol 40: 73–81.

Helweg-Larsen, J., T. L. Benfield, J. Eugen-Olsen et al. 1999. Effects of mutations in Pneu-mocystis carinii dihydropteroate synthase gene on outcome of AIDS-associated P. carinii pneumonia. Lancet 354: 1347–51.

Hope, W. W., L. Tabernero, D. W. Denning, M. J. Anderson 2004. Molecular mechanisms of primary resistance to flucytosine in Candida albicans. Antimicrob Agents Chemother 48: 4377–86.

Howard, S. J., I. Webster, C. B. Moore et al. 2006. Multi-azole resistance in Aspergillus fumigatus. Int J Antimicrob Agents 28: 450–3.

Ikeda, R., T. Sugita, E. S. Jacobson, T. Shinoda 2003. Effects of melanin upon susceptibility of Cryptococcus to antifungals. Microbiol Immunol 47: 271–7.

Iliades, P., S. R. Meshnick, I. G. Macreadie 2005a. Analysis of Pneumocystis jirovecii DHPS alleles implicated in sulfamethoxazole resistance using an Escherichia coli model system. Microb Drug Resist 11: 1–8.

Iliades, P., S. R. Meshnick, I. G. Macreadie 2005b. Mutations in the Pneumocystis jirovecii DHPS gene confer cross-resistance to sulfa drugs. Antimicrob Agents Chemother 49: 741–8.

Imhof, A., S. A. Balajee, D. N. Fredricks et al. 2004. Breakthrough fungal infections in stem cell transplant recipients receiving voriconazole. Clin Infect Dis 39: 743–6.

Johnson, E. M., D. W. Warnock, J. Luker et al. 1995. Emergence of azole drug resistance in Candida species from HIV-infected patients receiving prolonged fluconazole therapy for oral candidosis. J Antimicrob Chemother 35: 103–14.

Kahn, J. N., G. Garcia-Effron, M. J. Hsu et al. 2007. Acquired echinocandin resistance in a Candida krusei isolate due to modification of glucan synthase. Antimicrob Agents Chemother 51: 1876–8.

Kazanjian, P., W. Armstrong, P. A. Hossler et al. 2001. Pneumocystis carinii cytochrome b mutations are associated with atovaquone exposure in patients with AIDS. J Infect Dis 183: 819–22.

Kazanjian, P. H., D. Fisk, W. Armstrong et al. 2004. Increase in prevalence of Pneumocystis carinii mutations in patients with AIDS and P. carinii pneumonia, in the United States and China. J Infect Dis 189: 1684–7.

Kelly, S. L., D. C. Lamb, D. E. Kelly et al. 1997. Resistance to fluconazole and cross-resistance to amphotericin B in Candida albicans from AIDS patients caused by defective sterol delta5,6-desaturation. FEBS Lett 400: 80–2.

Kessl, J. J., P. Hill, B. B. Lange et al. 2004. Molecular basis for atovaquone resistance in Pneumocystis jirovecii modeled in the cytochrome bc(1) complex of Saccharomyces cerevisiae. J Biol Chem 279: 2817–24.

Kovacs, J. A., V. J. Gill, S. Meshnick, H. Masur 2001. New insights into transmission, diagnosis, and drug treatment of Pneumocystis carinii pneumonia. JAMA 286: 2450–60.

Kwon-Chung, K. J. 1994. Phylogenetic spectrum of fungi that are pathogenic to humans. Clin Infect Dis 19 Suppl 1: S1–7.

Laverdiere, M., R. G. Lalonde, J. G. Baril et al. 2006. Progressive loss of echinocandin activity following prolonged use for treatment of Candida albicans oesophagitis. J Antimicrob Chemother 57: 705–8.

Liebowitz, L. D., H. R. Ashbee, E. G. Evans et al. 2001. A two year global evaluation of the susceptibility of Candida species to fluconazole by disk diffusion. Diagn Microbiol Infect Dis 40: 27–33.

Lionakis, M. S., R. E. Lewis, H. A. Torres et al. 2005. Increased frequency of non-fumigatus Aspergillus species in amphotericin B- or triazole-pre-exposed cancer patients with positive cultures for aspergilli. Diagn Microbiol Infect Dis 52: 15–20.

Marr, K. A., R. A. Carter, F. Crippa et al. 2002. Epidemiology and outcome of mould infections in hematopoietic stem cell transplant recipients. Clin Infect Dis 34: 909–17.

Marr, K. A., C. N. Lyons, K. Ha et al. 2001. Inducible azole resistance associated with a heterogeneous phenotype in Candida albicans. Antimicrob Agents Chemother 45: 52–9.

Martins, M. D., J. H. Rex 1996. Resistance to antifungal agents in the critical care setting: problems and perspectives. New Horiz 4: 338–44.

Marty, F. M., L. A. Cosimi, L. R. Baden 2004. Breakthrough zygomycosis after voriconazole treatment in recipients of hematopoietic stem-cell transplants. N Engl J Med 350: 950–2.

Mathema, B., E. Cross, E. Dun et al. 2001. Prevalence of vaginal colonization by drug-resistant Candida species in college-age women with previous exposure to over-the-counter azole antifungals. Clin Infect Dis 33: E23–7.

Meis, J., M. Petrou, J. Bille et al. 2000. A global evaluation of the susceptibility of Candida species to fluconazole by disk diffusion. Global Antifungal Surveillance Group. Diagn Microbiol Infect Dis 36: 215–23.

Mellado, E., G. Garcia-Effron, L. Alcazar-Fuoli et al. 2004. Substitutions at methionine 220 in the 14alpha-sterol demethylase (Cyp51A) of Aspergillus fumigatus are responsible for resistance in vitro to azole antifungal drugs. Antimicrob Agents Chemother 48: 2747–50.

Mellado, E., G. Garcia-Effron, L. Alcazar-Fuoli et al. 2007. A new Aspergillus fumigatus resistance mechanism conferring in vitro cross-resistance to azole antifungals involves a combination of cyp51A alterations. Antimicrob Agents Chemother 51: 1897–904.

Meshnick, S. R. 1999. Drug-resistant Pneumocystis carinii. Lancet 354: 1318–9.

Messer, S. A., R. N. Jones, T. R. Fritsche 2006. International surveillance of Candida spp. and Aspergillus spp.: report from the SENTRY Antimicrobial Surveillance Program (2003). J Clin Microbiol 44: 1782–7.

Miller, C. D., B. W. Lomaestro, S. Park, D. S. Perlin 2006. Progressive esophagitis caused by Candida albicans with reduced susceptibility to caspofungin. Pharmacotherapy 26: 877–80.

Miller, R. F. 1999. Pneumocystis carinii infection in non-AIDS patients. Curr Opin Infect Dis 12: 371–7.

Moosa, M. Y., G. J. Alangaden, E. Manavathu, P. H. Chandrasekar 2002. Resistance to amphotericin B does not emerge during treatment for invasive aspergillosis. J Antimicrob Chemother 49: 209–13.

Morris, A., L. A. Kingsley, G. Groner et al. 2004a. Prevalence and clinical predictors of Pneumocystis colonization among HIV-infected men. AIDS 18: 793–8.

Morris, A., J. D. Lundgren, H. Masur et al. 2004b. Current epidemiology of Pneumocystis pneumonia. Emerg Infect Dis 10: 1713–20.

Nahimana, A., M. Rabodonirina, J. Bille et al. 2004. Mutations of Pneumocystis jirovecii dihydrofolate reductase associated with failure of prophylaxis. Antimicrob Agents Chemother 48: 4301–5.

Nahimana, A., M. Rabodonirina, G. Zanetti et al. 2003. Association between a specific Pneumocystis jiroveci dihydropteroate synthase mutation and failure of pyrimethamine/sulfadoxine prophylaxis in human immunodeficiency virus-positive and -negative patients. J Infect Dis 188: 1017–23.

Narasimhan, M., A. J. Posner, V. A. DePalo et al. 2004. Intensive care in patients with HIV infection in the era of highly active antiretroviral therapy. Chest 125: 1800–4.

Nascimento, A. M., G. H. Goldman, S. Park et al. 2003. Multiple resistance mechanisms among Aspergillus fumigatus mutants with high-level resistance to itraconazole. Antimicrob Agents Chemother 47: 1719–26.

Nguyen, M. H., J. E. Peacock, Jr., A. J. Morris et al. 1996. The changing face of candidemia: emergence of non-Candida albicans species and antifungal resistance. Am J Med 100: 617–23.

Nucci, M., A. L. Colombo 2002. Risk factors for breakthrough candidemia. Eur J Clin Microbiol Infect Dis 21: 209–11.

Ohmit, S. E., J. D. Sobel, P. Schuman et al. 2003. Longitudinal study of mucosal Candida species colonization and candidiasis among human immunodeficiency virus (HIV)-seropositive and at-risk HIV-seronegative women. J Infect Dis 188: 118–27.

Ostrosky-Zeichner, L., J. H. Rex, P. G. Pappas et al. 2003. Antifungal susceptibility survey of 2,000 bloodstream Candida isolates in the United States. Antimicrob Agents Chemother 47: 3149–54.

Park, S., R. Kelly, J. N. Kahn et al. 2005. Specific substitutions in the echinocandin target Fks1p account for reduced susceptibility of rare laboratory and clinical Candida sp. isolates. Antimicrob Agents Chemother 49: 3264–73.

Pepin, J., F. Sobela, N. Khonde et al. 2006. The syndromic management of vaginal discharge using single-dose treatments: a randomized controlled trial in West Africa. Bull World Health Organ 84: 729–38.

Perea, S., J. L. Lopez-Ribot, W. R. Kirkpatrick et al. 2001. Prevalence of molecular mechanisms of resistance to azole antifungal agents in Candida albicans strains displaying high-level fluconazole resistance isolated from human immunodeficiency virus-infected patients. Antimicrob Agents Chemother 45: 2676–84.

Perlin, D. S. 2007. Resistance to echinocandin-class antifungal drugs. Drug Resist Updat 10: 121–30.

Pfaller, M. A. 2000. Antifungal susceptibility testing: progress and future developments. Braz J Infect Dis 4: 55–60.

Pfaller, M. A., L. Boyken, R. J. Hollis et al. 2005a. In vitro susceptibilities of clinical isolates of Candida species, Cryptococcus neoformans, and Aspergillus species to itraconazole: global survey of 9,359 isolates tested by clinical and laboratory standards institute broth microdilution methods. J Clin Microbiol 43: 3807–10.

Pfaller, M. A., D. J. Diekema 2004. Twelve years of fluconazole in clinical practice: global trends in species distribution and fluconazole susceptibility of bloodstream isolates of Candida. Clin Microbiol Infect 10 Suppl 1: 11–23.

Pfaller, M. A., D. J. Diekema 2007. Epidemiology of invasive candidiasis: a persistent public health problem. Clin Microbiol Rev 20: 133–63.

Pfaller, M. A., D. J. Diekema, D. L. Gibbs et al. 2007. Results from the ARTEMIS DISK Global Antifungal Surveillance study, 1997 to 2005: an 8.5-year analysis of susceptibilities of Candida species and other yeast species to fluconazole and voriconazole determined by CLSI standardized disk diffusion testing. J Clin Microbiol 45: 1735–45.

Pfaller, M. A., D. J. Diekema, R. N. Jones et al. 2001. International surveillance of bloodstream infections due to Candida species: frequency of occurrence and in vitro susceptibilities to fluconazole, ravuconazole, and voriconazole of isolates collected from 1997 through 1999 in the SENTRY antimicrobial surveillance program. J Clin Microbiol 39: 3254–9.

Pfaller, M. A., D. J. Diekema, M. Mendez et al. 2006a. Candida guilliermondii, an opportunistic fungal pathogen with decreased susceptibility to fluconazole: geographic and temporal trends from the ARTEMIS DISK antifungal surveillance program. J Clin Microbiol 44: 3551–6.

Pfaller, M. A., D. J. Diekema, S. A. Messer et al. 2003. Activities of fluconazole and voriconazole against 1,586 recent clinical isolates of Candida species determined by Broth microdilution, disk diffusion, and Etest methods: report from the ARTEMIS Global Antifungal Susceptibility Program, 2001. J Clin Microbiol 41: 1440–6.

Pfaller, M. A., D. J. Diekema, M. G. Rinaldi et al. 2005b. Results from the ARTEMIS DISK Global Antifungal Surveillance Study: a 6.5-year analysis of susceptibilities of Candida and other yeast species to fluconazole and voriconazole by standardized disk diffusion testing. J Clin Microbiol 43: 5848–59.

Pfaller, M. A., D. J. Diekema, D. J. Sheehan 2006b. Interpretive breakpoints for fluconazole and Candida revisited: a blueprint for the future of antifungal susceptibility testing. Clin Microbiol Rev 19: 435–47.

Pfaller, M. A., R. N. Jones, G. V. Doern et al. 1999a. International surveillance of blood stream infections due to Candida species in the European SENTRY Program: species distribution and antifungal susceptibility including the investigational triazole and echinocandin agents. SENTRY Participant Group (Europe). Diagn Microbiol Infect Dis 35: 19–25.

Pfaller, M. A., R. N. Jones, G. V. Doern et al. 1998a. International surveillance of bloodstream infections due to Candida species: frequency of occurrence and antifungal susceptibilities of isolates collected in 1997 in the United States, Canada, and South America for the SENTRY Program. The SENTRY Participant Group. J Clin Microbiol 36: 1886–9.

Pfaller, M. A., R. N. Jones, G. V. Doern et al. 2000. Bloodstream infections due to Candida species: SENTRY antimicrobial surveillance program in North America and Latin America, 1997–1998. Antimicrob Agents Chemother 44: 747–51.

Pfaller, M. A., R. N. Jones, S. A. Messer et al. 1998b. National surveillance of nosocomial blood stream infection due to species of Candida other than Candida albicans: frequency of occurrence and antifungal susceptibility in the SCOPE Program. SCOPE Participant Group. Surveillance and Control of Pathogens of Epidemiologic. Diagn Microbiol Infect Dis 30: 121–9.

Pfaller, M. A., S. A. Messer, L. Boyken et al. 2004a. In vitro activities of voriconazole, posaconazole, and fluconazole against 4,169 clinical isolates of Candida spp. and Cryptococcus neoformans collected during 2001 and 2002 in the ARTEMIS global antifungal surveillance program. Diagn Microbiol Infect Dis 48: 201–5.

Pfaller, M. A., S. A. Messer, L. Boyken et al. 2005c. Global trends in the antifungal susceptibility of Cryptococcus neoformans (1990 to 2004). J Clin Microbiol 43: 2163–7.

Pfaller, M. A., S. A. Messer, L. Boyken et al. 2004b. Geographic variation in the susceptibilities of invasive isolates of Candida glabrata to seven systemically active antifungal agents: a global assessment from the ARTEMIS Antifungal Surveillance Program conducted in 2001 and 2002. J Clin Microbiol 42: 3142–6.

Pfaller, M. A., S. A. Messer, R. J. Hollis 1994. Strain delineation and antifungal susceptibilities of epidemiologically related and unrelated isolates of Candida lusitaniae. Diagn Microbiol Infect Dis 20: 127–33.

Pfaller, M. A., S. A. Messer, R. J. Hollis et al. 1999b. Trends in species distribution and susceptibility to fluconazole among blood stream isolates of Candida species in the United States. Diagn Microbiol Infect Dis 33: 217–22.

Pujol, C., M. A. Pfaller, D. R. Soll 2004. Flucytosine resistance is restricted to a single genetic clade of Candida albicans. Antimicrob Agents Chemother 48: 262–6.

Rangel-Frausto, M. S., T. Wiblin, H. M. Blumberg et al. 1999. National epidemiology of mycoses survey (NEMIS): variations in rates of bloodstream infections due to Candida species in seven surgical intensive care units and six neonatal intensive care units. Clin Infect Dis 29: 253–8.

Rex, J. H., M. A. Pfaller, J. N. Galgiani et al. 1997. Development of interpretive breakpoints for antifungal susceptibility testing: conceptual framework and analysis of in vitro–in vivo correlation data for fluconazole, itraconazole, and candida infections. Subcommittee on

Antifungal Susceptibility Testing of the National Committee for Clinical Laboratory Standards. Clin Infect Dis 24: 235–47.

Richter, S. S., R. P. Galask, S. A. Messer et al. 2005. Antifungal susceptibilities of Candida species causing vulvovaginitis and epidemiology of recurrent cases. J Clin Microbiol 43: 2155–62.

Riggle, P. J., C. A. Kumamoto 2006. Transcriptional regulation of MDR1, encoding a drug efflux determinant, in fluconazole-resistant Candida albicans strains through an Mcm1p binding site. Eukaryot Cell 5: 1957–68.

Rocha, E. M. F., G. Garcia-Effron, S. Park, D. S. Perlin 2007. A Ser678Pro Substitution in Fks1p Confers Resistance to Echinocandin Drugs in Aspergillus fumigatus. Antimicrob Agents Chemother 51: 4174–76.

Rogers, B., A. Decottignies, M. Kolaczkowski et al. 2001. The pleiotropic drug ABC transporters from Saccharomyces cerevisiae. J Mol Microbiol Biotechnol 3: 207–14.

Ruffini, D. D., S. A. Madhi 2002. The high burden of Pneumocystis carinii pneumonia in African HIV-1-infected children hospitalized for severe pneumonia. AIDS 16: 105–12.

Sabra, R., R. A. Branch 1990. Amphotericin B nephrotoxicity. Drug Saf 5: 94–108.

Saiman, L., E. Ludington, J. D. Dawson et al. 2001. Risk factors for Candida species colonization of neonatal intensive care unit patients. Pediatr Infect Dis J 20: 1119–24.

Samaranayake, Y. H., L. P. Samaranayake, P. C. Tsang et al. 2001. Heterogeneity in antifungal susceptibility of clones of Candida albicans isolated on single and sequential visits from a HIV-infected southern Chinese cohort. J Oral Pathol Med 30: 336–46.

Sandven, P. 1999. Detection of fluconazole-resistant Candida strains by a disc diffusion screening test. J Clin Microbiol 37: 3856–9.

Sanglard, D. 2002. Resistance of human fungal pathogens to antifungal drugs. Curr Opin Microbiol 5: 379–85.

Sanglard, D., F. C. Odds 2002. Resistance of Candida species to antifungal agents: molecular mechanisms and clinical consequences. Lancet Infect Dis 2: 73–85.

Sanguinetti, M., B. Posteraro, B. Fiori et al. 2005. Mechanisms of azole resistance in clinical isolates of Candida glabrata collected during a hospital survey of antifungal resistance. Antimicrob Agents Chemother 49: 668–79.

Sar, B., D. Monchy, M. Vann et al. 2004. Increasing in vitro resistance to fluconazole in Cryptococcus neoformans Cambodian isolates: April 2000 to March 2002. J Antimicrob Chemother 54: 563–5.

Selmecki, A., A. Forche, J. Berman 2006. Aneuploidy and isochromosome formation in drug-resistant Candida albicans. Science 31: 367–70.

Sobel, J. D. 2002. Pathogenesis of recurrent Vulvovaginal Candidiasis. Curr Infect Dis Rep 4: 514–519.

Sobel, J. D., S. E. Ohmit, P. Schuman et al. 2001. The evolution of Candida species and fluconazole susceptibility among oral and vaginal isolates recovered from human immunodeficiency virus (HIV)-seropositive and at-risk HIV-seronegative women. J Infect Dis 183: 286–93.

Sobel, J. D., S. E. Ohmit, P. Schuman et al. 2000. The evolution of Candida species and fluconazole susceptibility among oral and vaginal isolates recovered from human immunodeficiency virus (HIV)-seropositive and at-risk HIV-seronegative women. J Infect Dis 183: 286–293.

Standards, NCCL. 2002. Reference method for broth dilution antifungal susceptibility testing of yeasts, Approved Standard – Second Edition NCCLS document M27-A2. National Committee for Clinical Laboratory Standards. Wayne, PA.

Stein, C. R., C. Poole, P. Kazanjian, S. R. Meshnick 2004. Sulfa use, dihydropteroate synthase mutations, and Pneumocystis jirovecii pneumonia. Emerg Infect Dis 10: 1760–5.

Steinbach, W. J., D. K. Benjamin, Jr., D. P. Kontoyiannis et al. 2004. Infections due to Aspergillus terreus: a multicenter retrospective analysis of 83 cases. Clin Infect Dis 39: 192–8.

Stiefel, U., C. J. Donskey 2004. The role of the intestinal tract as a source for transmission of nosocomial pathogens. Curr Infect Dis Rep 6: 420–425.

Takahashi, T., M. Goto, T. Endo et al. 2002. Pneumocystis carinii carriage in immunocompromised patients with and without human immunodeficiency virus infection. J Med Microbiol 51: 611–4.

Tumbarello, M., E. Tacconelli, K. de Gaetano Donati et al. 1999. Candidemia in HIV-infected subjects. Eur J Clin Microbiol Infect Dis 18: 478–83.

Vanden Bossche, H., F. Dromer, I. Improvisi et al. 1998. Antifungal drug resistance in pathogenic fungi. Med Mycol 36(Suppl 1): 119–28.

Vermes, A., H. J. Guchelaar, J. Dankert 2000. Flucytosine: a review of its pharmacology, clinical indications, pharmacokinetics, toxicity and drug interactions. J Antimicrob Chemother 46: 171–9.

Vermitsky, J. P., T. D. Edlind 2004. Azole resistance in Candida glabrata: coordinate upregulation of multidrug transporters and evidence for a Pdr1-like transcription factor. Antimicrob Agents Chemother 48: 3773–81.

Visconti, E., E. Ortona, P. Mencarini et al. 2001. Mutations in dihydropteroate synthase gene of Pneumocystis carinii in HIV patients with Pneumocystis carinii pneumonia. Int J Antimicrob Agents 18: 547–51.

Wakiec, R., R. Prasad, J. Morschhauser et al. 2007. Voriconazole and multidrug resistance in Candida albicans. Mycoses 50: 109–15.

Walsh, T. J., A. H. Groll 1999. Emerging fungal pathogens: evolving challenges to immunocompromised patients for the twenty-first century. Transpl Infect Dis 1: 247–61.

Walsh, T. J., V. Yeldandi, M. McEvoy et al. 1998. Safety, tolerance, and pharmacokinetics of a small unilamellar liposomal formulation of amphotericin B (AmBisome) in neutropenic patients. Antimicrob Agents Chemother 42: 2391–8.

Wenzel, R. P. 1995. Nosocomial candidemia: risk factors and attributable mortality. Clin Infect Dis 20: 1531–4.

White, T. C., S. Holleman, F. Dy et al. 2002. Resistance mechanisms in clinical isolates of Candida albicans. Antimicrob Agents Chemother 46: 1704–13.

White, T. C., K. A. Marr, R. A. Bowden 1998. Clinical, cellular, and molecular factors that contribute to antifungal drug resistance. Clin Microbiol Rev 11: 382–402.

WHO. 2006. WHO expert consultation on cotrimoxazole prophylaxis in HIV infection. Geneva, World health Organization: 1–34.

Wingard, J. R. 2005. The changing face of invasive fungal infections in hematopoietic cell transplant recipients. Curr Opin Oncol 17: 89–92.

Wisplinghoff, H., T. Bischoff, S. M. Tallent et al. 2004. Nosocomial bloodstream infections in US hospitals: analysis of 24,179 cases from a prospective nationwide surveillance study. Clin Infect Dis 39: 309–17.

Wisplinghoff, H., H. Seifert, S. M. Tallent et al. 2003. Nosocomial bloodstream infections in pediatric patients in United States hospitals: epidemiology, clinical features and susceptibilities. Pediatr Infect Dis J 22: 686–91.

Wissmann, G., M. J. Alvarez-Martinez, S. R. Meshnick et al. 2006. Absence of dihydropteroate synthase mutations in Pneumocystis jirovecii from Brazilian AIDS patients. J Eukaryot Microbiol 53: 305–7.

Wong-Beringer, A., R. A. Jacobs, B. J. Guglielmo 1998. Lipid formulations of amphotericin B: clinical efficacy and toxicities. Clin Infect Dis 27: 603–18.

Xess, I., N. Jain, F. Hasan et al. 2007. Epidemiology of candidemia in a tertiary care centre of north India: 5-year study. Infection 35: 256–9.

Zar, H. J. 2001. Pneumocystis carinii pneumonia (PCP) in HIV-infected African children. SADJ 56: 617–9.

Chapter 10
Drug Resistance in African Trypanosomiasis

Enock Matovu and Pascal Mäser

Abstract African trypanosomes include the causative agents of sleeping sickness in humans and those affecting live stock. Vaccination being jeopardized by the ever-changing surface coats of bloodstream-form trypanosomes, chemotherapy is the mainstay in the control of infections. However, the drugs in use are old, cause severe side effects, and their efficacy is undermined by the emergence of drug-resistant trypanosomes. Reliable supply of drugs for the human disease is difficult to maintain since patients are unable to meet treatment costs. Fortunately the prospects for the control of trypanosomiasis have improved recently by drug donations from Sanofi-Aventis to the WHO and through support from the Bill and Melinda Gates Foundation. Here we review the current drugs against African trypanosomes, discuss the mechanisms of drug resistance, and address key issues for the control of trypanosomiasis in face of the limited options for chemotherapy.

10.1 The African Trypanosomiases

Trypanosomiasis is a vector-borne disease caused by pathogenic protozoa of the genus *Trypanosoma*. This together with the related *Leishmania* spp. is the most important of the order *Kinetoplastida*. The order is characterized by the presence of the kinetoplast, a unique structure made of mitochondrial DNA circles. These protozoans are important for the respective diseases associated with them: both trypanosomiasis and leishmaniasis are among the most neglected diseases affecting mainly the rural poor in tropical and sub-tropical regions. The major differences between the two lie in the transmission agents (tsetse/biting flies and sand flies, respectively) and the course of the diseases observed for each genus. While trypanosomiasis is invariably fatal if left untreated, the leishmaniases are a range of diseases from simple self-healing

E. Matovu (✉)
Department of Veterinary Parasitology and Microbiology, Faculty of Veterinary Medicine, Makerere University, Kampala Uganda
e-mail: matovue@vetmed.mak.ac.ug

A. de J. Sosa et al. (eds.), *Antimicrobial Resistance in Developing Countries*, DOI 10.1007/978-0-387-89370-9_10, © Springer Science+Business Media, LLC 2010

cutaneous lesions (*Leishmania major*) to the life-threatening visceral form (*L. donovani*). Dependence of these pathogens on a vector for transmission (biological or mechanical) dictates that the distribution of the diseases correlates with that of the intermediate hosts.

Trypanosomiasis transmitted by tsetse flies (*Glossina* spp.) is of great significance as regards animal productivity and human health in Africa. About 10 million square kilometers of arable land is infested by tsetse flies, hindering the development of livestock production. The human disease is highly debilitating and invariably fatal if untreated (WHO 1986), a factor that hampers the production and food security in sub-Saharan Africa. Of the African salivarian trypanosomes, *Trypanosoma congolense, T. vivax*, and *T. brucei brucei* are the most economically important and cause disease (nagana) in animals, while *T. b. rhodesiense* of Eastern and Southern Africa and *T. b. gambiense* of Western Africa cause sleeping sickness in humans (Mulligan 1970; Hoare 1972). The mechanically transmitted *T. evansi* is another species of economic importance in the arid and semiarid areas of Africa where it is responsible for "surra" that mainly affects camels.

Both animal and human African trypanosomiases (nagana and sleeping sickness, respectively) are transmitted alongside each other although the latter is restricted to tsetse fly-infested areas of Africa, while the former spreads beyond the tsetse fly belt. In addition to being biologically transmitted by the tsetse fly, *T. vivax* that is among the most important animal trypanosomes can be mechanically transmitted by other biting flies. This makes its existence beyond the tsetse fly belt possible. *T. evansi*, the cause of trypanosomiasis in horses, donkeys, and camels, is exclusively mechanically transmitted by biting flies, mainly *Tabanus* and *Stomoxys*. *T. equiperdum*, the cause of "dourine" in equines, is transmitted through coitus (Hoare 1972).

Thus, it would look as if the course of evolution started with trypanosome species, which solely relied on a vector for transmission, *T. vivax* providing a link to mechanically transmitted species since it can be transmitted both cyclically and mechanically. As the trypanosomes moved outside the tsetse belt, they had to adapt to mechanical transmission alone as is the case with *T. evansi*. This species is probably the most widespread among the most economically important trypanosomes. The evolution of trypanosomes appears to reach a climax at *T. equiperdum*, which is sexually transmitted.

10.2 Human African Trypanosomiasis Is a Zoonosis

Human African trypanosomiasis (HAT), also known as sleeping sickness, is endemic in 36 countries of sub-Saharan Africa between latitudes 14°N and 29°S where the vector (*Glossina*) occurs (Kuzoe 1993). There are two distinct forms of HAT attributed to two subspecies: *T. brucei gambiense*, which occurs in Western and Central Africa, and *T. brucei rhodesiense*, which is prevalent in

Eastern and Southern Africa. The two subspecies are morphologically indistinguishable and until recent years, there was no definite way of distinguishing them other than by some biochemical characteristics (e.g., isoenzyme analysis) and geographical distribution. It has now been unequivocally demonstrated that the presence of the human serum resistance-associated (*SRA*) gene is indicative of *T. b. rhodesiense* (Gibson et al. 2002; Xong t al. 1998), whereas a receptor-like flagella pocket glycoprotein (*TgsGP*) is specific to *T. b. gambiense* (Berberof et al. 2001). Indeed, the polymerase chain reaction can be used to diagnose the two subspecies in the laboratory (Radwanska et al. 2002a, b). Distinction of the two forms of HAT is based on differences in the disease evolution, although the final stages of infection in either case have similar clinical manifestations. Gambian sleeping sickness is characterized by a long asymptomatic stage, eventually succeeded by a subacute febrile illness, followed by late-stage chronic meningoencephalitis. If untreated, death follows several years after the onset of the disease. In contrast, the Rhodesian form of the disease progresses much more rapidly, with most deaths (48%) occurring within 6 months of the onset of illness (Welburn et al. 2001). HAT is zoonotic in nature and *T. brucei* spp. exist in animals with no apparent symptoms, thereby conforming to the concept of a reservoir. These animals act as constant source of parasites that are transmitted to humans, leading to continued disease outbreaks. To date, wild animals like the bushbuck, lion, hartebeest, hyena, kob, warthog, waterbuck, hippos, giraffe, and reedbuck are known to harbor human-infective trypanosomes (WHO 1986; Molyneux and Ashford 1983). With the ever-expanding human population, however, the role of wild animals in HAT transmission continues to dwindle and in such cases, domestic animals play a bigger role. Among the latter, cattle, sheep, dogs, and pigs also harbor the parasites.

10.3   Chemotherapy Is the Mainstay in the Control of Trypanosomiasis

Concerning the control of trypanosomiasis, there exist three possible areas of intervention, namely the host, the vector, and the parasite itself. Control by targeting the host is possible only in the case of animal trypanosomiasis where some breeds of cattle, e.g., N'dama and Muturu breeds, are known to be trypanotolerant (Molyneux and Ashford 1983). Thus, to increase production in trypanosomiasis-endemic areas, it is logical to rear such trypanotolerant breeds. In the past, the wild animal reservoir was controlled by hunting, which is but no longer practiced in the interest of biodiversity conservation. In the case of the human disease, however, the only alternative at host level is to remove people from high-risk areas as happened around the shores and islands of Lake Victoria, Uganda, early in the last century (Lester 1939). However, with increasing human populations, this is now less feasible.

As regards control of the vector, traditional methods include bush clearing and massive sprays of tsetse habitats with insecticides. These methods are crude in that they affect nontarget species as well, and the insecticides can find their way into various levels of the ecosystem leading to undesirable environmental effects. With the advent of environmental awareness, these approaches are proving unfashionable. The live bait technology, however, provides a safer way to apply insecticides to tsetse control. In this case, compounds with residual effect are applied as animal pour-ons that remain effective to knock off tsetse as the animal wanders into their habitats. Nevertheless, attention has now shifted to trapping that has no adverse effects on the environment. Various models of traps that may also be impregnated with insecticide have been developed, and the current trend is to involve the local communities in control as was reported in the southeast Uganda focus (Okoth et al. 1991). The biggest determination today is the efforts by the Pan African Tsetse and Trypanosomiasis Eradication Campaign (PATTEC), which is patronized by African heads of state. This approach will encompass intensive tsetse control by traditional methods (trapping, live bait, aerial sprays) to low levels that can be eliminated by the application of the sterile insect technique (SIT). SIT is a relatively new concept in tsetse control that saw elimination of the vector from Zanzibar Islands (Vreysen et al. 2000; Vreysen 2001). However, the practicability of SIT on mainland Africa, which is a vast terrain with many natural barriers (topographical and vegetational), remains to be seen.

Control at the parasite level is presented with a number of problems. Important of these is the phenomenon of antigenic variation of bloodstream trypanosomes (Vickerman 1978), which renders them elusive to the immune response. The surface coat made of variant surface glycoproteins (VSG) successively changes to different variant antigen types (VATs) as the disease progresses. The parasite can express a different VSG gene with frequencies ranging from 10^{-2} to 10^{-7} per cell per generation (Overath et al. 1994). The immune response, however, produces VAT-specific antibodies and in the event of a changed VAT, the parasitaemia continues to rise before a new type of antibody specific to this VAT is produced. This continuous development of antigenically distinct variants ensures that some trypanosomes always survive the development of an immune response to propagate the infection (Turner 1985). The immune response is therefore often a step behind the parasite in the course of the infection. The disease becomes chronic because there always exists a VSG coat against which the body has not yet produced an appropriate antibody (Overath et al. 1994). Hence trypanosome infections are characterized by waves of parasitaemia, the so-called "damped oscillations" (Hesse et al. 1995). These changes are controlled by genetic mechanisms and it remains to be determined whether the host immune response plays any role in the selection of the different VATs expressed at a time. Antigenic variation has dashed hopes for formulation of an effective antitrypanosome vaccine; current efforts are aimed at antidisease rather than at antiparasite vaccines. Thus, control of trypanosomes at parasite level is solely dependent on an effective chemotherapeutic intervention.

Chemotherapy is the use of chemicals (drugs) to rid infected host of parasites, while chemoprophylaxis refers to use of the same or other drugs to prevent transmission and limit the pathological effects of the disease. In the case of trypanosomiasis, such chemicals are called trypanocidal or trypanostatic depending on whether they kill or inhibit trypanosome multiplication, respectively. However, the term trypanocide is often used to embrace both classes of drugs. The success of trypanostatic drugs depends on the host immune system that has to eliminate the static population, if complete cure is to be achieved. Some trypanostatic drugs are said to inhibit multiplication by inducing the production of akinetoplastic trypanosomes that cannot undergo further multiplication (Chitambo et al. 1992).

Trypanocidal treatment is occasionally accompanied by high toxicity ranging from damage of tissues at the site of injection or hepatobiliary damage in the case of isometamidium chloride (Kinabo et al. 1991), to death in the case of melarsoprol, which claims 1–5% of patients due to reactive encephalopathies (WHO 1986). Administration of drugs should therefore follow proper diagnosis and confirmation of infection. There are several factors that can lead to chemotherapeutic failure. Most important of these are the following:

(a) Application of subtherapeutic doses.
(b) Maladministration of the drug: Different routes of administration are recommended for different drugs (orally, intravenously, or intramuscularly). Selecting the wrong route can cause failure.
(c) Resistance: If the parasites become nonresponsive to the recommended dose, the infection will survive the treatment.
(d) Natural tolerance. Where the parasites are naturally nonresponsive to the drug.

The major hindrance to chemotherapy is that there are few drugs on the market, most of which have been in use for over 50 years (Fig. 10.1). Because of economic reasons, the pharmaceutical industry has been reluctant to invest massively in research and development of new drugs. Both trypanosomiases and leishmaniases (the so-called neglected diseases) are diseases of the poor who cannot meet the costs of treatment. Thus, treatment of these ailments is dependent on handouts from the WHO that rely on donors. As such, only two drugs are available for early-stage human African trypanosomiasis (HAT): pentamidine and suramin for *T. b. gambiense* and *T. b. rhodesiense*, respectively. Two drugs are available for late-stage HAT: melarsoprol and difluoromethylornithine (DFMO), which is but ineffective against *T. b. rhodesiense*. Only one drug, DB289 or pafuramidine, has been undergoing clinical trials for early-stage *T. b. gambiense* HAT but was recently abandoned due to unexpected toxicity. The situation for AAT is even worse; several drugs have been abandoned due to resistance. Practically, only diminazene aceturate (Berenil®) and isometamidium chloride (Samorin®, Trypamidium®) are available for AAT; the

Early (stage I) HAT   Late (stage II) HAT

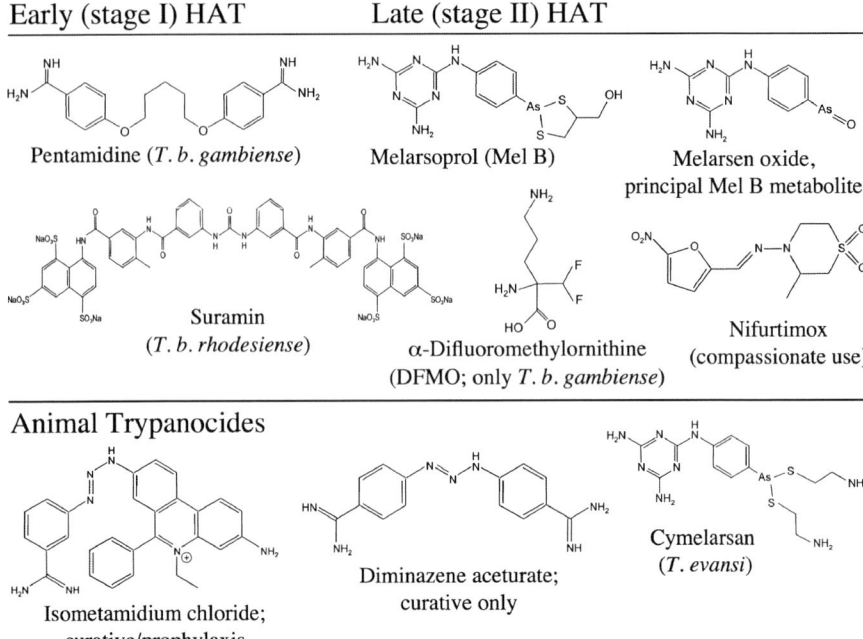

Pentamidine (*T. b. gambiense*)   Melarsoprol (Mel B)   Melarsen oxide,
                                                                principal Mel B metabolite

              Suramin   α-Difluoromethylornithine   Nifurtimox
          (*T. b. rhodesiense*)   (DFMO; only *T. b. gambiense*)   (compassionate use)

Animal Trypanocides

                                                                        Cymelarsan
                                        Diminazene aceturate;   (*T. evansi*)
    Isometamidium chloride;   curative only
    curative/prophylaxis

Fig. 10.1 Drugs in current use against African trypanosomes

third drug cymelarsan is used for the infection in camels. The inevitable outcome of continued use of the same compounds for decades is drug resistance that is largely responsible for the frequently observed chemotherapeutic failure. Reports of drug resistance are now widespread especially in the case of AAT; there are reports of resistant trypanosomes from all regions of Africa to all drugs in use (for HAT, there have been reports of melarsoprol treatment failure in Angola, the Democratic Republic of Congo, Uganda, and Sudan; Legros et al. 1999).

Reports of drug resistance are generated from isolated, small-scale studies that may not allow for effective monitoring for the spread of resistant trypanosomes. It should be noted that with increasing use of trypanocides, the inevitable emergence of resistant parasites continues to handicap chemotherapy as the mainstay for economic animal production in tsetse-infested areas. The principal factors fuelling the evolution of trypanocidal resistance can be described at four levels, namely the host, the vector, the drug itself, and the parasite. The host immune system plays an important role in treatment success as exemplified by DFMO. This drug is trypanostatic and requires a competent immune system to eliminate the arrested trypanosome population (De Gee et al. 1983). Thus, in the presence of immune depression due to concomitant infections, trypanosome exposure to the drug is increased, thereby favoring selection for resistance. Ironically, trypanosomiasis is an immunodepressant as well, a fact that may increase the chances of DFMO failure. Other host-related causes may be environmental, e.g., nutritional status, which may contribute to stress that renders the host unable to play its supportive role in chemotherapy. In principle, host-related factors lead to failure of treatment that may not be due to the presence of resistant parasites per se but to other reasons affecting the bioavailability of the drug in required concentrations, i.e., the drug pharmacokinetics. Some drugs are rapidly absorbed from the site of application and attain peak plasma levels within a short time. Diminazene aceturate was reported to reach peak plasma levels within 10–15 minutes and to be quickly excreted to below detection levels within 48 hours in dogs following intramuscular administration of 3.5 mg/kg (Aliu et al. 1993). In contrast, isometamidium is reportedly slowly released from tissues at the site of injection and was detected from the same tissues, liver, and kidney of cattle 6 weeks after treatment (Kinabo and Bogan 1988). These tissues act as slow-release devices and provide the basis for use of the drug for prophylaxis. Certain compartments like the central nervous system are partly or completely inaccessible to some drugs, e.g., diminazene aceturate was observed to be 3–4 times less in CSF than in plasma of treated goats (Mamman and Peregrine 1994). This would explain failure of diminazene treatment for animals in which the trypanosomes have crossed into the CNS (Jennings et al. 1979).

Metabolism of the drug is important where the active compound is not the native drug but its metabolite or where the drug is presented as a complex, e.g., with serum proteins (Kaminsky et al. 1990). One such example is the antimalarial proguanil, which is inactive but has to be converted into the active compound cycloguanil in the liver (Armstrong and Smith 1974). Thus, its success in treatment will depend on the role of the liver to convert the prodrug to the active substance.

Contribution by the vector is based on the notion that drug resistance is a genetically regulated characteristic. Indeed, the inheritance of experimental antibiotic resistance markers in *T. brucei* has previously been described (Gibson and Whittington 1993). Trypanosome populations are basically clonal, but there is evidence for genetic exchange between different trypanosomes within the tsetse fly (Jenni et al. 1986; Degen et al. 1995). The development of resistance may therefore rely on the degree of genetic exchange within the vector as well as the transmission intensity/efficiency, which in turn determines the rate of spread of resistant variants within the population.

At the level of the drug, several factors may fuel the development of resistance. Use of potential mutagens was widespread in AAT control with possible contribution to resistance. The drug homidium, for example, contains traces of ethidium bromide, a potent mutagenic agent. Although there is no documented evidence for the latter's initiation of mutagenesis in animals, it is speculated that its use may facilitate the emergence of mutants. Some of these mutants may be selected for and spread within the epidemic, especially if they favor survival in a drug-saturated environment. Underdosing is another important factor that may be more frequent in AAT where farmers may treat animals at their own jurisdiction. In this case, resistance is driven by persistent exposure of the parasites to suboptimal drug concentrations in host blood. Prophylactic treatment despite its utility in control may also suffer a similar fate. Isometamidium chloride (Samorin®, Trypamidium®), which is used for this purpose, stays in circulation for up to 3 months postinjection. Infections occurring toward the end of the prophylactic period when drug levels have declined will encounter sublethal doses, a situation that offers suitable conditions for the evolution of resistance. Traditional use of block treatment to control AAT increases the risk of unnecessary drug exposure of uninfected animals. This imposes a high selection pressure depending on the frequency of application. It is therefore not surprising that there are more reports of resistance in AAT than for the human disease. In all cases, appearance of new cases of resistance correlates with frequency of drug administration within a given focus. This supports the general belief that previous drug exposure is a significant factor in the evolution of resistance. Noteworthy is that most of the scenarios described above do not apply to HAT where patients are normally hospitalized and attended to by qualified health personnel. However, with the resurgence of epidemics, some of which were in areas of civil strife coupled with inadequate resources, cases of noncompliance are on the rise. This implies that some patients may not complete treatment regimens, leading to transmission of drug-fast trypanosomes. Thus, over the years, reports of treatment failure after melarsoprol are on the increase. In Uganda, relapse rates of up to 30% were reported; subsequent treatment of such patients with the same drug did not offer any relief (Legros et al. 1999).

10.4 Mechanisms of Drug Action

Although its mode of action is not well known, the primary target of melarsoprol was suggested to be the inhibition of trypanothione reductase, a key enzyme in detoxification processes in the trypanosomes (Walsh et al. 1991). However, Wang (1995) proposed inhibition of phosphofructokinase, a key enzyme in the glycolytic pathway.

Despite melarsoprol having been the drug of choice for late-stage HAT for a long time, its application was until recently based on empirical rather than rationally designed treatment protocols. It is only recently that some information on its pharmacokinetic properties has been generated. Burri et al. (1993) revealed that only 50 times lower drug level is attainable in the CSF as compared to the plasma level of treated patients. Furthermore, Keiser et al. (2000) demonstrated that melarsoprol is readily hydrolyzed into melarsen oxide, which possibly further dissociates into other metabolites all of which retain trypanocidal activity. Melarsen oxide was shown to be efficacious against rodent trypanosome infections, which does not render dissociation of melarsoprol a disadvantage to chemotherapy. Besides, the metabolites were evidenced to bind to plasma proteins; the resulting complexes may act as slow-release devices for prolonged drug challenge to the trypanosomes.

Melarsoprol use is hampered by a number of problems. The drug is highly toxic and in many cases the side effects are fatal. Some side effects include fever, chest pain, diarrhea, and necrosis at the injection site when the propylene glycol in which it is dissolved leaks into muscle tissue. But most important are the reactive encephalopathies, which occur in up to 10% (Pepin and Milord 1991) and claim 1–5% of treated patients (WHO 1986). Interestingly, melarsoprol uptake into the yeast *Saccharomyces pombe* and into mammalian cells appeared to occur through a thiamine transporter (Schweingruber 2004; Szyniarowski et al. 2006). Thiamine counteracted melarsoprol toxicity at much lower concentrations in mouse neuroblastoma cells than in bloodstream form *T. brucei* (Szyniarowski et al. 2006; Stoffel et al. 2006), suggesting that supplementation of the patients' diet with vitamin B1 may be a way to improve the tolerability of melarsoprol without jeopardizing its trypanocidal efficacy.

Difluoromethylornithine (DFMO) was originally manufactured as an anticancer agent but was also found to be effective against trypanosomes. The drug is an inhibitor of ornithine decarboxylase (ODC), a key enzyme in polyamine biosynthesis (Bacchi et al. 1980). A clinical study on DFMO carried out in Sudan in the early 1980s (Van Nieuwenhove et al. 1985) led to its registration for use against late-stage gambiense sleeping sickness. However, a similar trial against *T. b. rhodesiense* showed a cure rate of only one-third of treated patients (Bales et al. 1989). In a study performed in vitro using *T. b. rhodesiense* field isolates from southeast Uganda, the parasites responded to drug concentrations mainly above those reported in cerebrospinal fluids of treated patients, although *T. b. gambiense* included as controls were highly susceptible (Iten et al. 1995). Thus, *T. b. rhodesiense* was considered to be innately resistant (tolerant) to DFMO, presumably due to a shorter half-life of ODC compared to *T. b. gambiense* (Iten et al. 1997). But there could be factors beyond ODC turnover to explain the differential susceptibility of the *T. brucei* subspecies to DFMO.

Suramin is an aromatic oligo amide that was introduced in the 1920s. It is exclusively used against *T. b. rhodesiense* infections, probably due to its mode of application (intravenous) and possibility of inducing a fatal collapse if not carefully administered. The easier-to-administer drug for stage I disease, pentamidine, is less effective against *T. b. rhodesiense*. Despite its high efficacy against gambiense and rhodesiense sleeping sickness, its large molecules do not permit crossing of the blood–brain barrier into the CSF. Thus, it is of value only against early-stage disease. The mode of action of suramin has been reported to be inhibition of glycerol-3-phosphate oxidase and dehydrogenase (Fairlamb and Bowman 1977, 1980), as well as the inhibition of uptake of low-density lipoproteins and the consequent hampering of cell division (Vansterkenburg et al. 1993). Inhibitory effects of suramin on dihydrofolate reductase, L-α-glycerophosphate oxidase, dihydrofolate dehydrogenase, reverse transcriptase, thymidine kinase, trypsin, and RNA polymerase were also reported (Pepin and Milord 1994; Wang 1995).

The diamidines comprise the animal trypanocide diminazene aceturate, pentamidine used against early-stage *T. b. gambiense*, and pafuramidine.

Their modes of action appear to center around the kinetoplast DNA (kDNA) to which diminazene was reported to bind and subsequently inhibit replication (MacAdam and Williamson 1972). Experiments by Lanteri et al. (2004) demonstrated that DB75, the principal metabolite of pafuramidine, interferes with mitochondrial function by altering its membrane potential. More recently, studies with *L. panamensis* revealed that both diminazene and pentamidine inhibit topoisomerase II, although the former appeared to affect the host enzyme as well (Cortàzar et al. 2007). It is not clear whether the phenanthridine trypanocides (isometamidium chloride hydrochloride and ethidium bromide) have similar modes of action, although the latter is a well-known DNA intercalating agent.

10.5 Mechanisms of Drug Resistance

There are means by which the parasite can elude the toxic effects of a drug. One such mechanism is increased production of the target enzyme, in an attempt to "use up" the drug, thereby overcoming its effects. While episomal amplification of genes in response to drug pressure is typical for *Leishmania* (Ouellette et al. 1991), it has so far been observed in *T. brucei* spp. only once, when the gene for IMP dehydrogenase was amplified chromosomally upon selection for resistance to mycophenolic acid (Wilson et al. 1994). Alternatively, there may be structural changes within the target enzyme that change its affinity for a given drug (Diggens et al. 1970; Foote and Cowman 1994). This renders the drug ineffective while at the same time maintaining the physiological role of the enzyme. These changes are direct results of mutations within the genes in question. It was shown that mutations in the dihydrofolate reductase–thymidine synthetase gene rendered *Plasmodium falciparum* resistant to Fansidar (Reeder et al. 1996; Giraldo et al. 1998). Such a mechanism is yet to be reported for any trypanosomal drug target.

The predominant mechanism of drug resistance in African trypanosomes appeared to be failure of the accumulation of the drug by the parasite, either due to reduced uptake or due to enhanced export. Hawking (1936) had observed that arsenical-resistant trypanosomes removed less drug from the medium than do sensitive ones. Reduced net drug uptake was subsequently described for fluorescent compounds (Fulton and Grant 1955; Frommel and Balber 1987). Sutherland et al. (1992) showed that uptake of ^{14}C-labeled samorin was greater in drug-sensitive than resistant *T. congolense* and that samorin-resistant *T. congolense* eliminated the drug faster than do the sensitive ones (Sutherland and Holmes 1993). Members of the family of ATP-binding cassette (ABC) transporters (Upcroft 1994) whose role is transportation of foreign molecules from the cell are notorious for causing drug resistance in tumor cells and pathogens. ABC transporters are large membrane proteins that can function as ATP-dependent extrusion pumps conferring resistance to cytotoxic drugs by active efflux (Ouellette and Papadopoulou 1993). Indeed, this active drug efflux was reported to be a major mechanism for resistance of *L. talentaloe* to arsenite (Dey et al. 1994). ABC-type transporter genes have been identified in *T. brucei* (Mäser and Kaminsky 1998; Shahi et al. 2002) of which the multidrug resistance protein A (TbMRPA) has been extensively studied. The gene for this transporter, a 4.7-kb stretch, is located on chromosome 8 (www.genedb.org). Predictions based on sequence alignment and hydrophobicity profile of the protein and the closely related LtMRPA from *Leishmania* (formerly named LtPgpA, Ouellette et al. 1990) show that the proteins exhibit the hallmarks of ABC transporters, two regions each of six transmembrane domains and two ATP-binding cassettes, plus a kinetoplastid-specific sequence at the N-terminus (Fig. 10.2; Mäser et al. 2003). Recent experiments with laboratory trypanosome strains have elucidated the role of ATP transporters in drug resistance. Overexpression of the *TbMRPA* gene rendered trypanosomes arsenical-resistant (Shahi et al. 2002; Lüscher et al 2006). In yet another study, RNA interference targeted at *TbMRPA* expression was shown to confer hypersensitivity of trypanosomes to melarsoprol and melarsen oxide (Alibu et al. 2006). It is not clear whether these phenomena contribute to observed resistance phenotypes among field trypanosomes, although it can be postulated that strains with natural tendency to overexpress TbMRPA would stand a chance to survive drug challenge.

   Import of arsenicals and diamidines into trypanosomes is mediated by an unusual, aminopurine-specific adenosine transporter in trypanosomes called P2 (Carter and Fairlamb 1993). P2 is encoded by the gene *TbAT1*, as demonstrated by the functional expression in yeast (Mäser, Sütterlin, Kralli, and Kaminsky 1999) and by reverse genetics in *T. brucei* (Matovu et al. 2003). Loss of P2

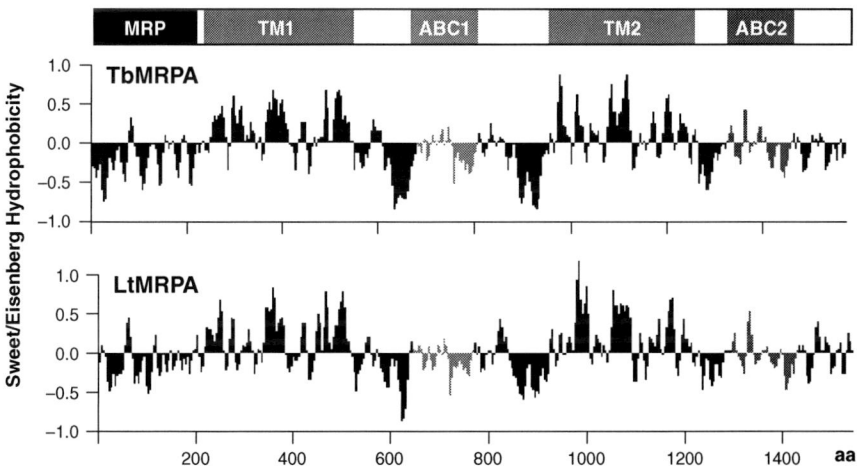

Fig. 10.2 Hydrophobicity profiles of TbMRPA and LtMRPA, ABC transporters involved in arsenical resistance. ABC, ATP-binding cassette; TM, transmembrane region with 6 TM domains; MRP, kinetoplastid-specific MRP N-terminus

function was associated with resistance to melarsoprol (Carter and Fairlamb 1993) and pentamidine (Carter et al. 1995). Genetic disruption of *TbAT1* in *T. b. brucei* caused only two- to threefold resistance to pentamidine and melarsoprol (Matovu et al. 2003), indicative of alternative routes for drug uptake. The residual pentamidine import in *tbat1*$^{-/-}$ trypanosomes fitted two-phase kinetics, indicating the presence of (at least) two additional pentamidine transport systems termed HAPT and LAPT for high- and low-affinity pentamidine transporter, respectively (de Koning 2001). Diminazene, in contrast, appeared to be predominantly accumulated through *TbAT1* (de Koning et al. 2004), explaining why *tbat1* null *T. brucei* are much more resistant to diminazene than to pentamidine (Matovu et al. 2003). A role for *TbAT1* in drug resistance in the field was underscored by the finding of mutations in the gene in drug-resistant trypanosomes. Three cases of homozygous loss of *TbAT1* have been described: in *T. b. gambiense* isolated from a melarsoprol-relapse patient in Angola (Matovu et al. 2001), in *T. b. rhodesiense* selected for melarsoprol resistance in vitro (Bernhard et al. 2007), and in *T. b. brucei* selected for resistance to furamidine (Lanteri et al. 2006). *TbAT1* is a polymorphic gene (Fig. 10.3). Two of the point mutations, alanine-178 to threonine and asparagine-286 to serine, correlated with drug resistance and interestingly, the same mutations were described from laboratory-selected mutants and from field isolates (Matovu et al. 1997; Maina et al. 2007; Fig. 10.3). These mutations were shown to render the *TbAT1* nonfunctional when expressed in *Saccharomyces cerevisiae* (Mäser et al. 1999). The presence of resistance mutations in *TbAT1* is readily detected by PCR- and RFLP-based tests (Mäser et al. 1999; Nerima et al. 2007). However, the rather weak melarsoprol resistance phenotype of *tbat1*$^{-/-}$ trypanosomes and the fact that the correlation between *TbAT1* alleles and melarsoprol treatment failures was not exactly perfect (Matovu et al. 2001) suggest additional, yet-to-be-identified determinants for the success of melarsoprol treatment.

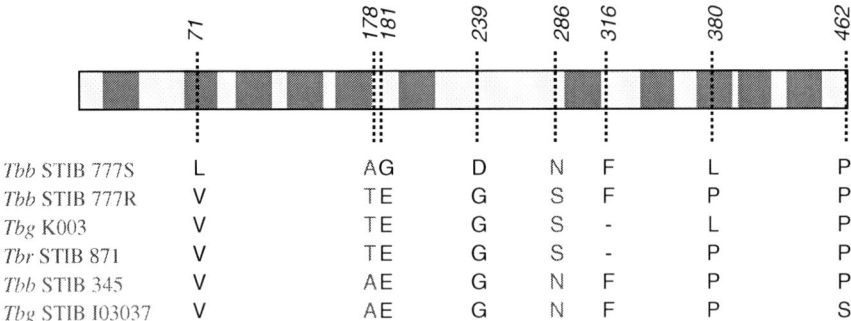

Fig. 10.3 *TbAT1* is polymorphic. Coding polymorphisms described from drug-sensitive (STIB 777S, 345 and 103037) and drug-resistant (STIB 777R, 871 and K003) *T. brucei* spp. isolates (Mäser et al. 1999; Matovu et al. 1997; Maina et al. 2007). Polymorphisms at positions 178 and 286 correlate with drug susceptibility; predicted transmembrane domains are indicated in *dark grey*

10.6 Facing the Challenge

Drug resistance is a reality facing control of African trypanosomiasis and there does not seem to be immediate solutions in the form of new drugs. Since the registration of DFMO in the early 1990s, only one new compound (DB289) went to phase III clinical evaluation stage but has been abandoned due to toxicity. Even then being a diamidine, it was likely to suffer the same fate of resistance in parasites where *TbAT1* mutations are already causing problems. Besides, the drug is incapable of crossing the blood–brain barrier, so it will not be used against the late-stage disease that is more affected by the problem of treatment failure. There are no other drugs in the pipeline, for both AAT and HAT. Thus, in the face of such challenges, integrated control of trypanosomiasis involving vector control and chemotherapy should be encouraged in order to limit overdependence on just a few drugs. The initiation of PATTEC is therefore a welcome approach that may see the problem of trypanosomiasis reduced to manageable levels. It will, however, require a concerted effort to maintain persistent challenge to the vector, given the vast terrain that has to be cleared of tsetse.

In the meantime, drug management will continue to play a central role in the control of the scourge. This can be achieved by rational drug use with the few available therapeutics at least to delay the spread of resistance. Present advocacy is for using drugs in combination rather than as monotherapies. The ideal would be to use drugs targeting different metabolic pathways. In such a way, it would be difficult for resistance to emerge since mutations at multiple sites would have to arise at the same time in order for the combination to select for such variants. Thus, in such a scenario, each of the drugs in combination would be protecting the other from emergence of resistant forms. The ultimate aim in combination chemotherapy is to achieve synergy, or at least additive effects between the combined drugs. For HAT, most compounds are known to exhibit high toxicity that is on its own a hindrance to effective case management. If selected compounds are proved to exhibit synergy, then it will mean using less of each drug in combination than would be the case in monotherapy. It would therefore be expected that the side effects would be drastically reduced, achieving faster patient recovery. Obviously, the drugs to be chosen for combination must be those that are not characterized by similar side effects. Fairlamb (2003) proposed a model for synergy between different drugs with reference to trypanothione metabolism (Fig. 10.4). This model was confirmed by subsequent RNAi experiments, which revealed that knockdown of trypanothione synthetase rendered procyclic forms more susceptible to arsenicals and nitro drugs (Arinayagam et al. 2005). Presently combination of nifurtimox and DFMO is in advanced stages of phase III trials and is likely to be recommended for late-stage HAT management. Interestingly, nifurtimox is registered as a drug against Chagas disease due to *T. cruzi* and to date has been used in a few HAT cases on compassionate basis, on individuals who failed to respond to multiple treatments with melarsoprol. The melarsoprol–nifurtimox combination was more doubted due to the similar side

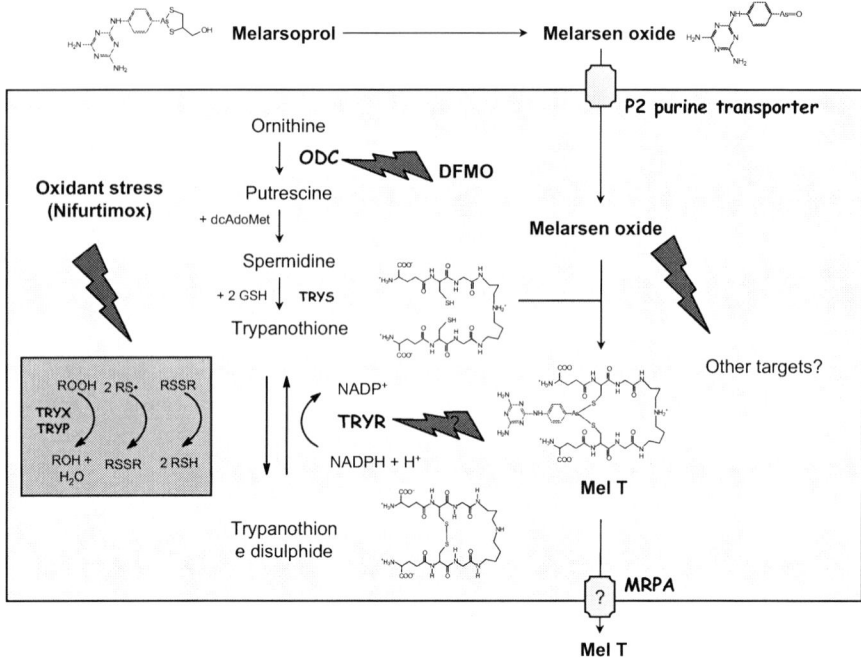

Fig. 10.4 Trypanocidal drug targets and predicted synergy between DFMO, melarsoprol, and nifurtimox with respect to trypanothione metabolism. DFMO inhibits ODC and depletes the cell of trypanothione. Melarsoprol is spontaneously converted to melarsen oxide in the host, which is imported into the trypanosome by P2 transport to react with trypanothione to form Mel T. Nifurtimox generates oxidation stress, forming DNA, lipid and hydrogen peroxide (ROOH) radicals (RSz and OHz), and disulfides (RSSR) all of which are reduced by trypanothione to form trypanothione disulphide. TRYR regenerates trypanothione, thereby maintaining the intracellular thiol–redox balance. TRYX, TRYP, and TRYS are the other potential targets. Abbreviations: DFMO, DL-α-difluoromethylornithine; MRPA, multidrug resistance protein; ODC, ornithine decarboxylase; TRPY, tryparedoxin peroxidase; TRYR, trypanothione reductase; TRYS, trypanothione synthetase; TRYX, tryparedoxine. Reproduced from Fairlamb (2003) with permission

effects observed for each of the drugs. Recent work has, however, demonstrated satisfactory treatment outcome when the combination was tried in the Congo. No relapse was observed out of 69 patients treated with the combination, while melarsoprol and nifurtimox monotherapies failed in 17 and 24 out of 70 cases, respectively (Bisser et al. 2007). What is now required is a larger scale evaluation of this combination, which could lead to its recognition as an alternative management option for HAT. Combination of DFMO and melarsoprol is yet to be explored, possibly because the former came into use in certain foci after high rates of melarsoprol relapse were observed. It therefore did not make sense in such foci to consider including melarsoprol in any combination.

The situation in AAT is even more uncertain: several drugs have been abandoned due to widespread resistance that was manifested much earlier than for HAT (Hawking 1963a, b). Reports from Sudan indicate resistance to suramin and quinapyramine by *T. evansi* infections in camels (Luckins et al. 1979; Boid et al. 1989; El Rayah et al. 1999), leaving cymelarsan as the only option. This will be useful only as long as P2 transporter mutants in *T. evansi* do not arise. Cymelarsan, just like melarsen oxide and diminazene aceturate, uses this transporter. Witola et al. (2004) demonstrated that RNAi of the *AT1* orthologue of *T. evansi*, which is highly similar to *TbAT1*, renders the trypanosomes diminazene resistant. Indeed, previous experiments had attributed resistance to reduction in the number of expressed P2 transporter molecules resulting in decreased uptake or changes in the drug-binding properties of the P2 transporter (Suswam et al. 2003).

Diminazene and isometamidium are almost the only drugs left to cater for the other economically important trypanosomes in sub-Saharan Africa. This is just by sheer luck due to properties associated with them. The advantages of these drugs are that (1) they affect different targets within the trypanosome and (2) their modes of import are different. While isometamidium appears to be principally taken up by simple diffusion, diminazene uptake is by the P2 transporter in which mutations inevitably lead to resistance. *TbAT1* knockout clearly indicated loss of diminazene susceptibility (Matovu et al. 2003), while mutant *TcoAT1* was reported to be associated with resistance in that species (Delespaux et al. 2006). By virtue of having different entry points and their targeting different pathways, diminazene and isometamidium cross-resistance is least expected. Indeed, this is a rare phenomenon in the field and as such the drugs are used as a sanative pair, each substituting the other whenever resistance is observed in a herd. The only challenge here is to be able to regularly monitor for resistance to either drug. This may be easier in the case of diminazene for which the P2 transporter is a known candidate with available techniques to detect mutants such as PCR/RFLP (Mäser et al. 1999; Matovu et al. 2001; Delespaux et al. 2006) and allele-specific PCR (Nerima et al. 2007). For isometamidium whose mechanisms of resistance are yet to be satisfactorily elucidated, such approaches will remain elusive and parasitological monitoring posttreatment is the only option.

References

Alibu, V.P., Richter C., Voncken, F., Marti, G., Shahi, S., Renggli, C.K., Seebeck, T., Brun, R. and Clayton, C. 2006. The role of *Trypanosoma brucei* MRPA in melarsoprol susceptibility. Mol. Biochem. Parasitol. 146, 38–44.

Aliu, Y.O., Mamman, M. and Peregrine, A.S. 1993. Pharmacokinetics of diminazene in female Boran (Bos indicus) cattle. J. Vet. Pharmacol. Therap. 16, 291–300.

Arinayagam, M.R., Oza, S.L., Guther, M.L.S. and Fairlamb, A.H. 2005. Phenotypic analysis of trypanothione synthetase knockdown in the African trypanosome. Biochem. J. 391, 425–432.

Armstrong, V.L. and Smith, C.C. 1974. Cyclization and N dealkylation of chloroguanide by rabbit and rat hepatic microsomes. Toxicol. Appl. Pharmacol. 29, 90.

Bacchi, C.J., Nathan, H.C., Hutner, S.H., McCann, P.P., and Sjerdsma, A. 1980. Polyamine metabolism: a potential therapeutic target in trypanosomes. Science 210, 332–334.

Bales, J.D., Harrison, S.M., Mbwabi, D.L. and Schechter, P.J. 1989. Treatment of arsenical refractory Rhodesian sleeping sickness in Kenya. Ann. Trop. Med. Parasitol. 83, 111–114.

Berberof, M., Perez-Morga, D. and Pays, E. 2001. A receptor-like flagellar pocket glycoprotein specific to *Trypanosoma brucei gambiense*. Mol. Biochem. Parasitol. 113, 127–138.

Bernhard, S., Nerima, B., Mäser, P. and Brun, R. 2007. Melarsoprol- and pentamidine-resistant *Trypanosoma brucei rhodesiense* populations and their cross-resistance. Int. J.Parasitol. 37, 1443–1448.

Bisser S., N'Siesi, F-X., Lejon, V., Preux, P-M., Van Nieuwenhove, S., Miaka, C. and Buscher P. 2007. Equivalence trail of melarsoprol and Nifurtimox monotherapy and combination therapy for the treatment of second stage trypanosome brucei gambiense sleeping sickness. J. Infect. Dis. 195, 322–329.

Boid, R., Jones, T.W. and Payne, R.C., 1989. Malic enzyme VII isoenzyme as an indicator of suramin resistance in *Trypanosoma evansi*. Exp Parasitol. 69, 317–323.

Burri, C., Baltz, T., Giroud, C., Doua, F., Welker, H.A. and Brun, R. 1993. Pharmacokinetic properties of the trypanocidal drug melarsoprol. Chemotherapy 39, 225–234.

Carter, N.S. and Fairlamb, A.H. 1993. Arsenical resistant trypanosomes lack an unusual adenosine transporter. Nature 361:173–175.

Carter, N.S., Berger, B.J. and Fairlamb, A.H. 1995. Uptake of diamidine drugs by the P2 nucleoside transporter in melarsen-sensitive and -resistant *Trypanosoma brucei brucei*. J. Biol. Chem. 270, 28153–28157.

Chitambo, H., Arakawa, A. and Ono, T. 1992. In vivo assessment of drug sensitivity of African trypanosomes using the akinetoplastic induction test. Res. Vet. Sci. 52, 243–249.

Cortàzar, T.M., Coombs, G.H. and Walker, J. 2007. *Leishmania panamensis*: comparative inhibition of nuclear DNA topoisomerase II enzymes from promastigotes and human macrophages reveals anti-parasite selectivity of fluoroquinolones, flavanoids and pentamidin. Exp. Parasitol. 116, 475–482.

Degen R., Pospichal H., Enyaru J. and Jenni L. 1995. Sexual compatibility among *Trypanosoma brucei* isolates from an epidemic area in south-eastern Uganda. Parasitol. Res. 81, 253–257.

De Gee, A.L.W., McCann, P.P. and Mansfield, J.M. 1983. Role of antibody in the elimination of trypanosomes after DL-α-difluoromethylornithine chemotherapy. J. Parasitol. 69, 818–822.

de Koning, H. 2001. Uptake of pentamidine in *Trypanosoma brucei* is mediated by three distinct transporters: implications for cross-resistance with arsenicals. Mol. Pharmacol. 59, 586–592.

de Koning, H.P., Anderson, L.F., Stewart, M., Burchmore, R.J., Wallace, L.J. and Barrett, M.P. 2004. The trypanocide diminazene aceturate is accumulated predominantly through the TbAT1 purine transporter: additional insights on diamidine resistance in African trypanosomes. Antimicrob. Agents Chemother. 48, 1515–1519.

Delespaux, V., Chitanga, S., Geysen, D., Goethals, A.P., Van den Bossche, P. and Geerts, S. 2006. SSCP analysis of the P2 purine transporter TcoAT1 gene of *Trypanosoma congolense* leads to a simple PCR-RFLP test allowing the rapid identification of diminazene resistant stocks. Acta Trop. 100, 96–102.

Dey, S., Papadopoulou, B., Haimeur, A., Roy, G., Grondin, K., Dou, D., Rosen, B.P. and Ouellette, M. 1994. High level arsenite resistance in *Leishmania tarentolae* is mediated by an active extrusion system. Mol. Biochem. Parasitol. 67, 49–57.

Diggens, S.M., Gutteridge, W.E. and Trigg, P.I. 1970. An altered dihydrofolate reductase associated with pyrimethamine resistant *P. bergei* produced in a single step. Nature 228, 579–580.

El Rayah, I.E., Kaminsky, R., Schmid, C. and El Malikh, K.H. 1999. Drug resistance in Sudanese *Trypanosoma evansi*. Vet. Parasitol. 80, 281–287.

Fairlamb, A.H. 2003. Chemotherapy of human African trypanosomiasis: current and future prospects. Trends Parasitol. 19, 488–494.

Fairlamb, A.H. and Bowman, I.B.R. 1977. *Trypanosoma brucei*: suramin and other trypanocidal compounds effects on glycerol-3-phosphate oxidase. Exp. Parasitol. 43, 353–361.

Fairlamb, A.H. and Bowman, I.B.R. 1980. Uptake of the trypanocidal drug suramin by bloodstream forms of *Trypanosoma brucei* and its effects on respiration and growth rate in vivo. Mol. Biochem. Parasitol. 1, 315–333.

Foote, S.J. and Cowman, A.F. 1994. The mode of action and the mechanism of resistance to antimalarial drugs. Acta Trop. 56, 157–171.

Frommel, T.O. and Balber, A.E. 1987. Flow cytofluorimetric analysis of drug accumulation by multidrug-resistant *Trypanosoma brucei brucei* and *T. b. rhodesiense*. Mol. Biochem. Parasitol. 26, 183–192.

Fulton, J.D. and Grant, P.T. 1955. The preparation of a strain of *Trypanosoma rhodesiense* resistant to stilbamidine and some observations on its nature. Ann. Trop. Med. Parasitol. 49, 377–387.

Gibson, W., Backhouse, T. and Griffiths, A. 2002. The human serum associated gene is ubiquitous and conserved in *Trypanosoma brucei rhodesiense* throughout East Africa. Infect. Genet. Evol. 1, 207–214.

Gibson, W. and Whittington, H. 1993. Genetic exchange in *Trypanosoma brucei*: selection of hybrid trypanosomes by introduction of genes conferring drug resistance. Mol. Biochem. Parasitol. 60, 19–26.

Giraldo, L.E., Acosta, M.C., Labrada, L.A., Praba, A., Montanegro-James, S., Saravia, N.G. and Krogstad, D.J. 1998. Frequency of the asn-108 and Thr-108 point mutations in the dihydrofolate reductase gene in *Plasmodium falciparum* from southwest Columbia. Am. J. Trop. Med. Hyg. 59, 124–128.

Hawking, F. 1936. The absorption of arsenical compounds and tartar emetic by normal and resistant trypanosomes and its relation to drug-resistance. J. Pharm. Exp. Therap. 59, 123–156.

Hawking, F. 1963a. Drug resistance of *Trypanosoma congolense* and other trypanosomes to quinapyramine, penanthridines, Berenil and other compounds in mice. Ann. Trop. Med. Parasitol. 57, 262–282.

Hawking, F. 1963b. History of chemotherapy. In Experimental chemotherapy Ed R.J. Schnitzer, and F. Hawking. New York, Academic Press, Vol I, pp. 2–3.

Hesse, F., Sezer, P.M., Mühlstädt, K. and Duszenko, M. 1995. A novel cultivation technique for long-term maintenance of bloodstream form trypanosomes in vitro. Mol. Biochem. Parasitol. 70, 157–166.

Hoare, C.A. 1972. The trypanosomes of mammals: a zoological monograph. Oxford: Blackwell.

Iten, M., Matovu, E., Brun, R. and Kaminsky, R. 1995. Innate lack of susceptibility of Ugandan *Trypanosoma brucei rhodesiense* to DL -α- difluorometylornithine (DFMO). Trop. Med. Parasitol. 46, 190–194.

Iten, M., Mett, H., Evans, A., Enyaru, J.C.K., Brun, R. and Kaminsky R. 1997. Alterations in ornithine decarboxylase characteristics account for tolerance of *Trypanosoma brucei rhodesiense* to DL-a-difluoromethylornithine. Antimicrob. Agents Chemoth. 41, 1922–1925.

Jenni, L., Marti, S., Schweizer, J., Betschart, B., Le Page, W.F., Wells, J.M., Tait, A., Paindavoine., Pays, E. and Steinert, M. 1986. Hybrid formation between African trypanosomes during cyclical transmission. Nature 322, 173–175.

Jennings, F.W., Whitelaw, D.D., Chizyuka, H.G.B., Holmes, P.H. and Urquhart, G. 1979. The brain as a source of relapsing *Trypanosoma brucei* infections in mice after chemotherapy. Int. J. Parasitol. 9, 381–384.

Kaminsky, R., Gumm, L.D., Zweygarth, E. and Chuma, F. 1990. A drug incubation infectivity test DIIT for assessing resistance in trypanosomes. Vet. Parasitol. 34, 335–343.

Keiser, J., Ericson, O. and Burri, C. 2000. Investigations on the metabolites of the trypano-cidal drug melarsoprol. Clin. Pharmacol. Therap. 67, 478–488.

Kinabo, L.D.B. and Bogan, J.A. 1988. Pharmacokinetic and histopathological investigations of isometamidium in cattle. Res. Vet. Sci. 44, 267–269.

Kinabo, L.D.B, McKellar, Q.A., and Eckersall, P.D. 1991. Isometamidium in pigs: disposition kinetics, tissue residues and adverse reactions. Res. Vet. Sci. 50, 6–13.

Kuzoe, F.A.S. 1993. Current situation of African trypanosomiasis. Acta Trop. 54, 153–162.

Lanteri, C., Stewart, M., Brock, J., Alibu, V., Meshnick, S., Tidwell, R. and Barrett, M. 2006. Roles for the *Trypanosoma brucei* P2 transporter in DB75 uptake and resistance. Mol. Pharmacol. 70, 1585–1592.

Lanteri, C.A., Trumpower, B.L., Tidwell, R.R. and Meshnick, S.R. 2004. DB75, a novel trypanocidalagent disrupts mitochondrial function in *Saccharomyces cerevisiae*. Antimicrob. Agents Chemoth. 48, 3968–3974.

Legros, D., Evans, S., Maiso, F., Enyaru, J.C.K. and Mbulamberi, D. 1999. Risk factors for treatment failure after melarsoprol for *Trypanosoma brucei gambiense* trypanosomiasis in Uganda. Trans. R. Soc. Trop. Med. Parasitol. 93, 439–442.

Lester, H.M.O. 1939. Certain aspects of trypanosomiasis in some African independencies. Trans. R. Soc. Trop. Med. Hyg. 33, 11–36.

Luckins, A.G., Boid, R., Rae, P., Mahmoud, M.M., Malik, K.H. and Gray, A.R. 1979. Serodiagnosis of infection with *T. evansi* in camels in Sudan. Trop. Anim. Health Prod. 11, 1–12.

Lüscher, A., Nerima, B. and Mäser, P. 2006. Combined contribution of TbAT1 and TbMRPA to drug resistance in *Trypanosoma brucei*. Mol. Biochem. Parasitol. 150, 364–366.

MacAdam, R.F. and Williamson, J. 1972. Drug effects on the fine structure of *Trypanosoma rhodesiense*: diamidines. Trans. R. Soc. Trop. Med. Hyg. 66, 897–904.

Mamman, M. and Peregrine, A.S. 1994. Pharmacokinetics of diminazene in plasma and cerebrospinal fluid of goats. Res. Vet. Sci. 57, 253–255.

Maina, N., Maina, K., Mäser, P. and Brun, R. 2007. Genotypic and phenotypic characterization of *Trypanosoma brucei gambiense* isolates from Ibba, South Sudan, an area of high melarsoprol treatment failure rate. Acta Trop. 104, 84–90.

Mäser, P., Sütterlin, C., Kralli, A. and Kaminsky, R. 1999. A nucleoside transporter from *Trypanosoma brucei* involved in drug resistance. Science 285, 242–244.

Mäser, P. and Kaminsky, R. 1998. Identification of three ABC transporter genes in *Trypanosoma brucei* spp. Parasitol. Res. 84, 106–111.

Mäser, P., Lüscher, A. and Kaminsky, R. 2003. Drug transport and drug resistance in African trypanosomes. Drug Resist. Update 6, 281–290.

Matovu, E., Iten, M., Enyaru, J.C.K., Schmid, C., Lubega, G.W., Brun, R. and Kaminsky, R. 1997. Susceptibility of Ugandan *Trypanosoma brucei rhodesiense* isolated from man and animal reservoirs to diminazene, isometamidium and melarsoprol. Trop. Med. Int. Health 2, 13–18.

Matovu, E., Geiser, F., Schneider, V., Mäser, P., Enyaru, J.C.K., Kaminsky, R., Gallati, S. and Seebeck, T. 2001. Genetic variants of the TbAT1 adenosine transporter from African trypanosomes in relapse infections following melarsoprol therapy. Mol. Biochem. Parasitol. 117, 73–81.

Matovu, E., Stewart, M.L., Geiser, F., Brun, R., Mäser, P., Wallace, L.J., Burchmore, R.J., Enyaru, J.C., Barrett, M.P., Kaminsky, R., Seebeck, T. and de Koning, H.P. 2003. The mechanisms of arsenical and diamidine uptake and resistance in *Trypanosoma brucei*. Euk. Cell. 2, 1003–1008.

Molyneux, D.H. and Ashford, R.W. 1983. The biology of *Trypanosoma* and *Leishmania*, Parasites of man and domestic animals. Taylor & Fransis, London.

Mulligan, H.W. (ed) 1970. The African trypanosomiases. London: George Allen & Unwin.

Nerima, B., Matovu, E., Lubega, G. and Enyaru, J. 2007. Detection of mutant P2 adenosine transporter TbAT1) gene in *Trypanosoma brucei gambiense* isolates from northwest

Uganda using allele-specific polymerase chain reaction. Trop. Med Int. Health 12, 1361–1368.

Okoth, J.O., Kirumira, E.K. and Kapaata, R. 1991. A new approach to community participation in tsetse control in the Busoga Sleeping sickness focus, Uganda. A preliminary report. Ann. Trop. Med. Parasitol. 85, 315–322.

Ouellette, M. and Papadopoulou, B. 1993. Mechanisms of drug resistance in *Leishmania*. Parasitol. Today 9, 150–153.

Ouellette, M., Fase-Fowler, F. and Borst, P. 1990. The amplified H-circle of methotrexate-resistant *Leishmania tarentolae* contains a novel P-glycoprotein. EMBO J. 9, 1027–1033.

Ouellette, M., Hettema, E., Wust, D., Fase-Fowler, F. and Borst, P. 1991. Direct and inverted DNA repeats associated with P-glycoprotein gene amplification in drug resistant *Leishmania*. EMBO J. 10, 1009–1016.

Overath, P., Chaudri, M., Steverdig, D. and Ziegelbauer, K. 1994. Invariant surface proteins in bloodstream forms of *Trypanosoma brucei*. Parasitol. Today 10, 53–58.

Pepin, J. and Milord, F. 1991. African trypanosomiasis and drug-induced encephalopathy; risk factors and pathogenesis. Trans. R. Soc. Trop. Med. Hyg. 85, 222–224.

Pepin, J. and Milord, F. 1994. The treatment of human African trypanosomiasis. Adv. Parasitol. 33, 1–47.

Radwanska, M., Claes, F., Magez, S., Magnus, E., Perez-Morga, D., Pays, E. and Büscher, P. 2002a. Novel primer sequences for polymerase chain reaction-based detection of *Trypanosoma brucei gambiense*. Am. J. Trop. Med. Hyg. 67, 289–295.

Radwanska, M., Chamekh, M., Vanhamme, L., Claes, F., Magez, S., Magnus, E., de Baetselier, P., Büscher, P. and Pays, E. 2002b. The serum resistance-associated gene as a diagnostic tool for the detection of *Trypanosoma brucei rhodesiense*. Am. J. Trop. Med. Hyg. 67, 684–690.

Reeder, J.C., Riekmann, K.H., Genton, B., Lorry, K., Wines, B. and Cowman, A.F. 1996. Point mutations in the dihydrofolate reductase and dihydropteroate synthetase genes and in vitro susceptibility to pyrimethamine and cycloguanil of *Plasmodium falciparum* isolates from Papua New Guinea. Am. J. Trop. Med. Hyg. 55, 209–213.

Schweingruber, M. 2004. The melaminophenyl arsenicals melarsoprol and melarsen oxide interfere with thiamine metabolism in the fission yeast *Schizosaccharomyces pombe*. Antimicrob. Agents Chemoth. 48, 3268–3271.

Shahi, S.K., Krauth-Siegel, R.L. and Clayton, C.E. 2002. Over-expression of the putative thiol conjugate transporter TbMRPA causes melarsoprol resistance in *Trypanosoma brucei*. Mol. Microbiol. 43, 1129–1138.

Stoffel, S., Rodenko, B., Schweingruber, A., Mäser, P., de Koning, H. and Schweingruber, M. 2006. Biosynthesis and uptake of thiamine vitamin B1. in bloodstream form *Trypanosoma brucei brucei* and interference of the vitamin with melarsen oxide activity. Int. J. Parasitol. 36, 229–236.

Suswam, E.A., Ross, C.A. and Martin, R.J. 2003. Changes in adenosine transport associated with melanophenyl arsenical (Mel CY. resiatnce in *Trypanosoma evansi*: down-regulation and affinity changes of the P2 transporter. Parasitology 127, 543–549.

Sutherland, I.A., Mounsey, A. and Holmes, P.H. (1992. Transport of isometamidium (Samorin) by drug resistant and drug sensitive *Trypanosoma congolense*. Parasitology 104, 461–467.

Sutherland, I.A. and Holmes, P.H. 1993. Alterations in drug transport in resistant *Trypanosoma congolense*, Acta Trop. 54, 271–278.

Szyniarowski, P., Bettendorff, L. and Schweingruber, M. 2006. The anti-trypanosomal drug melarsoprol competitively inhibits thiamin uptake in mouse neuroblastoma cells. Cell Biol. Toxicol. 22, 183–187.

Turner, M.J. 1985. The biochemistry of the surface antigens of African trypanosomes. Br. Med. Bull. 41, 137–143.

Upcroft, P. 1994. Multiple drug resistance in the pathogenic protozoa. Acta Trop. 56, 195–212.

Van Nieuwenhove, S., Schechter, P.J., Declercq, J., Bone, G., Burke, J. and Sjoerdsma, A. 1985. Treatment of gambiense sleeping sickness in the Sudan with Oral DFMO DL-a-difluoromethylornithine), an inhibitor of the ornithine decarboxylase; first field trial. Trans. Roy. Soc. Trop. Med. Hyg. 79, 692–698.

Vansterkenburg, E.L.M., Coppens, I., Wilting, J., Bos, O.J.M., Fischer, M.J.E., Janssen, L.H.M. and Opperdoes, F.R. 1993. The uptake of the trypanocidal drug suramin in combination with low-density lipoproteins by *Trypanosoma brucei* and its possible mode of action. Acta Trop. 54, 237–250.

Vickerman, K. 1978. Antigenic variation in trypanosomes. Nature 273, 613–617.

Vreysen, M.J. 2001. Principles of area-wide intergrated testse fly control using the sterile insect technique. Med. Trop. (Mars.) 61, 397–411.

Vreysen, M.J., Saleh, K.M., Ali, M.Y., Abdulla, A.M., Zhu, Z.R., Juma, K.G., Dyck, V.A., Msangi, A.R., Mkonyi, P.A. and Feldmann, H.U. 2000. *Glossina austeni* (Diptera: Glossinidae) eradicated on the island of Unguja, Zanzibar, using the sterile insect technique. J. Econ. Entomol. 93, 123–135.

Walsh, C., Bradley, M. and Nadeau, K. 1991. Molecular studies on trypanothione reductase, a target for anti-parasitic drugs. TIBS 16, 305–309.

Wang, C.C. 1995. Molecular mechanisms and therapeutic approaches to the treatment of African trypanosomiasis. Ann. Rev. Pharmacol. Toxicol. 35, 93–127.

Welburn, S.C., Coleman, P.G., Fevre, E., Maudlin, I. 2001. Sleeping sickness: a tale of two diseases. A review focussing on the contrasts between the epidemiology of *T.b. rhodesiense* and *T. b. gambiense*. Parasitol. Today 19, 19–24.

WHO. 1986. Epidemiology and control of African trypanosomiasis. Report of a WHO Expert Committee, WHO, Geneva. Technical Report Series 739, pp.7.

Wilson, K., Berens, R.L., Sifri, C.D. and Ullman, B. 1994. Amplification of the inosinate dehydrogenase gene in *Trypanosoma brucei gambiense* due to an increase in chromosome copy number. J. Biol. Chem. 269, 28979–28987.

Witola, W.H., Inoue, N., Ohasi, K. and Onuma, M. 2004. RNA-interference silencing of the adenosine transporter gene in *Trypanosoma evansi* confers resistance to diminazene aceturate. Exp. Parasitol. 107, 47–57.

Xong, H.V., Vanhamme, L., Chamekh, M., Chimfwembe, C.E., Van Den Abbele, J., Pays, A., Van Meirvenne, N., Hamers, R., De Baetselier, P. and Pays E. 1998. A VSG expression site-associated gene confers resistance to human serum in *Trypanosoma rhodesiense*. Cell, 95, 839–846.

Chapter 11
Antimicrobial Resistance in Enteric Pathogens in Developing Countries

Samuel Kariuki

Abstract Bacterial enteric infections exact a heavy toll on human populations, particularly among children and immunosuppressed individuals in developing countries, where malnutrition, HIV/AIDS and poor sanitation abound. Despite the explosion of knowledge on the pathogenesis of enteric diseases during the past two decades, the number of diarrhoeal episodes and human deaths reported especially among the poor populations in developing countries remain of apocalyptic dimensions. With several studies from developing countries showing worrying trends in multiple resistance among key enteric pathogens such as *Escherichia coli*, *Klebsiella*, *Salmonella* spp., *Vibrio cholerae* and *Shigella* spp. to nearly all commonly available antibiotics, it is imperative that this trend should be reversed. Many countries in the developing world lack a formal surveillance system for antibiotic resistance and often treatment is given empirically based on clinical diagnosis alone. Availability of antibiotics over the counter without prescription has not helped either. To minimize the negative effect on public health, a concerted effort is required in surveillance, public health education and awareness of dangers of resistance, and a clear policy on procurement and prudent use of antibiotics in these resource-poor settings.

11.1 Common Enteric Bacteria for Which Multidrug Resistance Is Documented

Diarrhoeal illness rarely requires antimicrobial treatment and can be prevented by improving living conditions; yet antimicrobials, which are widely available over the counter and through other unregulated outlets in a number of developing countries, remain a mainstay of empirical therapy. This widespread injudicious use diminishes the efficacy of affordable and available drugs, which poses a serious problem when antimicrobial treatment is needed (Wise et al. 1998; WHO 1999).

S. Kariuki (✉)
Kenya Medical Research Institute (KEMRI), Centre for Microbiology Research, Nairobi, Kenya
e-mail: skariuki@kemri.org

A. de J. Sosa et al. (eds.), *Antimicrobial Resistance in Developing Countries*, DOI 10.1007/978-0-387-89370-9_11, © Springer Science+Business Media, LLC 2010

Table 11.1 Antibacterial resistance among key enteric pathogens

Bacterium	Antibiotic
Still predictably sensitive	
Clostridium perfringens	Penicillin
Corynebacterium diphtheriae	Penicillin
Francisella tularensis	Streptomycin, gentamicin
Listeria monocytogenes	Penicillin
Becoming problematic	
Pasteurella multocida	Ampicillin, tetracyclines
Vibrio cholerae	Tetracyclines, fluoroquinolones
Yersinia pestis	Streptomycin, tetracyclines
Problematic	
Enterococcus spp.	Vancomycin, ampicillin, oxazolidonones[a]
Escherichia coli	Multidrug
Klebsiella pneumoniae	Multidrug
Non-typhoidal salmonellae	Multidrug
Salmonella Typhi	Multidrug
Shigella spp.	Multidrug

[a]Some untreatable infections.

Overuse of antimicrobials also wastes limited financial resources on ineffective medications that expose patients unnecessarily to potential toxicities, may prolong illness, and can increase risk of death (Legros et al. 1999). Enteric pathogens frequently acquire resistance when they are exposed to resistant normal flora that colonize intestinal tracts of humans and animals, or may have acquired resistance in other environments before infecting host species. Understanding the prevalent antimicrobial susceptibility patterns could inform guidelines for empirical therapy and also minimize injudicious antimicrobial use. Although there is a problem of antibiotic resistance in enteric pathogens, it is not uniformly bad. Some enteric pathogens remain predictably sensitive to a particular agent, and for others problems are developing but at a low prevalence; for a notorious few, however, there is certainly the spectre of untreatable infections (Table 11.1).

In the following section we endeavour to highlight some of the most critical areas in resistance among enteric pathogens of major clinical importance, especially in resource-poor settings in sub-Saharan Africa. We concentrate on enteric commensals, enteric pathogens and those pathogens with an enteric reservoir, and on problems that may be peculiar to these developing countries.

11.1.1 *Multidrug-Resistant* E. coli *from Clinical and Environmental Sources*

In a multicentre study to determine the prevalence of antibiotic-resistant faecal *E. coli* from adult volunteers from urban areas in Kenya, Mexico, Peru and the

Philippines, and non-urban locations in Curaçao, Mexico, Venezuela, Ghana, Zimbabwe and the Philippines (Nys et al. 2004), the mean age of the volunteers was 35 years; most of them were females. Ciprofloxacin resistance was in the range 1–63%: the highest percentage was found in the urban populations of Asia and South America. In Peru and the Philippines the prevalence of genta-micin resistance was >20%. Cefazolin resistance was the highest in the urban Philippines (25%). Higher prevalence of resistance to ampicillin, oxytetracy-cline and trimethoprim was found for urban areas compared with non-urban ones of Asia, Africa and South America, respectively (P-value <0.05). Anti-biotic resistance in faecal *E. coli* from these adult volunteers was emerging for cefazolin, gentamicin and ciprofloxacin and was high for the older drugs ampicillin, oxytetracycline, trimethoprim and chloramphenicol.

Ciprofloxacin resistance prevalence showed large variations between the populations and was higher in the urban areas of Asia and South America compared with the non-urban areas, but was similar in urban and non-urban regions in Africa. The high prevalence of resistance in Asia and South America might arise from the food supply. In both continents, poultry in particular are intensively raised and fluoroquinolones are commonly used during production; they are mixed through the water supply of the whole flock (WHO 1998). Similarly a study by van de Mortel et al. (1998) showed significant differences in the prevalence of resistance between non-urban and urban areas for ampicillin, oxytetracycline and trimethoprim. The differences in resistance between urban and non-urban areas were thought to be due to the availability of antibiotics, as in cities a large variety of often inexpensive antibiotics are available in pharmacies, over the counter and market stalls (Hart and Kariuki 1998; Okeke and Edelman 2001). Furthermore, crowding together with poor hygiene and poor sanitary facilities for sewage disposal in the cities might encourage the exchange of antibiotic-resistant bacteria in a population (Grenet 2004).

In a study by Bii et al. (2005) that characterized antibiotic resistance among diarrhoeagenic *E. coli* from children less than 5 years of age from Kenya isolates exhibited high-level multidrug resistance to WHO recommended antibiotics. Resistance rates to tetracycline, ampicillin and co-trimoxazole were 70.7, 65.9 and 68.3%, respectively, figures that were very similar to resistance prevalence among *E. coli* from healthy children. The presence of multiple virulence genes was associated with multidrug resistance; a selection and subsequent spread of this clone in the population would lead to a major public health concern. In another study conducted in Kenya, 74% of persons with bloody diarrhoea received antibiotics to which their isolate was not susceptible (Brooks et al. 2003), again indicating that antibiotic resistance among enteric pathogens within populations in Kenya was high and likely to jeopardize the effective treatment of infections using commonly available antibiotics.

In several studies worldwide it has been shown that *E. coli* from normal gut flora constitute an important reservoir of resistance genes (van den Bogaard and Stobberingh 2000). These bacteria may act as a reservoir for genes that encode multidrug-resistant (MDR) phenotypes which can be transferred to potential pathogens (van de Mortel et al. 1998). The rising prevalence of resistance against antibiotics, such as co-trimoxazole and ampicillin, could reflect the flow of MDR genes among gut-associated bacteria which would make a formidable reservoir for antibiotic resistance genes. For instance in their study, Bartoloni et al. (2006) used conjugation experiments to demonstrate that resistance towards ampicillin, tetracycline, trimethoprim, sulphamethoxazole and chloramphenicol could be transferred en bloc among commensal microflora suggesting that mobile genetic elements may be involved in the dissemination MDR phenotypes. In this respect, *E. coli* is capable of surviving in extra-intestinal environments and may acquire other MDR traits from soil and water bacteria.

To determine the interaction between resistance determinants in *E. coli* from different ecosystems, two comparative studies were conducted in two districts of central Kenya between 1999 and 2003 (Kariuki et al. 1997). Small-scale farmers in these areas grow food crops for subsistence and a majority keep less than 3 milking cows and goats, and 10–300 chickens for commercial purposes. In each study susceptibility of *E. coli* isolated from the faeces of healthy children attending mother and child well-clinics in selected hospitals and from the associated environments at their homes to commonly available antibiotics was evaluated. A total of 188 non-duplicate *E. coli* strains were obtained from 256 faecal specimens of children below 5 years of age, while 286 strains were isolated from environmental specimens from the homes of index cases. The environmental strains were from chicken droppings (214; 74.8%), rectal swabs of cattle (47; 16.4%), and from water sources (25; 8.7%) among which 18 strains were from boreholes while 4 strains were from rivers.

The first study found that *E. coli* isolates from children were less sensitive to the test drugs than the environmental isolates (P-value < 0.05). Of the *E. coli* isolates from children 164 (87.2%) were multidrug resistant, the commonest resistance being to ampicillin, chloramphenicol, co-trimoxazole and tetracycline. In contrast, only 26% (P-value <0.05) of *E. coli* isolates from chickens and none from cattle or water were multidrug resistant; resistance was usually to streptomycin or tetracycline. Resistance among *E. coli* isolates from chickens was mainly to tetracycline (72%), while the isolates were fully sensitive to most other antibiotics commonly used for treatment of patients including ampicillin, co-trimoxazole, chloramphenicol and gentamicin. However, all isolates were fully susceptible to third-generation cephalosporins and quinolones. This study strongly suggests that normal intestinal flora of children was more exposed to antibiotics than in livestock in the same setting.

In the second follow-up study conducted in 2003, a total of 344 *E. coli* isolates from healthy children were analysed for susceptibility to the panel of commonly available antimicrobials. High prevalence of multidrug resistance (88.6%) was still observed towards trimethoprim–sulphamethoxazole, ampicillin and tetracycline and this did not differ from prevalence observed in the 1999–2000 study. Resistance to other antimicrobials such as nalidixic acid and ciprofloxacin was rare. In contrast, only 17.4% (24/138) of environmental isolates (18 of these being *E. coli* from chickens) were multidrug resistant. It is plausible that the low prevalence of antibiotic-resistant *E. coli* strains from environmental sources may reflect the narrow range of antibiotics applied in poultry rearing and other farming activities in Kenya. In this survey 70% of the farmers interviewed indicated that they obtained antibiotics such as tetracycline, penicillins and sulphonamides from pharmacies without prescription for treatment of mastitis in cattle and diarrhoea in chickens. However, tetracycline was the most commonly used antibiotic in rearing chickens. High resistance to tetracycline among environmental isolates can therefore be attributed to improper use of this drug for prophylaxis in animal husbandry. Compared to previous studies in the same region in 1993 (Kariuki et al. 1997), the prevalence of resistance to commonly available antimicrobials in *E. coli* from children has risen from 85.5% in 1993 to 88.6% in 2003, but the difference was not statistically significant (P-value = 0.12). In contrast, the prevalence of MDR phenotype in *E. coli* from chickens reduced from 26% in 1993 to 17.4% in 2003 (P-value <0.05), probably a reflection of fewer farmers using antimicrobials for growth promotion. In both studies resistance in *E. coli* isolates was associated with the presence of 100–120 kb plasmids.

In a study on carriage of antibiotic resistance in *E. coli* from community and hospitalized patients in Indonesia, Duerink et al. (2007) observed that for most antibiotic classes, the most resistance was present in the group most exposed to antibiotics and the least resistance in the group least exposed to antibiotics. In the community, direct associations were observed between the use of specific antibiotics and resistance to those antibiotics, namely between β-lactam antibiotics and ampicillin resistance and sulphonamide use and trimethoprim–sulphamethoxazole resistance. Here, the majority of antibiotic therapy consisted of single therapy. Children, regardless of more frequent antibiotic use, were at greater risk of carriage of resistant *E. coli* than adults, perhaps because of the greater exposure to resistant micro-organisms.

11.1.2 Cefotaxime-Hydrolysing and Multidrug-Resistant Klebsiella *spp.*

Extended-spectrum beta-lactamase (ESBL)-producing strains of *Klebsiella pneumoniae* have caused major therapeutic problems worldwide since the majority are resistant to various antibiotics commonly used for treatment of most bacterial infections. In sub-Saharan Africa nosocomial infections caused by multidrug-resistant *K. pneumoniae* are becoming increasingly common. For instance, in a Kenyan referral hospital, all *K. pneumoniae* isolates obtained from sporadic infections in neonatal wards over a 6-month period were uniformly resistant to ampicillin, cephradine, cefuroxime, cefotaxime, carbenicillin, ceftazidime and tetracycline. However, they were susceptible to co-amoxyclav, ceftazidime, aztreonam, streptomycin, co-trimoxazole, gentamicin and nalidixic acid. Isolates had MIC of 24 and 1 µg/mL for cefotaxime and ceftazidime, respectively. The presence of clavulanic acid decreased the MIC of cefotaxime 750-fold to 0.032 µg/mL, resistance was shown to be a result of production of CTX-M-12-extended-spectrum β-lactamases (Kariuki et al. 2001).

In a separate study in Nigeria (Soge et al. 2006) CTX-M plasmids isolated from uropathogenic *K. pneumoniae* were large (58–320 kb) and carried the following genes: *aac(6′)*-Ib (aminoglycoside resistance) which included *aac(6′)*-Ib-cr (aminoglycoside–fluoroquinolone resistance), *aad*A2 (aminogly-coside resistance), *erm*(B) (macrolide–lincosamide–streptogramin B resistance), *bla*TEM-1 (ampicillin resistance), *tet*(A) (tetracycline resistance), *sul1* (sulphonamide resistance), *dfr* (trimethoprim resistance) and *intI*1, an integrase associated with class 1 integrons. Often, such high-level resistance in *K. pneumoniae* may lead to treatment failure using commonly available antibiotics. In Algeria, environmental isolates of *K. pneumoniae* resistant to extended-spectrum cephalosporins with a phenotype and genotype indicating CTX-M-15 enzyme production were found to be identical to *K. pneumoniae* clinical isolates recovered from urinary tract infection from hospitalized patients (Touati et al. 2007).

ESBL-producing *K. pneumoniae* have also caused major therapeutic pro-blems in many other developing countries including Iran (Liu et al. 2007) where 33% of ESBL strains were also resistant to ciprofloxacin and aminoglycosides, while in India (Shukla et al. 2004) ESBL-producing strains also showed resis-tance against amoxicillin, gentamicin, ciprofloxacin and amikacin in 93.3, 70, 10.4 and 26.1% of the isolates, respectively.

11.1.3 Multidrug-Resistant Non-Typhoidal Salmonella Infections

In literature there are an increasing number of reports of cases of multidrug-resistant foodborne *Salmonella enterica* infections in both developed and developing countries, with few options left for antimicrobial treatment. Exam-ples of increases in resistance in non-typhi *Salmonella* (NTS) in developing countries, particularly in sub-Saharan Africa, the Indian subcontinent and Southeast Asia, are exemplified by numerous outbreaks caused by multidrug-resistant NTS. These are frequently resistant to the newer quinolones both in hospitals and the community over wide geographical areas (Su et al. 2005; Kariuki et al. 2005). Although NTS infections are a common cause of self-limiting diarrhoea in healthy individuals, bacteraemia with or without focal infections can occur as a complication (Yang et al. 2002). In Africa, NTS are among the most common cause of bloodstream infections in children younger than 5 years (Graham 2002). In Kenya, for instance, NTS are second only to invasive pneumococcal disease in importance as a cause of bacteraemia in children less than 5 years of age, with a high mortality, especially in malnour-ished children. We have previously estimated that the local minimum incidence of community-acquired NTS is 166 per 100,000 per year for children below 5 years of age (Berkley et al. 2005). Also in Malawi, NTS was the most common blood culture isolate (40%) in hospitalized children, with a case fatality rate of 24% (Graham 2002). In the adult population, bacteraemia, relapses, and severe disease are a characteristic of NTS infection in the HIV-infected population (Cheesbrough et al. 1997). A major characteristic feature of NTS in AIDS patients in developing countries is the relapses that occur despite appropriate antibiotic therapy.

In Kenya over the last decade the two main serotypes of NTS-causing bacteraemia in children were *S.* Typhimurium and *S.* Enteritidis, which accounted for 70.8% of all NTS and these remained stably and almost equally distributed over this period. Genomic characterization of these two main serotypes by pulsed-field gel electrophoresis showed that within each serotype strains exhibited a minimal genetic diversity and that the strains appeared to be clonally related. During this period the NTS were multiply resistant to several commonly available antibiotics (Table 11.2) but remained fully susceptible to

Table 11.2 MICs and prevalence of resistance in NTS from paediatric admissions for antimicrobials tested during the 12-year study period

Antimicrobial agent MIC characteristic		Period of studyMIC in µg/mL		
		1994–1997 ($n=133$)	1998–2001 ($n=138$)	2002–2005 ($n=65$)
Amoxicillin	MIC range	0.75–256	0.5–256	0.5–>256
	MIC mode	256	256	2
	MIC50	256	1	2
	MIC90	256	256	4
	Resistant (%)	69.2	43.5	11
Co-trimoxazole	MIC range	0.05–32	0.03–32	0.05–32
	MIC mode	32	32	0.20
	MIC50	32	0.13	0.20
	MIC90	32	32	32
	Resistant (%)	68.4	36.2	13
Ciprofloxacin	MIC range	0.01–0.05	0.01–0.09	0.01–0.06
	MIC mode	0.01	0.01	0.01
	MIC50	0.01	0.01	0.01
	MIC90	0.02	0.02	0.02
	Resistant (%)	0	0	0
Cefotaxime	MIC range	0.03–1.5	0.06–4	0.02–0.4
	MIC mode	0.06	0.06	0.05
	MIC50	0.13	0.09	0.05
	MIC90	0.38	0.38	0.06
	Resistant (%)	0	0	0
Gentamicin	MIC range	0.13–16	0.1–24	0.1–8
	MIC mode	8	0.38	0.2
	MIC50	0.38	0.38	0.25
	MIC90	12	8	2
	Resistant (%)	12	7.2	4

cefotaxime and ciprofloxacin although MICs of these antibiotics had risen twofold to sixfold compared to fully susceptible strains.

The prevalence of NTS multiply resistant to all commonly available drugs including ampicillin, streptomycin, co-trimoxazole, chloramphenicol and tetracycline rose from 31% in 1993 to 42% at present, with concomitantly higher MICs of each drug, although the difference was not statistically significant (P-value >0.05). Molecular analysis of these isolates shows that resistance has been encoded on large self-transferable 100–110 kb plasmids among all the commonly isolated NTS serotypes. Higher resistance levels in the urban population may be due to higher usage in health clinics for treatment of diarrhoeal and respiratory infections and misuse of commonly available antimicrobials bought over the counter for self-treatment without prescription.

In contrast studies at a rural district hospital along the coast of Kenya have revealed a remarkable decreasing rate of prevalence of resistance especially to the most commonly available antimicrobials – amoxicillin, co-trimoxazole, gentamicin and chloramphenicol, respectively, during the 1994–2005 period (Kariuki et al. 2005). In an earlier study of susceptibility of Gram-negative bacteria from the same population, Bejon et al. (2005) initially observed the trend in decreasing resistance between 1998 and 2001. The decrease then was attributed to a change in blood culture sampling policy: from selective to all children admitted, thus including those with milder disease who are less likely to have received antibiotics before presentation. However, the trend in decreasing resistance levels in NTS has continued steadily after the year 2001 suggesting that changes in sampling criteria are not primarily responsible for the effect observed. Furthermore, from 1998, admission rates and laboratory processing of blood culture samples have not changed. Similarly in Uganda (Bachou et al. 2006) a study that sought to establish the magnitude of bacteraemia in severely malnourished children evaluated a total of 445 blood specimens that grew bacterial isolates; 58% were Gram-negative consisting of S. Typhimurium (27.6%) and S. Enteriditis (11.8%). The isolates were susceptible (> or = 80%) to ciprofloxacin, ceftriaxone and gentamicin, with low susceptibility to chloramphenicol, ampicillin (<50%) and co-trimoxazole (<25%).

During the 1980s, S. Typhimurium definitive phage type (DT) 104 with resistance to ampicillin, chloramphenicol, streptomycin, sulphonamides and tetracycline (ACSSuT-resistance type) emerged in the United Kingdom and subsequently disseminated throughout Europe and North America. By the 1990s, both DT 104 and non-DT 104 strains had acquired additional resistance to trimethoprim–sulphamethoxazole, ciprofloxacin and extended-spectrum cephalosporins (Threlfall 2000). This phenotype has now been widely disseminated to other regions of the world including South America where MDR blaCMY-2 S. Typhimurium has become the predominant phenotype of serovar Typhimurium in Mexico, and has caused severe enteric and systemic infections in children (Zaidi et al. 2007). Among food animals, the blaCMY-2 phenotype was only found in swine intestines, suggesting that the main selection pressure for MDR *S*. Typhimurium originated and persists in swine production systems in Mexico.

11.1.4 Emergence of Multidrug-Resistant S. enterica Serovar Typhi

Typhoid fever caused by *S. enterica* serovar Typhi (*S*. Typhi) is endemic in most parts of Central America (Fica et al. 1996), Southeast Asia (Mirza et al. 1996) and the Indian subcontinent (Nath et al. 2000), and recently increasing numbers of cases have been reported in Africa (Kariuki et al. 2000, Mills-Robertson et al. 2002). Globally, it is estimated that typhoid causes over 22 million cases of illness each year resulting in over 200,000 deaths (Crump et al. 2004). Although typhoid fever is largely considered an endemic disease, epidemics do occur frequently, often as a result of contamination or breakdown in water supplies and sanitation systems. In Tajikistan, for instance, recurrent outbreaks of multidrug-resistant (MDR) typhoid that evolved into the first reported epidemic of ciprofloxacin-resistant typhoid caused more than 24,000 cases and a case fatality rate of around 1% from 1996 to 1998 (Tarr et al. 1999). Over a 4-month period in 2004–2005, the Democratic Republic of the Congo reported to WHO more than 42,500 cases of typhoid, including 697 intestinal perforations and 214 deaths. Other recent outbreaks have been reported in Haiti (2003), Tajikistan (2003), Kochi and India (2006), among others (WHO 2006). Typhoid outbreaks caused by MDR serovar Typhi in most endemic parts Asia and the Indian subcontinent (Wain et al. 1997) have been well characterized. However, data from areas in Africa where there have been outbreaks are scarce. Since the first report of MDR serovar Typhi outbreaks, which occurred in Kenya in 1997–1999 when the prevalence of the MDR phenotype was 50–65%, continuous surveillance has shown that the prevalence of MDR serovar Typhi has been rising steadily and that, at present, 70–78% of all serovar Typhi isolates from blood cultures from the main referral hospital in Nairobi are MDR (Kariuki et al. 2004). This figure is much higher than the 52% prevalence for MDR serovar Typhi reported in outbreaks of typhoid in Ghana (Mills-Robertson et al. 2002) but is close to the high prevalence of MDR serovar Typhi noted in South African outbreaks (Coovadia et al. 1992). For resistant *S*. Typhi MICs for nalidixic acid and ciprofloxacin were, respectively, 5- and 10-fold higher than for sensitive strains.

Based on the increases in the prevalence of the MDR phenotype, it appears that MDR serovar Typhi strains have been spreading in many parts of Africa and are gradually replacing the fully sensitive strain type, probably due to their survival advantage over sensitive strains. This could be because the isolates, although showing raised quinolone MICs compared to fully sensitive isolates, are still within sensitive category on the basis of Clinical Laboratory Standards Institute (CLSI) (2005) criteria. Full resistance to the quinolones has normally been traced to sequential point mutations within the quinolone-resistance determining region (QRDR) of the chromosome of *S*. Typhi, with each mutation encoding higher levels of resistance. The increased quinolone MICs is not surprising since, within the last 5 years, the quinolones, especially norfloxacin and ciprofloxacin, have become the mainstay treatment choices for typhoid fever after resistance developed to previously commonly available antibiotics, including chloramphenicol and co-trimoxazole. The high prevalence of MDR serovar Typhi strains for which the nalidixic acid and ciprofloxacin MICs are high compared to those for sensitive strains means that soon resistance may emerge that renders these drugs ineffective, as has happened in Southeast Asia (Cooke et al. 2006), where even moderate rises in MICs have led to clinical treatment failure.

In Egypt (Srikantiah et al. 2006), incidence estimates of typhoid derived from hospital-based syndromic surveillance, which may not represent the population with typhoid fever, showed an estimated incidence of 59/100,000 persons/year. This study estimated that 71% of typhoid fever patients are managed by primary care providers. Of major concern was the high prevalence (29%) of multidrug-resistant (MDR) *S*. Typhi (resistant to chloramphenicol, ampicillin and co-trimoxazole). Similarly high incidence of multidrug-resistant *S*. Typhi was observed in studies in Nigeria (Itah and Uweh 2005) and Cameroon (Nkemngu et al. 2005) where one of the patients did not respond to treatment with ciprofloxacin, although the isolate was apparently susceptible to ciprofloxacin but resistant to nalidixic acid. It is important to note that nalidixic acid resistance has been identified as a key marker for reduced susceptibility to fluoroquinolones and sometimes this has led to treatment failures when the latter antibiotic was used to treat such typhoid cases. In several studies multidrug resistance in *S*. Typhi has also been associated with the presence of large transmissible plasmids in addition to point mutations in the QRDR in the chromosome of the bacterium (Hirose et al. 2002; Wain et al. 1997).

11.1.5 Emergence of Antibiotic-Resistant V. cholerae Outbreaks

Cholera is a widespread acute bacterial infection of the intestine caused by ingestion of food or water containing *V. cholerae*, toxigenic serogroups O1 or O139 (Salim et al. 2005). Symptoms include acute watery diarrhoea and vomiting which can result in severe dehydration or water loss. When left untreated, death can occur rapidly – sometimes within hours. Cholera is transmitted through contaminated food or drinking water, as well as from person to person through the faecal–oral route. Sanitary conditions in the environment play an important role since the *V. cholerae* bacterium survives and multiplies outside the human body and can spread rapidly where living conditions are crowded and water sources unprotected and where there is no safe disposal of faeces (WHO 2000). These conditions are common in poor countries and in crowded settings such as refugee camps. For example, in 1994 in a refugee camp in Goma, Democratic Republic of the Congo, a major epidemic took place. An estimated 58,000–80,000 cases and 23,800 deaths occurred within 1 month (Goma Epidemiology Group 1995). Cholera is now considered a reemerging disease because infections are appearing in novel communities or communities where the disease had been absent for many years, and the range of areas of endemicity is expanding. To date, nearly 200 serogroups of *V. cholerae* have been recorded, of which only the O1 and O139 strains have been associated with major epidemics due to their ability to express the cholera toxin (CT) which is rarely found in non-O1 and non-0139 serogroups (Anderson et al. 2004). In chronological order, seven cholera pandemics occurred in 1817–1823, 1826–1837, 1846–1862, 1864–1875, 1883–1896, 1899–1923 and 1961 to current. All the seven pandemics started in Asia with the first six starting in the Indian subcontinent. The seventh pandemic is believed to have originated in Indonesia and arrived in Africa and South America in the 1970s and in 1991, respectively (Karaolis et al. 1994). The first six pandemics were caused by the classical biotype which had been replaced by the O1 El Tor biotype (Ogawa and Inaba serovars) by 1961 (Kaper et al. 1995). The O139 serogroup first emerged as a pandemic strain in 1992, possibly through genetic exchange with *V. cholerae* O1 El Tor or with non-O1 strains (Faruque et al. 2004). Geographical areas once known to have experienced cholera epidemics can be characterized into three levels: cholera-free zones, endemic areas and epidemic areas. Cholera-free communities are defined as having no locally acquired infections. In cholera epidemic areas, the disease diminishes after an outbreak while in endemic areas, the disease does not disappear after an epidemic peak and returns in successive waves. Where cholera is endemic, cases tend to demonstrate distinct seasonal trends (Torres et al. 2001). After the sixth pandemic in 1950s, areas with true cholera endemicity had been reduced to southeastern India and Bangladesh. However, during the current seventh pandemic, the range of endemicity has expanded and includes vast areas on the African continent, and South and Central America (Borroto and Martinez-Piedra 2000).

Over the last two decades several outbreaks of cholera have been documented from Africa, and this is thought to be part of the ongoing seventh cholera pandemic. In a study to characterize 80 strains of *V. cholerae* O1 isolated during the Kenyan epidemic in 1998 and 1999, Scrascia et al. (2003) demonstrated that 61 strains from 25 outbreaks were ribotype B27, showed an identical and stable multiple resistances to chloramphenicol, streptomycin, co-trimoxazole and tetracycline, and produced the same cluster type of six randomly amplified DNA patterns. This uniformity of properties among outbreak strains from districts scattered over the entire area of the country provided strong genetic and epidemiological evidence that the predominant strains causing the epidemic had a clonal origin. Identification of strains, with traits typical of *V. cholerae* O1 strains active in

Somalia, from four outbreaks in North Eastern Province indicated that province as an epidemic zone where the Kenyan clone and Somali strains were overlapping and presumably competing. This epidemic has since spread into southern Sudan, Uganda and Tanzania. Ribotype B27 was first identified in Kolkata in 1993 and introduced into the western African country of Guinea-Bissau in 1994 (Dalsgaard et al. 1996). In 1995 and 1996, ribotype B27 was identified among *V. cholerae* O1 strains, causing cholera outbreaks in Senegal (Aidara et al. 1998). These reports suggest that this emerging ribotype has had a rapid spread into eastern Africa, with reports of sporadic outbreaks occurring in southern Sudan, parts of Somalia, western Kenya and along the coast as recently as 2006.

Antimicrobial resistance in *V. cholerae* was first observed in Tanzania (Mhalu et al. 1979) and later in Bangladesh (Glass et al. 1980). Resistance reflects the use or misuse of antibiotics in areas where frequent outbreaks of cholera occur or in cholera-endemic zones. Resistance to several or one of the following antibiotics has been observed: tetracycline, ampicillin, kanamycin, streptomycin, sulphona-mides, trimethoprim and gentamicin (Sack et al. 2004). In another recent surveillance study done in Angola, Ceccarelli et al. (2006) investigated *V. cholerae* O1 and *V. parahaemolyticus* clinical isolates, as well as *V. cholerae* O1 and *V. cholerae* non-O1 environmental isolates. All clinical isolates of *V. cholerae* O1 were resistant to ampicillin, chloramphenicol, trimethoprim, sulphamethoxazole and tetracycline. These multidrug-resistant isolates also contained a large con-jugative plasmid (p3iANG) with a set of three class 1 integrons harbouring *dfr*A15, *bla*P1 and *qac*H-*aad*A8 cassettes, which code for resistance to trimetho-prim, beta-lactams, quaternary ammonium compounds and aminoglycosides, clustered in a 19-kb region. Chloramphenicol (*cat*1), kanamycin (*aph*), sulpho-namide (*sul*2) and tetracycline (*tet*G) resistance genes were also carried on the plasmid within the same 19-kb region. A chromosomal integron containing the *dfr*A15 cassette was also revealed in *V. parahaemolyticus* strains. This study indicates that plasmids and integrons contributed mainly to the circulation of multiple-drug resistance determinants in *Vibrio* strains from Angola. Campos-Zahner et al. (2004) isolated multidrug-resistant environmental *V. cholerae* from different parts of Brazil between 1991 and 1999. The study reported that clinical and non-clinical O1 strains were more resistant to commonly used antibiotics than environmental non-O1 *V. cholerae* strains.

11.1.6 Outbreaks of Multidrug-Resistant Shigellosis in Africa and Other Developing Countries

Shigella dysenteriae type 1 was first isolated by Kiyoshi Shiga during a severe dysentery epidemic in Japan in 1896, when more than 90,000 cases were described with a mortality rate approaching 30% (Fasano 2000). Over the subsequent 50 years, the microbiology and epidemiology of *Shigella* species were clarified and the mechanisms by which the micro-organism causes disease have been intensively investigated. Humans are the only natural hosts for *Shigella* bacteria, and the transmission predominantly occurs by faecal–oral contact. In Africa, *Shigella* infections still predominates as a cause of sporadic bloody diarrhoea (Adeleye and Adetosoye 1993; Pitman et al. 1996), particularly *Sh. flexneri*. In their study in a rural population in Kenya, Brooks et al. (2003) observed that *Campylobacter* and *Shigella* were isolated with equal frequency from children less than 5 years old with bloody diarrhoea. In comparable studies of semi-urban Bolivian children less than 5 years old and Thai children 1–10 years of age with bloody diarrhoea, *Campylobacter* was isolated at least half as frequently as *Shigella* spp.

Shigellosis can occur in sporadic, epidemic and pandemic forms. Epidemics have been reported from Central American countries, Bangladesh, Sri Lanka, Maldives, Nepal, Bhutan, Myanmar and from the Indian subcontinent, Vellore, eastern India and Andaman and Nicobar islands. Plasmid profile of shigellae in Kolkata has shown a correlation between the presence of smaller plasmids and shigellae serotypes indicating epidemiological changes of the species (Sur et al. 2004). For a long time now antibiotics such as ampicillin, co-trimoxazole, chloramphenicol and tetracycline have been used for treatment of shigellosis, but their use is increasingly compromised by the emergence of resistance. For example, high prevalence of resistance to ampicillin (82%), chloramphenicol (73%), co-trimoxazole (88%) and tetracycline (97%) was detected in *Shigella* spp. isolated from children in Tanzania (Navia et al. 1999). Brooks et al. (2003) also found a high level of resistance to the antibiotics most commonly prescribed in Kenyan hospitals. Further investigation revealed that 74% of persons with bloody diarrhoea received antibiotics to which their isolate was not susceptible. Although these data were inadequate to assess the clinical impact of these findings (e.g. duration of bloody diarrhoea, mortality or bacterial shedding), nonetheless, strategies to improve prescription practices that use surveillance data to rationally guide more judicious antibiotic use warrant consideration if we are to preserve the effectiveness of commonly available antibiotics.

In another study on surveillance on antibiotic susceptibility in *Shigella* spp. (Kariuki et al. 1996) found all isolates multiply resistant to nearly all commonly available antibiotics including ampicillin, co-trimoxazole, co-amoxyclav, cefuroxime, chloramphenicol and tetracycline; however, no resistance to ciprofloxacin and minimal resistance to nalidixic acid were reported. Other studies also found little resistance to nalidixic acid and no resistance to ciprofloxacin among *Shigella* spp. in other parts of East Africa (Brooks et al. 2003). However, in areas where nalidixic acid has been introduced as the drug of choice to treat presumptive shigellosis, a marked increase in corresponding resistance has been observed (Mutwewingabo and Mets 1987). The ease with which antibiotics can be obtained without prescription may add further to selective pressure. Thus, although nalidixic acid is an attractive choice for treating bloody diarrhoea where antibiotic resistance limits other options, it should be used ideally only for illnesses most likely caused by *Shigella* or where *Shigella* infection could result in greater morbidity and increased risk of death, such as in immunosup-pressed individuals. If nalidixic acid is introduced for routine treatment of bloody diarrhoea, surveillance for antimicrobial susceptibility of local bacterial pathogens should be maintained.

11.1.7 Campylobacter *Species*

The bacteria *Campylobacter* spp. are part of the normal intestinal flora of poultry, cattle and a number of other food-producing and domestic animals, and are predominantly spread to humans in contaminated food. *Campylobacter jejuni* and, to a lesser extent, *C. coli* are responsible for most cases of campylo-bacteriosis, which is increasingly one of the most commonly detected bacterial enteric infections in poor resource settings in most developing countries (Samie et al. 2007). The incidence of human *Campylobacter* infections has increased markedly in both developed and developing countries worldwide and, more significantly, so has the rapid emergence of antibiotic-resistant *Campylobacter* strains, with evidence suggesting that the use of antibiotics, in particular the fluoroquinolones, as growth promoters in food animals and the veterinary industry is accelerating this trend. Although most infections are self-limiting and do not require antimicrobial chemotherapy, in a small proportion of individuals with severe or invasive disease treatment with erythromycin or a fluoroquinolone may be required. However, the prevalence of quinolone or macrolide-resistant *C. jejuni* and *C. coli* is increasing worldwide and several such isolates have been reported from Africa (Kinana et al. 2006).

Other developing world countries have also reported increasing incidences of multidrug-resistant *Campylobacter* infections, including resistance to fluoro-quinolones. For instance, multidrug resistance was detected in 27.5% of all *C. jejuni* isolated from a large farm in Estonia (Roasto et al. 2007), all of which were resistant to enrofloxacin. Multidrug resistance was significantly associated with enrofloxacin resistance ($P < 0.01$), and the use of enrofloxacin on the farms may have led to selection for multiresistant strains. In a separate study in Korea which investigated 232 retail stores for chicken raw meat (Hong et al. 2007), a total of 317 *Campylobacter* isolates examined showed that resistance to doxycycline was the most common (97.5%), followed by ciprofloxacin (95.9%), nalidixic acid (94.6%), tetracycline (94.6%), enrofloxacin (84.2%) and erythro-mycin (13.6%). A total of 93.4% showed multidrug (four or more antibiotics) resistance.

11.1.8 *Helicobacter Species and Gastritis*

Worldwide, *Helicobacter pylori* affects about 20% of persons below the age of 40 years and 50% of persons above the age of 60 years. Low socioeconomic status that is a hallmark in many resource-poor setting in developing countries correlate with *H. pylori* infection. Overall, there is a consistent pattern in most developing nations, where 70–90% of adults harbour the bacteria; most individuals acquired the infection as children, below the age of 10 years. A number of studies have shown that in developing countries >50% of children are infected by age of 10 years, the prevalence of infection rising to >80% in young adults (Misiewicz 1995). In contrast, in the majority of developed countries, children become infected at a rate of less than 1% a year (Graham et al. 1998.) Indeed, it is this significant difference in the rate of childhood acquisition of infection that is responsible for the differences in prevalence of *H. pylori* infection observed between developed and developing countries. A recent study in Soweto found 46% of children at 1 year and 100% of children at 12 years to be infected with *H. pylori* (Segal et al. 2001).

The adult prevalence starts at 20–50% in the developed world and 50–80% in sub-Saharan Africa (Skalsky 1995). A study in Ethiopia reported a prevalence of 81% of *H. pylori* infection in dyspeptic patients using histopathology and stool antigen test (Asrat et al. 2004). In a similar study in Kenya, Ogutu et al. (1998) reported a prevalence of 80.5% in dyspeptic patients with normal endoscopic findings, while Mohammed and Kumar (1991) found a prevalence of 84.6% in HIV-negative patients with upper gastrointestinal tract symptoms. These high prevalence rates of *H. pylori* infection indicate the seriousness of the infection in parts of the African continent.

Clarithromycin and metronidazole are the antibiotics most frequently used with amoxicillin for treatment of gastritis caused by *H. pylori*. Prevalence of resistance in *H. pylori* to these antibiotics is still low in Europe (0.8–9.1%) and in Asia (2–3%), and much higher in developing countries such as Kenya (18%). The situation is different after failure of a therapy using both clarithromycin and metronidazole, and up to 50% of strains may then harbour double resistance raising a major public health concern on effectiveness of commonly available antimicrobials (Lwai-Lume et al. 2004).

11.2 Conclusion

Since bacterial resistance is closely linked to antibiotic use it may be considered an inevitable antibiotic side effect. The potential of bacteria to spread and/or share the genetic resistance determinants in hospital wards, community and even globally makes the problem of resistance important not only for biomedical research but also for societal studies, health administrators and even governments. Human use of antibiotics has undoubtedly contributed to the resistance epidemic all over the world. Resistance in some bacterial species has driven the problem to the margin of there being no available clinically effective treatment for some infections. Although reports from several studies have alerted the medical and lay society for years about the problem of resistance, it has only been recently recognized as a dangerous epidemic. Several campaigns and projects to limit the antibiotic resistance are underway in the developed world at the moment. At the same time, of course, appropriate antibiotic use is a non-existing issue in most developing countries where people still buy a few doses of an antibiotic on a street market or over the counter without prescription. Although there is still a debate on the relationship between animal use of antibiotics and resistance in human pathogens, it is difficult to believe that low concentrations of antibiotics in contact with bacteria in our food and in the environment are harmless on the long term. From several studies especially in Europe it is estimated that about 20–50% of human and 40–80% agricultural antibiotic use is unnecessary or highly questionable. Studies showing epidemiological relationship between antibiotic consumption and bacterial resistance to antibiotics for several bacterial–drug combinations provide a scientific basis for the control measures in human medicine. In agriculture, a similar relationship is assumed and has been proven in a number of recent studies, but the results are less convincing.

Resistance to commonly available and affordable antimicrobials poses a major concern in the management of infections especially in resource-poor countries. Imprudent practices in the use of antibiotics by patients and for prophylaxis in animal husbandry undoubtedly contribute to the emergence of multidrug-resistant strains. Poor resource countries in particular are at greatest risk of drug-resistant infections as paucity of resources makes the purchasing of newer and more effective treatment agents difficult. Thus it would be important to carefully manage available therapeutic choices in order to ensure their effectiveness in the treatment of infections. Left unchecked, this problem will adversely affect our ability to treat and control infectious diseases.

References

Adeleye, I.A., and Adetosoye, A.I. 1993. Antimicrobial resistance patterns and plasmid survey of *Salmonella* and *Shigella* isolated is Ibadan, Nigeria. East Afr. Med. J. 70: 259–262.

Aidara, A., Koblavi, S., Boye, C.S., Raphenon, G., Gassama, A., Grimont, F., and Grimont, P.A.D. 1998. Phenotypic and genotypic characterization of *Vibrio cholerae* isolates from a recent cholera outbreak in Senegal: comparison with isolates from Guinea-Bissau. Am. J. Trop. Med. Hyg. 58: 163–167.

Anderson A.M., Varkey, J.B., Petti, C.A., Liddle, R.A., Frothingham, R., and Woods, C.W. 2004. Non-O1 *Vibrio cholerae* septicemia: case report, discussion of literature, and relevance to bioterrorism. Diagn. Microbiol. Infect. Dis. 49, 295–297.

Asrat, D., Nilsson, I., Mengistu, Y., Ashenafi, S., Ayenew, K., Al-Soud, W.A., Wadstrom, T., and Kassa, E. 2004. Prevalence of *Helicobacter pylori* infection among adult dyspeptic patients in Ethiopia. Ann. Trop. Med. Parasitol. 98: 181–189.

Bachou, H., Tylleskar, T., Kaddu-Mulindwa, D.H., and Tumwine, J.K. 2006. Bacteraemia among severely malnourished children infected and uninfected with the human immunodeficiency virus-1 in Kampala, Uganda. BMC Infect. Dis. 6: 160.

Bartoloni, A., Benedetti, M., Pallecchi, L., Larsson, M., Mantella, A., Strohmeyer, M., Bartalesi, F., Fernandez, C., Guzman, E., Vallejos, Y., Villagran, A.L., Guerra, H., Gotuzzo, E., Paradisi, F., Falkenberg, T., Rossolini, G.M., and Kronvalli, G. 2006. Evaluation of a rapid screening method for detection of antimicrobial resistance in the commensal microbiota of the gut. Trans. Roy. Soc. Trop. Med. Hyg. 100: 119–25.

Bii, C.C., Taguchi, H., Ouko, T.T., Muita, L.W., Wamae, N., and Kamiya, S. 2005. Detection of virulence-related genes by multiplex PCR in multidrug-resistant diarrhoeagenic *Escherichia coli* isolates from Kenya and Japan. Epidemiol. Infect. 133: 627–633.

Berkley, J.A., Lowe, B.S., Mwangi, I., Williams, T., Bauni, E., Mwarumba, S., Ngetsa, C., Slack, M.P., Njenga, S., Hart, C.A., Maitland, K., English, M., Marsh, K., and Scott, J.A. 2005. Bacteremia among children admitted to a rural hospital in Kenya. New Engl. J. Med. 352: 39–47.

Bejon, P., Mwangi, I., Ngetsa, C., Mwarumba, S., Berkley, J.A., Lowe, B.S., Maitland, K., Marsh, K., English, M., and Scott J.A. 2005. Invasive Gram-negative bacilli are frequently resistant to standard antibiotics for children admitted to hospital in Kilifi, Kenya. J. Antimicrob. Chemother. 56: 232–235.

Borroto, R.J., and Martinez-Piedra, R. 2000. Geographical patterns of cholera in Mexico, 1991–1996. Int. J. Epidemiol. 29: 764–772.

Brooks, J.T., Shapiro, R.L., Kumar, L., Wells, J.G., Phillips-Howard, P.A., Shi, Y.P., Vulule, J.M., Hoekstra, R.M., Mintz, E., and Slutsker, L. 2003. Epidemiology of sporadic bloody diarrhea in rural Western Kenya. Am. J. Trop. Med. Hyg. 68: 671–677.

Campos-Zahner, V., Avelar, K.E., Alves, R.M., Pereira, D.S., Vital, B.G., Freitas, F.S., Salles, C.S., and Karaolis, D.K. 2004. Genetic diversity and antibiotic resistance of clinical and environmental *Vibrio cholerae* suggests that many serogroups are reservoirs of resistance. Epidemiol. Infect. 135: 985–992.

Ceccarelli, D., Bani, S., Cappuccinelli, P., and Colombo, M.M. 2006. Prevalence of *aad*A1 and *dfr*A15 class 1 integron cassettes and SXT circulation in *Vibrio cholerae* O1 isolates from Africa. J. Antimicrob. Chemother. 58:1095–1107.

Cheesbrough, J.S., Taxman, B.C., Green, S.D., Mewa, F.I., and Numbi, A. 1997. Clinical definition for invasive Salmonella infection in African children. Pediatr. Infect. Dis. J. 16, 277–83.

Coovadia, A.Y.M., Gathiram, V., Bhamjee, A., Garratt, R.M., Mlisana, K., Pillay, N., Madlalose, T., and Short, M. 1992. An outbreak of multiresistant *Salmonella typhi* in South Africa. Q. J. Med. 82: 91–100.

Cooke, F.J., Wain, J., and Threlfall, E.J. 2006. Fluoroquinolone resistance in *Salmonella* Typhi. BMJ. 333:353–354.

Crump, J.A., Luby S.P. and Mintz, E.D. 2004. The global burden of typhoid fever. Bull WHO 82, 346–353.

Dalsgaard, A., Mortensen, H. F., Mølbak, F., Dias, Serichantalergs, F., and Echeverria, P. 1996. Molecular characterization of *Vibrio cholerae* O1 strains isolated during cholera outbreaks in Guinea-Bissau. J. Clin. Microbiol. 34: 1189–1192.

Duerink, D.O., Lestari, E.S., Hadi, U., Nagelkerke, N.D., Severin, J.A., Verbrugh, H.A., Keuter, M., Gyssens, I.C., and van den Broek, P.J. 2007. Determinants of carriage of resistant *Escherichia coli* in the Indonesian population inside and outside hospitals. J. Antimicrob. Chemother. 60: 377–384.

Faruque, S.M., Chowdhury, N., Kamruzzaman, M., Dziejman, M., Rahman, M.H., Sack, D. A., Nair, G.B., and Mekalanos, J.J. 2004. Genetic diversity and virulence potential of environmental *Vibrio cholerae* population in a cholera-endemic area. Proc. Natl. Acad. Sci. USA 101:2123–2128.

Fasano A., 2000. Intestinal infections: bacteria. In Pediatric Gastrointestinal Disease, ed. W.A. Walker, P. R. Durie, J. R. Hamilton. Ontario, BC Decker.

Fica, A.E., Prat-Miranda, S., Fernandez-Ricci, A., D'Ottone, K., and Cabello, F.C. 1996. Epidemic typhoid in Chile: analysis by molecular and conventional methods of *Salmonella typhi* strain diversity in epidemic (1977 and 1981) and non-epidemic 1990 years. J. Clin. Microbiol. 34:1701–1707.

Glass, R.I., Huq, I., Alim, A.R. and Yunus, M. (1980) Emergence of multiply antibiotic-resistance *Vibrio cholerae* in Bangladesh. J. Infect. Dis. 142, 939–942.

Goma Epidemiology Group. 1995. Public health impact of the Rwandan refugee crisis: what happened in Goma, Zaire in July 1994? Lancet 345:359–361.

Graham, D.Y., Breiter, J.R., Ciociola, A.A., Sykes, D.L., and McSorley, D.J. 1998. An alternative non-macrolide, non-imidazole treatment regimen for curing *Helicobacter pylori* and duodenal ulcers: ranitidine bismuth citrate plus amoxicillin. The RBC *H. pylori* Study Group. Helicobacter 3:125–131.

Graham, S.M. 2002. Salmonellosis in children in developing and developed countries and populations. Curr. Opin. Infect. Dis. 15:507–512.

Grenet, K. 2004. Antibacterial resistance, *Wayampis amerindians*, French Guyana. Emerg. Infect. Dis. 10:1150–1153.

Hart, C.A., and Kariuki, S. 1998. Antimicrobial resistance in developing countries. BMJ 317:649–650.

Hirose, K., Hashimoto, A., Tamura, K., Kawamura, Y., Ezaki, T., Sagara, H., and Watanabe, H. 2002. DNA sequence analysis of DNA gyrase and DNA topoisomerase IV quinolone resistance-determining regions of *Salmonella enterica* serovar Typhi and serovar Paratyphi A. Antimicrob. Agents Chemother. 46:3249–3252.

Hong, J., Kim, J.M., Jung, W.K., Kim, S.H., Bae, W., Koo, H.C., Gil, J., Kim, M., Ser, J., and Park, Y.H. 2007. Prevalence and antibiotic resistance of Campylobacter spp. isolated from chicken meat, pork, and beef in Korea, from 2001 to 2006. J. Food Prot. 70:860–866.

Itah, A.Y., and Uweh, E.E. 2005. Bacteria isolated from blood, stool and urine of typhoid patients in a developing country. Southeast Asian J. Trop. Med. Public Health 36:673–677.

Kaper, J.B., Morris, Jr, J.G. and Levine, M.M. 1995. Cholera. Clin. Microbiol. Rev. 8:48–86.

Karaolis, D.K., Lan, R., and Reeves P.R. 1994. Molecular evolution of the seventh-pandemic clone of *Vibrio cholerae* and its relationship to other pandemic and epidemic V. cholerae isolates. J. Bacteriol. 1176:6199–6206.

Kariuki, S., Muthotho, N., Kimari, J., Waiyaki, P., Hart, C.A., and Gilks, C.F. 1996. Molecular typing of multi-drug resistant *Shigella dysenteriae* type 1 by plasmid analysis and pulsed-field gel electrophoresis. Trans. Roy. Soc. Trop. Med. Hyg. 90:712–714.

Kariuki, S., Gilks, C.F., Kimari, J., Muyodi, J., Waiyaki, P., and Hart, C.A. 1997. Plasmid diversity of multi-drug-resistant *Escherichia coli* isolated from children with diarrhoea in a poultry-farming area in Kenya. Ann. Trop. Med. Parasitol. 91:87–94.

Kariuki, S., Gilks, C., Revathi, G. and Hart, C.A. 2000. Genotypic analysis of multidrug-resistant Salmonella enteric serovar Typhi, Kenya. Emerg. Infect. Dis. 6, 649–651.

Kariuki, S., Corkill, J.E., Revathi, G., Musoke, R., and Hart, C.A. 2001. Molecular characterization of a novel plasmid encoded cefotaximase (CTX-M-12) found in clinical *Klebsiella pneumoniae* isolates from Kenya. Antimicrob. Agents Chemother. 45:2141–2143.

Kariuki, S., Revathi, G., Muyodi, J., Mwituria, J., Munyalo, A., Mirza, S., and Hart, C.A. 2004. Characterization of multidrug-resistant typhoid outbreaks in Kenya. J. Clin. Microbiol. 42:1477–1482.

Kariuki, S., Revathi, G., Kariuki, N., Kiiru, J., Mwituria, J., and Hart C.A. 2005. Increasing prevalence of multidrug-resistant non-typhoidal salmonellae, Kenya, 1994–2003. Int. J. Antimicrob. Agents 25:38–43.

Kinana, A.D., Cardinale, E., Tall, F., Bahsoun, I., Sire, J.M., Garin, B., Breurec, S., Boye, C.S., and Perrier-Gros-Claude, J.D. 2006. Genetic diversity and quinolone resistance in *Campylobacter jejuni* isolates from poultry in Senegal. Appl. Environ. Microbiol. 72:3309–3313.

Legros, D., Paquet, C., Dorlencourt, F., and Le Saoult, E. 1999. Risk factors for death in hospitalized dysentery patients in Rwanda. Trop. Med. Int. Health 4:428–432.

Liu S.Y., Su L.H., Yeh Y.L., Chu C., Lai J.C., and Chiu C.H. 2007. Characterisation of plasmids encoding CTX-M-3 extended-spectrum beta-lactamase from Enterobacteriaceae isolated at a university hospital in Taiwan. Int J Antimicrob Agents. 29, 440–445.

Lwai-Lume, L., Ogutu, E.O., Amayo, E.O., and Kariuki, S. 2004. Drug susceptibility pattern of *Helicobacter pylori* in patients with dyspepsia at the Kenyatta National Hospital, Nairobi. East Afr. Med. J. 82:603–608.

Mohammed, R., and Kumar, S. 1991. Incidence of *Campylobacter pylori* in a consecutive series of surgical patients referred for endoscopy. J. R. Coll. Surg. Edinb. 36:422–423.

Misiewicz, J.J. 1995. Current insights in the pathogenesis of *Helicobacter pylori* infection. Eur. J. Gastroenterol. Hepatol. 7:701–703.

Mills-Robertson, F., Addy, M.E., Mensah, P., and Crupper, S.S. 2002. Molecular characterization of antibiotic resistance in clinical *Salmonella typhi* isolated in Ghana. FEMS Microbiol. Lett. 215:249–253.

Mirza, S.H., Beeching, N.J., and Hart, C.A. 1996. Multidrug resistant typhoid: a global problem. J. Med. Microbiol. 44:317–319.

Mhalu F.S., Mmari, P.W., and Ijumba, J. 1979. Rapid emergence of EI Tor *Vibrio cholerae* resistant to antimicrobial agents during first six months of fourth cholera in Tanzania. Lancet 1:345–347.

Mutwewingabo, A., and Mets, T. 1987. Increase in multiresistance of *Shigella dysenteriae* type 1 strain in Rwanda. East Afr. Med. J. 64:812–815.

Nath, G., Tikoo, A., Manocha, H., Tripathi, A.K., and Gulati A.K. 2000. Drug resistance in *Salmonella typhi* in North India with special reference to ciprofloxacin. J. Antimicrob. Chemother. 46:149–150.

Navia, M.M., Capitano, L., Ruiz, J., Vargas, M., Urassa, H., Schellemberg, D., Gascon, J., and Vila, J. 1999. Typing and characterization of mechanisms of resistance of Shigella spp. isolated from feces of children under 5 years of age from Ifakara, Tanzania. J. Clin. Microbiol. 37:3113–3117.

Nkemngu, N.J., Asonganyi, E.D., and Njunda, A.L. 2005. Treatment failure in a typhoid patient infected with nalidixic acid resistant *S. enterica* serovar Typhi with reduced susceptibility to ciprofloxacin: a case report from Cameroon. BMC Infect. Dis. 5:49.

Nys, S., Okeke, I.N., Kariuki, S., Dinant, G.J., Driessen, C., and Stobberingh, E.E. 2004. Antibiotic resistance of faecal *Escherichia coli* from healthy volunteers from eight developing countries. J. Antimicrob. Chemother. 54:952–955.

Ogutu, E.O., Kang'ethe, S.K., Nyabola, L., and Nyong'o, A. 1998. Endoscopic findings and prevalence of *Helicobacter pylori* in Kenyan patients with dyspepsia. East Afr. Med. J. 75:85–89.

Okeke, I. N., and Edelman, R. 2001. Dissemination of antibiotic-resistant bacteria across geographic borders. Clin. Infect. Dis. 33:364–369.

Pitman, C., Amali, R., Kanyerere, H., Siyasiya, A., Phiri,S., Phiri, A., Chakanika, I., Kampondeni, S., Chintolo, F.E., Kachenje, E., and Squire, S.B. 1996. Bloody diarrhea of adults in Malawi: clinical features, infectious agents, and antimicrobial sensitivities. Trans. R. Soc. Trop. Med. Hyg. 90:284–287.

Roasto, M., Juhkam, K., Tamme, T., Hörman, A., Häkkinen, L., Reinik, M., Karus, A., and Hänninen, M.L. 2007. High level of antimicrobial resistance in Campylobacter jejuni isolated from broiler chickens in Estonia in 2005 and 2006. J. Food Prot. 70:1940–1944.

Sack, A.D., Sack, R.B., Nair, G.B., and Siddique, A.K. 2004. Cholera. The Lancet 363:223–233.

Salim, A., Lan, R., and Reeves, R. 2005. *Vibrio cholerae* pathogenic clones. Emerg. Infect. Dis. 11:1758–1760.

Samie, A., Obi, C.L., Barrett, L.J., Powell, S.M., and Guerrant, R.L. 2007. Prevalence of *Campylobacter* species, *Helicobacter pylori* and *Arcobacter* species in stool samples from the Venda region, Limpopo, South Africa: Studies using molecular diagnostic methods. J. Infect. 54:558–566.

Scrascia, M., Forcillo, M., Maimone, F., and Pazzani, C. 2003. Susceptibility to rifaximin of *Vibrio cholerae* strains from different geographical areas. J. Antimicrob. Chemother. 52:303–305.

Segal, I., Ally, R., and Mitchell, H. 2001. Gastric cancer in sub-Saharan Africa. Eur. J. Cancer Prev. 10:479–482.

Shukla, I., Tiwari, R., and Agrawal, M. 2004. Prevalence of extended spectrum -lactamase producing *Klebsiella pneumoniae* in a tertiary care hospital. Indian J. Med. Microbiol. 22:87–91.

Skalsky, J.A. 1995. Dyspepsia and *Helicobacter pylori* infection in rural south west Cameroon. Trop. Doct. 25:92.

Soge, O.O., Adeniyi, B.A., and Roberts, M.C. 2006. New antibiotic resistance genes associated with CTX-M plasmids from uropathogenic Nigerian *Klebsiella pneumoniae*. J. Antimicrob. Chemother. 58:1048–1053.

Srikantiah, P., Girgis, F.Y., Luby, S.P., Jennings, G., Wasfy, M.O., Crump, J.A., Hoekstra, R.M., Anwer, M., and Mahoney, F.J. 2006. Population-based surveillance of typhoid fever in Egypt. Am. J. Trop. Med. Hyg. 74:114–119.

Sur, D., Ramamurthy, T., Deen, J., and Bhattacharya, S.K. 2004. Shigellosis: challenges & management issues. Indian J. Med. Res. 120:454–462.

Su, L.H., Wu, T.L., Chia, J.H., Chu, C., Kuo, A.J., and Chiu, C.H. 2005. Increasing ceftriaxone resistance in *Salmonella* isolates from a university hospital in Taiwan. J. Antimicrob. Chemother. 55:846–852.

Tarr, P.E., Kuppens, L., Jones, T.C., Ivanoff, B., Aparin, P.G., and Heymann, D.L. 1999. Considerations regarding mass vaccination against typhoid fever as an adjunct to sanitation and public health measures: potential use in an epidemic in Tajikistan. Am. J. Trop. Med. Hyg. 61:163–170.

Threlfall, E.J. 2000. Epidemic *Salmonella typhimurium* DT 104-a truly international multi-resistant clone. J. Antimicrob. Chemother. 46:7–10.

Torres, M.E., Pirez, M.C., Schelotto, F., Varela, G., Parodi, V., Allende, F., Falconi, E., Dell'Acqua, L., Gaione, P., Mendez, M.V., Ferrari, A.M., Montano, A., Zanetta, E., Acuna, A.M., Chiparelli, H., and Ingold, E. 2001. Etiology of children's diarrhea in Montevideo, Uruguay: associated pathogens and unusual isolates. J. Clin. Microbiol. 39:2134–2139.

Touati A., Benallaoua S., Djoudi F., Madoux J., Brasme L., and De Champs, C. 2007. Characterization of CTX-M-15-producing *Klebsiella pneumoniae* and *Escherichia coli* strains isolated from hospital environments in Algeria. Microb. Drug Resist. 13:85–89.

van den Bogaard, A.E., and Stobberingh, E.E. 2000. Epidemiology of resistance to antibiotics. Links between animals and humans. Int. J. Antimicrob. Agents 14:327–335.

van de Mortel, H.J., Jansen, E.J., Dinant, G.J., London, N., Palacios Pru, E., and Stobberingh, E.E. 1998. The prevalence of antibiotic-resistant faecal *Escherichia coli* in healthy volunteers in Venezuela. Infect. 26:292–297.

Wain, J., Hoa, N.T., Chinh, N.T., Vinh, H., Everett, M.J., Diep, T.S., Day, N.P., Solomon, T., White, N.J., Piddock, L.J., and Parry, C.M. 1997. Quinolone-resistant *Salmonella typhi* in Viet Nam: molecular basis of resistance and clinical response to treatment. Clin. Infect. Dis. 25:1404–1410.

Wise, R., Hart, T., Cars, O., Streulens, M., Helmuth, R., Huovinen, P., and Sprenger, M. 1998. Antimicrobial resistance is a major threat to public health. BMJ 317:609–610

World Health Organization. 1998. Use of quinolones in food animals and potential impact on human health. Report of a WHO meeting in Geneva. Switzerland.

World Health Organization. 1999. Containing antimicrobial resistance: review of the literature and report of a WHO workshop on the development of a global strategy for the containment of antimicrobial resistance. Geneva, Switzerland.

World Health Organization. 2000. WHO Report on Global Surveillance of Epidemic-prone Infectious Diseases. WHO/CDS/CSR/ISR/2000.1.

World Health Organization. 2006. Background document: The diagnosis, treatment and prevention of typhoid fever. Geneva, Switzerland.

Yang, Y.J., Huang, M.C., Wang, S.M., Wu, J.J., Cheng, C.P., and Liu, C.C. 2002. Analysis of risk factors for bacteremia in children with nontyphoidal *Salmonella* gastroenteritis. Eur. J. Clin. Microbiol. Infect. Dis. 21:290–293.

Zaidi, M.B., Leon, V., Canche, C., Perez, C., Zhao, S., Hubert, S.K., Abbott, J., Blickenstaff, K., and McDermott, P.F. 2007. Rapid and widespread dissemination of multidrug-resistant blaCMY-2 *Salmonella Typhimurium* in Mexico. J. Antimicrob. Chemother. 60:398–401.

Chapter 12
Bacterial-Resistant Infections in Resource-Limited Countries

Alessandro Bartoloni and Eduardo Gotuzzo

Abstract Considering that antibiotics play a crucial role in reducing morbidity and mortality due to bacterial infections, antibiotic resistance is a major problem in resource-limited countries (RLCs) where there is a high burden of infectious diseases, resistance rates are even higher than in industrialized countries, and therapeutical options are often unavailable or too expensive. Multidrug-resistant organisms – e.g. *Streptococcus pneumoniae, Salmonella* Typhi, *Shigella* spp., *Neisseria gonorrhoeae, Mycobacterium tuberculosis* – have been increasingly documented. Many factors contributing to antibiotic resistance in RLCs are strongly related to poverty: lack of knowledge or information of health-care professionals, lack of laboratory facilities, inadequate access to health system, lack of money available to pay for the appropriate amount of antibiotics, dispensation of drugs by untrained people, availability of substandard and counterfeit drugs, etc. Moreover, in RLCs, transmission of resistant bacteria is facilitated by person-to-person contact, through contaminated food, unsafe water or by vectors. An understanding of this complex and multifactorial scenario is crucial to develop any containment strategy based on the promotion of a correct use of antibiotics and infection control measures.

12.1 Introduction

The problem of antibiotic resistance represents a global public health problem but is particularly serious in resource-limited countries (RLCs) where bacterial infections remain among the major causes of morbidity and mortality, especially in childhood (WHO 2001a; WHO 2002).

The World Health Organization (WHO) states that infectious diseases are the world's leading cause of premature death. Infectious diseases account for 34% of deaths in RLCs. Of 57 million deaths in 2002, nearly 11 million were

A. Bartoloni (✉)
Infectious Diseases Unit, Department of Critical Care Medicine and Surgery, Careggi Hospital, University of Florence, Florence, Italy
e-mail: bartoloni@unifi.it

A. de J. Sosa et al. (eds.), *Antimicrobial Resistance in Developing Countries*,
DOI 10.1007/978-0-387-89370-9_12, © Springer Science+Business Media, LLC 2010

among children under 5 years of age, and 98% of them were in low- and middle-income countries and mostly due to infectious diseases (WHO 2007).

Bacterial agents play a mayor role among the leading infectious killers causing acute respiratory infections (ARIs), diarrhoeal diseases, tuberculosis and other less-frequent serious infections such as meningitis, typhoid fever, plague, melioidosis. Bacteria cause also other common infections such as urinary tract infections, sexually transmitted infections (STIs), skin and soft tissue infections (SSTIs) and are the major responsible for health-care associated infections.

ARIs constitute the leading cause of death of infectious aetiology, killing more than 3.5 million people a year. Most of these deaths (>80%) occur in RLCs (WHO 2007). *Streptococcus pneumoniae* is identified as the leading cause of bacterial pneumonia and pneumococcal bacteraemia is considered an important cause of child mortality (Levine et al. 2006). Globally, *S. pneumoniae* is estimated to cause 1 million deaths each year in children under 5 years old (Schrag, Beall and Dowell 2001). HIV infection increases risk for pneumococcal disease 20–40 fold (O'Dempsey et al. 1996; Madhi et al. 2000).

Diarrhoea, a typical disease of poverty, is one of the most common diseases afflicting children under 5 years of age and accounts for 15–30% of under five deaths in childhood (Kosek et al. 2003). In RLCs estimates of the annual incidence of diarrhoea vary from 3 to 10 episodes, while the incidence in industrialized countries is less than 1 per annum (Hart 2003). Bacterial enteropathogens such as salmonellae, shigellae, *Campylobacter jejuni*, enteroinvasive *Escherichia coli* (EIEC), Shiga-toxin (ST)-producing *E. coli* (STEC), enteroaggregative *E. coli* (EAEC) are the main cause of inflammatory diarrhoea, while enterotoxigenic *E. coli* (ETEC), enteropathogenic *E. coli* (EPEC), *Vibrio cholerae* are responsible for non-inflammatory diarrhoea.

Bacterial meningitis is a common medical emergency in many RLCs and is associated with high mortality rate and high risk of neurological sequelae. *Streptococcus pneumoniae*, *Haemophilus influenzae* and *Neisseria meningitidis* are the three major pathogens outside the neonatal period causing over 90% of acute bacterial meningitis. Neonatal meningitis may also be caused by these organisms but other bacteria such as *E. coli*, *Streptococcus agalactiae* (group B streptococcus), *Klebsiella* spp., *Salmonella* spp. and *Listeria monocytogenes* tend to predominate (Nel 2000).

WHO estimates that 340 million new cases of STIs have occurred worldwide in 1999. The largest number of new infections occurred in the region of South and Southeast Asia, followed by sub-Saharan Africa, Latin America and the Caribbean. However, the highest rate of new cases has occurred in sub-Saharan Africa (WHO 2001b). STIs represent a major public health problem in RLCs for the possible serious complications, particularly among women and children, and because they facilitate the transmission of HIV through sexual contact. Complications are partly due to lack of available diagnostic facilities and proper therapy. Bacterial agents such as *N. gonorrhoeae*, *Chlamydia trachomatis*, *Treponema pallidum*, *Haemophilus ducreyi*, *Ureaplasma urealyticum*, *Mycoplasma hominis* and *Klebsiella granulomatis* account for a large part of STIs.

SSTIs are among the main causes of consultation for health system members in RLCs. These infections may develop on normal skin but more commonly appear as a complication of wounds, arthropod bites, burns, dermatitis or other skin disease. Malnutrition and poor hygiene are predisposing factors in their development. The most common skin bacterial infections are caused by *Streptococcus pyogenes* and *Staphylococcus aureus*.

Staphylococcus aureus accounts also for 75–90% of pyomyositis, also known as "tropical piomyositis", an acute infection of skeletal muscle with localized abscess formation. This infection is endemic in tropical countries, accounting for 2.2–4% of surgical admission (Grose 2004).

Multidrug-resistant (MDR) tuberculosis is one of the most serious public health problem, but focussing this issue is beyond the scope of this chapter.

Bacteria are the most common pathogens of health-care-associated infections. Although very limited information is available from RLCs on health-care-associated infections, failures in the system of care are responsible for high infection rates, especially in hospital nurseries (Zaidi et al. 2005). Lack of essential equipment and supplies (soap, wash basin, clean water, etc.), failures in sterilization/disinfection of instruments, inadequate hand hygiene and glove use, overcrowded and understaffed nurseries, lack of knowledge, training and competency regarding infection control practice are the major critical points in the causation of health-care-associated infections in RLCs. The common nosocomial bacterial pathogens isolated in industrialized countries – coagulase-negative *Staphylococcus* spp., *S. aureus*, *Enterococcus* spp., *Pseudomonas aeruginosa*, *Klebsiella pneumoniae*, *Enterobacter* spp., *Acinetobacter* spp., etc. – are now of increasing importance also in RLCs (Zaidi et al. 2005).

Considering that antibiotics play a crucial role in reducing morbidity and mortality due to bacterial infections, antibiotic resistance is a major problem in RLCs where there is a high burden of infectious diseases, resistance rates are even higher than in industrialized countries, and therapeutical options are often unavailable or too expensive (WHO 1997; Byarugaba 2004; Lynch et al. 2007).

12.2 Micro-organisms

12.2.1 Streptococcus pneumoniae

Streptococcus pneumoniae colonizes the upper respiratory tract commonly in individuals, especially children, without causing disease, and is easily transmitted from person to person. The rapid pneumococcal acquisition observed in RLCs is associated with a high risk of serious infections early in infancy (Saha et al. 2003; Granat et al. 2007).

Streptococcus pneumoniae is the most frequently isolated respiratory pathogen in community-acquired pneumonia and also a major cause of meningitis, otitis media and sinusitis, resulting in high rates of morbidity and mortality in paediatric and adult patients. Among HIV-infected people, especially those living in RLCs with little access to highly active antiretroviral therapy (HAART), the burden of pneumococcal disease is greatly increased and associated with an extreme level of mortality (Klugman et al. 2007).

In the preantibiotic era, *S. pneumoniae* was defined as "Captain of the Men of Death" for the high case-fatality rate (25–35%) of pneumococcal pneumonia, especially if bacteraemia was present (80%) (Osler 1901). The advent of penicillin (at that time all *S. pneumoniae* were inhibited by 0.02 μg/ml or less) reduced dramatically mortality to rates below 10% and this drug has been considered the standard treatment for pneumococcal infection for nearly half a century. Although pneumococci with reduced susceptibility to penicillin were obtained in animal model shortly after the introduction of penicillin in 1940 (Schmidt and Sesler 1943), the first clinical isolate was not reported until 1967 (Hansman and Bullen 1967). In the late 1960s, strains with moderate penicillin resistance (MIC 0.1–1.0 μg/ml) were isolated in Australia and New Guinea (Hansman et al. 1971). In the 1970s, an epidemic of high-level penicillin-resistant pneumococci (MIC >1 μg/ml) that were also resistant to chloramphenicol and other drugs was reported in South Africa (Appelbaum et al. 1977). Patients with meningitis caused by those strains failed to respond to penicillin or chloramphenicol treatment. Despite limitations in available data from the majority of African, Latin American and Asian countries, the nonsusceptibility to penicillin among *S. pneumoniae* is a relevant and emerging problem in many countries (Table 12.1).

Resistance to beta-lactams in *S. pneumoniae* is an example of antibiotic target modification. It results from modifications of one or more of the five major penicillin-binding proteins (PBPs) affecting the affinity of these molecules for beta-lactam antibiotics (Hakenbeck and Coyette 1998). A variety of mosaic genes, resulting from genetic recombination, has been described, with the level and degree of resistance determined by the number and nature of recombinations. In general, modifications of more than one PBP are necessary for reduced susceptibility to penicillin, whereas high-level resistance (MIC≥2 μg/ml) usually implies alteration of all major PBPs (Barcus et al. 1995).

Considering the genetic bases of resistance (based on recombination events with foreign DNA rather than on single-point mutations or resistance plasmid acquisition), resistant pneumococcal infections result from the acquisition of resistant strains from the community rather than from the development of resistance during the course of antibiotic treatment (Schrag et al. 2001). Epidemiological data suggest that the emergence and diffusion of resistant pneumococci are the result of a clonal spread, and that asymptomatic nasopharyngeal carriers may represent an important reservoir of resistant strains in the community.

The clinical relevance of the *in vitro* penicillin nonsusceptibility remains difficult to measure (Feldman 2004; Nuermberger and Bishai 2004). Clinical and pharmacodynamic data suggest that pneumococcal resistance does not have impact on the outcome of antibiotic treatment for subjects with pneumococcal pneumonia (Nuermberger and Bishai 2004). Pulmonary infections due

Table 12.1 Percentage of nonsusceptibility to penicillin among *S. pneumoniae* in various countries

Region/country		Reference
Africa		
Gambia	7%	Adegbola et al. (2006)
Marocco	9%	Benbachir et al. (2001)
Mozambique	14%	Vallès et al. (2006)
Ivory Coast	22%	Benbachir et al. (2001)
Senegal	62%	Benbachir et al. (2001)
South Africa	74%	Felmingham et al. (2007)
Latin America		
Argentina	27%	Felmingham et al. (2007)
Brazil	34%	Felmingham et al. (2007)
Mexico	57%	Felmingham et al. (2007)
Asia		
Korea	80–81%	Song et al. (1999) and Inoue et al. (2004)
Vietnam	61%	Song et al. (1999)
Thailand	58–70%	Song et al. (1999) and Inoue et al. (2004)
Sri Lanka	41%	Song et al. (1999)
Taiwan	39%	Song et al. (1999)
Indonesia	21%	Song et al. (1999)
China	10%	Song et al. (1999)
Malaysia	9%	Song et al. (1999)
India	4%	Song et al. (1999)

to strains for which penicillin MICs are in the intermediate range may be treatable with penicillin, better if with higher dosage judging from pharmacokinetic/pharmacodynamic parameters, although cefotaxime or ceftriaxone are the preferred parenteral agents (Jacobs 2001; Feldman 2004). With regard to oral treatment, high-dose amoxicillin (1 g t.i.d. or 2 g b.i.d. for amoxicillin clavulanate) should be effective against ≥70% of penicillin-resistant pneumococci. Limited data are available about the impact of pneumococcal pneumonia due to isolates with high-level resistance to penicillin and the conflicting results do not allow to demonstrate worse outcomes following infection by resistant pneumococci (Metlay 2002). Unfortunately, although the burden of pneumococcal pneumonia in RLCs is significantly higher than in industrialized countries, the real impact of penicillin nonsusceptibility in that setting remains yet to be clearly assessed.

Different from pneumonia, treatment failures due to pneumococcal resistance have been documented for meningitis and otitis media (Dowell et al. 1999). As far as pneumococcal meningitis is concerned, the level of penicillin attained in the cerebral spinal fluid (CSF), around 1 μg/ml, does not sufficiently exceed the MIC of the penicillin nonsusceptible strain (Fernández Viladrich

2004). In case of moderately penicillin-resistant pneumococci, the third-generation cephalosporins (ceftriaxone and cefotaxime) remain active, while in case of penicillin-resistant and cefotaxime-resistant pneumococcal meningitis a regimen of high-dose cefotaxime or ceftriaxone plus vancomycin (or linezolid) is probably the best choice.

The phenomenon of resistance in *S. pneumoniae* is not limited to beta-lactam antibiotics. Resistance to trimethoprim–sulphamethoxazole (TMP–SMZ) has been reported worldwide, including RLCs (Table 12.2). Resistance to TMP–SMZ is conferred by alternative dihydrofolate reductase (DHFR) and dihydropteroate synthase (DHPS), encoded by *dhf* and *sul* genes mosaics and/or mutant alleles (Maskell et al. 1997; Adrian and Klugman 1998; Pikis et al. 1998). The common use of TMP–SMZ in RLCs is probably the key factor driving pneumococcal resistance. In HIV-infected subjects the impact of

Table 12.2 Percentage of nonsusceptibility to trimethoprim–sulphamethoxazole (TMP–SMZ), erythromycin (EM), fluoroquinolones (FQ) among *S. pneumoniae* in various countries

Region/country	Antibiotic			Reference
	TMP–SMZ	EM	FQ	
Africa				
Marocco	15%	4%		Benbachir et al. (2001)
Tunisia	20%	32%		Benbachir et al. (2001)
Senegal	29%	11%		Benbachir et al. (2001)
South Africa	37%		5%*	Jacobs et al. (2003)
Mozambique	45%	1%		Vallès et al. (2006)
Kenya	51%	1%	2.5%*	Jacobs et al. (2003)
Ivory Coast	60%	53%		Benbachir et al. (2001)
Gambia	95%			Adegbola et al. (2006)
Latin America				
Argentina		11%		Mendes et al. (2003)
Brazil	58%		3%*	Jacobs et al. (2003)
Colombia	44%	4%		Agudelo et al. (2006)
Mexico	60%	25–28%	6.5%*	Jacobs et al. (2003)
Asia				
Korea	67%	81%	3%[§]	Jacobs et al. (2003)
Vietnam	92%	74%		Song et al. (1999)
Thailand	67%	41%	3%[§]	Jacobs et al. (2003)
Sri Lanka	58%	27%		Song et al. (1999)
Taiwan	69%	90%		Song et al. (1999)
Indonesia	55%	48%		Song et al. (1999)
China	63%	35%	3%[§]	Jacobs et al. (2003)
Malaysia	21%	3%		Song et al. (1999)
India	39%	0%		Song et al. (1999)

*Ofloxacin.
[§]Levofloxacin.

TMP–SMZ prophylaxis is of particular concern (Madhi et al. 2000; Klugman et al. 2007), and the use of sulphadoxine/pyrimethamine to treat malaria can lead to an increase in TMP–SMZ-resistant pneumococci (Feikin et al. 2000; Vallès et al. 2006). The rate of resistance to TMP–SMZ is much greater in penicillin-resistant than penicillin-susceptible strains. WHO guidelines for the ARI management indicate TMP–SMZ as the recommended outpatient antibiotic treatment. Despite the great concern regarding the widespread of TMP–SMZ resistance in RLCs and the report of bacteriological failure of this agent in the treatment of pneumonia (Klugman 2002), no data are available about the real impact of this phenomenon on the clinical outcome of pneumonia due to nonsusceptible *S. pneumoniae*.

Macrolides (azithromycin, clarithromycin and erythromycin) represent first-line agents for community-acquired pneumonia due to their activity against most pathogens including atypical agents (Mandell et al. 2007). A number of studies show that increased macrolide consumption is connected to increased macrolide resistance (Moreno et al. 1995; Pihlajamäki et al. 2001; Bergman et al. 2006). In industrialized countries, macrolide resistance increased markedly in *S. pneumoniae* isolated in the last years, concomitant to penicillin resistance (Nuermberger and Bishai 2004). In RLCs, macrolides are less used than in the industrialized countries, because of their higher cost respect to that of other drugs such as amoxicillin or TMP–SMZ, and probably for this reason the prevalence of resistance is lower in these countries. However, the implementation of mass chemoprophylaxis against trachoma in a number of African countries may represent a risk for the emergence and spread of resistant pneumococci (Leach et al. 1997). Macrolide resistance in pneumococci is commonly associated with resistance to beta-lactams and TMP–SMZ and the mass use of azithromycin may have a negative impact on the efficacy of standard treatment for ARI (Schrag et al. 2001).

Two mechanisms are responsible for resistance of *S. pneumoniae* to macrolides in the vast majority of cases: a modification of the ribosomal target site (ribosomal methylation) conferred by the *erm*B gene, which may be constitutively or inducibly expressed, and an active drug efflux mechanism mediated by an efflux pump encoded by the *mef*A gene (Nuermberger and Bishai 2004). The first mechanism is associated with a higher level of resistance, and cross-resistance with clindamycin, while the second results in low- to middle-level resistance. The prevalence of the two mechanisms varies by geography but no data are available for RLCs.

In contrast to the scientific literature concerning penicillin, macrolide treatment failures resulting from antibiotic resistance have been documented (Baquero 1999; Hyde et al. 2001). From the analysis of the reported treatment failures it is evident that both mechanisms of macrolide resistance may have a clinical impact, and that the MIC level is an important predictor for the clinical outcome (Klugman 2002). Treatment failures have generally occurred in infections caused by isolates with MIC of 8 µg/ml or higher, but failures have also been reported in cases with macrolides MIC between 2 and 4 µg/ml (Lonks et al. 2002; Musher et al. 2002).

The very limited available data from RLCs indicate that resistance to fluoroquinolones among *S. pneumoniae* is still low in many countries, but an alarming increasing trend in resistance rates observed in some areas (Ho et al. 2001; Jones et al. 2003) and the clonal expansion of resistant isolates are a matter of concern and indicate the need to monitor the trends in development of resistance (McGee et al. 2001). The most common mechanism of clinically significant levels of fluoroquinolone resistance is through alterations of target enzymes (DNA topoisomerase II) involving DNA synthesis. These alterations are conferred by stepwise accumulation of mutations in genes that code for these enzymes and result in reduced drug affinity (Hooper 1999). The widespread use of fluoroquinolones has been associated with the emergence and diffusion of fluoroquinolone resistance (Goldstein and Garabedian-Ruffalo 2002) and surveillance studies in RLCs are important to monitor the evolution of antibiotic susceptibility of *S. pneumoniae* to fluoroquinolones.

Pneumococcal disease, especially when complicated by HIV infection and antibiotic resistance, represents a major public health issue in many RLCs. In this perspective, the implementation of conjugate pneumococcal vaccination in these countries is an urgent need (Levine et al. 2006).

12.2.2 **Staphylococcus aureus**

Staphylococcus aureus has long been recognized as one of the most important human pathogens. It colonizes and infects both hospitalized and healthy people in the community (Waldvogel 2000). It is responsible for a wide range of infections ranging from skin and soft tissue infections to bacteraemia, infections of the central nervous system, bone and joints, skeletal muscles, respiratory and urinary tracts, and infections associated with intravascular devices and foreign bodies. Before the introduction of penicillin in the early 1940s, invasive *S. aureus* infections had mortality over 90% (Smith and Vickers 1960). At that time, *S. aureus* strains were fully susceptible and penicillin treatment cured formerly untreatable infections (Abraham et al. 1941), but resistance, mediated by a plasmid-borne beta-lactamase capable of degrading the antibiotic, quickly emerged and spread. The same history happened to other antibiotics (streptomycin, tetracycline, chloramphenicol, erythromycin) soon after their introduction. The introduction of methicillin in 1961 was rapidly followed by reports of methicillin resistance in both hospital- and community-acquired *S. aureus* isolates (Appelbaum 2006). The resistance to methicillin results from the acquisition of the *mec*A gene, carried on a mobile genetic element known as staphylococcal cassette chromosome (SCC) *mec*, which encodes an altered low-affinity penicillin-binding protein 2A (PBP2A) (Hiramatsu et al. 2002). Methicillin-resistant *S. aureus* (MRSA) are cross-resistant to all other beta-lactam antibiotics and usually resistant to a wide range of antibiotics. Glycopeptide antibiotics (vancomycin and teicoplanin), linezolid, quinupristin–dalfopristin, daptomycin, tigecycline are often needed to treat infections caused by MRSA.

Since their emergence, MRSA have been associated with exposure to health-care settings where the prevalence of MRSA infections represents a challenge for infection control professionals and contributes to a high economic burden (Shorr 2007). However, the past few years have witnessed an extremely rapid increase in community-acquired MRSA infections (Saïd-Salim et al. 2003). Community-acquired MRSA are typically susceptible to most anti-staphylococcal antibiotics, have a different genetic background, and can be highly virulent (Zetola et al. 2005).

Data concerning resistance of *S. aureus* to antibiotics in RLCs are still extremely limited. In Africa, a study conducted between 1996 and 1997 documented rates of MRSA as 30% in Nigeria, 28% in Kenya, 21% in Cameroon, 17% in Ivory Coast, 14% in Morocco, 12% in Senegal, 8% in Tunisia and 5% in Algeria (Kesah et al. 2003). A more recent surveillance study conducted in a tertiary hospital of Tanzania showed a very low (2%) prevalence of methicillin resistance (Blomberg et al. 2004).

As far as Latin America is concerned, a report from the SENTRY surveillance program (1997–1999) indicates a rate of methicillin resistance among *S. aureus* isolates as 45% in Chile, 43% in Argentina, 34% in Brazil, 11% in Mexico, 9% in Colombia (Diekema et al. 2001). In a more recent study carried out in three major regional hospitals in Trinidad and Tobago, the prevalence of methicillin-resistant *S. aureus* was found to be 13% (Akpaka et al. 2006).

Data from the SENTRY surveillance program show much higher rates of methicillin resistance in two Asian countries, 74% in Hong Kong and 61% in Taiwan, respectively (Diekema et al. 2001).

12.2.3 Salmonella enterica *serotype Typhi and Paratyphi*

Salmonella enterica serotype Typhi (*S.* Typhi) and Paratyphi (*S.* Paratyphi) are Gram-negative bacilli restricted to humans, and cause typhoid fever, a severe systemic febrile illness. Transmission is person to person by faecal oral route with contaminated water and food. Typhoid fever was a seriously life-threatening disease in the preantibiotic era.

Salmonella Typhi and *S.* Paratyphi infections are recognized as a public health problem causing 21 million cases annually with 220,000 deaths worldwide (Crump et al. 2004). In fact, the infection has almost disappeared from developed world but it still remains endemic in many RLCs due to poor sanitary conditions.

Since its introduction 40 years ago, chloramphenicol has been the drug of choice for the treatment of typhoid fever. From 1972 resistance strains appeared in Southeast Asia and Latin America while ampicillin, TMP–SMZ and tetracycline still remained good therapeutic options (Parry 2004). At the end of 1980s and beginning of the 1990s, the spread of MDR strains of *S.* Typhi (resistant to ampicillin, TMP–SMZ, chloramphenicol, streptomycin, tetracycline) caused outbreaks in Asia and Africa associated with an increased severity of the disease. Decreased susceptibility to these antibiotic agents is often conferred by transferable, located into plasmid or other mobile genetic elements (e.g. transposons, gene cassettes inserted into integrons) (Parry 2004). Resistant genes are often carried on integrons that contain gene cassettes that can be transferred to other integrons, spreading resistance. Four classes of integrons have been identified to date, but class 1 integrons seems mainly involved with multiple antibiotic resistances.

Since the emergence of first-line antibiotic-resistant strains, quinolones have become drugs of choice for the treatment of typhoid fever. Unfortunately outbreaks of *S.* Typhi and *S.* Paratyphi resistant to nalidixic acid and with decreased susceptibility to fluoroquinolones appeared in Asia, mostly in Vietnam (Chau et al. 2007). Fluoroquinolones act by inhibiting and blocking the DNA replication. Reduced susceptibility to fluoroquinolones involves chromosomal genes and is associated with point mutations in the bacterial target gene encoding DNA gyrase *(gyrA* or *gyrB)* or topoisomerase IV (*parC* and *parE*) which are located in the quinolone resistance determining region (QRDR) (Parry 2004). Point mutations can occur also in the *marA* or *soxR* genes, affecting energy-depending efflux system.

Single-point mutations of *gyrA* confer resistance to nalidixic acid but only decreased susceptibility to fluoroquinolones. At the end of 1990s many clinical and microbiological failures of treatment with fluoroquinolones emerged in Asia, although isolates resulted susceptible to these drugs by disc testing with current breakpoints (Maskey et al. 2008). The MIC for fluoroquinolones was 10-fold higher in the nalidixic acid-resistant strains compared to wild type (0.125–1 µg/ml vs ≤0.03 µg/ml). Susceptibility to nalidixic acid is now considered the best predictor of clinical response to fluoroquinolones (Maskey et al. 2008; Parry 2004).

Resistance to expanded-spectrum beta-lactams is common among non-Typhi *Salmonella* isolates, but there are only anecdotal reports from India, Bangladesh, Nepal and Pakistan about resistance in *S.* Typhi and *S.* Paratyphi (Parry 2003). Azithromycin achieved excellent clinical results in the treatment of MDR and quinolone-resistant typhoid fever (Dolecek et al. 2008).

Data about resistant strains of *S.* Typhi and *S.* Paratyphi are available mainly from Asia where the spread of resistance is an emerging problem. A surveillance study conducted in seven Asian countries (Korea, Taiwan, Vietnam, Philippines, Singapore, Hong Kong and Sri Lanka) from 2002 to 2004 emerged high rates of resistance to ampicillin (44%), chloramphenicol (46%), TMP–SMZ (43%) and tetracycline (43%). MDR strains represented 41% of the total. All strains showed reduced susceptibility to ciprofloxacin while none was resistant to ceftriaxone. In Vietnam the proportion of MDR strains was 30% higher than in the other six countries (Chuang et al. 2008).

In contrast, a study carried out in Nepal between 1993 and 2003 showed an increased susceptibility to first-line antibiotics comparing the first half of the study (1993–1998) with the second half (1999–2003). Susceptibility increased for tetracycline from 77 to 92%, for amoxicillin from 72 to 94% and for chloramphenicol from 95 to 98%. However, isolates were all susceptible to ciprofloxacin until 1998 but from 1999 to 2003 resistance rates reached 13% (Maskey et al. 2008). A recent study conducted in India compared isolates collected in 2001 with those collected between 2005 and 2006. Resistance to nalidixic acid increased from 51 to 93% and ciprofloxacin-resistant enteric fever increased from 0.6 to 15%. This phenomenon can be attributed to the indiscriminate use of fluoroquinolones for empiric therapy of typhoid fever (Kumar et al. 2007).

As far as Africa is concerned, available data are very few. Data from a surveillance study carried out in Egypt during 2002 revealed that close to 30% of patients diagnosed with typhoid fever had MDR strains (Srikantiah et al. 2006). In Lagos, Nigeria, isolates of S. Typhi from hospitalized patients from 1997 to 2004 showed an increased resistance to first-line antibiotics, from 70% in 1997 to 85% in 2004. Resistant strains to nalidixic acid increased from 42 to 59% during this period (Akinyemi and Coker 2007). Data from Kenya, collected in three different districts from 2000 to 2002, reported high rates of MDR (from 85 to 100%) and increased MIC to ciprofloxacin resulted in almost 50% of cases (Kariuki et al. 2004).

To assess the trends in antibiotic susceptibility of S. Typhi, a retrospective study was performed in Dakar, Senegal, during 1987–1990 and 1997–2002 with, respectively, 135 and 97 isolates from patients with enteric fever. The study showed no resistance to quinolones, only one MDR isolate, and high susceptibility to first-line antibiotic agents (99.6% ampicillin, 99.6% TMP–SMZ, 99% chloramphenicol, 97% tetracycline). In this area it seems that these drugs are still appropriate for treatment of uncomplicated cases of enteric fever (Dromigny and Perrier-Gros-Claude 2003).

12.2.4 Shigella *spp.*

Among the bacterial causes of dysentery, *Shigella* spp. continue to be the most important, with a high infectivity rate and the development of antibiotic resistance. Four *Shigella* spp. are recognized as human pathogens: *Shigella dysenteriae, Shigella flexneri, Shigella boydii* and *Shigella sonnei*. The prevalence of *Shigella* species varies over time and in different geographical areas. *Shigella sonnei* is the most prevalent (>80%) in industrialized countries. In RLCs, *S. flexneri* is the most common. *Shigella* infections can lead to illness ranging from mild, self-limited diarrhoea to severe dysentery with frequent passage of blood and mucus, high fever, cramps, tenesmus, and in rare cases, bacteraemia. The severity of disease is determined in part by the infecting species; infections due to *S. dysenteriae* usually progress to dysentery, which may also occur in infections caused by *S. flexneri*, whereas *S. boydii* and *S. sonnei* generally cause a self-limited, watery diarrhoea. Antibiotic treatment is usually indicated for individuals with moderate or severe symptoms of shigellosis (Bhattacharya and Sur 2003). Most *Shigella* infections are treated empirically, and therefore an understanding of resistance patterns is important for management. Empirical treatment has been compromised in large part by emerging resistance and inadequate surveillance to monitor trends (Shapiro et al. 2001).

Resistance of *Shigella* to ampicillin, tetracycline, TMP–SMZ, and chloramphenicol has become widespread in Africa, even though these drugs are still used for first-line chemotherapy for dysentery in many parts of the continent (Iwalokun et al. 2001). In Nigeria, during 1990–2000, resistance to ampicillin increased from 70 to 90%, TMP–SMZ from 77 to 85%, chloramphenicol from 71 to 77%, streptomycin from 71 to 79% and nalidixic acid from 0 to 11%. On the contrary, resistance to tetracycline decreased from 89 to 79%. In Sudan *S. flexneri* isolates from diarrhoea were found resistant to ampicillin in 56% of cases, TMP–SMZ in 41%, chloramphenicol in 33% and tetracycline in 54% (Ahmed et al. 2000). In Kenya 85% of *S. flexneri* isolates from diarrhoea were resistant to ampicillin, 94% to TMP–SMZ, 91% to chloramphenicol and 100% to tetracycline (Brooks et al. 2003). In Eritrea 78% of *S. flexneri* isolates from diarrhoea were resistant to ampicillin, 81% to TMP–SMZ, 67% to chloramphenicol and 6% to nalidixic acid (Naik 2006); no isolate was resistant to ciprofloxacin. A recent study analysed antibiotic susceptibility of 98 *Shigella* isolates collected from eight centres in eight Asian countries from July 2001 to July 2004 in terms of species distribution (Kuo et al. 2008). The highest resistance rate was found for TMP–SMZ (81%), followed by tetracycline (74%) and ampicillin (53%). Five ceftriaxone-non-susceptible strains (from Taiwan, Hong Kong, Philippines) and 10 ciprofloxacin-non-susceptible strains (from Hong Kong, Philippines, Korea, Vietnam and Sri Lanka) were isolated. An Indian study on antibiotic resistance pattern of 166 shigellae strains isolated from stool samples of paediatric patients showed that all strains were susceptible to norfloxacin, but more than 90% strains were resistant to tetracycline and TMP–SMZ and 67% strains were resistant to ampicillin. Resistance to amoxicillin, chloramphenicol and nalidixic acid was found in 55, 46 and 29% strains, respectively (Dutta et al. 2002). A most recent study on clonal MDR *S. dysenteriae* type 1 strains associated with epidemic and sporadic dysentery in eastern India found that in addition to ampicillin, TMP–SMZ, tetracycline, chloramphenicol and nalidixic acid, all the recent strains were resistant to norfloxacin, lomefloxacin, pefloxacin and ofloxacin, and showed reduced susceptibility to ciprofloxacin (Pazhani et al. 2004). About 80% of the *Shigella* strains isolated from diarrhoeal patients between 1989 and 1998 in Vietnam were resistant to ampicillin, chloramphenicol, oxytetracycline, trimethoprim and sulphonamides. In contrast to neighbouring countries, low percentages of resistance were found to nalidixic acid and norfloxacin (3–5%) and no resistance was found to ciprofloxacin, indicating that nalidixic acid with its low cost and safety in children could be recommended for the treatment of shigellosis (Anh et al. 2001).

This clear trend of increasing multidrug-resistant *Shigella* in RLCs may be due to a worsening situation with regard to antibiotic overuse in both humans and animals. Of concern is the occurrence of isolates with intermediate sensitivity to newer quinolones in parallel with the emergence of quinolone resistance in *Salmonella* (Dutta et al. 2002). Although quinolone resistance is rare, given the wide availability and use of oral quinolones at a population level, it is generally believed that widespread occurrence of resistance is only a matter of time. According to the WHO revised guidelines for the control of shigellosis, ciprofloxacin is now the drug of choice for all patients with bloody diarrhoea, irrespective of their age (WHO 2005). Although the use of ciprofloxacin in paediatric patients is limited by concerns about the potential risk of damage to growing cartilage (Hampel et al. 1997), fluoroquinolones are generally safe for the treatment of shigellosis in children (Green and Tillotson 1997). Ceftriaxone and azithromycin are considered the second-line drugs for the treatment of MDR strains of *Shigella* in all age groups (WHO 2005). Because high rates of resistance to ampicillin, chloramphenicol, TMP–SMZ and tetracycline among *Shigella* isolates exist worldwide, these drugs are no longer recommended as empiric therapy for shigellosis (WHO 2005). In conclusion, empiric antibiotic therapy for shigellosis requires knowledge of the antibiotic resistance pattern of *Shigella* strains circulating locally. Physicians should be aware of the high multidrug resistance rate of *Shigella* spp., especially the increasing resistance to ceftriaxone and fluoroquinolones. *Shigella flexneri* is still the most prevalent species in most Asian countries and its higher antibiotic resistance rate compared with *S. sonnei* is a cause for concern. Continuous monitoring of resistance patterns at national and international levels is necessary to control the spread of resistance in *Shigella*.

12.2.5 Vibrio cholerae

Cholera still remains a threatening disease in RLCs and is endemic in many countries of Africa and south Asia with a typical seasonal periodicity (Sack et al. 2004). The infection is caused by *V. cholerae*, a motile Gram-negative rod that produces a toxin responsible for symptoms. The infection is caused by two serogroups of *V. cholerae*, serogroup O1, divided into three serotypes (Ogawa, Inaba and Hikojima) and two biotypes (classical and El Tor), and serogroup O139, first identified in Bangladesh in 1992. The transmission is by faecal oral route through contaminated water or food (especially undercooked seafood). There are no animal reservoirs apart from shellfish and plankton. Although notification is mandatory, cases of cholera are underreported especially from Asia. It is estimated that the burden of disease is about 6 million cases annually with 120,000 deaths, not including outbreaks (Griffith et al. 2006).

The disease is characterized by acute watery diarrhoea with sudden onset that causes severe dehydratation. Without treatment mortality rate is about 50% (Sack et al. 2004). Treatment is essentially based on rapid rehydratation, but it is demonstrated that antibiotic therapy can shorten the course of illness in severe cases. Tetracycline and TMP–SMZ are considered first-line drugs for the treatment of cholera. From 1977 multiply antibiotic-resistant *V. cholerae* strains (MARV) emerged in Tanzania and Bangladesh, and spread in India, East Africa, Thailand and Latin America. These isolates were resistant to ampicillin, tetracycline, TMP–SMZ, gentamicin, kanamycin and streptomycin (Sack et al. 2004, Okeke et al. 2005). Resistance genes are often located on conjugative plasmids or transposons. Furthermore, recently other genetic elements, like class I integrons and SXT constin, have been found to be involved in spreading of resistance. Class I integrons are often integrated in plasmids and allow resistance cassette associated genes to act like functional genes (Sack et al. 2004). The SXT constins, detected for the first time in 1992 in *V. cholerae* serogroup O139, is a conjugative, self-transmissible element embedded in the chromosomal gene *prfC* and determines resistance to TMP–SMZ, chloramphenicol and streptomycin (Iwanaga et al. 2004).

Quinolones are widely used as alternative drugs where first-line drugs are no more effective. *Vibrio cholerae* strains resistant to quinolones have been reported since 1998 in India and Africa. Point mutation in chromosomal genes *gyrA* (DNA gyrase) and *parC* (topoisomerase IV) and an increased activity of pump efflux are the most common mechanisms of resistance to quinolones (Okeke et al. 2007). Resistance to these drugs has also been attributed to plasmid-mediated *qnr* genes. Recently a new *qnr* determinant that confers resistance to ciprofloxacin carried by a class I integron has emerged in Brazil in *V. cholerae* O1 isolates (Fonseca et al. 2008). Data on resistance rates are mainly collected during outbreaks, with the exception of India and Bangladesh where longitudinal surveillance studies have been performed since the 1960s. A MARV O1 strain appeared in Bangladesh in 1979 and disappeared after 5 months and again a different MARV strain emerged in 1981. Abrupt emergence of multiply drug resistance occurred in 1991 with prevalence of resistance to tetracycline and TMP–SMZ reaching 90%. In 1993 O139 isolates, susceptible to tetracycline but resistant to TMP–SMZ and streptomycin and, moderately, to chloramphenicol, spread in Bangladesh, India and Pakistan (Sack et al. 2001).

In India *V. cholerae* O1 strains resistant to TMP–SMZ are documented since the end of the 1980s (resistant rates up to 82%). As far as serogroup O139 is concerned, during 1993 a tetracycline-sensitive strain replaced the O1 serogroup but tetracycline-resistant 0139 isolates appeared in 1996 (Narang et al. 2008). Since then, nalidixic acid has become the drug of choice. In 1994, *V. cholerae* O1 resistant to nalidixic acid appeared with increasing number of cases annually. During the epidemic of 2002 both MARV O1 and O139 were detected in the south of India, but only O1 serogroup showed resistance to fluoroquinolones (25%) while the O139 strains were all susceptible (Krishna et al. 2005). High incidence of *V. cholerae* O1 strains resistant to ciprofloxacin have been detected during 1995–1998 from hospitalized patients in Calcutta. In contrast *V. cholerae* O139 isolates in the same period showed susceptibility to ciprofloxacin, probably because O139 strains appeared in 1992 and have been less extensively treated with fluoroquinolones (Okuda et al., 2007). In a surveillance study conducted in Sevagram, India, from 1990 to 2005, 425 *V. cholerae* O1 and 86 *V. cholerae* O139 strains were collected and patterns of resistance estimated. O1 serogroup seemed not to have acquired resistance to tetracycline during this period, being relatively susceptible and the maximum rate of resistance was recorded in 1998 (17%). Resistance to nalidixic acid reached its higher level in 2003 (78%), while during the other years it was between 0 and 45%. Resistance to chloramphenicol fluctuated from 8 to 69%. O139 serogroup had different resistant patterns. Resistance to tetracycline was detected only in 1998 (59%) and 2002 (50%), while resistance to nalidixic acid had a rapid increase from 1999 (33%) to 2002 (100%). No resistance to ciprofloxacin was detected (Narang et al. 2008).

In Africa, cholera infections appeared in 1970 by biotype El Tor. Since then it has become endemic with seasonal outbreaks. Antibiotic susceptibility varies from countries and within outbreaks. At the end of 1970s *V. cholerae* isolates resistant to tetracycline were found in Tanzania, Kenya, and spread throughout Africa (Okeke et al. 2007). In a surveillance study enrolled in Maniçha, Mozambique, during 2002–2004, 91 samples of *V. cholerae* were collected, all belonging to O1 serogroup Ogawa. They showed high rates of resistance to tetracycline (92%), TMP–SMZ (97%), chloramphenicol (58%), while resistance to ampicillin and nalidixic acid was low (12 and 4%, respectively) (Mandomando et al. 2007a). Namibia experienced an outbreak of cholera from December 2006 to February 2007 with more than 250 cases of disease reported, but only nine isolates collected. All were O1 Inaba El Tor resistant to TMP–SMZ and streptomycin and susceptible to tetracycline, ampicillin, chloramphenicol and quinolones (Smith et al. 2008).

In Latin America, after 100 years of absence, an epidemic of cholera serogroup O1 began in Peru in 1991 and spread throughout the continent and in 1998 another epidemic occurred. The original strain was susceptible to all antibiotic while in 1998 different levels of resistance to tetracycline (11%), TMP–SMZ (25%) ampicillin (10%), streptomycin (10%), and nalidixic acid (30%) emerged (Ibarra and Alvarado, 2007).

12.2.6 **Campylobacter jejuni**

Campylobacter jejuni is considered to be one of the most significant bacterial causes of human gastroenteritis (Allos 2001). Annually, approximately 400 million cases of *Campylobacter*-associated gastroenteritis occur worldwide. *Campylobacter jejuni* is a ubiquitous pathogen that causes up to 30% of all cases of traveller's diarrhoea, particularly in Asia. In general, the occurrence of human *Campylobacter* gastroenteritis has been largely attributed to the consumption of contaminated food animal products, especially poultry, because of the high prevalence of *Campylobacter* in these animals. Other vehicles such as red meat, environmental water and unpasteurized milk are additional sources of infection. Person-to-person transmission is very rare (Allos 2001).

The majority of *C. jejuni* infections result in an acute, self-limited gastrointestinal illness. However, in a small number of subjects, it is followed by complications, including cholecystitis, pancreatitis and peritonitis, and by extraintestinal complications such as septicaemia, septic arthritis, osteomyelitis and endocarditis. Rarely, *C. jejuni* infections can cause autoimmune neuropathies, such as Guillain–Barré (GBS) and Miller-Fisher (MFS) syndromes. Approximately 1 in every 1000 *C. jejuni* infections will be followed by GBS (Allos 2001).

The range of *C. jejuni* disease outcomes, extending from acute inflammatory diarrhoea to the induction of the autoimmune neuropathies, could be associated with the differential expression of virulence factors in *C. jejuni* strains. Most cases of enteritis do not require treatment as they are of short duration, clinically mild and self-limiting. Antibiotic treatment, however, is necessary for systemic *Campylobacter* infections, infections in immune-suppressed patients and severe or long-lasting infections (Allos 2001).

Erythromycin is considered the drug of choice for treating *Campylobacter* gastroenteritis with ciprofloxacin and tetracycline as alternative drugs. Despite decades of use, the rate of resistance of *Campylobacter* to erythromycin remains quite low. Other advantages of erythromycin include its low cost, safety, ease of administration and narrow spectrum of activity. Unlike the fluoroquinolones and tetracyclines, erythromycin may be administered safely to children and pregnant women and is less likely than many agents to exert an inhibitory effect on other faecal flora (Allos 2001). The newer macrolides, azithromycin and clarithromycin, are also effective against *C. jejuni* infections, but they are more expensive than erythromycin and provide no clinical advantage. Since the 1990s, a significant increase in the prevalence of resistance to macrolides among *Campylobacter* spp. has been reported, and this is recognized as an emerging public health problem (Engberg et al. 2001). It has been suggested by some investigators that resistance to macrolides is mainly found in isolates of animal origin, especially *Campylobacter coli* from pigs and also *C. jejuni* from chickens (Aarestrup et al. 1997). Furthermore, macrolide resistance may develop during the course of antibiotic treatment in humans (Funke et al. 1994). It also seems possible that the selection pressure arising from the use of macrolides in human medicine might also affect resistance in *Campylobacter* (Phillips et al. 2004).

In the past few years, a rapidly increasing proportion of *Campylobacter* strains all over the world has been found to be fluoroquinolone resistant. Primary resistance to quinolone therapy in humans was first noted in the early 1990s in Asia and in Europe (Allos 2001). Not surprisingly, this coincided with initiation of the administration of the fluoroquinolone enrofloxacin to food animals in those areas.

All *Campylobacter* species are inherently resistant to vancomycin, rifampin and trimethoprim.

Serious systemic infections may be treated with an aminoglycoside such as gentamicin. *Campylobacter* species are also generally susceptible to chloramphenicol, clindamycin, nitrofurans and imipenem. High rates of resistance make tetracycline, amoxicillin, ampicillin, metronidazole and cephalosporins poor choices for the treatment of infections with *C. jejuni*.

The emergence of *Campylobacter* spp. that are resistant to ciprofloxacin has been reported in Thailand (Serichantalergs et al. 2007). An aetiologic study of acute bacterial dysentery conducted in children in two major hospitals in and around Bangkok shows that *Campylobacter* was the most common pathogen found: 90% of the *Campylobacter* isolates were resistant to ciprofloxacin but sensitive to macrolides (Bodhidatta et al. 2002).

For treatment of *Campylobacter* infections in Indonesia, erythromycin is most often used. A study that analysed 2812 bacterial pathogens isolated from patients with diarrhoea coming from several provinces of Indonesia, from 1995 to 2001, to determine their changing trends in response to eight antibiotics (ampicillin, TMP–SMZ, chloramphenicol, tetracycline, cephalotin, ceftriaxone, norfloxacin and ciprofloxacin) showed that 22–43% of *C. jejuni* strains developed resistance to ciprofloxacin since 1998 (Tjaniadi et al. 2003). No strains of *C. jejuni* were found resistant to erythromycin. In a north Indian rural community the pattern of antibiotic resistance of *Campylobacter* spp were ciprofloxacin 71%, erythromycin 6% and gentamicin 10%, while 30% of strains were multidrug resistant (Jain et al. 2005).

In Africa a study conducted on children with diarrhoea from a rural area of southern Mozambique showed that the resistance to quinolones was mainly observed in *Campylobacter* strains (11% of isolates) and that all the isolates were susceptible to erythromycin (Mandomando et al. 2007b). An increasing resistance to macrolides and quinolones was observed in South Africa (Samie et al. 2007). In southern Chile no *Campylobacter* strains were found to be resistant to ciprofloxacin, erythromycin and gentamicin, but 2% were resistant to tetracycline and all to aztreonam (Fernandez et al. 2000).

12.2.7 Neisseria gonorrhoeae

Neisseria gonorrhoeae can cause cervicitis, urethritis, proctitis and pelvic inflammatory disease. It can lead to long-term sequelae like infertility and ectopic pregnancy and it can be implicated in chronic pelvic pain. Remarkably it can cause increased susceptibility to transmission of HIV infection (Fleming and Wasserheit 1999). Among the aetiological agents of treatable STIs, *N. gonorrhoeae* stands out because of the extent to which antibiotic resistance compromises the effectiveness of individual case management and disease control programmes. *Neisseria gonorrhoeae* continues to develop resistance to both older, less expensive antibiotics (penicillins, tetracyclines, spectinomycin) and more recently introduced agents (quinolones).

Treatment of gonorrhoea has been greatly compromised by the emergence of penicillinase-producing *N. gonorrhoeae* (PPNG) in 1976. Between 1996 and 2001, the prevalence of PPNG in Hong Kong increased from 57 to 82% (Zheng et al. 2003). Penicillin resistance among gonococci currently ranges between 9 and 90% across Asia and exceeds 35% in sub-Saharan Africa and the Caribbean (Adegbola et al. 1997; Bogaerts et al. 1998; Guyot et al. 1998; Bhuiyan et al. 1999; Dillon et al. 2001a; Tapsall 2001). In the Americas, excluding the USA and Canada, data from 1995 on 10,500 isolates from 10 countries indicated a PPNG rate of about 10%, being less than 5% in 1990 (Ison et al. 1998).

Data from individual countries in this region confirm this trend. In Argentina, PPNG rose from 1.9% in 1980 to 28% in 1992–1994, declining to 14% in 1996. In Trinidad and Tobago 8% of 518 strains were reported to be penicillin resistant in 1997; 70% of them were PPNG (Swanston et al. 1997).

Tetracycline resistance is especially common in sub-Saharan Africa, where the low cost and wide distribution of these drugs have contributed to their use and abuse in the management of actual, presumed or expected infectious disease (Bogaerts et al. 1998; Tapsall 2001). Studies from various locations south of the Sahara showed high levels of resistance to penicillin (~70%) and tetracycline (up to 65%) in the early 1990s (Tapsall 2001). In 1997, GISP (gonococcal isolate surveillance programme) study in the USA reported that 26% of isolates were tetracycline resistant (Centers for Disease Control and Prevention 1997). In the rest of the Americas, regional data showed an increase in tetracycline-resistant *N. gonorrhoeae* (TRNG) from less than 5% in 1990 to nearly 15% in 1995 (Ison et al. 1998).

Spectinomycin retains its activity against *N. gonorrhoeae* in most parts of the world (Tapsall 2001). However, it will probably remain an alternative treatment rather than a recommended one because high levels of resistance developed when this antibiotic was widely used in the mid-1980s (Boslego et al. 1987). Widespread resistance has necessitated the replacement of penicillin and tetracycline with more expensive first-line drugs, to which resistance quickly emerged. In the Caribbean and South America, azithromycin resistance was found in 16–72% of isolates in different locations (Dillon et al. 2001b; Sosa et al. 2003). Even if demonstrated to be efficacious, its cost would be a limiting factor in using it to treat gonorrhoea.

The emergence of quinolone-resistant *N. gonorrhoeae* (QRNG) was first identified in Hawaii in 1991, at about the same time that it was recognized as a problem in Asia (Tapsall 2001; Rahman et al. 2002; Bala et al. 2003; Iverson et al. 2004). The proportion of less-sensitive strains is particularly high in China (36%), Hong Kong (44%), Japan (50%) and the Republic of Korea (52%). The highest proportions of fully quinolone-resistant isolates were seen in the Philippines (63%), Hong Kong SAR (48.8%), China (54.2%) and the Republic of Korea (11.2%). Lower rates, but still cause for concern, were found in Vietnam (8%), Singapore (7%) and New Caledonia (7.5%). In other centres there was a lower increase in fully resistant isolates. In Australia, resistant strains accounted for 3% of all isolates; most of these were from Sydney (Tapsall 2001). Both low- and high-level QRNG have been detected in areas of South and Southeast Asia, including New Delhi (Bala et al. 2003). In Taiwan, among 55 isolates collected from 1999 to 2003, ciprofloxacin resistance was found in 1 (25%) of 4 isolates obtained in 1999–2000 and in 27 of 29 isolates obtained in 2003 (Hsueh et al. 2005). Although quinolone resistance has emerged in Africa, Latin America, the Caribbean, and the Middle East, it appears less common than in Asia (Tapsall 2001).

The emergence of quinolone resistance increasingly leaves only less available drugs as fully reliable therapy. The high cost of third-generation cephalosporins in particular makes their use prohibitive in many RLCs. Ceftriaxone is the injectable agent and cefixime the oral cephalosporin most widely used. There is no documented case of treatment failure with these agents (Tapsall 2001), but recent reports from Japan and several other countries in the WHO Western Pacific Region (Australia; Brunei; China and Papua, New Guinea) suggest decreased susceptibility to cephalosporins, with the reported mechanism of resistance being alterations in $penA$ genes (Ito et al. 2004; The WHO Western Pacific Gonococcal Antibiotic Surveillance Programme 2008).

12.3 Surveillance

The burden imposed by antibiotic resistance on the human health of populations living in RLCs is certainly large but difficult to quantify with precision. Recently, Falagas and Karveli (2006) compiled a list of World Wide Web resources of data from surveillance studies on antibiotic resistance. They found 23 websites for 18 major international networks. Accessing these addresses it is evident that almost all RLCs have been insufficiently studied, with large areas from which no data are available. The limited data on antibiotic resistance for these countries are mostly from studies conducted on pathogens isolated during disease outbreaks or from community- or health-care-associated infections observed in the few health centres where high-quality laboratories are available (Guzmán-Blanco et al. 2000; Hart and Kariuki 1998; Rodríguez et al. 2001). Within the WHO global strategy for the containment of antibiotic resistance, surveillance is considered essential for providing information on the magnitude and trends in resistance and for monitoring the effect of interventions (WHO 2001a). Information from surveillance is important not only for clinicians to optimize syndromic (empirical) antibiotic treatment and for policy makers to define or update standard treatment guidelines, to design and monitor intervention strategies, to identify need for implementation of infection control measures but also for investigators with a research interest in the field of antibiotic resistance (WHO 2001a). In RLCs effective surveillance programs are difficult to be implemented for a number of reasons including the lack of laboratory facilities, and, where they do exist, the lack of quality control, reliable reagents, adequate supervision, personnel and periodic training (Shears 2001). In addition, RLCs lack other essential requirements for a correct surveillance, such as demographic data, databases and information dissemination channels.

12.3.1 Role of Surveillance Through Commensal Bacteria

There is increasing agreement about the importance of extending the surveillance of resistance to the commensal bacteria of humans and animals. Commensal bacteria, although not being a specific target, are continuously exposed to the selective pressure generated by antibiotic chemotherapy and may become a potential reservoir of resistant strains that can cause infections, and of resistant determinants that can be transferred to pathogenic bacteria (Levy et al. 1988; Leverstein-van Hall et al. 2002; Alliance for the Prudent Use of Antibiotics-APUA). Therefore, some members of the commensal microbiota, such as the faecal *E. coli*, are considered as sensitive indicator bacteria for surveillance of antibiotic resistance, to measure the selective pressure generated by antibiotic usage, to evaluate the impact of modifications in antibiotic prescription policies and to predict the emergence of resistance in pathogens (Lester et al. 1990; Calva et al. 1996; Bartoloni et al. 1998, 2006a, 2008a; Okeke et al. 2000; Osterblad et al. 2000; Sörberg et al. 2002; Mathai et al. 2008). Reliable and low-cost rapid screening methods have been used in various epidemiological settings showing to represent a valid tool to conduct large-scale surveillance studies and to monitor resistance control programs in a cost-effective manner (Lester et al. 1990; Bartoloni et al. 2006b).

12.4 Antibiotic Use and Bacterial Resistance

It is generally accepted that the principal cause for the emergence and spread of bacterial resistance has been the selective pressure generated by widespread antibiotic use in clinical practice. The consumption of antibiotics has increased substantially since they were first introduced and in most RLCs they now constitute the single largest group of drug purchased. Misuse of antibiotics is common in RLCs and may be expressed as inappropriate prescription (Hui et al. 1997; Bosu and Ofori-Adjei 1997; Larsson et al. 2000; Kristiansson et al. 2008) or incorrect dosage (Gilson 1993; Uppal et al. 1993).

Prescribers probably have the major responsibility for inappropriate drug use in RLCs. There are many causes which may explain over-prescription and inappropriate prescription in RLCs including lack of education and training for health-care professionals and lack of laboratory facilities. Both of these latter reasons imply inadequacy of the diagnostic process needed to reach a correct diagnosis. The uncertainty about diagnosis and the most appropriate drug, and fear of a poor outcome for the patient may lead to over-prescription of antibiotics. The problem is probably more complex and antibiotic usage behaviours of prescribers are influenced not only by knowledge but also by cultural preferences and beliefs and by the health administration system (Radyowijati and Haak 2003; Sterky et al. 1991; Rowe et al. 2005; Kristiansson et al. 2008). A few studies demonstrated that even physicians with adequate knowledge on use of antibiotics prescribed them inappropriately (Gani et al. 1991; Paredes et al. 1996) and that health systems that do not provide diagnostic facilities free of charge or allow for a restricted number of free visits may contribute to increase antibiotic prescription (Nizami et al. 1996; Bosu and Ofori-Adjei 1997; Sterky et al. 1991; Rowe et al. 2005; Kristiansson et al. 2008). Moreover, irrational prescribing may be due to the lack of time for a correct clinical decision as it happens in settings where there is pressure and demands of caring for large number of ill subjects. Other rational reasons, at least from the prescriber's point of view, which may influence physicians' prescribing behaviour are economic factors. Financial incentives for physicians, especially in private sectors, may be obtained from the patient or from the pharmaceutical companies by prescribing and/or dispensing a specific medicine (Gani et al. 1991; Dong et al. 1999; Radyowijati and Haak 2003; Shao-Kang et al. 1998).

In RLCs, antibiotics can be purchased directly from pharmacies without prescriptions. This phenomenon may be due to a variety of reasons, including lack of appropriate legislation on dispensing antibiotics and economic incentives for pharmacists. Moreover, unauthorized staff, including family members or children, lacking appropriate knowledge may attend costumers and dispense antibiotics (Serkkola 1990).

Antibiotic usage patterns in RLCs are also strongly influenced by patients' cultural perceptions on illness, health, and power of drugs (Radyowijati and Haak 2003). There is a widespread perception that antibiotics are "wonder drugs", never harmful and capable of healing any sort of illness (Kunin et al. 1987) and patient may request specific drugs or their particular formulations, e.g. capsules rather than tablets, or injections rather than pills. Prescribers' and dispensers' perception regarding patient hopes and demands is considered an influencing factor for the prescribing and dispensing behaviour (Gani et al. 1991; Wolff 1993; Indalo 1997; Radyowijati and Haak 2003).

Self-medication is common in RLCs (Tomson and Sterky 1986; Larsson et al. 2000). Antibiotics may be purchased directly from street-vendors or markets and used in accordance with an earlier recommendation from a physician or on advice of friends or relatives (Dua, Kunin and White 1994; Lansang et al. 1990; Radyowijati and Haak 2003). The quantity of drug purchased is often small and in accordance with the customer's ability to pay (Bartoloni et al. 1998; Lansang et al. 1990). Yet another problem in many RLCs is the availability of substandard and counterfeit drugs (Okeke et al. 2005; WHO 2001a). Although the link between quality of antibiotics and bacterial resistance has not been documented, the exposure to suboptimal levels of drugs could result in both therapeutic failure and selection of drug-resistant strains (Okeke et al. 2005; WHO 2001a). The inappropriate use of antibiotics is certainly a complex and multifactorial problem in RLCs, and a proper understanding of this phenomenon should be considered a basic need to guide the development of any model for action.

12.5 Knowledge Gaps

The high prevalence of resistance in RLCs is likely due to several factors, including both inappropriate use of antibiotics and unique environmental conditions, such as crowding and poor sanitation. However, the recent findings of unexpectedly high levels of acquired resistance in commensal bacteria from human populations living in geographically remote areas, and with very low levels of antibiotic exposure and limited exchanges with the exterior, raise the question about the role of other factors in favouring the emergence and spreading of antibiotic resistance (Walson et al. 2001; Bartoloni et al. 2004; Grenet et al. 2004; Bartoloni et al. 2009). The molecular characterization of resistant isolates from two remote human communities in Bolivia and Peru allowed to document a remarkable variety of acquired resistance genes, not dissimilar from those circulating in antibiotic-exposed areas, as well as clonal and plasmid heterogeneity (Pallecchi et al. 2007; Bartoloni et al. 2009). These findings are not consistent with the hypothesis of an occasional introduction of a few resistant strains from antibiotic-exposed settings, followed by the dissemination and replacement of the existing susceptible population. They would rather suggest either a sustained flow of diverse resistant strains from the exterior (despite the very limited exchanges) or the occurrence of substantial horizontal gene transfer following the occasional introduction of resistance strains from the exterior into the remote settings. Whichever the mechanism responsible for this genetic diversity, what remains to be defined are the reasons for maintaining the high prevalence of resistance in the absence of a significant selective pressure generated by antibiotic consumption. Further studies are needed to clarify why, in certain settings, the spread and maintenance of antibiotic resistance can occur regardless of the selective pressure generated by the use of antibiotics. An understanding of the mechanism underlying similar phenomena is crucial to design any program aimed at controlling antibiotic resistance and further studies are needed to address these knowledge gaps.

12.6 Conclusion

In conclusion, the issue of antibiotic use and bacterial resistance is complex and multifactorial and represents only a subset of the problems responsible for the low quality of health care in RLCs. Many factors contributing to antibiotic resistance in RLCs are strongly related to poverty: lack of knowledge or information of health-care professionals, lack of laboratory facilities, inadequate access to health system, lack of money available to pay for the appropriate amount of antibiotics, dispensation of drugs by untrained people, availability of substandard and counterfeit drugs, etc. Moreover, in RLCs, transmission of resistant bacteria is facilitated by person-to-person contact, through contaminated food, unsafe water or by vectors. Governments have a critical role in developing containment strategies and should address the problem within the wider priorities of strengthening health systems and disease control and prevention programs. Considering the difficulties of many RLCs, where the problem of antibiotic resistance is even more serious, there is a urgent need of international significant financial and technical support.

Acknowledgments The authors thank Lucia Pallecchi and Marianne Strohmeyer for their helpful comments and Annalisa Cavallo, Costanza Fiorelli, Sarah Jacopini, Jacopo Nocentini, Sonia Vicidomini for critical reviewing of published informations.

References

Aarestrup, F. M., Nielsen, E. M., Madsen, M., and Engberg, J. 1997. Antmicrobial susceptibility patterns of thermophilic *Campylobacter* spp. from humans, pigs, cattle, and broilers in Denmark. Antimicrob. Agents Chemother. 41:2244–50.

Abraham, E. P., Gardener, A. D., and Chain E. 1941. Further observation on penicillin. Lancet. 16:177–89.

Adegbola, R. A., Sabally, S., Corrah T., West B., and Mabey, D. 1997. Increasing prevalence of penicillinase-producing *Neisseria gonorrhoeae* and the emergence of high-level, plasmid-mediated tetracycline resistance among gonococcal isolates in The Gambia. Trop. Med. Int. Health. 2:428–32.

Adegbola, R. A., Hill, P. C., Secka, O., Ikumapayi, U. N., Lahai, G., Greenwood, B. M., and Corrah, T. 2006. Serotype and antimicrobial susceptibility patterns of isolates of Streptococcus pneumaniae causing invasive disease in The Gambia 1996–2003. Trop. Med. Int. Health. 117:1128–35.

Adrian, P. V., and Klugman, K. P. 1998. Mutations in the dihydrofolate reductase gene of trimethoprim-resistant isolates of *Streptococcus pneumoniae*. Antimicrob. Agents. Chemother. 41:2406–13.

Agudelo, C. I., Moreno, J., Sanabria, O. M., Ovalle, M. V., Di Fabio, J. L., Castañeda, E., and Grupo Colombiano de Trabajo en Streptococcus pneumonite. 2006. *Streptococcus pneumoniae*: evolution de los serotipos y los patrones de susceptibilidad antimicrobiana en aislamientos invasores en 11 anos de vigilancia en Colombia 1994–2004. Biomedica. 26: 234–49.

Ahmed, A., Osman, H., Mansour, A., Musa, H. A., Ahmed, A. B., Karrar, Z., and Hassan, H. S. 2000. Antimicrobial agent resistance in bacterial isolates from patients with diarrhoea and urinary tract infection in the Sudan. Am. J. Trop. Med. Hyg. 63:259–63.

Akinyemi, K. O., and Coker, A. O. 2007. Trends of antibiotic resistance in Salmonella enterica serovar typhi isolated from hospitalized patients from 1997 to 2004 in Lagos, Nigeria Indian. J. Med. Microbiol. 25:436–7.

Akpaka, P. E., Kissoon, S., Swanston, W. H., and Monteil, M. 2006. Prevalence and antimicrobial susceptibility pattern of methicillin resistant *Staphylococcus aureus* isolates from Trinidad & Tobago. Ann. Clin. Microbiol. Antimicrob. 3:5–16.

Alliance for the Prudent Use of Antibiotics (APUA). Available at http://www.tufts.edu/med/apua/

Allos, B. M. 2001. *Campylobacter jejuni* infections: update on emerging issues and trends. Clin. Infect. Dis. 32:1201–6.

Anh, N. T., Cam, P. D., and Dalsgaard, A. 2001. Antimicrobial resistance of *Shigella* spp isolated from diarrheal patients between 1989 and 1998 in Vietnam. Southeast Asian J. Trop. Med. Public. Health. 32:856–62.

Appelbaum, P. C., Bhamjee, A., Scragg, J. N., Hallett, A. F., Bowen, A. J., and Cooper, R. C. 1977. *Streptococcus pneumoniae* resistant to penicillin and chloramphenicol. Lancet. 12: 995–7.

Appelbaum, P. C. 2006. MRSA-the tip of the iceberg. Clin. Microbiol. Infect. 12 (Suppl. 2): 3–10.

Bala, M., Ray, K., and Kumari, S. 2003. Alarming increase in ciprofloxacin- and penicillin-resistant *Neisseria gonorrhoeae* isolates in New Delhi, India. Sex Transm. Dis. 30:523–5.

Baquero, F. 1999. Evolving resistance patterns of *Streptococcus pneumoniae*: a link with long-acting macrolide consumption? J. Chemother. 11:35–43.

Barcus, V. A., Ghanekar, K., Yeo, M., Coffey, T. J., and Dowson, C. G. 1995. Genetics of high level penicillin resistance in clinical isolates of *Streptococcus pneumoniae*. FEMS Microbiol. Lett. 126:299–303.

Bartoloni, A., Cutts, F., Leoni, S., Austin, C. C., Mantella, A., Guglielmetti, P., Roselli, M., Salazar, E., and Paradisi, F. 1998. Patterns of antimicrobial use and antimicrobial resistance among healthy children in Bolivia. Trop. Med. Int. Health. 3:116–23.

Bartoloni, A., Bartalesi, F., Mantella, A., Dell'Amico, E., Roselli, M., Strohmeyer, M., Barahona, H. G., Barrón, V. P., Paradisi, F., and Rossolini, G. M. 2004. High prevalence of acquired antimicrobial resistance unrelated to heavy antimicrobial consumption. J. Infect. Dis. 189 (7):1291–4.

Bartoloni, A., Pallecchi, L., Benedetti, M., Fernandez, C., Vallejos, Y., Guzman, E., Villagran, A. L., Mantella, A., Lucchetti, C., Bartalesi, F., Strohmeyer, M., Bechini, A., Gamboa, H., Rodríguez, H., Falkenberg, T., Kronvall, G., Gotuzzo, E., Paradisi, F., and Rossolini, G. M. 2006a. Multidrug-resistant commensal *Escherichia coli* in children, Peru and Bolivia. Emerg. Infect. Dis. 12:907–13.

Bartoloni, A., Benedetti, M., Pallecchi, L., Larsson, M., Mantella, A., Strohmeyer, M., Bartalesi, F., Fernandez, C., Guzman, E., Vallejos, Y., Villagran, A. L., Guerra, H., Gotuzzo, E., Paradisi, F., Falkenberg, T., Rossolini, G. M., and Kronvall. G. 2006b. Evaluation of a rapid screening method for detection of antimicrobial resistance in the commensal microbiota of the gut. Trans. R. Soc. Trop. Med. Hyg. 100:119–25.

Bartoloni, A., Pallecchi, L., Fiorelli, C., Di Maggio, T., Fernandez, C., Villagran, A. L., Mantella, A., Bartalesi, F., Strohmeyer, M., Bechini, A., Gamboa, H., Rodriguez, H., Kristiansson, C., Kronvall, G., Gotuzzo, E., Paradisi, F., and Rossolini, G. M. 2008a. Increasing resistance in commensal Escherichia coli, Bolivia and Peru. Emerg. Infect. Dis. 142:338–40.

Bartoloni, A., Pallecchi, L., Rodríguez, H., Fernandez, C., Mantella, A., Bartalesi, F., Strohmeyer, M., Kristiansson, C., Gotuzzo, E., Paradisi, F., and Rossolini, G. M. 2009. Antibiotic resistance in a very remote Amazonas community. Int. J. Antimicrob. Agents 33(2):125–9.

Benbachir, M., Benredjeb, S., Boye, C. S., Dosso, M., Belabbes, H., Kamoun, A., Kaire, O., and Elmdaghri, N. 2001. Two-year surveillance of antibiotic resistance in *Streptococcus pneumoniae* in four African cities. Antimicrob. Agents. Chemother. 452:627–9.

Bergman, M., Huikko, S., Huovinen, P., Paakkari, P., and Seppälä, H. 2006. Macrolide and azithromycin use are linked to increased macrolide resistance in *Streptococcus pneumoniae*. Antimicrob. Agents Chemother. 50:3646–50.

Bhattacharya, S. K., and Sur, D. 2003. An evaluation of current shigellosis treatment. Expert Opin. Pharmacother. 4:1315–20.

Bhuiyan, B. U., Rahman, M., Miah, M. R., Nahar, S., Islam N., Ahmed M., Rahman K. M., and Albert, M. J. 1999. Antimicrobial susceptibilities and plasmid contents of *Neisseria gonorrhoeae* isolates from commercial sex workers in Dhaka, Bangladesh: emergence of high-level resistance to ciprofloxacin. J. Clin. Microbiol. 37:1130–6.

Blomberg, B., Mwakagile, D. S., Grassa, W. K., Maselle, S. Y., Mashurano, M., Digranes, A., Harthug, S., and Langeland, N. 2004. Surveillance of antimicrobial resistance at a tertiary hospital in Tanzania. BMC Public Health. 4:45.

Bodhidatta, L., Vithayasai, N., and Eimpokalarp, B. 2002. Bacterial enteric pathogens in children with acute dysentery in Thailand: increasing importance of quinolone-resistant *Campylobacter*. Southeast Asian J. Trop. Med. Public Health. 33:752–7.

Bogaerts, J., Van Dyck E., Mukantabana, B., Munyabikali, J. P., and Martinez Tello, W. 1998. Auxotypes, serovars, and trends of antimicrobial resistance of *Neisseria gonorrhoeae* in Kigali, Rwanda (1985–93). Sex. Transm. Infect. 74:205–9.

Boslego, J. W., Tramont, E. C., Takafuji, E. T., Diniega, B. M., Mitchell, B. S., and Small. 1987. Effect of spectinomycin use on the prevalence of spectinomycin-resistant and of penicillinase-producing *Neisseria gonorrhoeae*. N. Engl. J. Med. 317:272–8.

Bosu, W. K., and Ofori-Adjei, D. 1997. Survey of antibiotic prescribing pattern in government health facilities of the Wassa west district of Ghana. East. Afr. Med. J. 74: 138–42.

Brooks, J. T., Shapiro, R. L., Kumar, L., Wells, J. G., Phillips-Howard, P. A., Shi, Y. P., Vulule, J. M., Hoekstra, R. M., Mintz, E., and Slutsker, L. 2003. Epidemiology of sporadic bloody diarrhea in rural western Kenya. Am. J. Trop. Med. Hyg. 68:671–7.

Byarugaba D. K. 2004. A view on antimicrobial resistance in developing countries and responsible risk factors. Int. J. Antimicrob. Agents. 24:105–10.

Calva, J. J., Sifuentes-Osornio, J., and Cerón, C. 1996. Antimicrobial resistance in fecal flora: longitudinal community-based surveillance of children from urban Mexico. Antimicrob. Agents. Chemother. 40:1699–702.

Centers for Disease Control and Prevention. 1997. Sexually Transmitted Disease Surveillance. Division of STD Prevention. Centers for Disease Control and Prevention, Atlanta

Chau, T. T., Campbell, J. I., Galindo, C. M., Van Minh Hoang, N., Diep, T. S., Nga, T. T., Van Vinh Chau, N., Tuan, P. Q., Page, A. L., Ochiai, R. L., Schultsz, C., Wain, J., Bhutta, Z. A., Parry, C. M., Bhattacharya, S. K., Dutta, S., Agtini, M., Dong, B., Honghui, Y., Anh, D. D., Canh do, G., Naheed, A., Albert, M. J., Phetsouvanh, R., Newton, P. N., Basnyat, B., Arjyal, A., La, T. T., Rang, N. N., Phuong le, T., Van Be Bay, P., von Seidlein, L., Dougan, G., Clemens, J. D., Vinh, H., Hien, T. T., Chinh, N. T., Acosta, C. J., Farrar, J., and Dolecek, C. 2007. Antimicrobial drug resistance of Salmonella enterica serovar typhi in Asia and molecular mechanism of reduced susceptibility to the floroquinolones. Antimicrob. Agents Chemother. 51:4315–23.

Chuang, C. H., Su, L. H., Perera, J., Carlos, C., Tan, B. H., Kumarasinghe, G., So, T., Van, P.H., Chongthaleong, A., Hsueh, P. R., Liu, J. W., Song, J. H., and Chiu, C. H. 2008. Surveillance of antimicrobial resistance of Salmonella enterica serotype Typhi in seven Asian countries. Epidemiol. Infect. 12:1–4.

Crump, J., Luby, S., and Mintz, E. 2004. The global burden of typhoid fever. Bull. World Health Organ. 82:346–53.

Diekema, D. J., Pfaller, M. A., Schmitz, F. J., Smayevsky, J., Bell, J., Jones, R. N., Beach, M., and SENTRY Participants Group. 2001. Survey of infections due to Staphylococcus species: frequency of occurrence and antimicrobial susceptibility of isolates collected in the United States, Canada, Latin America, Europe, and the Western Pacific region for the SENTRY Antimicrobial Surveillance Program, 1997–1999. Clin. Infect. Dis. 32 (Suppl. 2):S114–32.

Dillon, J. A., Li, H., Sealy, J., Ruben, M., and Prabhakar, P. 2001a. Antimicrobial susceptibility of *Neisseria gonorrhoeae* isolates from three Caribbean countries: Trinidad, Guyana, and St. Vincent. Sex. Transm. Dis. 28:508–14.

Dillon, J. A., Rubabaza, J. P., Benzaken, A .S., Sardina, J. C., Li, H., Bandiera, M. G., and dos Santos Fernando Filho, E. 2001b. Reduced susceptibility to azithromycin and high percentages of penicillin and tetracycline resistance in *Neisseria gonorrhoeae* isolates from Manaus, Brazil, 1998. Sex. Transm. Dis. 28:521–6.

Dolecek, C., Tran, T. P., Nguyen, N. R., Le, T. P., Ha, V., Phung, Q. T., Doan, C. D., Nguyen, T. B., Duong, T. L., Luong, B. H., Nguyen, T. B., Nguyen, T. A., Pham, N. D., Mai, N. L., Phan, V. B., Vo, A. H., Nguyen, V. M., Tran, T. T., Tran, T. C., Schultsz, C., Dunstan, S. J., Stepniewska, K., Campbell, J. I., To, S. D., Basnyat, B., Nguyen, V. V., Nguyen, V. S., Nguyen, T. C., Tran, T. H., and Farrar J. 2008. A multi-center randomised controlled trial of gatifloxacin versus azithromycin for the treatment of uncomplicated typhoid fever in children and adults in Vietnam. PLoS ONE. 3(5):2188.

Dong, H., Bogg, L., Rehnberg, C., and Diwan, V. 1999. Association between health insurance and antibiotics prescribing in four countries in rural China. Health Policy. 48:29–45.

Dowell, S. F., Butler, J. C., Giebink, G. S., Jacobs, M. R., Jernigan, D., Musher, D. M., Rakowsky, A., and Schwartz, B. 1999. Acute otitis media: management and surveillance in an era of pneumococcal resistance-a report from the Drug-resistant *Streptococcus pneumoniae*. Therapeutic Working Group. Pediatr. Infect. Dis. J. 18:1–9.

Dromigny, J. A., and Perrier-Gros-Claude, J. D. 2003. Antimicrobial resistance of Salmonella enterica serotype Typhi in Dakar, Senegal. Clin. Infect. Dis. 37:465–6.

Dua V., Kunin C. M., and White L. V. 1994. The use of antimicrobial drugs in Nagpur, India. A window on medical care in a developing country. Soc. Sci. Med. 38:717–24.

Dutta, S., Rajendran, K., Roy, S., Chatterjee, A., Dutta, P., Nair, G. B., Bhattacharya, S. K., and Yoshida, S. I. 2002. Shifting serotypes, plasmid profile analysis and antimicrobial resistance pattern of *Shigellae* strains isolated from Kolkata, India during 1995–2000. Epidemiol. Infect. 129:235–43.

Engberg, J., Aarestrup, F. M., Taylor, D. E., Gerner-Smidt, P., and Nachamkin, I. 2001. Quinolone and macrolide resistance in *Campylobacter jejuni* and *C. coli*: resistance mechanisms and trends in human isolates. Emerg. Infect. Dis. 7:24–34.

Falagas, M. E., and Karveli, E. A. 2006. World Wide Web resources on antimicrobial resistance. Clin. Infect. Dis. 43:630–3.

Feikin, D. R., Dowell, S. F., Nwanyanwu, O. C., Klugman, K. P., Kazembe, P. N., Barat, L. M., Graf, C., Bloland, P. B., Ziba, C., Huebner, R. E., and Schwartz, B. 2000. Increased carriage of trimethoprim/sulfamethoxazole-resistant *Streptococcus pneumoniae* in Malawian children after treatment for malaria with sulfadoxine/pyrimethamine. J. Infect. Dis. 181:1501–5.

Feldman, C. 2004. Clinical relevance of antimicrobial resistance in the management of pneumococcal community-acquired pneumonia. J. Lab. Clin. Med. 143:269–83.

Felmingham, D., Cantón, R., and Jenkins, S. G. 2007. Regional trends in beta-lactam, macrolide, fluoroquinolone and telithromycin resistance among *Streptococcus pneumoniae* isolates 2001–2004. J. Infect. 552:111–8.

Fernandez, H., Mansilla, M., and Gonzalez, V. 2000. Antimicrobial susceptibility of C. jejuni subsp. Jejuni assessed by E-test and double dilution agar method in Southern Chile. Mem. Inst. Oswaldo Cruz. 95:247–9.

Fernández Viladrich, P. 2004. Management of meningitis caused by resistant *Streptococcus pneumoniae*. Management of multiple drug-resistant infections, ed. S. H. Gillespie, pp. 31–48. Human press, Totowa, NJ.

Fleming, D. T., and Wasserheit, J. N. 1999. From epidemiologic synergy to public health policy and practice: the contribution of other sexually transmitted diseases to sexual transmission of HIV infection. Sex. Transm. Infect. 75:3–17.

Fonseca, E. L., dos Santos Freitas, F. F., Vieira, V. V., and Vicente A. C. P. 2008. New qnr gene cassettes associated with superintegron repeats in Vibrio cholerae O1. Emerg. Infect. Dis. 14:1129–31.

Funke, G., Baumann, R., Penner, J. L., and Altwegg, M. 1994. Development of resistance to macrolide antibiotics in an AIDS patient treated with clarithromycin for *Campylobacter jejuni* diarrhea. Eur. J. Clin. Microbiol. Infect. Dis. 13:612–5.

Gani, L., Arif, H., Widjaja, S. K., Adi, R., Prasadja, H., Tampubolon, L. H., Lukito, E., and Jauri, R. 1991. Physicians' prescribing practice for treatment of acute diarrhoea in young children in Jakarta. J. Diarrhoeal. Dis. Res. 9:194–9.

Gilson, L. 1993. Health-care reform in developing countries. Lancet. 342:800.

Goldstein, E. J., and Garabedian-Ruffalo, S. M. 2002. Widespread use of fluoroquinolones versus emerging resistance in pneumococci. Clin. Infect. Dis. 35:1505–11.

Granat, S. M., Mia, Z., Ollgren, J., Herva, E., Das, M., Piirainen, L., Auranen, K., and Mäkelä P H. 2007. Longitudinal study on pneumococcal carriage during the first year of life in Bangladesh. Pediatr. Infect. Dis. J. 26:319–24.

Green, S., and Tillotson, G. 1997. Use of ciprofloxacin in developing countries. Pediatr. Infect. Dis. J. 16:150–9.

Grenet, K., Guillemot, D., Jarlier, V., Moreau, B., Dubourdieu, S., Ruimy, R., Armand-Lefevre, L., Bau, P., and Andremont, A. 2004. Antibacterial resistance, Wayampis Amerindians, French Guyana. Emerg. Infect. Dis. 10:1150–3.

Griffith, D. C., Kelly-Hope, L. A., and Miller, M. 2006. Review of reported outbreaks worldwide, 1995–2005. Am. J. Trop. Med. Hyg. 75:973–7.

Grose, C. 2004. Pyomiositis and bacterial myositis. In Textbook of paediatric infectious diseases, 5th ed., eds. R. D. Feigin, J. D. Cherry, G. J. Demmler and S. L. Kaplan, pp. 737–41. Saunders, Philadelphia:.

Guyot, A., Jarrett, B., Sanvee, L., and Dore D. 1998. Antimicrobial resistance of *Neisseria gonorrhoeae* in Liberia. Trans. R. Soc. Trop. Med. Hyg. 92:670–4.

Guzmán-Blanco, M., Casellas, J. M., and Sader, H. S. 2000. Bacterial resistance to antimicrobial agents in Latin America. The giant is awakening. Infect. Dis. Clin. North. Am. 14:67–81.

Hakenbeck, R., and Coyette, J. 1998. Resistant penicillin-binding proteins. Cell. Mol. Life Sci. 54:332–40.

Hampel, B., Hullmann, R., and Schmidt, H. 1997. Ciprofloxacin in pediatrics: worldwide clinical experience based on compassionate use – safety report. Pediatr. Infect. Dis. J. 16: 127–9.

Hansman, D., and Bullen, M. 1967. A resistant pneumococcus. Lancet. 2:264–5

Hansman, D., Glasgow, H., Surt, J., Devitt, H. L., and Douglas, R. 1971. Increased resistant to penicillin of pneumococci isolated from man. N. Eng. J. Med. 284:175–7.

Hart, C. A., and Kariuki, S. 1998. Antimicrobial resistance in developing countries. BMJ 317: 647–50.

Hart, C. A. 2003. Introduction to acute infective diarrhoea in *Manson's tropical Diseases,* eds G. C Cook and A. I. Zumla, pp. 907–9.

Hiramatsu, K., Katayama, Y., Yuzawa, H., and Ito, T. 2002. Molecular genetics of methicillin-resistant *Staphylococcus aureus.* Int. J. Med. Microbiol. 292:67–74.

Ho, P. L., Yung, R. W., Tsang, D. N., Que, T. L., Ho, M., Seto, W. H., Ng, T. K., Yam, W. C., and Ng, W. W. 2001. Increasing resistance of *Streptococcus pneumoniae* to fluoroquinolones: results of a Hong Kong multicentre study in 2000. J. Antimicrob. Chemother. 48:659–65.

Hooper, D. C. 1999. Mechanisms of fluoroquinolones resistance. Drug Resist. Update. 2:38–55.

Hsueh, P. R., Tseng, S. P., Teng, L. J., Ho, S. W., Hughes, R. A., and Cornblath, D. R. 2005. Guillain-Barré syndrome. Lancet. 366:1653–66.

Hui, L., Li, X. S., Zeng, X. J., Dai, Y. H., and Foy, H. M. 1997. Patterns and determinants of use of antibiotics for acute respiratory tract infection in children in China. Pediatr. Infect. Dis. J. 16:560–4.

Hyde, T. B., Gay, K., Stephens, D. S., Vugia, D. J., Pass, M., Johnson, S., Barrett, N. L., Schaffner, W., Cieslak, P. R., Maupin, P. S., Zell, E. R., Jorgensen, J. H., Facklam, R. R., and Whitney, C. G. 2001. Macrolide resistance among invasive *Streptococcus pneumoniae* isolates. JAMA 286:1857–62.

Ibarra, J. O., and Alvarado, D. E. 2007. Antimicrobial resistance of clinical and environmental strains of *Vibrio cholerae* isolated in Lima- Peru during epidemics in 1991 and 1998. Braz. J. Infect. Dis. 11:100–5.

Indalo, AA. 1997. Antibiotic sale behaviour in Nairobi: a contributing factor to antimicrobial drug resistance. East Afr. Med. J. 74:121–3.

Inoue, M., Lee, N. Y., Hong, S. W., Lee, K., and Felmingham, D. 2004. PROTEKT 1999–2000: a multicentre study of the antibiotic susceptibility of respiratory tract pathogens in Hong Kong, Japan and South Korea. Int. J. Antimicrob. Agents. 23: 44–51.

Ison, C. A., Dillon, J. A., and Tapsall, J. W. 1998. The epidemiology of global resistance among Neisseria gonorrhoeae and Haemophilus ducreyi. Lancet. 351 (Suppl. 3):8–11.

Ito, M., Yasuda, M., Yokoi, S., Ito, S., Takahashi, Y., Ishihara, S., Maeda, S., and Deguchi, T. 2004. Remarkable increase in central Japan in 2001–2002 of Neisseria gonorrhoeae isolates with decreased susceptibility to penicillin, tetracycline, oral cephalosporins, and fluoroquinolones. Antimicrob. Agents Chemother. 48:3185–7.

Iverson, C. J., Wang, S. A., Lee, M. V., Ohye, R. G., Trees, D. L., and Knapp, J. S. 2004. Fluoroquinolone resistance among Neisseria gonorrhoeae isolates in Hawaii, 1990–2000: role of foreign importation and increasing endemic spread. Sex. Transm. Dis. 31:702–8.

Iwalokun, B. A., Gbenle, G. O., Smith, S. I., Ogunledun, A., Akinsinde, K. A., and Omonigbehin, E. A. 2001. Epidemiology of shigellosis in Lagos, Nigeria: trends in antimicrobial resistance. J. Health Popul. Nutr. 19:183–90.

Iwanaga, M., Toma, C., Miyazato, T., Insisiengmay, S., Nakasone, N., and Ehara M. 2004. Antibiotic resistance conferred by a class I integron and SXT constin in Vibrio cholerae O1 strains isolated in Laos. Antimicrob. Agents Chemother. 48:2364–9.

Jacobs, M. R. 2001. Optimisation of antimicrobial therapy using pharmacokinetic and pharmacodynamic parameters. Clin. Microbiol. Infect. 711:589–96.

Jacobs, M. R., Felmingham, D., Appelbaum, P. C., Grüneberg, R. N., and the Alexander Project Group. 2003. The Alexander Project 1998–2000: susceptibility of pathogens isolated from community-acquired respiratory tract infection to commonly used antimicrobial agents. J. Antimicrob. Chemother. 52:229–246.

Jain, D., Sinha, S., Prasad, K. N., and Pandey, C. M. 2005. Campylobacter species and drug resistance in a north Indian rural community. Trans. R. Soc. Trop. Med. Hyg. 99:207–14.

Jones, M. E., Blosser-Middleton, R. S., Thornsberry, C., Karlowsky, J. A., and Sahm, D. F. 2003. The activity of levofloxacin and other antimicrobials against clinical isolates of Streptococcus pneumoniae collected worldwide during 1999–2002. Diagn. Microbiol. Infect. Dis. 47:579–86.

Kariuki, S., Revathi, G., Muyodi, J., Mwituria, J., Munyalo, A., Mirza, S., and Hart, C. A. 2004. Characterization of multidrug-resistant typhoid outbreaks in Kenya. J. Clin. Microbiol. 1477–82.

Kesah, C., Ben Redjeb, S., Odugbemi, T. O., Boye, C. S., Dosso, M., Ndinya Achola, J. O., Koulla-Shiro, S., Benbachir, M., Rahal, K., and Borg, M. 2003. Prevalence of methicillin-resistant Staphylococcus aureus in eight African hospitals and Malta. Clin. Microbiol. Infect. 9:153–6.

Klugman, K. P. 2002. Bacteriological evidence of antibiotic failure in pneumococcal lower respiratory tract infections. Eur. Respir. J. Suppl. 36:3s–8s.

Klugman, K. P, Madhi, S. A., and Feldman, C. 2007. HIV and pneumococcal disease. J. Clin. Microbiol. 4112:5582–7.

Kosek, M., Bern, C., and Guerrant, R. L. 2003. The global burden of diarrhoeal disease, as estimated from studies published between 1992 and 2000. Bull. World Health Organ. 81:197–204.

Krishna, B. V. S., Patil, A. B., and Chandrasekar, M. R. 2006. Fluoroquinolone-resistant Vibrio cholerae isolated during a cholera outbreak in India. Trans. R. Soc. Trop. Med. Hyg. 100: 224–226.

Kristiansson, C., Reilly, M., Gotuzzo, E., Rodriguez, H., Bartoloni, A., Thorson, A., Falkenberg, T., Bartalesi, F., Tomson, G., and Larsson, M. 2008. Antibiotic use and health-seeking behaviour in an underprivileged area of Perú. Trop. Med. Int. Health. 13:434–41.

Kumar, R., Gupta, N. S., and Shalini 2007. Multidrug-resistant typhoid fever. Indian. J. Pediatr. 74:39–42.

Kunin, C. M., Lipton, H. L., Tupasi, T., Sacks, T., Scheckler, W. E., Jivani, A., Goic, A., Martin, R. R., Guerrant, R. L., and Thamlikitkul, V. 1987. Social, behavioral, and practical factors affecting antibiotic use worldwide: report of Task Force 4. Rev. Infect. Dis. 9 (Suppl 3):270–85.

Kuo, C. Y., Su, L. H., Perera, J., Carlos, C., Tan, B. H., Kumarasinghe, G., So, T., Van, P. H., Chongthaleong, A., Song, J. H., and Chiu, C. H. 2008. Antimicrobial susceptibility of *Shigella* isolates in eight Asian countries, 2001–2004. J. Microbiol. Immunol. Infect. 412: 107–11.

Lansang, M. A., Lucas-Aquino, R., Tupasi, T. E., Mina, V. S., Salazar, L. S., Juban, N., Limjoco, T. T., Nisperos, L. E., and Kunin, C. M. 1990. Purchase of antibiotics without prescription in Manila, the Philippines. Inappropriate choices and doses. J. Clin. Epidemiol. 43:61–7.

Larsson, M., Kronvall, G., Chuc, N. T., Karlsson, I., Lager, F., Hanh, H. D., Tomson, G., and Falkenberg, T. 2000. Antibiotic medication and bacterial resistance to antibiotics: a survey of children in a Vietnamese community. Trop. Med. Int. Health. 5:711–21.

Leach, A. J., Shelby-James, T. M., Mayo, M., Gratten, M., Laming, A. C., Currie, B. J., and Mathews, J. D. 1997. A prospective study of the impact of community-based azithromycin treatment of trachoma on carriage and resistance of *Streptococcus pneumoniae*. Clin. Infect. Dis. 24:356–62.

Lester, S. C., del Pilar Pla, M., Wang, F., Perez Schael, I., Jiang, H., and O'Brien, T. F. 1990 The carriage of Escherichia coli resistant to antimicrobial agents by healthy children in Boston, in Caracas, Venezuela, and in Qin Pu, China. N. Engl. J. Med. 323(5):285–9.

Leverstein-van Hall, M. A., Box, A. T., Blok, H. E., Paauw, A., Fluit, A. C., and Verhoef, J. 2002. Evidence of extensive interspecies transfer of integron-mediated antimicrobial resistance genes among multidrug-resistant Enterobacteriaceae in a clinical setting. J. Infect. Dis. 186:49–56.

Levine, O. S., O'Brien, K. L., Knoll, M., Adegbola, R. A., Black, S., Cherian, T., Dagan, R., Goldblatt, D., Grange, A., Greenwood, B., Hennessy, T., Klugman, K. P., Madhi, S. A., Mulholland, K., Nohynek, H., Santosham, M., Saha, S. K., Scott, J. A., Sow, S., Whitney, C. G., and Cutts, F. 2006. Pneumococcal vaccination in developing countries Lancet. 367: 1880–2.

Levy, S. B., Marshall, B., Schluederberg, S., Rowse, D., and Davis, J. 1988. High frequency of antimicrobial resistance in human fecal flora. Antimicrob. Agents Chemother. 32:1801–6.

Lonks, J. R., Garau, J., Gomez, L., Xercavins, M., Ochoa de Echagüen, A., Green, I. F., Reiss, P. T., and Medeiros, A. A. 2002. Failure of macrolide antibiotic treatment in patients with bacteremia due to erythromycin-resistant *Streptococcus pneumoniae*. Clin. Infect. Dis. 35:556–64.

Lynch, P., Pittet, D., Borg, M. A., and Mehtar, S. 2007. Infection control in countries with limited resources. J. Hosp. Infect. 65:148–50.

Madhi, S. A., Petersen, K., Madhi, A., Wasas, A., and Klugman, K. B. 2000. Impact of human immunodeficiency virus type 1 on the disease spectrum of *Streptococcus pneumoniae* in South African children. Pediatr. Infect. Dis. J. 19:1141–7.

Mandell, L. A., Wunderink, R. G., Anzueto, A., Bartlett, J. G., Campbell, G. D., Dean, N. C., Dowell, S. F., File, T. M. Jr., Musher, D. M., Niederman, M. S., Torres, A., and Whitney, C. G. 2007. Infectious Diseases Society of America/American Thoracic Society consensus guidelines on the management of community-acquired pneumonia in adults. Clin. Infect. Dis. 44 (Suppl. 2):S27–72.

Mandomando, I., Espasa, M., Vallès, X., Sacarlal, J., Sigaúque, B., Ruiz, J., and Alonso P. 2007a. Antimicrobial resistance of Vibrio cholerae O1 serotype Ogawa isolated in Manhiça District Hospital, southern Mozambique. J. Antimicrob. Chemother. 60:662–4.

Mandomando, I. M., Macete, E. V., Ruiz, J., Sanz, S., Abacassamo, F., Vallès, X., Sacarlal, J., Navia, M. M., Vila, J., Alonso, P. L., and Gascon, J. 2007b. Etiology of diarrhea in children younger than 5 years of age admitted in a rural hospital of southern Mozambique. Am. J. Trop. Med. Hyg. 76:522–7.

Maskell, J. P., Sefton, A. M., and Hall, L. M. 1997. Mechanism of sulfonamide resistance in clinical isolates of *Streptococcus pneumoniae*. Antimicrob Agents Chemother. 41(10): 2121–6.

Maskey, A. P., Besnyat, B., Thwaites, G. E., Campell, J. I., Farra, J. J., and Zimmermann, M. D. 2008. Emerging trends in enteric fever in Nepal: 9124 cases confirmed by blood colture 1993–2003. Trans. R. Soc. Trop. Med. Hyg. 102:91–5.

Mathai, E., Chandy, S., Thomas, K., Antoniswamy, B., Joseph, I., Mathai, M., Sorensen, T. L., and Holloway, K. 2008. Antimicrobial resistance surveillance among commensal Escherichia coli in rural and urban areas in Southern India. Trop. Med. Int. Health. 13:41–5.

McGee, L., McDougal, L., Zhou, J., Spratt, B. G., Tenover, F. C., Gorge, R., Hakenbeck, R., Hryniewicz, W., Lefévre, J. C., Tomasz, A., and Klugman, K. P. 2001 Nomenclature of major antimicrobial-resistant clones of *Streptococcus pneumoniae* defined by the pneumococcal molecular epidemiology network. J. Clin. Microbiol. 39:2565–71.

Mendes, C., Marin, M. E., Quiñones, F., Sifuentes-Osornio, J., Siller, C. C., Castanheira, M., Zoccoli, C. M., López, H., Súcari, A., Rossi, F., Angulo, G. B., Segura, A. J., Starling, C., Mimica, I., and Felmingham D. 2003. Antibacterial resistance of community-acquired respiratory tract pathogens recovered from patients in Latin America: results from the PROTEKT surveillance study 1999–2000. Braz. J. Infect. Dis. 71:44–61.

Metlay, J. P. 2002. Update on community-acquired pneumonia: impact of antibiotic resistance on clinical outcomes. Curr. Opin. Infect. Dis. 152:163–7.

Moreno, S., García-Leoni, M. E., Cercenado, E., Diaz, M. D., Bernaldo de Quirós, J. C., and Bouza, E. 1995. Infections caused by erythromycin-resistant *Streptococcus pneumoniae*: incidence, risk factors, and response to therapy in a prospective study. Clin. Infect. Dis. 20:1195–200.

Musher, D. M., Dowell, M. E., Shortridge, V. D., Flamm, R. K., Jorgensen, J. H., Le Magueres, P., and Krause, K. L. 2002. Emergence of macrolide resistance during treatment of pneumococcal pneumonia. N. Engl. J. Med. 346:630–1.

Naik, D. G. 2006. Prevalence and antimicrobial susceptibility patterns of *Shigella* species in Asmara, Eritrea, northeast Africa. J. Microbiol. Immunol. Infect. 395:392–5.

Narang, P., Mendiratta, D. K., Deotale, V. S., and Narang R. 2008. Changing patterns of *Vibrio cholerae* in Sevagram between 1990 and 2005. Indian J. Med. Microbiol. 26: 40–4

Nel, E. J. 2000. Neonatal meningitis: mortality, cerebrospinal fluid, and microbiological findings. Trop. Pediatr. 46:237–9.

Nizami, S. Q., Khan, I. A., and Bhutta, Z. A. 1996. Drug prescribing practices of general practitioners and paediatricians for childhood diarrhoea in Karachi, Pakistan. Soc. Sci. Med. 42:1133–9.

Nuermberger, E. L., and Bishai, W. R. 2004. Management of community-acquired pneumonia caused by drug-resistant *Streptococcus pneumoniae*. In *Management of multiple drug-resistant infections*. ed S. H Gillespie , pp 3–29. Human press, Totowa, NJ.

O'Dempsey, T. J., McArdle, T. F., Morris, J., Lloyd-Evans, N., Baldeh, I., Laurence, B. E., Secka, O., and Greenwood, B. M. 1996. A study of risk factors for pneumococcal disease among children in a rural area of West Africa. Int. J. Epidemiol. 25:885–93.

Okeke, I. N., Fayinka, S., and Lamikanra, A. 2000. Antibiotic resistance trends in *Escherichia coli* from apparently healthy Nigerian students (1986–1998). Emerg. Infect. Dis. 6:393–6.

Okeke, I. N., Klugman, K. P., Bhutta, Z. A., Duse, A. G., Jenkins, P., O'Brien, T. F., Pablos-Mendez, A., and Laxminarayan R. 2005. Antimicrobial resistance in developing countries. Part II: strategies for containment. Lancet Infect. Dis. 5:568–80.

Okeke, I. N., Aboderin, O. A., Byarugaba, D. K., Ojo, K. K., and Opintan J. A. 2007. Growing problem of multidrug-resistant enteric pathogens in Africa. Emerg. Infect. Dis. 13:1640–6

Okuda, J., Ramamurthy, T., and Yamasaki, S. 2007. Antibacterial activity of ciprofloxacin against clinical strains of Vibrio cholerae O139 recently isolated from India. Yakugaku Zasshi. 127:903–4.

Osler, W. 1901. The principles and practice of medicine. 4th ed. D. Appleton and Company, New York.

Osterblad, M., Hakanen, A., Manninen, R., Leistevuo, T., Peltonen, R., Meurman, O., Huovinen, P., and Kotilainen, P. 2000. A between-species comparison of antimicrobial resistance in enterobacteria in fecal flora. Antimicrob. Agents Chemother. 44:1479–84.

Pallecchi L., Lucchetti C., Bartoloni A., Bartalesi F., Mantella A., Gamboa H., Carattoli A., Paradisi F., and Rossolini G. M. 2007. Population structure and resistance genes in antibiotic-resistant bacteria from a remote community with minimal antibiotic exposure. Antimicrob. Agents Chemother. 51:1179–84

Paredes, P., de la Peña, M., Flores-Guerra, E., Diaz, J., and Trostle, J. 1996. Factors influencing physicians' prescribing behaviour in the treatment of childhood diarrhoea: knowledge may not be the clue. Soc. Sci. Med. 42:1141–53.

Parry, C. M. 2003. Antimicrobial drug resistance in Salmonella enterica. Curr. Opin. Infect. Dis. 16:467–2.

Parry, C. M. 2004. Management of multiple drug-resistant *Salmonella* infections. In *Management of multiple drug-resistance infections*, ed. S. H. Gillespie , pp. 189–202. Humana Press, Totowa, NJ.

Pazhani, G. P., Sarkar, B., Ramamurthy, T., Bhattacharya, S. K., Takeda, Y., and Niyogi, S. K. 2004. Clonal multidrug-resistant *Shigella dysenteriae* type 1 strains associated with epidemic and sporadic dysenteries in Eastern India. Antimicrob. Agents Chemother. 48: 681–4.

Phillips, I., Casewell, M., Cox, T., De Groot, B., Friis, C., Jones, R., Nightingale, C., Preston, R., and Waddell, J. 2004. Does the use of antibiotics in food animals pose a risk to human health? A critical review of published data. J. Antimicrob. Chemother. 53:28–52.

Pihlajamäki, M., Kotilainen, P., Kaurila, T., Klaukka, T., Palva, E., and Huovinen, P. 2001. Macrolide-resistant *Streptococcus pneumoniae* and use of antimicrobial agents. Clin. Infect. Dis. 33:483–8.

Pikis, A., Donkersloot, J. A., Rodriguez, W. J., and Keith, J. M. 1998. A conservative amino acid mutation in the chromosome-encoded dihydrofolate reductase confers trimethoprim resistance in *Streptococcus pneumoniae*. J. Infect. Dis. 178:700–6.

Radyowijati, A., and Haak, H. 2003. Improving antibiotic use in low-income countries: an overview of evidence on determinants. Soc. Sci. Med. 57:733–44.

Rahman, M., Sultan, Z., Monira, S., Alam, A., Nessa, K., Islam, S., Nahar, S., Shama-A-Waris, Alam Khan, S., Bogaerts, J., Islam, N., and Albert, J. 2002. Antimicrobial susceptibility of *Neisseria gonorrhoeae* isolated in Bangladesh (1997 to 1999): rapid shift to fluoroquinolone resistance. J. Clin. Microbiol. 40:2037–40.

Rodríguez, A. J., Niño Cotrina, R. A., Neyra Pérez, C., Rodríguez, C. N., Barbella, R., Lakatos, M., Molina, N., García, A., Duque, C., and Meijomil, P. 2001. Comparative study of antimicrobial resistance of Escherichia coli strains isolated from urinary tract infection in patients from Caracas and Lima. Antimicrob. Chemother. 47:903–4.

Rowe A. K, de Savigny D., Lanata C. F., and Victora C. G. 2005. How can we achieve and maintain high-quality performance of health workers in low-resource settings? Lancet. 366:1606.

Saha, S. K., Baqui, A. H., Darmstadt, G. L., Ruhulamin, M., Hanif, M., El Arifeen, S., Santosham, M., Oishi, K., Nagatake, T., and Black, R. E. 2003. Comparison of antibiotic resistance and serotype composition of carriage and invasive pneumococci among Bangladeshi children: implications for treatment policy and vaccine formulation. J. Clin. Microbiol. 4112:5582–7.

Saïd-Salim, B., Mathema, B., and Kreiswirth, B. N. 2003. Community-acquired methicillin-resistant *Staphylococcus aureus*: an emerging pathogen. Infect. Control. Hosp. Epidemiol. 24:451–5.

Sack, D. A., Lyke, C., McLaughlin, C., and Suwanvanichkij, V. 2001. Antimicrobial resistance in shigellosis, cholera and campylobacteriosis. WHO/CDS/CSR/DRS/2001.8. WHO Document Production Services, Geneva, Switzerland.

Sack, D. A., Sack, R. B., Nair G. B., and Siddique, A. K. 2004. Cholera. Lancet. 363:223–33.

Samie, A., Ramalivhana, J., and Igumbor, E. O. 2007. Prevalence, haemolytic and haemoagglutination activities and antibiotic susceptibility profiles of *Campylobacter spp.* isolated from human diarrhoeal stools in Vhembe District, South Africa. J. Health Popul. Nutr. 25:406–13.

Schmidt, L. H., and Sesler, C. L. 1943. Development of resistance to penicillin by pneumococci. Proc. Soc. Exp. Biol. Med 52:353–7.

Schrag, S. J., Beall, B., and Dowell, S. 2001. Resistant pneumococcal infections: the burden of disease and challenger in monitoring and controlling antimicrobial resistance. Da: a background document for the WHO global strategy for containment of antimicrobial resistance.

Serichantalergs, O., Dalsgaard, A., Bodhidatta, L., Krasaesub, S., Pitarangsi, C., Srijan, A., and Mason C. J. 2007. Emerging fluoroquinolone and macrolide resistance of *Campylobacter jejuni* and *Campylobacter coli* isolates and their serotypes in Tai children from 1991 to 2000. Epidemiol. Infect. 135:1299–306.

Serkkola A. 1990. Medicines, pharmacy and family: Triplicity of self-medication in Mogadishu, Somalia. Occasional papers #11. University of Helsinki, Institute of Development Studies, Helsinki.

Shao-Kang, Z., Sheng-Ian, T., You-de, G., and Bloom, G. 1998. Drug prescribing in rural health facilities in China: implications for service quality and cost. Trop. Doct. 28:42–48.

Shapiro, R. L., Kumar, L., Phillips-Howard, P., Ochieng, J. B., Mintz, E., Wahlquist, S., Waiyaki, P., and Slutsker, L. 2001. Antimicrobial resistant bacterial diarrhoea in rural western Kenya. J. Infect. Dis. 183:1701–4.

Shears, P. 2001. Antibiotic resistance in the tropics. Epidemiology and surveillance of antimicrobial resistance in the tropics. Trans. R. Soc. Trop. Med. Hyg. 95:127–30.

Shorr, A. F. 2007. Epidemiology of staphylococcal resistance. Clin. Infect. Dis. 45 (Suppl. 3):S171–6.

Smith, A. M., Keddy, K. H., and De Wee, L. 2008. Characterization of cholera outbreak isolates from Namibia December 2006 to February 2007. Epidemiol. Infect. 136(9):1207–9.

Smith, I. M., and Vickers, A. B. 1960. Natural history of 338 treated and untreated patients with staphylococcal septicaemia (1936–1955). Lancet. 18:1318–22.

Song, J. H., Lee, N. Y., Ichiyama, S., Yoshida, R., Hirakata, Y., Fu, W., Chongthaleong, A., Aswapokee, N., Chiu, C. H., Lalitha, M. K., Thomas, K., Percra, J., Yee, T. T., Jamal, F., Warsa, U. C., Vinh, B. X., Jacobs, M. R., Appelbaum, P. C., and Pai, C. H. 1999. Spread of drug-resistant *Streptococcus pneumoniae* in Asian countries: Asian Network for Surveillance of Resistant Pathogens ANSORP Study. Clin. Infect. Dis. 286:1206–11.

Sörberg, M., Farra, A., Ransjö, U., Gårdlund, B., Rylander, M., Wallén, L., Kalin, M., and Kronvall, G. 2002. Long-term antibiotic resistance surveillance of gram-negative pathogens suggests that temporal trends can be used as a resistance warning system. Scand. J. Infect. Dis. 34:372–8.

Sosa, J., Ramirez-Arcos, S., Ruben, M., Li, H., Llanes, R., Llop, A., and Dillon, J. A. 2003. High percentages of resistance to tetracycline and penicillin and reduced susceptibility to azithromycin characterize the majority of strain types of *Neisseria gonorrhoeae* isolates in Cuba, 1995–1998. Sex. Transm. Dis. 30:443–8.

Srikantiah, P., Girgis, F. Y., Luby, S. P., Jennings, G., Wasfy, M. O., Crump, J. A., Hoekstra, R. M., Anwer, and M., Mahoney, F. J. 2006. Population-based surveillance of typhoid fever in Egypt. Am. J. Trop. Med. Hyg. 74:114–9.

Sterky, G., Tomson, G., Diwan, V. K., and Sachs L. 1991. Drug use and the role of patients and prescribers. J. Clin. Epidemiol. 44 (Suppl 2): 67S–72S.

Swanston, W. H. Ali, C., Mahabir, B. S., Prabhakar, P., Basraj, S., and George, J. 1997. Antibiotic susceptibility of *Neisseria gonorrhoeae* in Trinidad and Tobago. West Indian Med. J. 46:107–10.

Tapsall, J. 2001. Antimicrobial resistance in *Neisseria gonorrhoeae*. Department of Communicable Disease Surveillance and Response, WHO, Geneva, Switzerland.

The WHO Western Pacific Gonococcal Antimicrobial Surveillance Programme. 2008. Annual Report of the Australian Gonococcal Surveillance Programme 2007. Commun. Dis. Intell. 32 (2):227–31.

Tjaniadi, P., Lesmana, M., Subekti, D., Machpud, N., Komalarini, S., Santoso, W., Simanjuntak, C. H., Punjabi, N., Campbell, J. R., Alexander, W. K., Beecham, H. J. 3rd., Corwin, A. L., and Oyofo, B. A. 2003. Antimicrobial resistance of bacterial pathogens associated with diarrheal patients in Indonesia. Am. J. Trop. Med. Hyg. 68:666–70.

Tomson G., and Sterky G. 1986. Self-prescribing by way of pharmacies in three Asian developing countries. Lancet. 2:620–2.

Uppal, R., Sarkar, U., Giriyappanavar, C., and Kacker, V. 1993. Antimicrobial drug use in primary health care. J. Clin. Epidemiol. 46:671–3.

Vallès, X., Flannery, B., Roca, A., Mandomando, I., Sigaúque, B., Sanz, S., Schuchat, A., Levine, M., Soriano-Gabarró, M., and Alonso, P. 2006. Serotype distribution and antibiotic susceptibility of invasive and nasopharyngeal isolates of *Streptococcus pneumoniae* among children in rural Mozambique, Trop. Med. Int. Health. 11:358–66.

Waldvogel, F. A. 2000. *Staphylococcus aureus*. In Principles and practice on infectious diseases, eds G. L. Mandell, J. E. Bennett, R. Dolin, pp. 1069–2092. 5th ed., Edinburgh: Churchill, Livingston.

Walson J. L., Marshall B., Pokhrel B. M., Kafle K. K., and Levy S. B. 2001. Carriage of antibiotic-resistant fecal bacteria in Nepal reflects proximity to Kathmandu. J. Infect. Dis. 185:1542.

Wolff M. J. 1993. Use and misuse of antibiotics in Latin America. Clin. Infect. Dis. 17 (Suppl 2):S346–51.

World Health Organization. 1997. Resistance to antimicrobial agents. Wkly. Epidemiol. Rec. 72:333–40.

World Health Organization. 2001a. WHO Global strategy for containment of antimicrobial resistance. WHO/CDS/CSR/DRS/2001.2. WHO Document Production Services, Geneva, Switzerland.

World Health Organization. 2001b. Global prevalence and incidence of selected curable sexually transmitted diseases: overview and estimates. WHO/HIV_AIDS/2001.02. WHO Document Production Services, Geneva, Switzerland.

World Health Organization. 2002. Communicable Disease Surveillance and Response CSR/WHO/2002. WHO Document Production Services, Geneva, Switzerland.

World Health Organization. 2005. Guidelines for the control of shigellosis, including epidemics due to *Shigella dysenteriae* type 1. WHO Document Production Services, Geneva, Switzerland.

World Health Organization. 2007. The top ten causes of death. Fact sheet N 310. WHO Document Production Services, Geneva, Switzerland.

Zaidi, A. K., Huskins, W. C., Thaver, D., Bhutta, Z. A., Abbas, Z., and Goldmann, D. A. 2005. Hospital-acquired neonatal infections in developing countries. Lancet. 365:1175–88.

Zetola N., Francis J. S., Nuermberger E. L., and Bishai W. R. 2005. Community-acquired methicillin-resistant *Staphylococcus aureus*: an emerging threat. Lancet Infect. Dis. 5:275–86.

Zheng, H. P., Cao, W. L., Wu, X. Z., and Yang, L. G. 2003. Antimicrobial susceptibility of *Neisseria gonorrhoeae* strains isolated in Guangzhou, China, 1996–2001. Sex. Transm. Infect. 79:399–402.

Chapter 13
Prevalence of Resistant Enterococci in Developing Countries

Lorena Abadía-Patiño

Abstract Glycopeptide resistance in enterococci is a problem of worldwide distribution, with a wider incidence in industrialized countries. This incidence is reflected as both nosocomial illnesses in the United States, because of the extensive use of antibiotic therapy, and community-borne diseases in Europe, due to the use of avoparcin as a growth promoter in animal husbandry. Vancomycin-resistant strains, however, have started to emerge in the developing world, where VRE carriers in the hospital may infect themselves upon receiving either third-generation cephalosporins to combat Gram-negative infections or vancomycin, for the treatment of severe methicillin-resistant *Staphylococcus* infections. Detecting VRE carriers is then of paramount importance, and although several techniques for that purpose do exist, no gold standard has been proposed to certify the accuracy or sensitivity of the assays, and it all depends on the inoculum in the intestine.

13.1 Epidemiology in Developing Countries

13.1.1 East Asia and Pacific

In Korea, although VRE strains have been isolated since 1992 (Kim and Song 1998), the VRE prevalence index remained low until 1997. Of 5,275 enterococci isolated between 1995 and 1997, eight were VRE (Shin et al. 2003). Between 1998 and 2000 the VRE index abruptly escalated to 325 isolates (Park et al. 2007). In a study done on Korean patients between 1998 and 2005, no VRE were detected in 1998, but in 2005, *Enterococcus faecalis* and *E. faecium* exhibited 16 and 12% glycopeptide resistance, respectively (Yang et al. 2007). In the Ajou University Hospital, 57 *E. faecium vanA* were isolated. The strains were studied according to ISs distribution in the Tn*1546*. Two types of Tn*1546* were

L. Abadía-Patiño (✉)

Departamento de Biomedicina, Instituto de Investigaciones en Biomedicina y Ciencias Aplicadas, Universidad de Oriente, Cumaná. Edo. Sucre, Venezuela

e-mail: grurbact_iibcaudo@yahoo.com

A. de J. Sosa et al. (eds.), *Antimicrobial Resistance in Developing Countries*, DOI 10.1007/978-0-387-89370-9_13, © Springer Science+Business Media, LLC 2010

found (Park et al. 2007). The type II strains had a higher transfer frequency, due to the deletion caused by the IS*1216 V* at the 3' end, which may be influencing the rapid spread of VRE in the Ajou University Hospital. The Tn*1546* genetic rearrangements are due to the ISs, which function as excellent molecular markers (Woodford et al. 1998). There are reports that describe the isolation of VRE in east Asian regions and countries, including Japan, Korea, and Taiwan (Lauderdale et al. 2002; Yu et al. 2003). Since the initial isolation of VRE in Japan in 1996, the isolation frequency has increased (Fujita et al. 1998). Little information is available regarding the prevalence of VRE in mainland China, in spite of the fact that glycopeptides have been used there for decades.

In a study done in the Prince of Wales Hospital in Hong Kong, 498 strains of enterococci were isolated between 1997 and 1998, of which only 12 were VRE; it was not a clonal dissemination and the incidence was sporadic (Cheng et al. 2002). VRE isolates were studied between 2001 and 2005 (Zheng et al. 2007). Thirteen strains of *E. faecium vanA* were isolated in the course of the study, unrelated to each other or to human or animal strains. The first case of community-borne endophthalmitis caused by VRE was reported in Taiwan (Tang et al. 2007); in the past 10 years few cases have been reported worldwide (Han et al. 1997; Esmaeli et al. 2003; Scott et al. 2003).

In Malaysia, *E. faecalis (vanA)* and *E. casseliflavus (vanC2)* are the most prevalent species in chickens sold at the retail level (Radu et al. 2001). In a study done on chicken samples in Sri Serdang, Selangor, the principal species with high glycopeptide resistance were *E. faecalis* and *E. hirae*. It was not possible to establish a correlation between the plasmid profiles and antibiotic resistance (Toosa et al. 2001). The difference in prevalence of the species reflects the ability of the enterococcal clones to colonize the animal ecology under specific environmental conditions, which generate different evolutionary selection pressures (Aznar et al. 2004; Taneja et al. 2004). Although VRE is principally associated with nosocomial infections, a few cases have been reported in the community (Aznar et al. 2004; Taneja et al. 2004). In Malaysia, the first case of community-borne skin and soft tissue infection was reported in a male patient with a buccal abscess caused by *E. faecium* of the *vanA* genotype, detected at the University of Malaysia Medical Center, Kuala Lumpur, Malaysia (Raja et al. 2005).

13.1.2 Europe and Central Asia

The first report of a case in Turkey came in 2001, caused by *E. faecium,* isolated by hemoculture (Basustaoglu et al. 2001). In 2002, the first outbreak of 20 clonally disseminated strains of VRE was reported in Turkey (Colak et al. 2002). Of 23 enterococci isolated in the Bayindir-Ankara hospital, 34.8% were of high-level and 60.9% of low-level glycopeptide resistance, indicating an increase in the rate of resistance in that hospital (Coleri et al. 2004). In that same year, another study was done without finding any VRE (Oncu et al. 2004).

In 1996, 10 years after the first isolation of the highly resistant *E. faecium* in the world, a study was done in Polish hospitals in which no VRE strain was detected (Hryniewicz et al. 1998). Between 1996 and 1997, the number of isolates detected in the Gdańsk hospital evidenced the first nosocomial outbreak in Poland (Hryniewicz et al. 1999a; Samet et al. 1999), which was subsequently genotyped as *vanA*. In the Hematology Unit of a Polish hospital, VanA and VanB phenotypes of VRE strains were isolated and found to have caused outbreaks in the cities of Kraków, Poznań and Warsaw (Kawalec et al. 2001, 2004). A total of 291 *E. faecalis* isolates from Polish hospitals were evaluated over a period of 10 years, from 1996 to 2005 (Ruiz-Garbajosa et al. 2006), from different clinical samples. Thirty *E. faecalis* samples were found to be VRE, of which 5.5% were *vanA* and 4.8% *vanB* (Kawalec et al. 2007).

The first isolate of VanA VRE found in Serbia in this century (Stošovic et al. 2004) was detected in a female patient in the Cardiovascular Disease Clinic in Belgrade. In Serbia, avoparcin is not used due to its side effects in animal husbandry, and hospital use of vancomycin has been restricted because of its high cost. It was suspected that the strain was imported, but the patient had not been out of Serbia, which suggests that it originated in the hospital environment. The origin of this strain is still unknown. Among the enterococci isolated from unsterile locations in Hungary, Poland, the Czech Republic, and Slovakia, the VRE prevalence rate ranged between 1 and 3% (Krcméry et al. 1998; Hryniewicz 1999b).

The first signs of VRE appearance and increase occurred in the Department of Hematology and Oncology of the University Hospital of Olomuoc and Ostrava in 1999, when the prevalence index was determined to be 10% (Kollar 1999; Urbásková 1999). As a consequence of this outbreak, a 3-year nationwide study was initiated in six university hospitals, from January 1997 until January 2000 (Bilikova et al. 2001), which found a 5.3% VRE rate.

In a community study in the Czech Republic, in which 5,283 rectal swabs were taken, 558 enterococci were isolated, of which only 1.6% were found to be VRE (Kolař et al. 2004). Of this 1.6% of fecal carriers, the species identified were VanA phenotype *E. faecium*, VanB phenotype *E. faecalis*, and VanC phenotypes *E. gallinarum* and *E. casseliflavus*, with *E. casseliflavus* being most prevalent; this represents a risk of spread from the community to the hospital environment. This is not the first study done in the country; nosocomial cases of VRE had been previously reported (Bergerova 1997; Kolař et al. 2002). In Europe, the VRE prevalence rate varies from 2 to 28%, principally VanA/ VanB mediated (Endtz et al. 1997).

A group of investigators from the Department of Veterinary Microbiology, Infections and Parasitic Diseases at the University of Trakia in Bulgaria analyzed 520 samples from farm chickens between February 2001 and July 2002, and found no VRE. The principal species in the intestines of the chickens was *E. faecium* (87.27%). Only one isolate of *E. faecalis* had intermediate susceptibility to vancomycin (16 mm diffusion with 30 µg disk) and grew in the vancomycin screening, although its genotype was not confirmed by PCR (Urumova et al. 2005). Probably the absence of VRE in the avian samples was due to the short time during which avoparcin was used in Bulgaria. An indirect proof of antibiotics as growth factors is the presence of VRE in healthy humans, animals, and environmental sources. This is evidence that the presence of VRE in commensal flora can be spread through the food chain and be a pathway of dissemination (Hawkey 1986).

It has been demonstrated that in week-old chickens the enterococcal flora is predominated by *E. faecalis* and *E. faecium*. After the second week, species diversity changes to *E. faecium, E. durans*, and *E. hirae* (Urumova et al. 2005). In Bulgaria, avoparcin was used for a very short time in the early 1990s, principally in the growth phase of chickens, and was then substituted with bacitracin, virginiamycin, and bambermycin. Various studies have shown that the apparition of VRE in the intestinal flora of animals raised with antibiotics as growth factors has been observed even after the postapplication period (Butaye et al. 1999; van den Bogaard et al. 2002). The prevalence of VRE varies widely in different regions of the world (Cetinkaya et al. 2000; Lauderdale et al. 2002; Hsueh et al. 2005). VRE is widely distributed in Taiwan, all of Asia and Europe, as a community reservoir. This has been demonstrated in chicken samples in Taiwan, where 28 VRE were isolated from 30 samples due to the wide use of avoparcin in farm animals (Lauderdale et al. 2002).

As VRE data from Russia were not available until early in this century, a group of investigators from the Institute of Antimicrobial Chemotherapy, Smolensk State Medical Academy, decided to analyze 362 clinical enterococci strains (Dekhnich 2001) isolated between 2000 and 2001 from 15 hospitals in Moscow, Ryazan, Smolensk, St. Petersburg, Krasnodar, N. Novgorod, Kazan, Ekaterinburg, Krasnoyarsk, Novosibirsk, and Tomsk, from patients coming from the Surgical, Neonatal, General Medicine, and Intensive Care Units. Not one VRE was detected, which indicates that glycopeptide resistance did not seem to be a problem at that time in Russia. Shortly after, in June 2002 in Russia, the first cases of *E. gallinarum* VanB were isolated from a patient in a hematology center. In the study 31 VRE were obtained from 3,578 samples, with a prevalence of intestinal carriers of 3.5%. The distribution was as follows: One *E. faecalis*, 23 *E. faecium* VanA, three *E. faecium*, and four *E. gallinarum* VanB (Mironova et al. 2006).

13.1.3 Latin America and the Caribbean

The prevalence of VRE in Latin America has remained low, less than 2% (Low et al. 2001). The majority of the strains isolated in Latin America are of the *vanA* genotype (Dalla Costa et al. 1998; Marin et al. 1998; Panesso et al. 2002), although Chile and Argentina have already reported cases of *vanB* (Miranda et al. 2003). In

August 1996, the first glycopeptide-resistant *E. faecium* of the VanD phenotype was isolated in Brazil (Dalla Costa et al. 1998). No other VRE strain was isolated in that hospital nor was there any source of infection identified. *E. faecium* VanA was subsequently isolated (Zanella et al. 1999). In 2000, the first *E. faecalis* VRE was isolated in a tertiary care hospital in Porto Alegre, in southern Brazil (Cereda et al. 2002). A vigilance program was immediately established in all hospitals by the infection committees, through which in February 2002 a patient carrying *E. gallinarum* VanA (Camargo et al. 2004a) was detected. The patient had been colonized by VanA phenotype *E. faecalis*; it was supposed that there had been an in vivo transfer of the *vanA* operon. Since the appearance of the first VRE strains in Brazil, their spread has continued to grow in the entire country, as has been reported in the international literature. The said dissemination was noted principally in the southeast, central, and southern regions of Brazil (Dalla Costa et al. 2000; Cereda et al. 2001; Reis et al. 2001; Almeida et al. 2004; Camargo et al. 2004b; Furtado et al. 2005). Before long, cases of the same clone appeared in hospitals in the northeast of Brazil (Vilela et al. 2006; Palazzo et al. 2006). The rapid dissemination of the resistant *vanA* genotype was due to the association with Tn*1546*- and Tn1546-like elements, which are carried by plasmids (Arthur et al. 1996) at the moment of conjugation between VRE strains and vancomycin-sensitive enterococci. Antibiotics were used as growth promoters in Brazil for a period of 10 years. During that time, VRE was not found in chicken samples (Leme et al. 2000). In another study done with chickens in the central west region of Brazil in August and September 2005, neither *vanA*- nor *vanB*- carrying strains were found (Batista Xavier et al. 2006).

In the city of Mendoza, Argentina, the presence of VanB phenotype VRE was first reported in 1996 (Marin et al. 1998), as was the first detection of *E. faecalis* (Casellas et al. 1996). The following year three VRE strains with identical restriction profiles were detected in samples from the Hospital Aeronáutico de Buenos Aires using pulsed field electrophoresis (Targa et al. 1997). In a hospital in Córdova, Argentina, a study was done of the prevalence of VRE carriers (Llittvik et al. 2006). A total of 235 rectal swabs were performed on 147 ICU patients, detecting 12.2% VRE, 94.4% of which identified as *E. faecium*, and 5.6% *E. gallinarum* (Llittvik et al. 2006). The WHONET Antimicrobial Surveillance Program Report, Argentina, 2002, showed an increase of VRE from 0.8% in 1998 to 11% in 2002. In a male diabetic patient in the Evita General Hospital for Acute Patients, after 53 days of hospitalization and treatment with third-generation cephalosporins, lincosamides, and glycopeptides, the first strain of VanA phenotype *E. gallinarum* was isolated. Due to this unusual isolation, surveillance rectal swabs were performed on the other ICU patients between August 2000 and February 2001. A total of 256 rectal swabs were analyzed from 124 patients: 35 VRE were obtained, 16 of them *E. gallinarum*, all but one being carriers of the *vanA* genotype (Corso et al. 2005).

An investigative analysis of the enterococci resistance profile was done in a public hospital in Guadalajara, Mexico (Morfín et al. 1992). The study only detected vancomycin resistance in *E. faecium* (Morfín et al. 1999). Few cases were ever reported in Mexico before the 1990s (Sifuentes-Osornio et al. 1996; McDonald et al. 1999). All the isolates of *E. faecium* and *E. faecalis* tested sensitive to vancomycin at an oncology hospital in Mexico City (Cornejo-Juarez et al. 2005). In a 2-year prospective study (January 1998 to December 1999) done at the National Pediatric Hospital in Mexico City, 97 strains of enterococci were processed. Seven strains, or 7.2% of the samples, were vancomycin-resistant *E. faecium*, five VanA and two VanB (Calderón-Jaimes et al. 2003), an increase from previously reported figures (Miranda et al. 2001). From May 2004 until April 2005 in the Instituto Nacional de Ciencias Médicas y Nutrición "Salvador Zubirán" (INCMNSZ) the rate of resistance to glycopeptides was 0.27%. In May 2005, the first VRE was detected at the INCMNSZ, and within 12 months the rate had risen to 6.23%, 23 times greater than the previous year, according to a retrospective study done at the same institution (Cuellar-Rodríguez et al. 2007). At the Public Hospital in Guadalajara, 681 strains of enterococci (*E. faecalis* and *E. faecium*) isolated between 1991–1993 and 1996–1999 were studied. It was not until 1999 that two vancomycin-resistant *E. faecium* were detected among the 52 strains isolated (Morfín et al. 1999).

In 1998, the first VanA VRE strains were isolated from 23 patients from different infection sites at the University Hospital at Medellin, Colombia (Panesso et al. 2002). A multicentric study was subsequently done in which five *E. faecalis* and seven *E. faecium* glycopeptide-resistant strains were isolated from urine, peritoneal and pleural fluid, hemoculture, and surgical wounds. All isolates of *E. faecalis* amplified VanB ligase and those of *E. faecium* amplified VanA ligase. The said study was done between 2001 and 2002 in five Colombian cities, but VRE was isolated only in Bogotá (Arias et al. 2003). Colombia has particular characteristics that exert direct influence over the emergence and dissemination of bacterial resistance, such as the availability of "over-the-counter" compounds, a high referral rate between medical institutions, and a lack of multicentric surveillance data (Arias et al. 2003).

In April 1999 the National Bacterial Resistance Network was established in Quito, Ecuador, with data from 13 hospitals (Zurita et al. 2001). Glycopeptide-resistant *E. faecium* was isolated beginning in August 2001.

The first three cases of VRE detected in Uruguay were isolated from a urinary colonization, a polymicrobial intra-abdominal infection, and an infectious endocarditis in the Pasteur Hospital and the Hospital de Clínicas. In the Pasteur Hospital, the first case occurred in 2000 and the second in 2003 (Bazet et al. 2005). The prevention of VRE-caused infections is the first step in controlling VRE-related morbimortality. VRE surveillance studies of hemodialysis and ICU patients are not done in all developing countries.

The few studies that have been done show the absence of VRE carriers, and this may be due to the detection methodology employed. Previous studies have shown greater VRE isolation in carrier patients when the rectal swabs or feces are inoculated into enriched broth before placing them onto screen plates (Novicki et al. 2004). Countries in which VRE cases have not been detected are not being efficient in searching for possible carrier patients, in the belief that the problem does not exist in their hospitals. Since 2006, Venezuela has experienced the appearance of various patients who have died from VRE in Caracas, Maracaibo, and Valencia (Perdomo et al. 2006; Pineda et al. 2007; Montilla et al. 2007; Ruiz et al. 2007; Silva et al. 2007; Vásquez et al. 2007).

It was not until September 2000 that the first case of VRE appeared in Peru. A retrospective descriptive study was done in August and September 2002 to analyze the possible cases of VRE during this period. Fourteen vancomycin-resistant *E. faecium* were found at different infection sites. Two chains of clonal dissemination were found in the ICU between December 2001 and January 2002 (Paredes and Hernán 2007).

Cuba had not reported a single case of VRE in its hospital centers (Quiñones 2002; Nodarse 2005), but in 2004, 99 enterococci taken from 12 Cuban hospitals between October 2000 and September 2001 were analyzed and found to have VRE distributed as follows: 1.2% *E. faecalis* and 10% *E. faecium* (Quiñones et al. 2005).

In a study done in Chile no VRE strains were detected (Giglio et al. 1999). In contrast to European findings, few VRE strains of animal origin have been isolated in Latin America (Knudtson and Hartman 1993; Thal et al. 1995). Although avoparcin is not officially used in the United States or Canada, VRE has been found in those countries in chickens and turkeys (Coque et al. 1996), and there could be a risk of dissemination in the food chain (Tenover 1998).

13.1.4 Middle East and North Africa

There is a low incidence of VRE in the Middle East (Udo et al. 2002). Nevertheless, in Iran, a high number of VanA VRE carriers have begun to be isolated (Feizabadi et al. 2004). The VanB VRE phenotype was isolated from a urine sample from an ICU patient in 2004 (Emaneini et al. 2005). An Iranian study (Afkhamzadehm 2004) had shown that there was a 3.78 times greater risk of colonization in hemodialyzed patients than in other patients. The problem of isolating strains of VRE highly resistant to almost all antibiotics available in the market is that it implies the use of costly new antibiotics. Of 120 enterococci isolated from urine in the El Ahari Children's Medical Center in Teheran, Iran, 7% of isolates were VRE with the following distribution: *E. faecalis* 38%, *E. faecium* 25%, *E. mundtii* 25%, and *E. raffinosus* 12% (Fatholahzadeh et al. 2006). The diversity of highly VRE-resistant species (*E. faecalis, E. faecium, E. avium, E. durans, E. hirae, E. mundtii,* and *E. raffinosus*) isolated in Iran, combined with the isolation of the latter two for the first time in other regions of the Middle East, suggests a rapid dissemination of resistance determinants. The isolation of these two species reflects a change in the diversity of VRE strains in Iranian hospitals in contrast to first reports from the region (Udo et al. 2002; Feizabadi et al. 2004.). Other studies in different parts of the world have also described this change in diversity (Ruoff et al. 1990; Prakash et al. 2005). In another study done in Iran, 6.2% of the patients with underlying illnesses were colonized by VRE, without clinical manifestations of VRE infection. Of all risk factors described in the literature, only two were statistically significant for this study: consumption of antibiotics 2 months before the study, and hospitalization in the previous year (Assadian et al. 2007). This is the first study that reports the prevalence of VRE in Iran.

In a study of prevalence in five hospitals in Kuwait in 2003, *E. faecalis* was shown to be the principal cause of infections in 85.3% of isolates, 2.5% of which were VRE (Udo et al. 2003). In Lebanon, *E. faecalis* and *E. faecium* are the two most prevalent species in ICU patients (Zouain and Araj 2001).

As in Egypt there are not sufficient data on the resistance of enterococci to glycopeptides, an investigative group of the Department of Clinical Pathology of the University of Cairo and the Dar Al Fouad Hospital reviewed the patterns of susceptibility of all Gram-positive isolates obtained from the hemocultures from 1999 to 2000, finding 4% VRE strains (Kholy et al. 2002).

13.1.5 South Africa

The first two cases of VRE infection were described in 1997, from a VanA phenotype strain (Budavari et al. 1997; Derby et al. 1998). Subsequent to these cases, a VRE screening was done on patients in four Johannesburg hospitals, collecting a total of 184 rectal swabs; 10.9% of patients were colonized by VRE. The strains were identified by species isolated in the following order: *E. faecium* vanA, *E. faecium* vanB, *E. gallinarum* vanC1, and *E. avium* vanA (von Gottberg et al. 2000). This study permitted the macrorestriction profile analysis of VanA and VanB strains of *E. faecium* that had possibly caused infectious outbreaks in high-risk patients in these hospitals. *E. faecium* VanA is widely described in outbreaks worldwide, although the importance of *E. faecium* VanB has also been reported (Boyce et al. 1994; McNeeley et al. 1998). Dissemination of inter- and intrahospital clonal VanA and VanB strains was shown to be similar to that in other hospitals (Murray et al. 1990). No *E. faecalis* VanB strain was identified, perhaps due to the low resistance level. This prevalence rate refers only to high-risk patients in tertiary care hospitals in Johannesburg, in contrast to smaller or rural hospitals with different patterns of antibiotic use.

13.2 Conclusion

The lack of surveillance studies in developing countries is a problem. Outbreaks occur soon after the first cases are detected, and the strains, being rapidly disseminated intra- and interhospitally, establish themselves in different hospital services. Detection of carrier patients is vital in order to be able to implement necessary measures to avoid patient autoinfection, or infection via medical or paramedical personnel. The alarm regarding the dissemination of resistance to glycopeptides is due to (i) serious limitations being placed on the therapeutic options for treatment of serious VRE infections; (ii) in vivo dissemination of *vanA* or *vanB* operons in methicillin-resistant *S. aureus* strains; (iii) different selective pressures depending on country and hospital; and (iv) scant success of prevention strategies to contain the resistance once VRE is established.

References

Afkhamzadehm, A. 2004. Risk factors of vancomycin-resistant enterococci colonization in patients admitted in Namazi hospital-Shiraz 2004. Shiraz: Shiraz University of Medical Sciences.

Almeida, R. T., Filho, M., Silveira, R., Rodrigues, C. A. N., Filho, I. P., Nascimento, J. E., Ferreira, R. S., Moraes, II R. F., Boelens, L. M. P., Belkum, H., Felipe, A. V. 2004. Molecular epidemiology and antimicrobial susceptibility of enterococci from Brazilian Intensive Care Units. Braz. J. Infect. Dis. 8:197–205.

Arias, C. A, Reyes, J., Zúñiga, M., Cortés, L., Cruz, C., Rico C. L., and D. Panesso on behalf of the Colombian Antimicrobial Resistance Group (RESCOL). 2003. Multicentre surveillance of antimicrobial resistance in enterococci and staphylococci from Colombian hospitals, 2001–2002. J. Antimicrob. Chemother. 51:59–68.

Arthur, M., Reynolds, R., and Courvalin, P. 1996. Glycopeptide resistance in enterococci. Trends Microbiol. 4:401–407.

Assadian, O., Askarian, M., Stadler, M., and Shaghaghian, S. 2007. Prevalence of vancomycin-resistant enterococci colonization and its risk factors in chronic hemodialysis patients in Shiraz, Iran. BMC Infect. Dis. 7:52–56.

Aznar, E., Buendía, B., Garcia-Peñuela, E., Escudero, E., Alarcon, T., and Lopez-Brea, M. 2004. Community-acquired urinary tract infection caused by vancomycin-resistant *Enterococcus faecalis* clinical isolate. Rev. Esp. Quimioter. 17:263–265.

Basustaoglu, A., Aydogan, H., Beyan, C., Yalcin, A., and Unal, S. 2001. First glycopeptide-resistant *Enterococcus faecium* isolate from blood culture in Ankara, Turkey. Emerg. Infect. Dis. 7:160–161.

Bazet, C., Blanco, J., Seija, V., and Palacio, R. 2005. Enterococos resistentes a vancomicina. Un problema emergente en Uruguay. Rev. Med. Uruguay 21:151–158.

Bergerova, T., and Turkova, S. 1997. First finding of vancomycin-resistant enterococci in the Faculty Hospital, Plzeň. Klin. Mikrobiol. Inf. Lek. 3:287–288.

Bilikova, E., Hanzen, J., Roidova, A., Trupl, J., Lamosova, J., Liskova, A., Macekova, J., Szvenyova, Z., Svetlansky, I., and Krcméry, V. 2001. Bacteremia due to vancomycin resistant enterococci in Slovakia. J. Antimicrob. Chemother. 47:362–364.

Boyce, J. M., Opal, S. M., Chow, J. W., Zervos, M. J., Potter-Bynoe, G., Sherman, C. B., Romulo, R. L. C., Fortna, S., and Medeiros, A. A. 1994. Outbreak of multidrug-resistant *Enterococcus faecium* with transferable *vanB* class vancomycin resistance. J. Clin. Microbiol. 32:1148–1153.

Budavari, S. M., Saunders, G. L., Liebowitz, L. D., Khoosal, M., and Crewe-Brown, H. H. 1997. Emergence of vancomycin-resistant enterococci in South Africa. S. Afr. Med. J. 87:1557.

Butaye, P., Devriese, L. A., Goosens, H., Leven, M., and Haesebrousk, F. 1999. Enterococci with acquired vancomycin resistance in pigs and chickens of different age groups. Antimicrob. Agents Chemother. 43:365–366.

Calderón-Jaimes, E., Arredondo-García, J. L., Aguilar-Ituarte, F., and García-Roca, P. 2003. *In vitro* antimicrobial susceptibility in clinical isolates of *Enterococcus* species. Salud Pública de México. 45(2):96–101.

Camargo, I. L. B. C., Barth, A. L., Pilger, K., Seligman, B.G.S., Machado, A.R.L., and Darini, A.L.C. 2004a. *Enterococcus gallinarum* carrying the *vanA* gene cluster: first report in Brazil. Braz. J. Med. Biol. Res. 37:1669–1671.

Camargo, I. L. B. C., Del Peloso, P., Da Costa, L. F., Goldman, G. H., and Darini, A. L. C. 2004b. Identification of an unusual *vanA* element in glycopeptide-resistant *Enterococcus faecium* in Brazil following international transfer of a bone marrow transplant patient. Can. J. Microbiol. 50:767–770.

Casellas, J. M., Tome, G., Goldberg, M., Gilardoni, M., Rolon, M. J., Lopez-Furst, M. J., Marcopido, M. 1996. Primer aislamiento en Argentina de *Enterococcus faecalis* resistente a vancomicina. La faceta del Cindim. 4:22.

Cereda, R. F., Sader, H. S., Jones, R. N., Sejas, L., Machado, A. M., Zanatta, Y. P., Rego, S. T. M. S., Medeiros, E. A. S. 2001. *Enterococcus faecalis* resistant to vancomycin and teicoplanin (VanA phenotype) isolated from a bone narrow transplanted patient in Brazil. Braz. J. Infect. Dis. 5:40–46.

Cereda, R. F., Gales, A. C., Silbert, S., Jones, R. N., and Sader, H. S. 2002. Molecular typing and antimicrobial susceptibility of vancomycin resistant *Enterococcus faecium* in Brazil. Infect. Contr. Hosp. Epidemiol. 23:19–22.

Cetinkaya, Y., Falk, P., and Mayhall, C. G. 2000. Vancomycin-resistant enterococci. Clin. Microbiol. Rev. 13:686–707.

Cheng, A. F., Char, T. S., and Ling, J. M. 2002. Are multiply resistant enterococci a common phenomenon in Hong Kong? J. Antimicrob. Chemother. 50:761–763.

Colak, D., Nass, T., Gunseren, F., Fortineau, N., Ogunc, D., Gultekin, M., and Nordmann, P. 2002. First outbreak of vancomycin-resistant enterococci in a tertiary hospital in Turkey. J. Antimicrob. Chemother. 50:397–401.

Coleri, A., Cokmus, C., Ozcan, B., Akcelik, M., and Tukel, C. 2004. Determination of antibiotic resistance and resistance plasmids of clinical *Enterococcus* species. J. Gen. Appl. Microbiol. 50:213–219.

Coque, T. M., Tomayko, J. F., Ricke, S. C., Okhyusen, P. C., and Murray, B. E. 1996. Vancomycin-resistant enterococci from nosocomial, community and animal sources in the United States. Antimicrob. Agents Chemother. 40:2605–2609.

Cornejo-Juarez, P., Velásquez-Acosta, C., Díaz-Gonzalez, A., and Volkow-Fernandez P. 2005. Tendencia del perfil de sensibilidad antimicrobiana de los aislamientos de sangre en un hospital oncológico (1998–2003). Salud Pública Mex. 47:288–293.

Corso, A., Faccone, D., Gagetti, P., Togneri, A., Lopardo, H., Melano, R., Rodríguez, V., Rodriguez, M., and Galas, M. 2005. First report of VanA *Enterococcus gallinarum* dissemination within an intensive care unit in Argentina. Inter. J. Antimicrob. Agents 25:51–56.

Cuellar-Rodríguez, J., Galindo-Fraga, A., Guevara, V., Pérez-Jiménez, C., Espinosa-Aguilar, L., Rolón, A. L., Hernández-Cruz, A., López-Jácome, E., Bobadilla-del-Valle, M., Martínez-Gamboa, A., Ponce-de-León, A., and Sifuentes-Osornio, J. 2007. Vancomycin resistant enterococci, Mexico City. Emerg. Infect. Dis. 13(5):798–799.

Dalla Costa, L. M., Souza, D. C., Martins, L. T. F., Zanella, R. C., Brandileone, M. C., Bokermann, S., Sader, H. S., and Souza, H. A. P. H. M. 1998. Vancomycin-resistant *Enterococcus faecium*: first case in Brazil. Braz. J. Infect. Dis. 2:160–2163.

Dalla Costa, L. M., Reynolds, P. E., Souza, H. A. P. H. M., Souza, D. C., Palepou, M. F. I., and Woodford, N. 2000. Characterization of a divergent VanD-type resistance element from the first glycopeptide-resistant strain of *Enterococcus faecium* isolated in Brazil. Antimicrob. Agents Chemother. 44:3444–3446.

Dekhnich, A., Edelstain, I., Narezkina, A., and Stratchounski, L. 2001. Antimicrobial resistance of nosocomial strains of *Enterococcus* spp. in Russia. Poster Nr. P 1207. http://www.antibiotic.ru/files/pdf/en/eccmid_arnsesru.pdf

Derby, P., Allan, B., Lambrick, M., and Gay-Elisha, B. 1998. Detection of glycopeptide-resistant enterococci using susceptibility testing and PCR. S. Afr. J. Epidemiol. Infect. 13:66–69.

Emaneini, M., Farhad, B., Hashemi, M. A., Aligholi, M., and Fatholahzadeh, B. 2005. Detection of *vanB* genotype enterococci in Iran. Inter. J. Antimicrob. Agents 26:95–99.

Endtz, H. P., van den Braak, N., van Belkum, A. Kluytmans, J. A., Koeleman, J. G., Spanjaard, L., Voss, A., Weersink, A. J., Vandenbroucke-Grauls, C. M., Buiting, A. G., van Duin, A., and Verbrugh, H. A. 1997. Fecal carriage of vancomycin-resistant enterococci in hospitalized patients and those living in the community in the Netherlands. J. Clin. Microbiol. 35:3026–3031.

Esmaeli, H., Holz, E. R., Ahmadi, M. A., Kathren, R. A., and Raad, I. I. 2003. Endogenous endophtalmitis secondary a vancomycin-resistant enterococci infection. Retina 23:118–119.

Fatholahzadeh, B., Hashemi, F. B., Emaneini, M., Aligholi, M., Nakhjavani, F. A., Kazemi, B., 2006. Detection of vancomycin resistant enterococci (VRE) isolated from urinary tract infections (UTI) in Tehran, Iran. DARU. 14(3):141–146.

Feizabadi, M. M., Asadi, S., Aliahmadi, A., Parvin, M., Parastan, R., Shayegh, M., and Etemadi, G. 2004. Drug resistant patterns of enterococci recovered from patients in Tehran during 2000–2003. Int. J. Antimicrob. Agents 24:521–522.

Fujita, N., Yoshimura, M., Komori, T., Tanimoto, K., and Ike, Y. 1998. First report of the isolation of high-level vancomycin-resistant *Enterococcus faecium* from a patient in Japan. Antimicrob. Agents Chemother. 42:2150.

Furtado, G. H. C., Martins, S. T., Coutinho, A. P., Wey, S. B, and Medeiros, E. A S. 2005. Prevalence and factors associated with rectal vancomycin-resistant enterococci colonization in two intensive care units in São Paulo, Brazil. Braz. J. Infect. Dis. 9:64–69.

Giglio, M. S., Farías, O., Lafourcade, M., and Pinto, M. E. 1999. Vigilancia de susceptibilidad de cocáceas grampositivas a betalactámicos, glicopéptidos y otros antimicrobianos. Rev. Med. Chil. 127:8.

Han, D. P., Wisniewski, S. R., and Wilson, A. 1996. Spectrum and susceptibilities of micro-biologic isolates of endophthalmitis vitrectomy study. Am. J. Ophthalmol. 122:1–17.

Hawkey, P. M. 1986. Resistant bacteria in the normal human flora. J. Antimicrob. Chemother. 18(Suppl. C):133–139.

Hryniewicz, W., Zareba, T., and Kawalec, M. 1998. Susceptibility patterns of *Enterococcus* spp. isolated in Poland during 1996. Int. J. Antimicrob. Agents 10:303–307.

Hryniewicz, W., Szczypa, K., Bronk, M. Samet, A., Hellmann, A., and Trzciński, K. 1999a. First report of vancomycin-resistant *Enterococcus faecium* isolated in Poland. Clin. Microbiol. Infect. 5:503–505.

Hryniewicz, W. 1999b. Antibiotic resistant in Central Europe. Abstract on the 9th ECCMID, Berlin. Clin. Microbiol. Infect. 8(Suppl. 1):234.

Hsueh, P. R., Chen, W. H., Teng, L. J., and Luh. K. T. 2005. Nosocomial infections due to methicillin-resistant *S. aureus* and vancomycin-resistant enterococci at a university hospital in Taiwan from 1991 to 2003: resistant trends, antibiotics usage and *in vitro* activities of newer antimicrobial agents. Int. J. Antimicrob. Agents 26:43–49.

Kawalec, M., Gniadkowski, M., Kedzierska, Skotnicki, J. A., Fiett, J., and Hryniewicz, W. 2001. Selection of a teicoplanin-resistant *Enterococcus faecium* mutant during an outbreak caused by vancomycin-resistant enterococci with the VanB phenotype. J. Clin. Microbiol. 39:4274–4282.

Kawalec, M., Ziókowski, R., Ozorowski, T., Kedzierska, J., Baran, M., and Hryniewicz, W. 2004. 14th European Congress Clinical Microbiology Infectious Diseases. Abstract P812. Clin. Microbiol. Infect. 10(Suppl.3):207.

Kawalec, M., Pietras, Z., Daniowicz, E., Jakubczak, A., Gniadkowski, M., Hryniewicz, W., and Willems, R. J. L. 2007. Clonal structure of *Enterococcus faecalis* isolated from Polish hospitals: characterization of epidemic clones. J. Clin. Micrbiol. 45(1):147–153.

Kim, J. M., and Song, Y. G. 1998. Vancomycin-resistant enterococcal infections in Korea. Yonsei Med. J. 39:562–568.

Kholy, A. E., Baseem, H., Hall, G. S., Procop, G. W., and Longworth, D. L. 2002. Antimicrobial resistance in Cairo, Egypt 1999–2000: a survey of five hospitals. J. Antimicrob. Chemother. 51:625–630.

Knudtson, L. M., and Hartman, P. A. 1993. Antibiotic resistance among enterococcal isolates from environmental and clinical sources. J. Food Protect. 56:489–492.

Kolař, M., Vagnerova, I., Latal, T., Urbanek, K., Typovska, H., Hubacek, J., Papajik, T., Raida, L., and Faber, E. 2002. The occurrence of vancomycin-resistant enterococci in hematological patients in relation to antibiotic use. Microbiologica 25:205–212.

Kolař, M., Čekanova, L., Vagnerova, I., Kesselova, M., Sauer, P., Koukalova, D., and Hejnar, P. 2004. Molecular-biological analysis of vancomycin-resistant enterococci isolated from a community in the Czech Republic. Biomed. Papers 148(2):167–169.

Kollar, M. 1999. Susceptibility of enterococci to vancomycin. Report from a University Hospital hematology ward. Clin. Microbiol. 5:13–17.

Krcméry, V. Jr., Hryiewicz, W., Ludwig, E., Radzilowski, H., Stratchoynski, C., Navachin, S. M. 1998. Epidemiology of antimicrobial resistant in Central/Eastern Europe. J. Pub. Health. 1:13–30.

Lauderdale, T. L., McDonald, L. C., Shiau, Y. R., Chen, P.-C., Wang, H.-Y., Lai, J.-F., and Ho, M. 2002. Vancomycin-resistant enterococci from humans and retail chickens in Taiwan with unique phenotype VanB-vanA genotype incongruence. Antimicrob. Agents Chemother. 46:525–527.

Leme, I. L., Piantino, A. J., Bottino, J. A., and Pignatari, A. C. 2000. Glycopeptides susceptibility among enterococci isolated from a poultry farm in São Paulo, Brazil (1996/1997). Braz. J. Microbiol. 31:53–57.

Llittvik, A. M., López, T. N., González, S. E., Fernández, C. M., and Pavan, J. V. 2006. Colonization with vancomycin-resistant enterococci (VRE) in intensive care unit patients in Cordoba City, Argentina. Rev. Argent. Microbiol. 38:28–30.

Low, D. E., Keller, N., Barth, A., and Jones, R. N. 2001. Clinical prevalence, antimicrobial susceptibility, and geographic resistance patterns of enterococci: results from the SENTRY Antimicrobial Surveillance Program, 1997–1999. Clin. Infect. Dis. 32(Suppl.2):S133–S145.

Marin, M. E., Mera, J. R., Arduino, R. C., Correa, A. P., Coque, T. M., Stamboulian, D., and Murray, B. E. 1998. First report of vancomycin-resistant *Enterococcus faecium* isolated in Argentina. Clin. Infect. Dis. 26:235–236.

McDonald, L. C., Garza, L. R., and Jarvis, W. R. 1999. Proficiency of clinical laboratories in and near Monterrey, Mexico, to detect vancomycin-resistant enterococci. Emerg. Infect. Dis. 5:143–146.

McNeeley, D. F., Brown, A. E., Noel, G. J., Chung, M., and de Lancastre, H. 1998. An investigation of vancomycin-resistant *Enterococcus faecium* within the pediatric service of a large urban medical center. Pediatr. Infect. Dis. J. 17:184–188.

Miranda, G., Lee, L., Kelly, C., Solórzano, F., Solórzano, F., Leaños, B., Muñoz, O., and Patterson, J. E. 2001. Antimicrobial resistance from enterococci in a pediatric hospital. Plasmids in *Enterococcus faecalis* isolates with high-level gentamicin and streptomycin resistance. Arch. Med. Res. 32:159–163.

Miranda, G., Corso, A., Melano, R., Arismendi, P., Rodríguez, M., and Garbervetsky, L. 2003. First isolation of vancomycin-resistant *Enterococcus faecium* with vanB genotype in Argentina: presentation of two cases. Rev. Argent. Microbiol. 35(1):41–44.

Mironova, A., Cherkashin, E., Kliasova, G., Tishkov, V., Brilliantova, A., Rezvan, S., and Sidorenko, S. 2006. First detection of vancomycin-resistant enterococci in Russia: genetic background. 16th European Congress of Clinical of Clinical Microbiology and Infectious Diseases. Nice. France. Abstract 1819.

Montilla, N., León, Y., Payares, D., Paraqueimo, M., Machado, Y., Ojeda, X., Marcano, D., and Vacampenhaud, M. 2007. *Enterococcus faecium* resistente a vancomicina con genotipo vanA en el hospital Dr. Domingo Luciani. Caracas, Venezuela. Abstract 06. XVI Jornadas Nacionales de Infectología. XIII Jornadas Nororientales de Infectología. Margarita, Estado Nueva Esparta, Venezuela. Bol. Venez. Infectol. 18(2):53.

Morfín, O. R., Heredia, C. J., Pinto, T. D., Rodríguez, C. J. J, and Rodríguez, E. 1992. *E. faecalis* and *E. faecium* resistance patterns, México. Abstract C-260. En programa y abstractos de la 92ª American Society of Microbiology General Meeting. 464.

Morfín, R., Esparza, S., Atilano, G., Pinto, D., Heredia, J., Rodríguez, J. J., and Rodríguez, E. 1999. Tendencias de resistencia en enterococos: 1991–1999. Enf. Infec. Microbial. 19(5):222–226.

Murray, B. E., Singh, K. V., Heath, J. D., Sharma, B. R., and Weinstock, G. M. 1990. Comparison of genomic DNAs of different enterococcal isolates using restriction endonucleases with infrequent recognition sites. J. Clin. Microbiol. 28:2059–2063.

Nodarse, R. 2005. Informe Corto. Rev. Cub. Med. Mil. 51:63–67.

Novicki, T. J., Schapiro, J. M., Ulness, B. K., Sebeste, A., Busse-Johnston, L., Swanson, K. M., Swanzy, S. R., Leisenring, W., and Limaye, A. P. 2004. Convenient selective differential broth for isolation of vancomycin-resistant *Enterococcus* from fecal material. J. Clin. Microbiol. 42:1637–1640.

Oncu, S., Punar, M., and Eraksoy, H. 2004. Susceptibility patterns of enterococci causing infections. Tohoku J. Exp. Med. 202:23–29.

Palazzo, I. C. V., Camargo, I. B. C., Zanella, R. C., and Darini, A. L. C. 2006. Evaluation of clonality on enterococci isolated in Brazil carrying Tn*1546*–like elements associated to *vanA* plasmids FEMS Microbiol. Lett. 258:29–36.

Panesso, D., Ospina, S., Robledo, J., Vela, M. C., Peña, J., Hernández, O., Reyes, J., and Arias, C. A. 2002. First characterization of a cluster of VanA-type glycopeptide resistant *Enterococcus faecium*, Colombia. Emerg. Infect. Dis. 8(9):961–965.

Paredes, F., and Hernán, W. 2007. Características clínico-epidemiológicas de 14 casos con aislamiento clínico de enterococo resistente a vancomicina en el Hospital Nacional Edgardo Rebagliati Martins. Rev. Med. Hered. 18(2):68–75.

Park, I. J., Lee, W. G., Lim, Y. A., and Cho, S. R. 2007. Genetic rearrangements of Tn*1546*–like elements in vancomycin-resistant *Enterococcus faecium* isolates collected from hospitalized patients over a seven-year period. J. Clin. Microbiol. 45(2):3903–3908.

Perdomo, Y., Borrero, L., González, S., Castillo, Z., Castillo, O., and Sánchez, C. 2006. *Enterococcus faecium* resistente a vancomicina, genotipo *vanA*, en la ciudad de valencia. Reporte del primer caso. Abstract 141. VII Congreso Venezolano de Infectología Dr. Belisario Gallegos. XIV Jornadas Guayanesas de Infectología. Puerto Ordaz. Estado Bolívar. Venezuela.

Pineda, M., Perozo-Mena, A., Lleras, A., Bonilla, X., Mendez, A., González, M., and Villalobos, H., 2007. Primer reporte de *Enterococcus* resistente a vancomicina en el estado Zulia. Abstract BA-16. XXXI Jornadas Venezolanas de Microbiología. "Dr. Sócrates Medina y Elba Aracelis Padrón. Ciudad Bolívar, Estado Bolívar. Venezuela.

Prakash, V. P., Rao, S. R., and Parija, S. C. 2005. Emergence of unusual species of enterococci causing infections, South India. BMC Infect. Dis. 5:14.

Quiñones, D. 2002. Fenotipos de resistencia antimicrobiana y genes de resistencia en *Enterococcus* sp de importancia clínica en Cuba. Rev. Latinoam. Microbiol. 44(4):150.

Quiñones, D., Goñib, P., Rubioc, M., Duranc, E., and Gómez-Lusb, R. 2005. Enterococci spp. isolated from Cuba: species frequency of occurrence and antimicrobial susceptibility profile. Diagn. Microbiol. Infect. Dis. 51(1):63–67.

Radu, S., Toosa, H., Rahim, R. A., Reezal, A., Ahmad, M., Hamid, A. N., Rusul, G., and Nishibuchid, M. 2001. Occurrence of the *vanA* and *vanC2/C3* genes in *Enterococcus* species isolated from poultry sources in Malaysia. Diagn. Microbiol. Infect. Dis. 39:145–153.

Raja, N. S., Karunakaran, R., Ngeow, Y. F., and Awang, R. 2005. Community-acquired vancomycin-resistant *Enterococcus faecium*: a case report from Malaysia. J. Med. Microbiol. 54:901–903.

Reis, A. O., Cordeiro, J. C., Machado, A. M., and Sader, H. S. 2001. *In vitro* antimicrobial activity against vancomycin-resistant enterococci isolated in Brazilian hospitals. Braz. J. Infect. Dis. 5:243–251.

Ruiz, N., Vásquez, Y., Gayoso, E., Moy, F., Guzmán, M., Hernández, M., and Spadola, E. 2007. *Enterococcus faecium* resistente a vancomicina. Reporte del primer caso en el hospital militar "Dr. C. Arvelo" y revisión de la literatura. Abstract 07. XVI Jornadas Nacionales de Infectología. XIII Jornadas Nororientales de Infectología. Margarita, Estado Nueva Esparta, Venezuela. Bol. Venez. Infectol. 18(2):53.

Ruiz-Garbajosa, P., Bonten, M. J., Robinson, D. A., Top, J., Nallapareddy, S. R., Torres, C., Coque, T. M., Canton, R., Baquero, F., Murray, B. E., Del Campo, R., and Willems, R. J. 2006. Multilocus sequence typing scheme for *Enterococcus faecalis* reveals hospital-adapted genetic complexes in a background of high rates of recombination. J. Clin. Microbiol. 44:2220–2228.

Ruoff, K. L., de la Maza, L., Murtagh, M. J., Spargo, J. D., and Ferraro, M. J. 1990. Species identities of enterococci isolated from clinical specimens. J. Clin. Microbiol. 28:435–443.

Samet, A., Bronk, M., Hellmann, A., and Kur, J. 1999. Isolation and epidemiological study of vancomycin-resistant *Enterococcus faecium* from patients of a haematological unit in Poland. J. Hosp. Infect. 41:137–143.

Scott, I. U., Loo, R. H., Flynn, H. W. Jr., and Miller, W. 2003. Endophtalmitis caused by *Enterococcus faecalis*: antibiotic selection and treatment outcomes. Opthalmol. 110:1573–1577.

Shin, J. W., Yong, D., and Kim, M. S. 2003. Sudden increase of vancomycin-resistant enterococcal infections in a Korean tertiary care hospital: possible consequences of increased use of oral vancomycin. J. Infect. Chemother. 9:62–67.

Sifuentes-Osornio, J., Ponce-de-León, A., Muñoz-Trejo, T., Villalobos-Zapata, Y., Ontiveros-Rodriguez, C., and Gómez-Roldan, C. 1996. Antimicrobial susceptibility patterns and high-level gentamicin resistance among enterococci isolated in a Mexican tertiary care center. Rev. Invest. Clin. 48:91–96.

Silva, M., Pitteloud, J., Villarroel, E., Figueredo, A., Payares, D., Sánchez, D., Martín, A., Carvajal, A., López, L., Villarroel, E., Khalil, R., Núñez, M., González, E., Pacheco, C., and Sojo, G. 2007. Aislamiento de *Enterococcus faecium* resistente a vancomicina: características clínicas y epidemiológicas de los pacientes. Hospital Universitario de Caracas. 2005–2007. Abstract 46. XVI Jornadas Nacionales de Infectología. XIII Jornadas Nororientales de Infectología. Margarita, Estado Nueva Esparta, Venezuela. Bol. Venez. Infectol. 18(2):72.

Stošovic, B., Stepanovic, S., Donabedian, S., Tošic, T., and Jovanovic, M. 2004. Vancomycin resistant *Enterococcus faecalis* in Serbia. Emerg. Infect. Dis. 10(1):157–158.

Taneja, N., Rani, P., Emmanuel, R., and Sharma, M. 2004. Significance of vancomycin resistant enterococci from urinary specimens at a tertiary care centre in northern India. Indian J. Med. Res. 119:72–74.

Tang, C. W., Cheng, C. K., and Lee, T. S. 2007. Comunity-adquired bled-related endophtalmitis caused by vancomycin-resistant enterococci. Can. J. Opthalmol. 42:477–478.

Targa, L., Carbone, E., and Gallego, V. 1997. Brote epidémico por *Enterococcus faecium* resistente a la vancomicina en un hospital de Buenos Aires. I Congreso Internacional de Infectología y Microbiología Clínica SADI-SADEBAC, Resumen H330, Buenos Aires, Argentina.

Tenover, F. C. 1998. Laboratory methods of surveillance of vancomycin-resistant enterococci. Clin. Microbiol. News Lett. 20:1–5.

Thal, L.A., Chow, J. W., Mahayani, R., H. Bonilla, H., Perri, M. B., Donabedian, S. A., Silverman, J., Taber, S., and Zervos, M. J. 1995. Characterisation of antimicrobial resistance in enterococci of animal origin. Antimicrob. Agents Chemother. 39:2112–2115.

Toosa, H., Son, R., Rasul, G., Reezal, A., Rahim, R., Ahmad, N., and Ling, O. 2001. Detection of vancomycin resistant *Enterococcus* spp (VRE) from poultry. Malays. J. Med. Sci. 8(1):53–58.

Udo, E. E., Al-Sweih, N., and Chugh, T. D. 2002. Antibiotic resistance of enterococci isolated at a teaching hospital in Kuwait. Diagn. Microbiol. Infect. Dis. 43:233–238.

Udo, E. E., Al-Sweih, N, Phillips, O. A., and Chugh, T. D. 2003. Species prevalence and anti-bacterial resistance of enterococci isolated in Kuwait hospitals. J. Med. Microbiol. 52:163–168.

Urbásková, P. 1999. Survey of resistance of enterococcal isolates to antibiotics in Czech Republic. Clin. Microbiol. 6:113–118.

Urumova, V., Lyutzkanov, M., and Petrov, M. 2005. *In vitro* study of antimicrobial sensitivity of *Enterococcus isolates* from farm birds. Trakia. J. Sci. 3(5):40–45.

van den Bogaard, A. E., Willems, R., London, N., Top, J., and Stobberingh, E. E. 2002. Antibiotic resistance and faecal enterococci in poultry, poultry farmers and poultry slaughterers. J. Antimicrob. Chemother. 49:497–505.

Vásquez, Y., Guzmán, M., Ruiz, N., Gayoso, E., Moy, F., Hernández, M., Spadola, E., and Córdova, J. 2007. Endocarditis por *Enterococcus faecalis* en paciente trasplantado renal en el Hospital Militar "Dr. Carlos Arvelo" a propósito de un caso. Abstract 47. XVI Jornadas Nacionales de Infectología. XIII Jornadas Nororientales de Infectología. Margarita, Estado Nueva Esparta, Venezuela. Bol. Venez. Infectol. 18(2):72.

Vilela, M. A., de Souza, S. L., Palazzo, I. C. V., Ferreira, J. C., de Morais M. A. Jr, da Costa, A. L., de Morais, M. M. C. 2006. Identification and molecular characterization of VanA-type vancomycin-resistant *Enterococcus faecalis* in Northeast of Brazil. Mem. Inst. Oswaldo. Cruz. 101(7):716–719.

von Gottberg, A., van Nierop, W., Dusé, A., Kassel, M., McCarthy, K., Brink, A., Meyers, M., Smego, R., and Koornhof, H. 2000. Epidemiology of glycopeptide-resistant enterococci colonizing high-risk patients in hospitals in Johannesburg, Republic of South Africa. J. Clin. Microbiol. 38(2):905–909.

Woodford, N., Adebiyi, A. M., Palepou, M. F., and Cookson, B. D. 1998. Diversity of VanA glycopeptide resistance elements in enterococci from humans and nonhuman sources. Antimicrob. Agents Chemother. 42:502–508.

Batista Xavier, D., Moreno, F. E., and Titze-de-Almeida, R. 2006. Absence of VanA- and VanB-containing enterococci in poultry raised on non intensive production farms in Brazil. Appl. Environ. Microbiol. 72(4):3072–3073.

Yang, J., Lee, D., Kim, Y., Kang, B., Kim, K., and Ha, N. 2007. Occurrence of the *van* genes in *Enterococcus faecalis* and *Enterococcus faecium* from clinical isolates in Korea. Arch. Pharm. Res. 30(3):329–336.

Yu, H. S., Seol, S. Y., and Cho, D. T. 2003. Diversity of Tn*1546*-like elements in vancomycin-resistant enterococci isolated from humans and poultry in Korea. J. Clin. Microbiol. 41:2641–2643.

Zanella, R. C., Valdetaro, F., Lovgren, M., Tyrrel, G. J, Bokermann, S, Almeida, S. C. G, Vieira, V. S. D, and Brandileone, M. C. C. 1999. First confirmed case of a vancomycin-resistant *Enterococcus faecium* with VanA phenotype from Brasil: isolation from a meningitis case in São Paulo. Microbiol. Drug. Resist. 5:159–162.

Zheng, B., Tomita, H., Xiao, Y. H., Wang, S., Li, Y., and Ike, Y. 2007. Molecular characterization of vancomycin-resistant *Enterococcus faecium* isolates from Mainland China. J. Clin. Microbiol. 45(9): 2813–2818.

Zouain, M. G., and Araj, G. F. 2001. Antimicrobial resistance of enterococci in Lebanon. Int. J. Antimicrob. Agents17:209–213.

Zurita, J., Ayabaca, J., Pavón, L., Espinosa, Y., Narváez, I., and Grupo REDNARBEC. 2001. Se detectan *Enterococcus faecium* resistentes a vancomicina en dos hospitales de Quito. Rev. Ecuat. Med. Crit. 2(2): http://www.medicosecuador.com/medicina_critica/rev_vol2_num2/se_detectana.html

Chapter 14
Antimicrobial Resistance in Gram-Negative Bacteria from Developing Countries

Soraya Sgambatti de Andrade, Ana Cristina Gales, and Helio Silva Sader

Abstract Antimicrobial resistance is a worldwide problem, with severe implications for developing countries, where a high infectious disease burden coexists with rapid emergence and spread of antimicrobial resistance. Furthermore, cost constrains the broader use of newer, more expensive therapeutic agents and the application of appropriate infection control measures. In the present chapter, we address the main antimicrobial resistance problems affecting hospital- and community-acquired bacterial infections. The majority of the data reviewed here was obtained from multinational surveillance programs and the chapter is structured according to organisms or organism groups.

14.1 *Salmonella* spp. and *Shigella* spp.

Salmonella gastroenteritis is one of the most prevalent and widely distributed foodborne diseases. This infection constitutes a major public health burden in many developing countries. Although the majority of non-typhoidal human salmonellosis is usually associated with self-limited foodborne gastroenteritis, life-threatening bacteremia can occur, particularly in developing regions and certain patient populations, such as neonates, elderly patients, and immuno-compromised hosts.

The WHO Global Salm-Surv is a global network of laboratories involved in surveillance, isolation, identification, and antimicrobial resistance testing of *Salmonella*. This program was launched by the World Health Organization (WHO) in 2000 as an integrated laboratory-based surveillance. At the end of 2005, the WHO Salm-Surv had more than 800 members from 142 countries. Data can be procured from the WHO Global Salm-Surv Country Databank, which is a web-based databank maintained by the Danish Institute for Food

S.S. de Andrade (✉)
Infectious Diseases Division, Laboratório Especial de Microbiologia Clínica,
Department of Medicine, Universidade Federal de São Paulo, São Paulo, Brazil
e-mail: soraya.andrade@lemc.com.br

A. de J. Sosa et al. (eds.), *Antimicrobial Resistance in Developing Countries*,
DOI 10.1007/978-0-387-89370-9_14, © Springer Science+Business Media, LLC 2010

and Veterinary Research. Analysis of this data provides the frequency of worldwide and regional *Salmonella* serotypes.

The WHO Global Salm-Surv revealed that *Salmonella enterica* serovar Enteritidis was the most common human serotype worldwide in 2000–2002, followed by *S.* Typhimurium (see Galanis et al. 2006). However, the distribution of human *Salmonella* serotypes varied according to the geographic region studied. *S.* Enteritidis (26% of all isolates) and *S.* Typhimurium (25% of all isolates) were the most common serotypes reported from Africa in 2002 (Fig. 14.1). In Latin America and the Caribbean, *S.* Enteritidis comprised 31% of all human clinical isolates in that study year. The predominance of *S.* Enteritidis was also verified in Asia, where almost 40% of all isolates belonged to this serotype. Of note, only 6% of all isolates in this region were *S.* Typhimurium, and *S.* Typhi ranked only ninth in the Asian countries included in the study.

The WHO Global Salm-Surv has recently published a progress report (see WHO Global Salm-Surv 2007). During the 2000–2004 period, results from this databank indicated that the frequency of *S.* Typhimurium increased in Africa (31% of all isolates), followed by *S.* Enteritidis (19% of all isolates). In contrast, *S.* Enteritidis continued to predominate in Asia, Central and South America, and the Caribbean.

The SENTRY Antimicrobial Surveillance Program revealed that *Salmonella* spp. caused 2.0% of all bloodstream infections in Latin America between 1997 and 2000 (see Gales et al. 2002b). Among the isolates with defined types, *S.* Typhi (17.4%) was the most common pathogen followed by *S.* Typhimurium (4.9%). In this study, however, nearly 70% of the studied *Salmonella* spp. isolates were not identified to species or serotyped by the participant medical centers. So, the frequency of *S.* Enteritidis and other serotypes may have been underestimated.

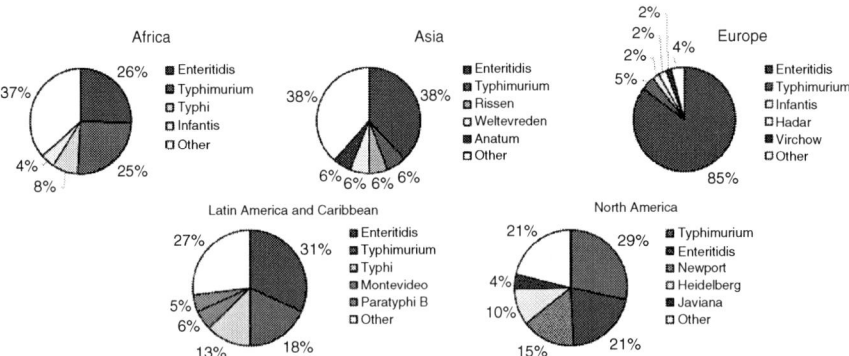

Fig. 14.1 Proportion of most common serotypes of reported human *Salmonella* isolates by region, 2002. World Health Organization Global Salm-Surv (Adapted from Galanis et al., 2006)

For decades, ampicillin, chloramphenicol, and trimethoprim/sulfamethoxazole have been considered first-choice agents for the treatment of severe salmonellosis. However, clinical use of these antimicrobials in patients from developing countries has been jeopardized by decreased susceptibility rates among clinical isolates of *Salmonella* spp. Multidrug-resistant (MDR) strains of *Salmonella* have now been frequently reported in many countries, and rates of MDR phenotypes have increased considerably in recent years. Rao et al. (1993) reported that 78.4% of the *S.* Typhi bloodstream isolates collected from infected patients in India, isolated in the early 1990s, were already resistant to ampicillin, chloramphenicol, and trimethoprim/sulfamethoxazole. Moreover, decreased susceptibility rates of quinolones among *S. enterica* serovar Typhi have also been observed in some regions. In Latin America, susceptibility rates to ampicillin among *Salmonella* spp. varied from 92.4% (bloodstream infections, 1997–2002) to 87.1% (gastroenteritis and bloodstream infections, 2003) (Table 14.1) (see Gales et al. 2002b; Streit et al. 2006).

The emergence of clinical isolates with elevated minimum inhibitory concentration (MIC) values to quinolones has been associated with point mutations in the quinolone resistance-determining region (QRDR) and/or active efflux. Single *gyrA* mutations mediate resistance to nalidixic acid and

Table 14.1 Activity of 10 selected antimicrobial compounds tested against 314 bacterial strains recovered from gastroenteritis infections in patients in Latin America (SENTRY Antimicrobial Surveillance Program, 2003)

Organism/antimicrobial agent (no. tested)	MICs (µg/ml)				
	50%	90%	Range	% susceptible	% resistant
Salmonella spp. (170)					
Ampicillin	2	>16	≤1–>16	87.1	12.9
Amoxicillin/clavulanate	≤1	8	≤1–>16	94.1	2.4
Cefuroxime	4	8	≤0.12–>16	56.5	1.8
Ceftriaxone	≤0.25	≤0.25	≤0.25–>32	98.8	0.6
Nalidixic acid	4	>32	1–>32	87.1	12.9
Ciprofloxacin	≤0.03	0.12	≤0.03–0.5	100.0	0.0
Tetracycline	≤2	>8	≤2–>8	81.2	18.8
Trimethoprim/ sulfamethoxazole	≤0.5	≤0.5	≤0.5–>2	91.1	8.9
Shigella spp. (144)					
Ampicillin	>16	>16	≤1–>16	26.4	73.6
Amoxicillin/clavulanate	8	16	≤1–>16	72.9	0.7
Cefuroxime	4	4	0.5–8	96.5	0.0
Ceftriaxone	≤0.25	≤0.25	≤0.25	100.0	0.0
Nalidixic acid	2	4	≤0.5–>32	99.3	0.7
Ciprofloxacin	≤0.03	≤0.03	≤0.03–>4	99.3	0.7
Tetracycline	>8	>8	≤2–>8	43.1	56.2
Trimethoprim/ sulfamethoxazole	>2	>2	≤0.5–>2	34.7	65.3

Adapted from Streit et al. (2006).

increase the MIC to broader spectrum fluoroquinolones, while a further *parC* mutation may lead to resistance to fluoroquinolones. In this manner, decreased susceptibility to ciprofloxacin can be monitored by nalidixic acid phenotypic results, which can be particularly helpful as an indicator of mutations occurring in the quinolone resistance-determining region (QRDR). A recent study evaluated more than 1800 isolates in various Asian countries, collected between 1993 and 2005 (see Chau et al. 2007). The authors found that rates of multidrug resistance and nalidixic acid resistance have significantly increased in the last years of the study. In Vietnam, for example, nalidixic acid resistance rates for *S.* Typhi dramatically increased from 4% (1993) to 97% (2005) among the studied isolates. Biedenbach et al. (2006) found that resistance to nalidixic acid was much higher among Latin American isolates of *Salmonella* spp. in comparison to North American isolates (15.0 versus 6.3%, respectively). In addition, the frequency of this resistance phenotype was variable between countries, with higher rates in Mexico (50.0%) and Brazil (33.6%) compared with 1.8–5.6% in the other participating Latin American countries (see Biedenbach et al. 2006). In that study, many *Salmonella* serotype Enteritidis isolates resistant to nalidixic acid detected from a medical center in Brazil had an identical *gyrA* mutation. This resistance pattern is worrisome since ciprofloxacin is considered a first-line agent to treat such infections. While such strains may remain susceptible in vitro to ciprofloxacin, therapeutic failures may arise if the clinical isolate presents resistance to nalidixic acid.

Reports of ciprofloxacin resistance among *Salmonella* spp. strains suggest that third-generation cephalosporins can be one alternative for empiric management of these infections in developing countries. However, resistance to these β-lactam agents has already arisen, usually due to the acquisition of mobile β-lactamase genes, often associated with outbreaks. The description of extended-spectrum β-lactamases (ESBLs) in *Salmonella* spp. compromises the clinical use of third-generation cephalosporins in serious infections. In addition, plasmids harboring ESBL genes may also carry elements for other antimicrobial resistances including tetracycline, gentamicin, and trimethoprim/sulfamethoxazole. These enzymes have been increasingly identified in *Salmonella* spp., with a variety of enzymes including TEM, SHV, and CTX-M types. Resistant *Salmonella* spp. harboring ESBLs have been characterized molecularly and reported in many developing countries, such as Argentina, Brazil, China, and Turkey.

A report published at the Bulletin of the World Health Organization showed that *Shigella flexneri* was the most common species in the developing countries studied, followed by *S. sonnei*. *S. dysenteriae* was seen most often in South Asia and sub-Saharan Africa (see Kotloff et al. 1999). These data are corroborated by surveillance studies by the Brazilian Shigella Surveillance Program, which found *S. flexneri* as the predominant (53%) serogroup, followed by *S. sonnei* (44%) (see Peirano et al. 2006). High resistance rates to ampicillin (73.6%), tetracycline (56.2%), and trimethoprim/sulfamethoxazole (65.3%) among *Shigella* spp. isolates have also been reported by the SENTRY Latin American study (Table 14.1). Fluoroquinolone-resistant *Shigella* spp. continues to be an uncommon but emerging phenotype, with cases identified from Asian nations such as Bangladesh, India, and Nepal (see Talukder et al. 2004; Taneja 2007).

14.2 Enterobacteriaceae

Enterobacteriaceae is a large family of bacteria that causes both nosocomial and community-acquired infections. This family includes many more familiar pathogens, such as *Salmonella* spp. and *Escherichia coli*, as well as *Citrobacter* spp., *Enterobacter* spp., *Klebsiella* spp., *Morganella* spp., *Proteus* spp., *Providencia* spp., *Serratia* spp., and others. The resistance mechanisms mostly related to β-lactam resistance among these pathogens are the overexpression of AmpC β-lactamases and/or the production of extended-spectrum β-lactamases (ESBL), and, more recently, the emergence of carbapenem-hydrolyzing enzymes.

Many reports have documented the emergence of ESBLs in developing countries. In Bolivia and Peru, *E. coli* harboring $bla_{CTX-M-2}$ and $bla_{CTX-M-15}$ were detected in stool from healthy children after a surveillance study (see Pallecchi et al. 2004). Further studies in these two Latin American countries demonstrated increasing ESBL rates, mainly associated with the dissemination of CTX-M-type ESBL determinants among the commensal *E. coli* strains studied (see Pallecchi et al. 2004). In Argentina, a single isolate of *Klebsiella pneumoniae* presented nine antimicrobial resistance mechanisms comprising two ESBLs (PER-2 and CTX-M-2), TEM-1-like, OXA-9-like, AAC(3)-IIa, AAC(6′)-Ib, ANT(3″)-Ia and resistance determinants to tetracycline and chloramphenicol (see Melano et al. 2003). CTX-M-3 was found to be disseminated among members of the Enterobacteriaceae family in many hospitals in Poland (see Baraniak et al. 2002). In Lebanon, ESBL-producing isolates have been identified by phenotypic tests, but the authors did not further characterize the types of enzyme (see Daoud et al. 2006).

Although β-lactam and fluoroquinolones are distinct classes of antimicrobial agents, with distinct mechanisms of action, ESBL-producing Enterobacteriaceae are usually resistant to fluoroquinolones. In many cases, the plasmid harboring CTX-M-15 β-lactamase also carries *aac(6′)Ib-cr*, a gene encoding an aminoglycoside acetylase that modifies ciprofloxacin and several other fluoroquinolones, leading to resistance to these compounds.

The frequency of occurrence and antimicrobial susceptibility profile of uropathogens collected from hospitalized patients in Latin America between 1997 and 2000 were evaluated by the SENTRY Antimicrobial Surveillance Program (see Gales et al. 2002a). A total of 1961 isolates were included, and *E. coli* and *Klebsiella* spp. were the most frequent uropathogens collected from Latin American medical centers. There was little variation in rates of ESBL-producing *E. coli* according to the study period: from 5.9% (1997–1999) to 5.2% (2000). In contrast, 45.9% of all *Klebsiella* spp. isolates collected in 2000 could be considered ESBL producers, in comparison to the 1997–1999 period (28.9% of ESBL producers). *E. coli and K. pneumoniae* isolates exhibited high ciprofloxacin resistance rates, nearly 20% (Gales et al., 2002a).

The SENTRY Program also described the frequencies of occurrence and antimicrobial susceptibility profiles of bloodstream isolates collected from January 1997 to December 2000 in the Latin American participating centers (see Sader et al. 2002). A total of 7207 isolates were collected from this geographic region. *E. coli* ranked second as the most frequent isolated pathogen from bloodstream infections (17.2%). *K. pneumoniae* (9.2%) and *Enterobacter* spp. (5.6%) also ranked among the top 10 pathogens in this study. This study showed high frequencies of ESBL-producing *E. coli* (6.7%) and *K. pneumoniae* (47.3%).

Bell et al. (2003) have evaluated 587 *Enterobacter cloacae* isolates from the Asia-Pacific region, collected for the SENTRY Program. A total of eight countries/regions participated in the study: Australia, Hong Kong, Japan, Mainland China, Philippines, Singapore, South Africa, and Taiwan. Rates of isolates screened as possible ESBL producers varied according to the country/ region of isolation, as shown in Table 14.2.

Fluoroquinolones, i.e., ciprofloxacin, gatifloxacin, and levofloxacin, are frequently employed for the treatment of urinary tract infections (UTI). *E. coli* and *Klebsiella* spp. were also the most frequent pathogens causing UTI reported by two distinct studies conducted in Turkey and Senegal, respectively (see Dromigny et al. 2003; Yuksel et al. 2006). Norfloxacin resistance rates among *E. coli* and *K. pneumoniae* isolates from Senegal were 14.2 and 2.9%, respectively. Interestingly, *K. pneumoniae* causing UTI in Turkey exhibited 100% susceptibility to ciprofloxacin, while resistance rate to this compound among *E. coli* was 12.0%. Gales et al. (2000) have demonstrated that most of urinary *E. coli* resistant to ciprofloxacin isolated in Latin America presented double mutations in *gyrA* and single mutations in *parC*, resulting in altered topoisomerases as previously observed.

The Study for Monitoring Antimicrobial Resistance Trends (SMART) evaluated the in vitro susceptibilities of 6150 gram-negative bacilli (86% Enterobacteriaceae) isolated from patients with intra-abdominal infections in different geographic areas. Medical centers were located in the Asia-Pacific region, North America, Europe, Latin America, and Middle East/Africa. This study concluded that the tested fluoroquinolones were not very active against *E. coli*. For instance, the lowest susceptibility rates for ciprofloxacin and levofloxacin among this pathogen were observed in the Asia-Pacific region (64.5 and 66.0%, respectively), followed by Latin America (72.1 and 74.3%, respectively) (see Rossi et al. 2006).

Recently, Castanheira et al. (2007b) reported the presence of a quinolone resistance gene (*qnrA1*) in an *E. coli* isolated from a urinary specimen of an 80-year-old female who was hospitalized in Porto Alegre, Brazil. This isolate showed resistance to most β-lactams (except cefepime and carbapenems), ciprofloxacin, levofloxacin,

Table 14.2 Prevalence of reduced third-generation cephalosporin susceptibility and ESBL production among *Enterobacter cloacae* strains accessed by country and year – SENTRY study Asia-Pacific

Country or locale	1998			1999			2000			2001			4-year totals		
	No. of strains	Percentage of CRO or CAZ R[a]	Percentage of ESBL positive	No. of strains	Percentage of CRO or CAZ R[a]	Percentage of ESBL positive	No. of strains	Percentage of CRO or CAZ R[a]	Percentage of ESBL positive	No. of strains	Percentage of CRO or CAZR[a]	Percentage of ESBL positive	No. of strains	Percentage of CRO or CAZ R[a]	Percentage of ESBL positive
Australia	33	45	6	45	36	2	55	36	2	45	31	7	178	37	4
Hong Kong	8	75	25	9	22	0	10	40	0	9	44	0	36	44	6
Japan	22	9	5	32	47	0	33	36	6	14	43	0	101	35	3
Mainland China	25	48	28	13	92	54	0			0			38	61	37
Philippines	19	58	37	27	56	33	37	57	43	11	45	9	94	55	35
Singapore	14	48	29	2	0	0	9	89	89	2	0	0	27	52	44
South Africa	7	29	14	6	67	0	18	39	33	23	48	17	54	44	20
Taiwan	8	63	38	2	50	0	10	30	0	39	56	21	59	53	19
Total	136	43	20	136	47	13	172	44	19	143	43	11	587	44	16

CRO, ceftriaxone; CAZ, ceftazidime; R, resistant.

a. Resistant

Adapted from Bell et al. (2003).

gatifloxacin, streptomycin, and chloramphenicol. This *qnr* was harbored on a 41-kb conjugative plasmid, which also carried a FOX-type cephalosporinase and a class 1 integron with the *aadB* and *catB3* cassettes, conferring resistance to aminoglycosides and chloramphenicol, respectively.

In another study, performed in a Taiwanese hospital, the authors determined the prevalence of plasmid-mediated quinolone resistance determinants among 526 non-replicate clinical isolates of *E. cloacae* (see Wu et al. 2007). The authors observed that 16.3% of all isolates were *qnr*-positive. In addition, *qnr* was present in 39.6% of the 149 ESBL producers and in 78.6% of the 56 metallo-β-lactamase (IMP-8) producers. These findings are worrisome due to the possibility of horizontal transfer of plasmids harboring *qnr* along with other resistance genes, which would increase the frequency of resistant isolates and jeopardize not only the future use of fluoroquinolones but also the use of other unrelated antimicrobial compounds.

Although carbapenem susceptibility rates are generally high among Enterobacteriaceae in developing countries, there are reports of the emergence of carbapenemase-producing Enterobacteriaceae in such regions. Carbapenemases are enzymes that hydrolyze, at least partially, carbapenems such as imipenem, meropenem, and ertapenem, broad-spectrum penicillins, and cephalosporins. The carbapenemases described in Enterobacteriaceae so far includes NMC-A from *E. cloacae*; SME from *S. marcescens*; IMI-1 from *E. cloacae*; KPC originally from *K. pneumoniae*; and the metallo-β-lactamases IMP and VIM types. VIM-11-producing *K. pneumoniae* have been detected in Turkey (see Yildirim et al. 2007); IMP-1- and KPC-2-producing *K. pneumoniae*, as well as IMP-1-producing *E. cloacae*, have been detected in Brazil (see Lincopan et al. 2005; Castanheira et al. 2006; Monteiro et al. 2007); KPC-2-producing *K. pneumoniae* have been detected in Colombia (see Villegas et al. 2006) and China (see Mendes et al. 2008); IMP-8-producing *E. cloacae* have been detected in Taiwan (see Wu et al. 2007).

14.3 *Pseudomonas aeruginosa* and *Acinetobacter* spp.

Nonfermentative gram-negative bacilli (NFB) are primarily opportunists, mainly causing infections in seriously ill, hospitalized patients, immunocompromised hosts, and patients with cystic fibrosis. NFB can be isolated from the environment and are intrinsically resistant to many commonly used antimicrobial agents. Dramatic increases in the prevalence of multidrug-resistant NFB have influenced the empiric antimicrobial therapy for treatment of serious infections caused by such pathogens. The knowledge of their epidemiology and antimicrobial susceptibility patterns are of crucial importance to guide therapeutic strategies. Among the NFB, *Pseudomonas aeruginosa* is the organism most frequently isolated by routine clinical microbiology laboratory followed by *Acinetobacter* spp. and *Stenotrophomonas maltophilia*. Among the Brazilian medical centers participating in the SENTRY Program, *P. aeruginosa* and *Acinetobacter* spp. ranked as the first and fourth most common pathogens causing nosocomial pneumonia (unpublished data from the SENTRY Antimicrobial Surveillance Program). Although all other NFB are less frequently isolated, the prevalence of infections caused by multidrug-resistant *P. aeruginosa* and *Acinetobacter* spp. strains has been increasing in many developing countries such as those in the Latin American region, as represented in Fig. 14.2 (see Andrade et al. 2003). In South Africa, the National Antimicrobial Surveillance Forum has reported that carbapenem resistance in *P. aeruginosa* ($N = 382$) and *Acinetobacter baumannii* ($N = 190$) isolated from private institutions varied between 42 and 45% and 32 and 33%, respectively (see Brink et al. 2007). Wang and Chen (2005) have shown that there was no significant increase in the imipenem resistance rates among *Acinetobacter* spp. isolated throughout a 7-year period in China (1996–2002). In contrast, these authors have demonstrated an obvious decrease in imipenem activity against *P. aeruginosa*. Sader et al. (2007) have reported very high resistance rates among all antimicrobials, except tigecycline (MIC_{90}, 1 µg/ml; 98.6% susceptible) and polymyxin B (MIC_{90}, 1 µg/ml; 99.6% susceptible) among *Acinetobacter* spp. complex strains recently isolated from the Asia-Pacific region. Imipenem (MIC_{50}, 4 µg/ml) and amikacin (MIC_{50}, >32 µg/ml) were active against only 56.8 and 41.3% of these isolates. Against *P. aeruginosa* strains isolated from the same geographic region, polymyxin B (MIC_{90}, 2 µg/ml; 99.6% susceptible) was the only antimicrobial agent with reasonable activity (Sader et al. 2007).

Among these organisms, the production of chromosomally encoded AmpC enzymes coupled with decreased outer membrane permeability has been responsible for resistance to broad-spectrum β-lactam agents. In the beginning of the 1990s, carbapenem resistance due to the production of mobile metallo-β-lactamases (MBL) was very rare and basically restricted to Japan. Since then, MBL have become increasingly more prevalent in various geographic regions, and their resistance mechanism seems to be spreading more rapidly (see Sader et al. 2005; Poirel et al. 2007). New types of MBL were described for the first time in developing nations such as the São Paulo metallo-β-lactamase (SPM) in

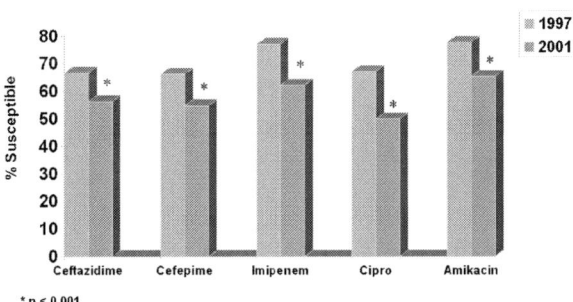

Fig. 14.2 Percentage of *Pseudomonas aeruginosa* isolates collected from Latin American medical centers to selected antimicrobial agents (SENTRY Antimicrobial Surveillance Program, 1997–2001). (Adapted from Andrade et al., 2003)

Brazil and the Seoul metallo-β-lactamases (SIM) in Korea among *P. aeruginosa* and *Acinetobacter* spp., respectively. In addition, variants of previously described MBL were also detected in these regions (see Sader et al. 2005; Poirel et al. 2007). Clinical isolates harboring IMP-1, IMP-16, VIM-2 or SPM-1 have been isolated from some Latin American countries by the SENTRY Program (Sader et al., 2005). VIM-6-producing *P. aeruginosa* was also detected from India (13 strains), Indonesia (2 strains), and 1 strain each from Korea and the Philippines (see Castanheira et al. 2007a). Recently, 54 of 282 (19.1%) *P. aeruginosa* isolates collected from Indian medical centers were detected as MBL producers. A great variety of VIM-type enzymes (VIM-2, -5, -6,-11, and -18) were detected in this collection (see Castanheira et al. 2008). Numerous studies have shown that the high carbapenem resistance rates found in some institutions are due to the dissemination of MβL-producing clones. The SENTRY study has evaluated imipenem resistance rates and screened for the presence of MβLs by phenotypic tests in isolates from Argentina, Brazil, Chile, Mexico, and Venezuela (see Sader et al. 2005). More than 70% of the screen-positive *P. aeruginosa* isolates were sent by a single medical center located in São Paulo, Brazil, in which the imipenem resistance rates among *P. aeruginosa* were very high (Table 14.3 and Fig. 14.2).

Table 14.3 Origin of metallo-β-lactamase-producing isolates during the study period (2001–2003) and imipenem resistance rates in the respective Latin American medical centers in the year 2002

	P. aeruginosa		*Acinetobacter* spp.	
Country/ medical center #	No. of isolates with positive MβL screen tests	Imipenem resistance[a] (no. of isolates tested)	No. of isolates with positive MβL screen tests	Imipenem resistance[a] (no. of isolates tested)
Argentina				
039	0	0% (35)	24	86% (30)
040	0	8% (48)	1	20% (5)
Brazil				
046	0	9% (66)	0	0% (11)
048[b]	38	65% (121)	7	13% (38)
057	0	2% (45)	0	0% (12)
101	8	29% (45)	0	0% (25)
Chile				
042	0	0% (12)	0	0% (6)
043[b]	0	13% (64)	0	0% (23)

a. Imipenem resistance (MIC, $\geq$ 16 μg/ml) rates in 2002 at the medical center.
b. *P. fluorescens* producing IMP-1 (one isolate) and VIM-2 (one isolate) were detected in medical centers 048 and 043, respectively.
Adapted from Sader et al. (2005).

Class D β-lactamases, oxacillinases (OXA), correspond to another important group of β-lactamases capable of hydrolyzing carbapenems. Some of these OXA-type carbapenemases are widely dispersed in *P. aeruginosa* and especially in *A. baumannii*. Although most of the OXA-type carbapenemases show only weak carbapenemase activity, carbapenem resistance may result from a combined action of an OXA-type carbapenemase and a secondary resistance mechanism such as porin deficiencies or overexpression of efflux pumps. Carbapenem-resistant *A. baumannii* isolates producing OXA-23 have been identified worldwide, some of them as scattered reports but most of them as sources of outbreaks, such as in Korea, French Polynesia, Turkey, and Brazil (see Poirel et al. 2007). Table 14.4 summarizes the OXA-type carbapenemase-producing bacteria isolated in developing nations. In addition, the SENTRY Program has recently detected *Acinetobacter* spp. harboring OXA-23 in China, Hong Kong, India, Korea, Singapore, and Thailand, while *Acinetobacter* spp. harboring OXA-58 have been detected in China and Thailand (see Bell et al. 2007).

Since MβL- and OXA-type enzymes hydrolyze all β-lactams, except for aztreonam, MBL-producing *P. aeruginosa* and *Acinetobacter* spp. are noted with trepidation. Very few therapeutic options are available for treatment of serious infections caused by these organisms and patients are mainly restricted to drugs of last resort, such as polymyxins. In general, polymyxin B and colistin remain active against clinical isolates of *P. aeruginosa* and *Acinetobacter* spp., including isolates resistant to carbapenems. However, as the usage of the polymyxins increases, emergence of resistance to these potentially toxic agents becomes an obvious concern.

Table 14.4 OXA-type carbapenemase-producing bacteria isolated in developing nations

Enzyme	Host	Country
OXA-23	*A. baumannii*	Brazil, China, Korea, Romania, Singapore, Thaiti
OXA-23	*A. junii*	China
OXA-27	*A. baumannii*	Singapore
OXA-49	*A. baumannii*	China
OXA-50	*P. aeruginosa*	India, Republic of South Africa, Thailand
OXA-58	*A. baumannii*	Argentina, Greece, Romania, Turkey
OXA-64	*A. baumannii*	Republic of South Africa
OXA-65	*A. baumannii*	Argentina
OXA-66	*A. baumannii*	Hong Kong, Poland, Singapore
OXA-69	*A. baumannii*	Singapore, Turkey
OXA-70	*A. baumannii*	Hong Kong
OXA-72	*A. baumannii*	Thailand
OXA-78	*A. baumannii*	Turkey

Adapted from Walther-Rasmussen and Høiby (2006); Poirel et al. (2007).

14.4 *Neisseria gonorrhoeae*

N. gonorrhoeae remains one of the most common causes of sexually transmitted (STD) diseases in developing countries (see Tapsall 2001; Okeke et al. 2005). Rates of gonorrhea vary greatly among countries in the developed and developing world. The highest rates are in South and Southeast Asia, sub-Saharan Africa, and Latin America (see Tapsall 2001). Unlike many STDs, gonorrhea is treatable, but treatment success has been greatly jeopardized by the emergence of resistant isolates. Resistance also affects STD control programs. The appearance of penicillin resistance in the 1970s was rapidly followed by the emergence of tetracycline and spectinomycin resistance. *N. gonorrhoeae* exhibits high rates of resistance to penicillins and tetracycline worldwide, and resistance to multiple agents is common in several geographic regions. In the WHO Western Pacific Region, 2004 data are available for about 10,000 isolates from 15 countries. The proportion of isolates resistant to penicillins by any mechanism ranged from 7% in New Caledonia to 100% in Laos People's Democratic Republic. Particularly high levels of resistance were recorded in Brunei (85.6%), China (75.2%), Korea (77.4%), Philippines (51.4%), and Singapore (50.5%) (see WHO 2006). Penicillin resistance rates exceed 35% in sub-Saharan Africa and Latin American region (see Tapsall 2001; Dillon et al. 2006). Tetracycline resistance is especially common in sub-Saharan Africa, where the low cost and wide distribution of these drugs have contributed to their use and abuse. In this region, tetracycline resistance increased from 20 to 40–65% in the early 1990s (see Van Dyck et al. 2001). Widespread resistance has led to the substitution of penicillin and tetracycline by azithromycin and quinolones as first-line drugs for treatment of gonorrhea (see Tapsall 2001). In South America and the Caribbean region, azithromycin resistance rates varied from 16 to 72% depending on the region reported (see Dillon et al. 2006). Quinolone resistance is most commonly reported in the Asia-Pacific region. According to the WHO report, the highest proportions of quinolone-resistant isolates were seen in the Hong Kong Special Administrative Region (99.3%), China (99.2%), Laos People's Democratic Republic (96.0%), Japan (91.6%), Vietnam (83.9%), and the Republic of Korea (88.2%) (see WHO 2006).

Although *N. gonorrhoeae* continues to develop resistance to older, less expensive antimicrobials and to more recently introduced agents, no resistance to the third-generation cephalosporins was reliably documented until the 1990s (see Tapsall 2001). After first reports from Japan, from 2000 onward, a small number of isolates with altered susceptibility to third-generation cephalosporins have been reported each year in WHO WPR surveys in various countries. At different time periods, Australia, Cambodia, Brunei, China, Japan, Korea, Malaysia, New Zealand, Papua New Guinea, and Singapore have reported their presence (see WHO 2006). The reduced susceptibility is associated with the presence of a number of mosaic *penA* genes and these gonococci are often multi-resistant due to the aggregation of different resistance mechanisms.

14.5 *Neisseria meningitidis*

Neisseria meningitidis is the etiologic agent of meningococcal disease, most commonly meningococcal bacteremia and meningitis, in industrialized countries and in the developing world. These two clinically overlapping syndromes may occur simultaneously, but meningitis alone occurs most frequently (see CDC 2003). Meningococcal disease differs from the other leading causes of bacterial meningitis because of its potential to cause large-scale epidemics. Historically, these epidemics have been typically caused by serogroup A, to a lesser extent, serogroup C. In Africa, the highest incidence rates of serogroup A meningococcal disease occur in a region of sub-Saharan Africa extending from Sudan in the east to Gambia in the west; this region comprehends 15 countries comprising more than 260 million people and has been referred to as the "meningitis belt" (see CDC 2003; Stephens et al. 2007). Patients with invasive meningococcal infection must be treated with effective antibiotics due to the severity of meningococcal disease. Historically, penicillin has been the first therapeutic option for treatment of meningococcal disease. However, strains with reduced susceptibility to penicillin have been reported in many parts of world including Latin American (Argentina, Colombia, Costa Rica, Cuba, and Venezuela), African, and Asian countries. Such strains have been shown to have alterations in penicillin-binding protein 2 (PBP2), which is encoded by the *penA* gene, result of the formation of mosaic genes derived through transformation with DNA from commensal *Neisseria* species. Although the high-level resistance to penicillin due to production of β-lactamase was reported in Africa, these strains are still extremely rare. Resistance to other antimicrobial agents that may be used for therapy of meningococcal infections or for prophylaxis of case contacts has been reported in several countries. This includes resistance to chloramphenicol, sulfonamides, tetracycline, and rifampin (see Jorgensen et al. 2005).

Jorgensen et al. (2005) have evaluated the susceptibility profile of 442 *N. meningitidis* isolates collected from 15 countries to help in the establishment of CLSI breakpoints. Most of these isolates were collected from the United States and other developed countries, and only 17 strains were isolated from developing countries [Bangladesh (2 strains), Burkina Faso (4 strains), Cameroon (1 strain), Chad (1 strain), Chez Republic (1 strain), Gambia (1 strain), Hong Kong (1 strain), Niger (2 strains), Saudi Arabia (1 strain), Singapore (1 strain), and Sudan (2 strains)]. Elevated penicillin and ampicillin MICs (≥ 0.12 and ≥ 0.25 µg/ml, respectively) occurred in 14.3 and 8.6% of strains and were associated with polymorphisms of the *penA* gene. None of the 442 isolates produced β-lactamase. Two strains were resistant to chloramphenicol due to the production of chloramphenicol acetyltransferase mediated by the gene *catP*. Crawford et al. (2005) have also evaluated this collection and detected 13 tetracycline-resistant strains. All 13 tetracycline-resistant isolates belonged to serogroup A, clonal complex ST-5, and were recovered from five African countries (Cameroon, Chad, Dhaka, Niger, and Sudan) and four US states between 1999 and 2002. Tetracycline resistance in this collection was associated with the drug efflux mechanism encoded by *tet*(B), which did not lead to elevated minocycline MICs.

Fluoroquinolones are used for prophylaxis of case contacts. Reduced susceptibility to ciprofloxacin has been described due to mutations in the gene *gyrA* encoding the gyrase A, the fluoroquinolone target, and hyperexpression of naturally occurring efflux systems. Only two strains of *N. meningitidis* with reduced susceptibility to ciprofloxacin were detected by Jorgensen, Crawford, and Fiebelkorn. In contrast, 4 of 13 strains of *N. meningitidis* serogroup A showing resistance to ciprofloxacin (MICs of 0.25 μg/ml) were isolated during a recent epidemic (April 2005–2006) in New Delhi. Unhappily, the genetic relatedness of these isolates was not evaluated. The report of decreased susceptibility to ciprofloxacin is a cause for real concern and will have implications in chemoprophylaxis since ciprofloxacin is used to eradicate meningococci from nasopharyngeal carriers.

Manchanda and Bhalla (2006) reported eight serogroup A *N. meningitidis* non-ceftriaxone-susceptible isolates in India showing MICs ranging from 0.25 to 8 μg/ml. This is of great concern, as ceftriaxone is used for treatment of bacterial meningitis. However, the identification of such isolates has been questioned since Gram staining and latex agglutination tests are not considered sufficient for full characterization of such isolates.

References

Andrade, S. S., Jones, R. N., Gales, A. C., and Sader, H. S. 2003. "Increasing prevalence of antimicrobial resistance among *Pseudomonas aeruginosa* isolates in Latin American medical centres: 5 year report of the SENTRY Antimicrobial Surveillance Program (1997–2001)". *Journal of Antimicrobial Chemotherapy* **52**:140–141.
Baraniak, A., Fiett, J., Sulikowska, A., Hryniewicz, W., and Gniadkowski, M. 2002. "Countrywide spread of CTX-M-3 extended-spectrum β-lactamase-producing microorganisms of the family Enterobacteriaceae in Poland." *Antimicrobial Agents and Chemotherapy* **46**:151–159.
Bell, J. M., Turnidge, J. D., and Jones, R. N. 2003. "Prevalence of extended-spectrum b-lactamase-producing *Enterobacter cloacae* in the Asia-Pacific region: Results from the SENTRY Antimicrobial Surveillance Program, 1998 to 2001." *Antimicrobial Agents and Chemotherapy* **47**:3989–3993.
Bell, J. M., Turnidge, J. D., Walters, L. J., and Jones, R. N. 2007. "Emergence and widespread dissemination of OXA-23, -24 and -58 carbapenemases among *Acinetobacter* spp. in Asia-Pacific Surveillance Program (2006)", Abstract C2-1510a. 47th Interscience Conference on Antimicrobial Agents and Chemotherapy (ICAAC), Chicago, IL, 228.
Biedenbach, D. J., Toleman, M., Walsh, T. R., and Jones, R. N. 2006. "Analysis of *Salmonella* spp. with resistance to extended-spectrum cephalosporins and fluoroquinolones isolated in North America and Latin America: report from the SENTRY Antimicrobial Surveillance Program (1997–2004)." *Diagnostic Microbiology and Infectious Diseases* **54**:13–21.
Brink, A., Moolman, J., da Silva, M. C., and Botha, M. 2007. "Antimicrobial susceptibility profile of selected bacteraemic pathogens from private institutions in South Africa." *South African Medical Journal* **97**:273–279.

Castanheira, M., Bell, J. M., Jones, R. N., Mathai, D., and Turnidge, J. D. 2008. "High carbapenem resistance among *Pseudomonas aeruginosa* from India: A national epidemic of multiple metallo-β-lactamase clones (VIM-2, -5, -6, -11 and new VIM-18)", Abstract P900. 18th European Congress of Clinical Microbiology and Infectious Diseases (ECC-MID), Barcelona, Spain.

Castanheira, M., Bell, J. M., Turnidge, J. D., Mendes, R. E., Sader, H. S., and Jones, R. N. 2007a. "Dissemination and genetic context analysis of bla_{VIM-6} among *Pseudomonas aeruginosa* in Asia-Pacific Nations", Abstract C1-109a. 47th International Conference on Antimicrobial Agents and Chemotherapy (ICAAC), Chicago, IL.

Castanheira, M., Mendes, R. E., Picao, R. C., Pinto, F. P., Machado, A. M. O., Walsh, T. R., and Gales, A. C. 2006. "Genetic analysis of a multidrug-resistant (MDR) *Enterobacter cloacae* producing IMP-1 metallo-b-lactamase (MBL)," Abstract C1-63." 46th Interscience Conference on Antimicrobial Agents and Chemotherapy (ICAAC), San Francisco, CA, 72.

Castanheira, M., Pereira, A. S., Nicoletti, A. G., Pignatari, A. C., Barth, A. L., and Gales, A. C. 2007b. "First report of plasmid-mediated qnrA1 in a ciprofloxacin-resistant *Escherichia coli* strain in Latin America." *Antimicrobial Agents and Chemotherapy.* 51:1527–1529.

CDC. 2003. Centers for Disease Control and Prevention: National Center for Infectious Disease and World Health Organization: Department of Communicable Disease Surveillance and Response. Manual for the laboratory identification and antimicrobial susceptibility testing of bacterial pathogens of public health importance in the developing world: *Haemophilus influenzae, Neisseria meningitidis, Streptococcus pneumoniae, Salmonella* serotype Typhi, *Shigella* and *Vibrio cholerae*. Chapter IV P 29–43. WHO/CDS/CSR/RMD/2003.6. Available from http://www.who.int/csr/resources/publications/drugresist/WHO_CDS_CSR_RMD_2003_6/3n/.

Chau, T. T., Campbell, J. I., Galindo, C. M., Van Minh Hoang, N., Diep, T. S., Nga, T. T., Van Vinh Chau, N., Tuan, P. Q., Page, A. L., Ochiai, R. L., Schultsz, C., Wain, J., Bhutta, Z. A., Parry, C. M., Bhattacharya, S. K., Dutta, S., Agtini, M., Dong, B., Honghui, Y., Anh, D. D., Canh do, G., Naheed, A., Albert, M. J., Phetsouvanh, R., Newton, P. N., Basnyat, B., Arjyal, A., La, T. T., Rang, N. N., Phuong le, T., Van Be Bay, P., von Seidlein, L., Dougan, G., Clemens, J. D., Vinh, H., Hien, T. T., Chinh, N. T., Acosta, C. J., Farrar, J., and Dolecek, C. 2007. "Antimicrobial drug resistance of *Salmonella enterica* Serovar Typhi in Asia and molecular mechanism of reduced susceptibility to the fluoroquinolones." *Antimicrobial Agents and Chemotherapy* 51:4315–4323.

Crawford, S. A., Fiebelkorn, K. R., Patterson, J. E., Jorgensen, and J. H. 2005. "International clone of *Neisseria meningitidis* serogroup A with tetracycline resistance due to *tet(B)*." *Antimicrobial Agents and Chemotherapy* 49:1198–1200.

Daoud, Z., Moubareck, C., Hakime, N., and Doucet-Populaire, F. 2006. "Extended-spectrum β-lactamase producing Enterobacteriaceae in Lebanese ICU patients: Epidemiology and patterns of resistance." *The Journal of General and Applied Microbiology* 52:169–178.

Dillon, J. A., Ruben, M., Li, H., Borthagaray, G., Marquez, C., Fiorito, S., Galarza, P., Portilla, J. L., Leon, L., Agudelo, C. I., Sanabria, O. M., Maldonado, A., and Prabhakar, P. 2006. "Challenges in the control of gonorrhea in South America and the Caribbean: monitoring the development of resistance to antibiotics." *Sexually Transmitted Diseases* 33:87–95.

Dromigny, J. A., Ndoye, B., Macondo, E. A., Nabeth, P., Siby, T., and Perrier-Gros-Claude, J. D. 2003. "Increasing prevalence of antimicrobial resistance among Enterobacteriaceae uropathogens in Dakar, Senegal: A multicenter study." *Diagnostic Microbiology and Infectious Diseases* 47:595–600.

Galanis, E., Lo Fo Wong, D. M., Patrick, M. E., Binsztein, N., Cieslik, A., Chalermchikit, T., Aidara-Kane, A., Ellis, A., Angulo, F. J., and Wegener, H. C. 2006. "Web-based surveillance and global *Salmonella* distribution, 2000–2002." *Emerging Infectious Diseases* 12:381–388.

Gales, A. C., Jones, R. N., Gordon, K. A., Sader, H. S., Wilke, W. W., Beach, M. L., Pfaller, M. A., and Doern, G. V. 2000. "Activity and spectrum of 22 antimicrobial agents tested against urinary tract infection pathogens in hospitalized patients in Latin America: Report from the second year of the SENTRY Antimicrobial Surveillance Program (1998)." *Journal Antimicrobial Chemotherapy* **45**:295–303.

Gales, A. C., Sader, H. S., and Jones, R. N. 2002a. "Urinary tract infection trends in Latin American hospitals: Report from the SENTRY antimicrobial surveillance program (1997–2000)." *Diagnostic Microbiology and Infectious Diseases* **44**:289–299.

Gales, A. C., Sader, H. S., Mendes, R. E., and Jones, R. N. 2002b. "*Salmonella* spp. isolates causing bloodstream infections in Latin America: Report of antimicrobial activity from the SENTRY Antimicrobial Surveillance Program (1997–2000)." *Diagnostic Microbiology and Infectious Diseases* **44**:313–318.

Jorgensen, J. H., Crawford, S. A., and Fiebelkorn, K. R. 2005. "Susceptibility of *Neisseria meningitidis* to 16 antimicrobial agents and characterization of resistance mechanisms affecting some agents." *Journal of Clinical Microbiology* **43**:3162–3171.

Kotloff, K. L., Winickoff, J. P., Ivanoff, B., Clemens, J. D., Swerdlow, D. L., Sansonetti, P. J., Adak, G. K., and Levine, M. M. 1999. "Global burden of *Shigella* infections: Implications for vaccine development and implementation of control strategies." *Bulletin of the World Health Organization* **77**:651–666.

Lincopan, N., McCulloch, J. A., Reinert, C., Cassettari, V. C., Gales, A. C., and Mamizuka, E. M. 2005. "First isolation of metallo-β-lactamase-producing multiresistant *Klebsiella pneumoniae* from a patient in Brazil." *Journal of Clinical Microbiology* **43**:516–519.

Manchanda, V., and Bhalla, P. 2006. "Emergence of non-ceftriaxone-susceptible *Neisseria meningitidis* in India." *Journal of Clinical Microbiology* **44**:4290–4291.

Melano, R., Corso, A., Petroni, A., Centron, D., Orman, B., Pereyra, A., Moreno, N., and Galas, M. 2003. "Multiple antibiotic-resistance mechanisms including a novel combination of extended-spectrum beta-lactamases in a *Klebsiella pneumoniae* clinical strain isolated in Argentina." *Journal of Antimicrobial Chemotherapy* **52**:36–42.

Mendes, R. E., Bell, J. M., Turnidge, J. D., Yang, Q., Yu, Y., Sun, Z., and Jones, R. N. 2008. "Carbapenem-resistant isolates of *Klebsiella pneumoniae* in China and detection of a conjugative plasmid (bla_{KPC-2} + $qnrB4$) and a bla_{IMP-4}." *Antimicrobial Agents and Chemotherapy* **52**:798–799.

Monteiro, J., Henriques, A. P. C., Santos, A. F., Matos, D. G. C., Perano, G., Asensi, M. D., and Gales, A. C. 2007. Carbapenem-resistant *Klebsiella pneumoniae* outbreak: Emergence of KPC-2-producing strains in Brazil, Abstr C2-1929. 47th Interscience Conference on Antimicrobial Agents and Chemotherapy (ICAAC), Chicago, IL, 141.

Okeke, I. N., Laxminarayan, R., Bhutta, Z. A., Duse, A. G., Jenkins, P., O'Brien, T. F., Pablos-Mendez, A., and Klugman, K. P. 2005. "Antimicrobial resistance in developing countries. Part I: Recent trends and current status." *The Lancet Infectious Diseases* **5**:481–493.

Pallecchi, L., Malossi, M., Mantella, A., Gotuzzo, E., Trigoso, C., Bartoloni, A., Paradisi, F., Kronvall, G., and Rossolini, G. M. 2004. "Detection of CTX-M-type β-lactamase genes in fecal *Escherichia coli* isolates from healthy children in Bolivia and Peru." *Antimicrobial Agents and Chemotherapy* **48**:4556–4561.

Peirano, G., Souza, F. S., and Rodrigues, D. P. 2006. "Frequency of serovars and antimicrobial resistance in *Shigella* spp. from Brazil." *Memórias do Instituto Oswaldo Cruz* **101**:245–250.

Poirel, L., Pitout, J. D., and Nordmann, P. 2007. "Carbapenemases: Molecular diversity and clinical consequences." *Future Microbiology* **2**:501–512.

Rao, P. S., Rajashekar, V., Varghese, G. K., and Shivananda, P. G. 1993. "Emergence of multidrug-resistant *Salmonella typhi* in rural southern India." *American Journal of Tropical Medicine and Hygiene* **48**:108–111.

Rossi, F., Baquero, F., Hsueh, P. R., Paterson, D. L., Bochicchio, G. V., Snyder, T. A., Satishchandran, V., McCarroll, K., DiNubile, M. J., and Chow, J. W. 2006. "In vitro susceptibilities of aerobic and facultatively anaerobic Gram-negative bacilli isolated from patients with intra-abdominal infections worldwide: 2004 results from SMART (Study for Monitoring Antimicrobial Resistance Trends)." *Journal of Antimicrobial Chemotherapy* **58**:205–210.

Sader, H. S., Bell, J. M., Jones, R. N., Fritsche, T. R., and Turnidge, J. D. 2007. Spectrum and potency of tigecycline tested against non-enteric Gram-negative bacilli collected across the Asia-Pacific region (2006), Abstract 531. 45th Infectious Disease Society of America (IDSA), San Diego, CA, 160–161.

Sader, H. S., Castanheira, M., Mendes, R. E., Toleman, M., Walsh, T. R., and Jones, R. N. 2005. "Dissemination and diversity of metallo-β-lactamases in Latin America: Report from the SENTRY Antimicrobial Surveillance Program." *International Journal of Antimicrobial Agents* **25**:57–61.

Sader, H. S., Jones, R. N., Andrade-Baiocchi, S., and Biedenbach, D. J. 2002. "Four-year evaluation of frequency of occurrence and antimicrobial susceptibility patterns of bacteria from bloodstream infections in Latin American medical centers." *Diagnostic Microbiology and Infectious Diseases* **44**:273–280.

Stephens, D. S., Greenwood, B., and Brandtzaeg, P. 2007. "Epidemic meningitis, meningococcaemia, and *Neisseria meningitidis*." *Lancet* **369**:2196–2210.

Streit, J. M., Jones, R. N., Toleman, M. A., Stratchounski, L. S., and Fritsche, T. R. 2006. "Prevalence and antimicrobial susceptibility patterns among gastroenteritis-causing pathogens recovered in Europe and Latin America and *Salmonella* isolates recovered from bloodstream infections in North America and Latin America: Report from the SENTRY Antimicrobial Surveillance Program (2003)." *International Journal of Antimicrobial Agents* **27**:378–386.

Talukder, K. A., Khajanchi, B. K., Islam, M. A., Dutta, D. K., Islam, Z., Safa, A., Khan, G. Y., Alam, K., Hossain, M. A., Malla, S., Niyogi, S. K., Rahman, M., Watanabe, H., Nair, G. B., and Sack, D. A. 2004. "Genetic relatedness of ciprofloxacin-resistant *Shigella dysenteriae* type 1 strains isolated in south Asia." *Journal of Antimicrobial Chemotherapy* **54**:730–734.

Taneja, N. 2007. Changing epidemiology of shigellosis and emergence of ciprofloxacin-resistant *Shigellae* in India. *Journal of Clinical Microbiology* **45**:678–679.

Tapsall, J. W. 2001. Antimicrobial resistance in *Neisseria gonorrhoeae*. Available at: http://www.who.int/drugresistance/Antimicrobial_resistance_in_Neisseria_gonorrhoeae.pdf. Geneve, Switzerland: Department of Communicable Diseases Surveillance and Response.

Van Dyck, E., Karita, E., Abdellati, S., Dirk, V. H., Ngabonziza, M., Lafort, Y., and Laga, M. 2001. "Antimicrobial susceptibilities of *Neisseria gonorrhoeae* in Kigali, Rwanda, and trends of resistance between 1986 and 2000." *Sexually Transmitted Diseases* **28**:539–545.

Villegas, M. V., Lolans, K., Correa, A., Suarez, C. J., Lopez, J. A., Vallejo, M., and Quinn, J. P. 2006. First detection of the plasmid-mediated class A carbapenemase KPC-2 in clinical isolates of *Klebsiella pneumoniae* from South America. *Antimicrobial Agents and Chemotherapy* **50**:2880–2882.

Walther-Rasmussen J, and Høiby N. 2006. OXA-type carbapenemases. *Journal of Antimicrobial Chemotherapy* **57**:373–83.

Wang, H., and Chen, M. 2005. "Surveillance for antimicrobial resistance among clinical isolates of Gram-negative bacteria from intensive care unit patients in China, 1996 to 2002." *Diagnostic Microbiology and Infectious Diseases* **51**:201–208.

WHO. 2006. World Health Organization. "Surveillance of antibiotic resistance in *Neisseria gonorrhoeae* in the WHO Western Pacific Region, 2004." *Communicable Diseases Intelligence* **30**: 129–132.

WHO Global Salm-Surv. 2007. Progress report (2000–2005): Building capacity for laboratory-based foodborne disease surveillance and outbreak detection and response (http://www.who.int/salmsurv/links/GSSProgressReport2005.pdf). Accessed November 2007.

Wu, J. J., Ko, W. C., Tsai, S. H., and Yan, J. J. 2007. "Prevalence of plasmid-mediated quinolone resistance determinants *qnrA, qnrB*, and *qnrS* among clinical isolates of

Enterobacter cloacae in a Taiwanese hospital." *Antimicrobial Agents and Chemotherapy* **51**:1223–1227.

Yildirim, I., Ceyhan, M., Gur, D., Mugnaioli, C., and Rossolini, G. M. 2007. "First detection of VIM-1 type metallo-β-lactamase in a multidrug-resistant *Klebsiella pneumoniae* clinical isolate from Turkey also producing the CTX-M-15 extended-spectrum β-lactamase." *Journal of Chemotherapy* **19**:467–468.

Yuksel, S., Ozturk, B., Kavaz, A., Ozcakar, Z. B., Acar, B., Guriz, H., Aysev, D., Ekim, M., and Yalcinkaya, F. 2006. "Antibiotic resistance of urinary tract pathogens and evaluation of empirical treatment in Turkish children with urinary tract infections." *International Journal of Antimicrobial Agents* **28**:413–416.

Chapter 15
Resistance in Reservoirs and Human Commensals

Michael Feldgarden

Abstract The role of commensal bacteria in the spread of antibiotic resistance is recognized as a vital component in understanding how to preserve the power of antibiotics. Excepting tuberculosis, the majority of bacterially caused illness results from infection by commensal organisms, such as *Escherichia coli*, streptococci, and staphylococci. Thus, better knowledge of the biology of antibiotic resistance in commensal bacteria is a crucial component to treating these life-threatening diseases. The two roles that commensal bacteria play as reservoirs of antibiotic resistance genes and as drug-resistant opportunistic pathogens are discussed. Then, the patterns of resistance in commensal *E. coli* in the developing world are surveyed. Finally, various methods for studying antibiotic resistance are discussed, along with project design to determine the effects of commensals on resistance in clinical isolates.

15.1 Introduction

Increasingly, the role of commensal bacteria in the spread of antibiotic resistance is recognized as a vital component in understanding how to preserve the power of antibiotics. In the United States, for example, the National Institute of Allergy and Infectious Disease has funded the Reservoirs of Antibiotic Resistance project in recognition of the need to better understand the biology of commensal bacteria (www.roarproject.org). In the United States, roughly 75% of clinical isolates are commensal organisms that can also exist as opportunistic pathogens (*E. coli, Enterococcus, Staphylococcus*, and *Streptococcus*; Styers et al. 2006).

As serious as the commensal problem is in the developed world, it is far more serious in the developing world. Forty-six percent of childhood mortality is due to diarrhea and pneumonia, much of which is bacterial in cause (Wardlaw et al. 2006). Most of these disease-causing bacteria are either related to commensals

M. Feldgarden (✉)
The Alliance for the Prudent Use of Antibiotics; Genome Sequencing and Analysis Program, The Broad Institute, Boston, MA, USA
e-mail: feldgard@broad.mit.edu

A. de J. Sosa et al. (eds.), *Antimicrobial Resistance in Developing Countries*,
DOI 10.1007/978-0-387-89370-9_15, © Springer Science+Business Media, LLC 2010

or are themselves commensals in other organisms or even in humans (e.g., extraintestinal pathogenic *E. coli*; Johnson and Russo 2002). Thus, better knowledge of the biology of antibiotic resistance in commensal bacteria is a crucial component to treating these life-threatening diseases.

In this review, I will first offer a (tentative) definition of *commensal*. I will then briefly discuss the two roles that commensal bacteria play as reservoirs of resistance. Then, the patterns of resistance in commensal *E. coli* in the developing world will be described and discussed. Finally, the advantages and disadvantages of various methods for studying antibiotic resistance will be discussed, along with suggestions for the design of projects that can rigorously assess the effects of commensals on resistance in clinical isolates.

15.2 Defining Commensals

One of the most conceptually difficult terms to define in microbiology is *commensal*. Many species of bacteria contain members that can coexist with human hosts without causing disease, as well as members which cause disease. To complicate matters further, many strains within a species are able to switch between pathogenic and commensal life histories, depending on the ecological context. For example, at a global level, *E. coli* strains belonging to the ST69 complex account for roughly 7% of *E. coli* isolated from humans (Manges et al. 2001). This clonal complex is also capable of behaving as a highly virulent pathogen when it gains access to the urinary tract because it possesses specific adaptations for survival and reproduction in that habitat. This clone appears to be well adapted to commensal and pathogenic strategies. Additionally, many strains are virulent in some hosts and not others. Shiga toxin-producing *E. coli* cause severe gastrointestinal illness in humans, but, in cattle, are unable to cause disease (Acheson 2000). Even within the same host species, a given strain can vary in its ability to cause disease, as is the case with travelers' diarrhea, which sickens naïve, unexposed hosts.

One useful distinction is to distinguish between "pathogenic commensals" that can cause disease when a patient is vulnerable or the bacterium gains access to a sterile site and "non-pathogenic commensals" which are incapable of causing disease, such as some lactobacilli (Alekshun and Levy 2006). Another distinction is to classify bacteria based on the source of isolation. Commensals can be defined as those isolates collected from asymptomatic individuals and from "typical" colonization sites, such as *E. coli* from feces and *Streptococcus* from nasal passages, or those isolated from environmental or non-human animal sources.

While these definitions may be arbitrary, for purposes of this discussion, we require a working definition of commensals. Here, I will define a commensal as an organism which typically does not cause disease, although some individual strains within a commensal species may be able to cause disease. A commensal isolate is defined as an isolate which has not been acquired from a diseased animal or person, even though that isolate could cause disease in an atypical site, and which would not cause disease in humans following typical colonization. Under this definition, an *E. coli* isolate that has the capacity to be a virulent bloodstream pathogen would be considered a commensal since the bloodstream is not the primary habitat of *E. coli*, whereas a Shiga toxin-producing *E. coli* isolate would not be considered a commensal, even though it does not cause disease in bovines, because this isolate will cause disease following intestinal colonization in humans.

15.3 Commensals as Reservoirs of Antibiotic Resistance: A Conceptual Overview

As discussed above, the meaning of the phrase "commensal organism" has been a subject of much debate, and there has been little resolution of this issue. Most host-associated organisms fall along a continuum of pathogenicity (Alekshun and Levy 2006). Some, such as the *Shigella* complexes of *E. coli* and diarrheagenic *E. coli*, are essentially pathogens. Others, such as the extraintestinal pathogenic *E. coli* (e.g., urinary tract infections) and the vancomycin-resistant enterococci, are commensals in their usual habitat (the intestinal tract), but when present in non-commensal sites are opportunistic pathogens. In opportunistic pathogens, virulence will be dependent on the complement of virulence-associated loci. Consequently, commensalism should be viewed as a sliding scale or a continuum, and not as a binary state.

When commensalism is viewed as a continuum and not an "either-or" state, there are two ways in which commensals can function as reservoirs of antibiotic resistance for pathogens:

1. Commensals may act as a potential source of antibiotic resistance genes that can be transferred to pathogens via horizontal gene transfer.
2. Resistant commensals which are able to behave as opportunistic pathogens will increase in frequency due to selection for resistance. This would increase the likelihood that a given opportunistic infection would be antibiotic resistant and thus difficult to treat.

These are not mutually exclusive roles. As the proportion of resistant commensals increases, the likelihood that successful gene transfer will occur also increases via two mechanisms. First, if pre-existing resistant clones are favored, then that environment will also favor resistant transferrants. Second, as resistant commensals increase in number, there are more potential resistant donors. Many examples of horizontal transfer of antibiotic resistance genes from commensals to pathogens have been documented since the 1970s (Neu et al. 1973; Klare et al. 2001; Oppegaard et al. 2001; Winokur et al. 2001; Witte et al. 2002). Given widespread antibiotic resistance in many organisms, it is not clear, at this point, to what extent the further generation of novel resistance genotypes via horizontal gene transfer would be the primary response to selection versus the clonal spread of pre-existing resistance genotypes.

15.4 Survey of Resistance in Commensals

15.4.1 Resistance in African Commensals

While relatively few studies have been conducted on commensal organisms in Africa, those that have indicate that resistance among commensals is rising and that there are ample reservoirs of resistance that experience regular human contact. Okeke and colleagues tested *E. coli* isolates collected from healthy Nigerian students between 1986 and 1998 for resistance to seven antimicrobials (Okeke et al. 2000). Trimethoprim resistance was high throughout the 12 years (~40%), while five of the six other antimicrobials showed dramatic increases in resistance, with sulfonamide and ampicillin resistance higher than 70%. Disturbingly, by 1998, all isolates were tetracycline resistant, whereas in 1986, less than 40% had been tetracycline resistant. The only antimicrobial that was effective was nalidixic acid, which at the time of the study was used rarely in Nigeria. Also, the percentage of isolates resistant to six drugs increased from 1.6% in 1986 to 15.9% in 1998. A comparison of resistance in *E. coli* across eight developing countries observed similar patterns in Ghana, Kenya, and Zimbabwe: the frequency of resistance to ampicillin, chloramphenicol, oxyte-tracycline, and trimethoprim ranged between 45 and 96%, while the frequency of resistance to rarely used or recently introduced antibiotics, cefazolin, ciprofloxacin, and gentamicin, was less than 10% in all three countries (Nys et al. 2004).

Resistance in African animals is typically much less frequent than in humans. Interestingly, *E. coli* isolated from children versus chickens in close contact with rural Kenyan farming communities showed very different patterns of resistance (Kariuki et al. 1999). The human *E. coli* isolates were overwhelmingly multidrug resistant (85.7%), while the poultry isolates were either sensitive or resistant to only tetracycline (96.8%). Pulsed field gel electrophoresis indicated that these two groups of isolates were genetically distinct, which suggests that the segregation of clones by host can have a profound effect on the prevalence of resistance. While the poultry data suggest that resistance is typically lower in non-human animals, an older study of African baboons, which unfortunately did not use standard susceptibility typing methods, found that wild baboons had very low frequencies of resistance phenotypes, while those baboons asso-ciated with humans (i.e., that picked through human garbage dumps) had higher levels of resistance (Routman et al. 1985). Additionally, the individual resistance profiles for the human-associated baboons were very similar to those observed in the local human population. While poultry in close contact with humans carry lower frequencies of resistant organisms than humans, baboons appear to have acquired resistant organisms from humans quite readily. This suggests that the influence of a given host species as a potential reservoir of resistant organisms and resistance genes can be quite variable.

15.4.2 Resistance in Asian and Middle Eastern Commensals

Surveys of commensals in Asia, as in Africa, have focused on *E. coli*. As in Africa, resistance is quite high. Unlike Africa, however, there is extensive and widespread use of antibiotics in agriculture and aquaculture. A broad survey of chloramphenicol-resistant heterotrophs isolated from aquacultural settings in Malaysia, Thailand, and Vietnam indicates that aquaculture is a significant reservoir of drug-resistant *E. coli* (Huys et al. 2007). While the total number of heterotrophs was not reported, total counts of resistant organisms in sediment and water averaged 10^6 CFU/g and 10^5 CFU/ml, respectively. Given that 40% of heterotrophic-resistant isolates were *E. coli*, commensals in aquaculture are significant environmental reservoirs of antibiotic resistance.

Poultry production is also highly associated with resistance. *E. coli* isolates from Thai chicken farms and abattoirs were highly resistant to many antibiotics (>60%), including ampicillin, nalidixic acid, and tetracycline. Florfenicol resistance exceeded 50%, and, most disturbingly, enrofloxacin and ciprofloxacin resistance were common at 28 and 15%, respectively (Hanson et al. 2002). Poultry production was also implicated in resistance in Saudi Arabia, with poultry *E. coli* isolates possessing higher levels of resistance to those antibiotics used in poultry production than either clinical specimens or those isolates acquired from poultry workers (Al-Ghamdi et al. 1999). However, for those drugs not used in poultry production, the frequency of resistance was lower in poultry isolates than in human clinical isolates, with isolates from poultry workers possessing intermediate levels of resistance. Finally, recent data from China indicate that poultry are a significant reservoir of cephalosporin resistance (Liu et al. 2007). One observation of particular concern is that 2.4% of *E. coli* isolates from Chinese poultry contained the CTX-Mβ-lactamase gene. The widespread use of antibiotics in tandem with high levels of resistance in Asian agriculture provides compelling evidence that agriculture is a significant factor in maintaining high levels of resistance among commensals.

Resistance is also high among human commensals. As early as 1973, 15% of *E. coli* isolated from healthy Thai volunteers were resistant to two or more antibiotics out of five tested (Koonkhamlert and Sawyer, 1973). Since then, resistance to older drugs has risen and become quite frequent, while resistance to newer drugs, such as second- and third-generation cephalosporins, gentamicin, and ciprofloxacin, is rising (Lester et al. 1990; Mansouri and Shareifi 2002; Nys et al. 2004). One factor that does appear to affect the carriage of resistant commensals is proximity to accessible health care (Walson et al. 2001).

15.4.3 Resistance in Latin American Commensals

Of all the developing regions, Latin America has the most data regarding commensals and resistance. What makes much of the human commensal research in Latin America unique is that several groups have examined resistance in remote indigenous communities with very low levels of antibiotic exposure and rare contact with high-use populations. An early study in Amazonian Brazil suggested that fecal *E. coli* from healthy children in remote communities possessed high levels of multidrug resistance, with roughly 30% of isolates resistant to at least three out of four antimicrobials (Nascimento et al. 1999). In French Guyana, resistance in both the predominant fecal flora and the subdominant fecal flora ranged between 25 and 96% in remote Wayampis Amerindian villages to all antibiotics except for later-generation cephalosporins and fluoroquinones which still retained high efficacy (Grenet et al. 2004). While some members of the community had received antibiotic therapy, there was no correlation between therapy and the carriage of resistant organisms; dwelling in a house with a recipient of antibiotic therapy also did not affect the likelihood of carriage of resistant organisms.

A similar investigation of a remote rural Bolivian community that received minimal antibiotic exposure also yielded results similar to those observed in French Guyana (Bartoloni et al. 2004). Resistant isolates were readily isolated from fecal samples, and the majority of these isolates were resistant to one or more antibiotics. Resistance to most classes of antibiotics was common with the exception of later-generation antibiotics and all fluoroquinolones. These resistant isolates were genetically characterized and possessed many loci commonly associated with human clinical settings, such as class I and II integrons and common plasmid replicons (Pallecchi et al. 2007a). Because the resistant isolates and resistant genes were highly diverse, this suggests that multiple introductions of resistant strains have occurred, as opposed to the de novo evolution of resistance. In addition, there appeared to be limited resistance gene transfer among clonal background which would indicate that resistance in this community is largely due to the clonal spread of introduced strains, as opposed to horizontal gene transfer of antibiotic resistance loci. These remote communities provide compelling evidence that antibiotic use can have far-reaching consequences.

Environmentally obtained commensal isolates indicate another source of antibiotic resistance: linkage with metal resistance. Heterotrophic isolates from mining tailing ponds were also resistant to many older antibiotics, such as ampicillin, chloramphenicol, streptomycin, and tetracycline (Ball et al. 2007). Heavy-metal resistance loci are often encoded on plasmids or transposons that also encode for drug resistance, so this association should not be surprising. However, environmental pollution should also be viewed as a potential selective pressure favoring the evolution of resistance (Barkay et al. 2003). Environmental commensals can also serve as reservoirs of resistance. *E. coli* collected from dust and air samples in Mexico City, Mexico, indicated that multidrug-resistant strains were prevalent in human-associated environments (Rosas et al. 1997, 2006).

Human-associated commensals in Latin America also show high levels of resistance. In oral streptococci, resistance was common to seven drugs in 205 isolates from healthy Mexicans and Cubans not taking antibiotics (Diaz-Meija et al. 2002). Despite marked differences in antibiotic usage between the two countries (Cuba is a low-use country, while Mexico is a high-use country), resistance frequency to any single antibiotic did not differ significantly, suggesting that other selection pressures help to maintain the resistance phenotypes. The Cuban isolates were less likely to be resistant to many (>3) antibiotics and were usually resistant to one to three drugs, whereas the Mexican isolates were either susceptible to all or most drugs or resistant to most or all drugs. The genetic elements responsible for the acquisition of resistance might be different between the two countries and stem from the different antibiotic usage regimens in these two countries.

Other studies find very high frequencies of resistance to older antibiotics in human fecal *E. coli* (Bartoloni et al. 1998, 2006; van de Mortel et al. 1998). Most of the resistance loci involved in resistance are commonly occurring loci with a global distribution, such as *bla*TEM, *tet(A)*, *tet(B)*, *sul1*, *sul2*, and *cat1* (Bartoloni et al. 2006). However, the widespread use of antibiotics in Latin America may be functioning as an evolutionary "amplifier" of resistance genes of clinical importance. The first reported observation of the *sul3* gene was in Latin America, which was then observed in Canada 1 year later (Infante et al. 2005; Blahna et al. 2006). The rising prevalence of CTX-M extended spectrum β-lactamases in *E. coli* from healthy children in Bolivia and Peru is another serious emerging threat. In particular, CTX-M-15, which imparts resistance to the critically important antibiotic cefepime, as well as many other cephalosporins, was not observed in Bolivia in 2002, but 3 years later was present in commensal *E. coli* (Pallecchi et al. 2007b). In Peru, similar rapid emergences of other CTX-Mβ-lactamases were observed (Pallecchi et al. 2007b). Given that resistant fecal *E. coli* can easily colonize travelers visiting developing countries, these emerging resistance loci should be viewed as global health problems, and not local or regional problems (Murray et al. 1990).

15.4.4 Summary of Resistance Patterns in Developing Regions

Overall, resistance among commensals has risen to extremely high frequencies, particularly to older, and commonly available, antibiotics. The evolution of resistance appears to be partially attributable to heavy local use enabling the evolution of novel resistant strains and the selection of previously resistant strains. Another critical component is the immigration of resistant strains into habitats and communities where antibiotic use is very low. Much of the research examining resistance in commensals in developing countries has used phenotypic assays. This presents a significant problem in tracking the spread of resistance genes and resistant organisms. Not only is clonal typing information absent, but the high frequencies of resistance further compound the problem, since many isolates are phenotypically similar, making it impossible to determine the origin of isolates. Likewise, the prevalence of resistance in "naïve" communities also means that clonal spread cannot be definitively ascertained simply by examining the presence of resistance in these communities. Consequently, molecular approaches are needed to understand the evolution and spread of resistance.

15.5 Molecular Approaches

Molecular techniques can be used to track the clonal background of both commensal isolates (i.e., those portions of the genome that are not directly involved with resistance) and resistance alleles and genes. Here, I will review the advantages and disadvantages of various molecular approaches, taking into consideration cost, time, resolution, and population genetics concerns. Depending on the particular question being addressed, combining techniques may be the optimal approach for many studies.

15.5.1 DNA Sequencing

DNA sequencing has several advantages. First, by definition, sequencing DNA is the most comprehensive technique for determining genotypes. Second, it can be applied to determine alleles of resistance loci (e.g., the various TEM alleles) and clonal backgrounds, using the method known as multi-locus sequence typing ("MLST"; Maiden et al. 1998) which typically involves sequencing several hundred base pairs of DNA at multiple loci scattered around the genome. Commonly used MLST methods are available at www.mlst.net and pubmlst.org. Third, it is portable among facilities because the data are informationally simple (four nucleotides in a unitary string). Fourth, sequence data can be readily adapted for population genetic analyses that can be used for the evolution of resistance clones (Didelot and Falush 2007). Finally, sequence data also lend themselves to molecular analyses that could identify regions of resistance loci that affect function (Chattopadhyay et al. 2007; Sokurenko et al. 2004).

DNA sequencing has two major disadvantages. The first is cost. Most sequencing facilities can perform MLST at a cost of roughly $40–$80 US per isolate, so large numbers of isolates often exceed the resources of many investigators. The other disadvantage is a lack of temporal resolution. When two isolates differ by only one nucleotide out of approximately 3000 examined, it is difficult to determine when that divergence occurred: it may have occurred yesterday, or, in the case of *E. coli*, roughly 200 years ago. Since the antibiotic era did not even begin until the 1930s, we are essentially attempting to measure an Olympic sprinter with an hour glass. Hopefully, sequencing costs will continue to decrease, enabling more data collection per isolate. In the near future, genomic data coupled with DNA microarray technology, which is currently used to detect DNA variation in humans, may be adapted for use in microbiology (Gibson and Muse 2005).

15.5.2 Pulsed Field Gel Electrophoresis

Pulsed field gel electrophoresis (PFGE) assesses the overall genome architecture. The procedure is similar to standard gel electrophoresis, but the fragments generated are much larger (20–750 kb). The isolate is digested by a restriction enzyme that cuts at eight base recognition sites and then electrophoresed through a gel under alternating (hence "pulsed") current for 18–36 h. PFGE is inexpensive, although it does require equipment typically not used for other laboratory procedures, and fast. The level of discrimination is finer than MLST, as demonstrated by the Center for Disease Control's use of PFGE to discriminate among *E. coli* O157:H7 isolates, most of which have identical MLST profiles, in its PulseNet program.

One problem PFGE faces is that it requires a great deal of skill, and results among laboratories can vary. Additionally, fragment calling can be highly subjective and also depends on the quality of visualization equipment. Determining relatedness of isolates with PFGE is also difficult. Gain or loss of fragment can result from a variety of causes, including single base alterations of a recognition site, acquisition of prophage, pathogenicity islands and plasmids, and inversions. As a result, two strains that share more fragment with each other than with a third strain might not be closest relatives.

15.5.3 PCR Methods

All PCR methods have the common feature that they are fast and cheap, typically costing at most $1 US per strain per PCR reaction. However, they differ greatly in their reproducibility. RAPD PCR uses 10 base primers to randomly amplify unknown DNA fragments, which are then sized using gel electrophoresis (Williams et al. 1990). This method suffers from several defects. Like PFGE, it is unclear how the sharing of fragments matches strain relatedness. Fragments can also be gained or lost for the same reasons as with PFGE. In *E. coli*, the presence of fragments can vary based on minor changes in experimental conditions, such as DNA concentration and purity, reagents, PCR tubes, and PCR cycler (Johnson and O'Bryan 2000; pers. obs.). Related techniques, including REP-PCR and ERIC-PCR, can share similar problems.

PCR methods that amplify specific genes are typically more robust since the reactions are run to exhaustion under defined conditions. Virulence factor typing, where genes involved in pathogenesis are amplified, often in multiplex reactions, has been widely used in studies of commensal *E. coli*. This method, which involves a handful of PCR reactions, is cheap and fast, and also provides information about pathogenic potential, and provides some limited information about relatedness. PCR methods are also widely used to determine the presence of antibiotic resistance (Aarts et al. 2006).

15.6 Future Directions

Understanding resistance in commensals can inform intervention programs to limit the evolution and spread of resistance. At this point, it is not clear whether we need further studies documenting the high frequencies of resistance phenotypes in commensal populations (i.e., "cataloguing the apocalypse"). Such surveillance-oriented projects need to include basic molecular characterization of resistance genes, especially in light of the emergence of potent resistance loci in developing countries. When possible, these studies ideally would type the clonal backgrounds in which resistance occurs.

A second avenue should be to conduct "natural experiments." For example, in the United States, the National Antimicrobial Resistance Monitoring System (NARMS) tracks the spread of resistance from the farm to the diseased patient along each step of the pathway by sampling agricultural isolates, slaughterhouse and supermarket isolates, and isolates associated with disease. Different geographic regions are sampled annually, and isolates are stored for future analysis. Similar projects should be conducted in other countries, particularly those where florfenicol is widely used in agriculture, and this drug can serve as a biomarker of agriculturally influenced antibiotic resistance. Other natural experiments could involve detailed, long-term comparisons of commensal isolates in individuals who have or have not received antibiotic therapy with isolates implicated in disease at local health centers. Given the large immuno-challenged population in many developing countries due to HIV, malaria, and tuberculosis, resistance in commensals may severely limit the ability to effectively treat commensal-associated disease.

While this review has focused exclusively on resistance in commensals, the boundary between commensals and pathogens, to a considerable extent, is arbitrary. As discussed earlier, many commensals are quite proficient at causing disease when presented with the opportunity (or necessity) to do so. In light of the considerable burden treatment failure can place on developing countries, the author hopes that funding agencies and governments recognize the importance of understanding the biology of antibiotic resistance in commensal bacteria.

Acknowledgments This work was supported in part by The Alliance for the Prudent Use of Antibiotics (NIH grant no. U24 AI 050139).

References

Aarts, H.J.M., Guerra, B., and Malorny, B. 2006. "Molecular methods for detection of antibiotic resistance." In: Antimicrobial Resistance in Bacteria of Animal Origin, ed. F.M. Aaarestrup, pp. 37–48. Washington, DC: ASM Press.

Acheson, D.W. 2000. "How does *Escherichia coli* O157:H7 testing in meat compare with what we are seeing clinically?" *Journal of Food Protection* 63: 819–821.

Alekshun, M.N. and Levy, S.B. 2006. "Commensals upon us." *Biochemical Pharmacology* 71, 893–900.

Al-Ghamdi, M.S., El-Morsy, F., Al-Mustafa, Z.H., Al-Ramadhan, M., and Mohammad, H. 1999. "Antibiotic resistance of *Escherichia coli* isolated from poultry workers, patients and chicken in the eastern province of Saudi Arabia." *Tropical Medicine and International Health* 4: 278–283.

Ball, M.M., Carrero, P., Castro, D., and Yarzabal, L.A. 2007. "Mercury resistance in bacteria strains isolated from tailing ponds in a gold mining area near El Callao Bolivar State, Venezuela." *Current Microbiology* 54: 149–154.

Barkay, T., Miller, S.M., and, Summers, A.O. 2003. "Bacterial mercury resistance from atoms to ecosystems." *FEMS Microbiology Reviews* 27: 355–384.

Bartoloni, A., Cutts, F., Leoni, S., Austin, C.C., Mantella, A., Guglielmetti, P., Roselli, Salazar, M., and Paradisi, F. 1998. "Patterns of antimicrobial use and antimicrobial resistance among healthy children in Bolivia." *Tropical Medicine and International Health* 3: 116–123.

Bartoloni, A., Bartalesi, F., Mantella, A., Dell'Amico, E., Roselli, M., Strohmeyer, M., Gamboa Barahona, H., Prieto Barron, V., Paradesi, F., and Rossolini, G.M. 2004. "High prevalence of acquired antimicrobial resistance unrelated to heavy antimicrobial consumption." *Journal of Infectious Diseases* 189: 1291–1294.

Bartoloni, A., Pallecchi, L., Bendetti, M., Fernandez, C., Vallejos, Y., Guzman, E., Villagran, A.L., Mantella, A., Lucchetti, C., Bartalesi, F., Strohmeyer, M., Bechini, A., Gamboa, H., Rodriguez, H., Falkenberg, T., Kronvall, G., Gotuzzo, E., Paradisi, F., and G.M. Rossolini. 2006. "Multidrug-resistant commensal *Escherichia coli* in children, Peru and Bolivia." *Emerging Infectious Diseases* 12: 907–913.

Blahna, M.T., Zalewski, C.A., Reuer, J., Kahlmeter, G., Foxman, B., and Marrs, C.F. 2006. "The role of horizontal gene transfer in the spread of trimethoprim-sulfamethoxazole resistance among uropathogenic *Escherichia coli* in Europe and Canada." *Journal of Antimicrobial Chemotherapy* 57: 666–672.

Chattopadhyay, S., Feldgarden, M., Weissman, S.J., Dykhuizen, D.E., van Belle, G., and Sokurenko, E.V. 2007. "Haplotype diversity in "source-sink" dynamics of *Escherichia coli* urovirulence." *Journal of Molecular Evolution* 64: 204–214.

Diaz-Meija, J.J., Carbajal-Saucedo, A., and Amabile-Cuevas, C.F. 2002. "Antibiotic resistance in oral commensal streptococci from healthy Mexicans and Cubans: resistance prevalence does not mirror antibiotic usage." *FEMS Microbiology Letters* 217: 173–176.

Didelot, X. and Falush, D. 2007. "Inference of bacterial microevolution using multilocus sequence data." *Genetics* 175: 1251–1266.

Gibson, G. and Muse, S.V. 2005. A Primer of Genome Science, Second Edition. Sunderland, MA: Sinauer Associates.

Grenet, K., Guillemot, D., Jarlier, V., Moreau, B., Dubourdieu, S., Ruimy, R., Armand-Lefevre, L., Bau, P., and Andremont, A. 2004. "Antibacterial resistance, Wayampis Amerindians, French Guyana". *Emerging Infectious Diseases* 10: 1150–1153.

Hanson, R., Kaneene, J.B., Padungtod, P., Hirokawa, K., and Zeno, C. 2002. "Prevalence of *Salmonella* and *E. coli*, and their resistance to antimicrobial agents, in farming communities in northern Thailand." *Southeast Asian Journal of Tropical Medicine and Public Health* **33** S3: 120–126.

Huys, G., Bartie, K., Cnockaert, M., Hoang, Oanh, D.T., Phuong, N.T., Somsiri, T., Chinabut, S., Yusoff, F.M., Shariff, M., Giacomini, M., Teale, A., and Swings, J. 2007. "Biodiversity of chloramphenicol-resistant mesophilic heterotrophs from Southeast Asian aquaculture environments." *Research in Microbiology* **158**: 228–235.

Infante, B., Grape, M., Larsson, M., Kristiansson, C., Pallecchi, L., Rossolini, G.M., and Kronvall, G. 2005. "Acquired sulphonamide resistance genes in faecal *Escherichia coli* from healthy children in Bolivia and Peru." *International Journal of Antimicrobial Agents* **25**: 308–312.

Johnson, J.R. and T.T. O'Bryan. 2000. "Improved repetitive-element PCR fingerprinting for resolving pathogenic and nonpathogenic phylogenetic groups within *Escherichia coli*." *Clinical and Diagnostic Laboratory Immunology* **7**: 265–273.

Johnson, J.R. and Russo, T.A. 2002. "Extraintestinal pathogenic *Escherichia coli*: "the other bad *E. coli*."" *Journal of Laboratory and Clinical Medicine* **139**, 155–162.

Kariuki, S., Gilks, C., Kimari, J., Obanda, A., Muyodi, J., Waiyaki, P., and Hart, C.A. 1999. "Genotype analysis of *Escherichia coli* strains isolated from children and chickens living in close contact." *Applied and Environmental Microbiology* **65**, 472–476.

Klare, I., Werner, G., and Witte, W. 2001. "Enterococci. Habitats, infections, virulence factors resistances to antibiotics, of resistance determinants." *Contributions to Microbiology* **8**: 108–122.

Koonkhamlert, C. and Sawyer, W.D. 1973. "Drug-resistant *Escherichia coli* and *Klebsiella–Enterobacter* in healthy adults in Thailand and the effect of antibiotic administration." *Antimicrobial Agents and Chemotherapy* **4**: 198–200.

Lester, S.C., del Pilar Pia, M., Wang, F., Perez Schael, I., Jiang, H., and O'Brien, T.F. 1990. "The carriage of *Escherichia coli* resistant to antimicrobial agents by healthy children in Boston, in Caracas, Venezuala, and in Qin Pu, China." *New England Journal of Medicine* **323**: 285–289.

Liu, J.H., Wei, S.Y., Ma, J.Y., Zeng, Z.L., Lu, D.H., Yang, G.X., and Chen, Z.L. 2007. "Detection and characterisation of CTX-M and CMY-2 beta-lactamases among *Escherichia coli* isolates from farm animals in Guangdong Province of China." *International Journal of Antimicrobial Agents* **29**: 576–581.

Maiden, M.C., Bygraves, J.A., Feil, E., Morelli, G., Russell, J.E., Urwin, R., Zhang, Q., Zhou, J., Zurth, K., Caugant, D.A., Feavers, I.M., Achtman, M., and Spratt, B.G. 1998. "Multilocus sequence typing: a portable approach to the identification of clones within populations of pathogenic microorganisms." *Proceedings of the National Academy of Sciences U.S.A.* **95**: 3140–3145.

Manges, A.R., Johnson, J.R., Foxman, B., O'Bryan, T.T., Fullerton, K.E., and Riley, L.W. 2001. "Widespread distribution of urinary tract infections caused by a multidrug-resistant *Escherichia coli* clonal group." *New England Journal of Medicine* **345**, 1007–1013.

Mansouri, S. and Shareifi, S. 2002. "Antimicrobial resistance pattern of *Escherichia coli* causing urinary tract infections, and that of human fecal flora, in the South-east of Iran." *Microbial Drug Resistance* **8**: 123–128.

Murray, B.E., Mathewson, J.J., DuPont, H.L., Ericsson, C.D., and Reeves, R.R. 1990. "Emergence of resistant fecal *Escherichia coli* in travelers not taking prophylactic antimicrobial agents." *Antimicrobial Agents and Chemotherapy* **34**: 515–518.

Nascimento, A.M.A, Campos, C.E.P., Campos, E.P., Azevedo, J.L., and Chartone-Souza, E. 1999. "Re-evaluation of antibiotic and mercury resistance in *Escherichia coli* populations isolated in 1978 from Amazonian rubber tree tappers and Indians." *Research in Microbiology* **150**: 407–411.

Neu, H.C., Huber, P.J., and Winshell, E.B. 1973. "Interbacterial transfer of R factor in humans." *Antimicrobial Agents Chemotherapy* **3**: 542–544.

Nys, S., Okeke, I.N., Kariuki, S., Dinant, G.J., Driessen, C., and Stobberingh, E.E. 2004. "Antibiotic resistance of faecal *Escherichia coli* from healthy volunteers from eight developing countries." *Journal of Antimicrobial Chemotherapy* **54**: 952–955.

Okeke, I.N., Fayinka, S.T., and Lamikanra, A, 2000. "Antibiotic resistance in *Escherichia coli* from Nigerian students, 1986–1998." *Emerging Infectious Diseases* **6**: 393–396.

Oppegaard, H., Steinum, T.M., and Wasteson Y. 2001. "Horizontal transfer of a multi-drug resistance plasmid between coliform bacteria of human and bovine origin in a farm environment." *Applied Environmental Microbiology* **67**: 3732–3734.

Pallecchi, L., Lucchetti, C., Bartoloni, A., Bartalesi, F., Mantella, A., Gamboa, H., Carattoli, A., Paradisi, F., and Rossolini, G.M. 2007a. "Population structure and resistance genes in antibiotic-resistant bacteria from a remote community with minimal antibiotic exposure." *Antimicrobial Agents and Chemotherapy* **51**: 1179–1184.

Pallecchi, L, Bartoloni, A., Fiorelli, C., Mantella, A., Di Maggio, T., Gamboa, H., Gotuzzo, E., Kronvall, G., Paradisi, F., and Rossolini, G.M. 2007b. "Rapid dissemination and diversity of CTX-M extended spectrum[R]-lactamase genes in commensal *Escherichia coli* isolates from healthy children from low-resources settings in Latin America." *Antimicrobial Agents and Chemotherapy* **51**: 2720–2725.

Rosas, I., Salinas, E., Yela, A., Calva, E., Eslava, C., and Cravioto, A. 1997. "*Escherichia coli* in settled-dust and air samples collected in residential environments in Mexico City." *Applied and Environmental Microbiology* **63**: 4093–4095.

Rosas, I., Salinas, E., Martinez, L., Calva, E., Eslava, C., Cravioto, A., and Amabile-Cuevas, C.F. 2006. "Urban dust fecal pollution in Mexico City: antibiotic resistance and virulence factor of *Escherichia coli*." *International Journal of Environmental Health* **209**: 461–470.

Routman, E., Miller, R.D., Phillips-Conroy, J., and Hartl, D.L. 1985. "Antibiotic resistance and population structure in *Escherichia coli* from free-ranging African yellow baboons." *Applied and Environmental Microbiology* **50**: 749–754.

Sokurenko, E.V., Feldgarden, M., Trintchina, E., Weissman, S.J., Avagyan, S., Chattopadhyay, S., Johnson, J.R., and Dykhuizen, D.E. 2004. "Selection footprint in the FimH adhesin shows pathoadaptive niche differentiation in *Escherichia coli*." *Molecular Biology and Evolution* **21**: 1373–1383.

Styers, D., Sheehan, D.J., P. Hogan, and Sahm, D.F. 2006. "Laboratory-based surveillance of current antimicrobial resistance patterns and trends among *Staphylococcus aureus*: 2005 status in the United States." *Annals of Clinical Microbiology and Antimicrobials* **5**: 2.

Van de Mortel, H.J., Jansen, E.J., Dinant, G.J., London, N., Palacios Pru, E., and Stobberingh, E.E. 1998. "The prevalence of antibiotic-resistant faecal *Escherichia coli* in healthy volunteers in Venezuala." *Infection* **26**: 292–297.

Walson, J.L., Marshall, B., Pokhrel, B.M., Kafle, K.K., and Levy, S.B. 2001. "Carriage of antibiotic-resistant fecal bacteria in Nepal reflects proximity to Kathmandu." *Journal of Infectious Diseases* **184**: 1163–1169.

Wardlaw, T., Salama, P., Johansson, E.W., and Mason, E. 2006. "Pneumonia: the leading killer of children." *Lancet* **368**: 1048–1050.

Williams, J.G., Kubelik, A.R., Livak, K.J., Rafalski, J.A., and Tingey, S.V. 1990. "DNA polymorphisms amplified by arbitrary primers are useful as genetic markers." *Nucleic Acids Research* **18**: 6531–6535.

Winokur, P.L., Vonstein, D.L., Hoffman, L.J., Uhlenhopp, E.K., and Doem, G.V. 2001. "Evidence for transfer of CMY-2 AmpC beta-lactamase plasmids between *Escherichia coli* and *Salmonella* isolates from food animals and humans." *Antimicrobial Agents and Chemotherapy* **45**: 2716–2722.

Witte, W., Klare, I., and Werner, G. 2002. "Molecular ecological studies on spread of antibiotic resistance genes." *Animal Biotechnology* **13**: 57–70.

Part III
Antimicrobial Use and Misuse

Chapter 16
Determinants of Antimicrobial Use: Poorly Understood–Poorly Researched

Hilbrand Haak and Aryanti Radyowijati

Abstract Resistance of microbes against available antimicrobials has grown in past decades, and should be expected to continue to do so in the coming years, unless effective strategies are developed against it. This is especially the case in developing countries, where regulatory systems are less effective and health systems are weaker. Antimicrobials have been part of daily life for a long time in developing countries and local 'cultures' have been built around their indications, efficacy, dosing and duration of use. Hence, poor use of these agents cannot simply be 'regulated', 'reformed' or 'trained' away. Poor antimicrobial use requires careful study of its determinants and to use the findings to define promising interventions. Available literature on determinants is limited and, hence, there is a huge gap in our understanding of what actually drives poor antimicrobial use. Given this poor understanding, it should not come as a surprise that improving antimicrobial use in developing countries has had limited success. This chapter will explore the literature on factors that influence antimicrobial use by providers, dispensers and the general population in developing countries. Antimicrobial prescribing clearly received more research attention than antimicrobial dispensing, and antimicrobial use by populations. Especially the practices and perceptions of dispensers are little studied, despite the key role dispensers have in antimicrobial use. Based on the available information, proposals are made to improve our understanding of determinants, and to develop more effective interventions into antimicrobial use.

16.1 Introduction

There has been an increasing concern over the resistance of microbes against available antimicrobials. While this problem is serious in industrialised countries, it may be more so in developing countries. In developing countries antimicrobials are widely available to the public, without effective controls that are common in

H. Haak (✉)
Consultants for Health and Development, Leiden, The Netherlands
e-mail: haakh@chd-consultants.nl

A. de J. Sosa et al. (eds.), *Antimicrobial Resistance in Developing Countries*,
DOI 10.1007/978-0-387-89370-9_16, © Springer Science+Business Media, LLC 2010

most industrialised countries. Despite legislation against it, these agents can be purchased without a prescription in many countries in Africa, Asia, Latin America and the Eurasian region. Antimicrobials are available from a variety of sources, including hospitals and pharmacies (Bartoloni et al. 1998), licensed medicine stalls and drugstores (Hardon 1987; Calva 1996; Bartoloni et al. 1998), and roadside stalls and hawkers (Bartoloni et al. 1998; Okeke et al. 1999). Formal, informal, illegal, and clandestine aspects are present in all systems (Van der Geest 1984), and there are various incentives for and barriers to appropriate antimicrobial prescribing and use. Prescriptions are not only made by trained health-care providers but also by drug dispensers with a varying degree of training (Haak 1988). It is often believed that this lack of control is the main cause of emergence and spread of resistance.

When exploring the literature on antimicrobial use, it seems that there is a common understanding that people's behaviour is largely determined by what they know about correct practices. As a result, there has been a fair amount of training activities to improve practices, assuming that people have insufficient knowledge on how antimicrobials should be used. Hence, it is felt that they should be taught how to do better. However, prescribing, dispensing and use of drugs are determined by a complex set of reasons, motivations, and a mixture of medical, psycho-social, cultural, economic and even geopolitical factors. In addition, access and non-access to drugs, economic factors, patient pressure, promotional activities of pharmaceutical companies and others may play important roles (WHO 1999). Antimicrobial agents do not escape this reality, and the term 'antibiotic culture' is sometimes used (Haak and Hardon 1988; Wolff 1993), implying that for every ill-defined symptom, antimicrobials are thought to offer a solution. Folk beliefs and traditions may also influence antimicrobial use in many cultures (Haak 1988; Nyazema 1992; Denno et al. 1994; Dua et al. 1994; Duong et al. 1997).

16.2 Who Decides on Antimicrobial Use?

Several authors have argued that 'self-medication' is the key determinant of improper antimicrobial use (Obaseiki-Ebor et al. 1987). As consumers are believed not to have correct knowledge on antimicrobial use, 'self-medication' is assumed to be undesirable. 'Self-medication' is practised widely, and appears to be by far the most common medical response (Van der Geest 1982; Hardon 1991). In a community study in Brazil, antimicrobials were the group of medicines that were most often used in 'self-medication' (Haak 1988). In a Nigerian community, all members admitted having used an antimicrobial in recent times, and most of them admitted that they had treated themselves once or more with antimicrobials before consulting a physician (Obaseiki-Ebor et al. 1987). Surveys of pharmacy sales confirm high rates of 'self-medication' with antimicrobials. In Nigerian pharmacies, oral or injectable antimicrobials were the second most frequent treatments sold for dysentery and diarrhoea (Igun 1987). Motives for 'self-medication' with antimicrobials include the need to save money, and the desire to treat 'confirmed' or suspected bacterial infections. For example, in China, 18–70% of children with ARI were 'self-medicated' with antimicrobials (Hui et al. 1997). In cases such as sexually transmitted infections, where social stigmata may play a role 'self-medication' offers anonymity (Schorling et al. 1991).

The idea of trying to control 'self-medication' has fascinated many in research and medical community, but the main problem is that it is far from clear what should be considered 'self-medication'. The World Health Organization (WHO 1998) does not distinguish whether a prescription was presented when buying the drug for 'self-medication'. As a result, repeat prescriptions, and prescriptions for others may be presented for drugs that are taken at own initiative. More importantly, a doctor or a nurse cannot ensure that patients take medicines as instructed. Any drug, including antimicrobials, may be used differently than intended. Given these methodological difficulties, it is doubtful that measures to control 'self-medication' will lead to improved antimicrobial use. While 'self-medication' may not be helpful in defining strategies to fight antimicrobial resistance, it is certainly useful to define the sources of advice in antimicrobial use. In contrast to industrialised countries, use of antimicrobials in developing countries may be advised by many different actors, having different degrees of medical training or no training at all. These 'advisory agents' include physicians, pharmacists, pharmacy clerks, paramedics, traditional healers, family members and friends. Each of these 'agents' may have different determining factors for their advice on antimicrobial usage.

16.2.1 Lay Networks

Informal care networks have an important influence on the use of medicines. Drug store customers in the Philippines, India, Mexico and Brazil based their decisions to buy antimicrobials on the advice given by friends or relatives (Lansang et al. 1990; Schorling et al. 1991; Dua et al. 1994; Calva 1996). In the Philippines, nearly 50% of mothers mentioned sources of information that were outside the formal health-care network. They claimed that they or their relatives 'knew what to do.'

Lay networks and professional health workers are not necessarily mutually exclusive. Professional health workers may live in the community in which they practice, and their practices may be influenced by folk traditions and perceptions (Soroffman 1992).

16.2.2 Pharmacies

Pharmacies are often the first source of advice for patients who seek care (Lansang et al. 1990; Hui et al. 1997). In most developing countries pharmacists are legally barred from diagnosis and prescription, but in practice they often do. They may even have special rooms for physical examinations and injections (Serkkola 1990; Van Staa 1993). In some instances they have commercial contacts with medical clinics or private doctors. Pharmacy personnel in the Philippines and Mexico give patients advice to buy antimicrobials (Lansang et al. 1990; Calva 1996), while in Egypt pharmacy staff simply refill old bottles of antimicrobials, in most cases, without asking for a prescription (Khallaf et al. 1991). Mothers in poor Brazilian urban slums were reported to seek treatment from pharmacies because it is cheaper and less time-consuming (Schorling et al. 1991).

16.2.3 Physicians

Physicians influence antimicrobial use in three distinct ways: by giving recommendations to obtain antimicrobials (unwritten or 'verbal' prescriptions), by issuing written prescriptions or by prescribing and directly dispensing/selling drugs (Calva 1996). Physicians' prescribing habits may be an important determining factor in 'self-medication' (Hardon 1987), as they are often seen as authority figures and their written prescriptions are valued. According to Hardon (1991), doctors may even have a role of 'legitimising' popular choices of pharmaceuticals, and the consequences of their antimicrobial prescribing practices may extend to other actors. In India, 75% of pharmacy clients based their antimicrobials purchasing decision on an earlier prescription by a physician (Dua et al. 1994). In Mexico, antimicrobial therapy was found to be up to seven times more likely if a sick person had seen a physician (Bojalil and Calva 1994). In the Philippines, community members rank antimicrobials third for the treatment of diarrhoea and ARI, while physicians rank them second (Hardon 1987). Furthermore, in a Mexican study, 61% of all episodes of diarrhoea were treated with one or more antimicrobials that had been prescribed by a physician (Bojalil and Calva 1994). Similar findings are reported from other countries (Lansang et al. 1990; Calva 1996; Bartoloni et al. 1998).

16.2.4 Traditional Medical Practitioners

Traditional medical practitioners generally practice without formal supervision. They often lack access to medical technology or other diagnostic services and they rarely receive training in antimicrobial prescribing. Their information on indications, contraindications and side effects of drugs tends to come from informal, non-medical sources or sometimes from pharmaceutical representatives.

Some traditional healers apply western medicine, including the use of antimicrobials (Wolffers 1987; Singh and Raje 1996). Competition with western doctors may be stiff and the ability to prescribe antimicrobials is believed to attract patients. For example, an ayurvedic healer in India used penicillin injections in the treatment of serious infections, such as skin ulcerations, pulmonary tuberculosis, abscesses and conjunctivitis. Because his patients demanded antimicrobials, the ayurvedic healer was unable to eliminate penicillin from his practice (Burghart 1988).

16.3 Determinants of Antimicrobial Prescribing, Dispensing and Consumption

Because of the different sources of advice in antimicrobial use, determinants of antibiotic use will be discussed according to the different 'advisory agents'.

16.3.1 Antimicrobial Use by Consumers

16.3.1.1 Popularity Antimicrobial Folk Culture

Antimicrobial agents enjoy high popularity in managing health problems, and people have established criteria for types of antimicrobials for defined health problems. In Nigeria, ampicillin and tetracycline are believed to be suitable for the treatment of virtually any ailment, especially STDs, cough, stomach upset and diarrhoea (Obaseiki-Ebor et al. 1987). In Pakistan, any antimicrobial drug is felt to be useful to treat respiratory tract infections (Sturm et al. 1997). Vietnamese drug purchasers believe that antimicrobials are indicated for inflammations, infections, diarrhoea and fever (Duong et al. 1997). Mothers in Ghana use them to treat cough, fever and rhinorrhoea (Denno et al. 1994).

Lay people's indications for antimicrobial use are usually broader than those used by medical doctors, and often more focussed on symptoms. Use of antimicrobials may therefore be stopped as soon as the target symptom has been resolved (Etkin et al. 1990). Methods used in traditional medicine may be imitated, for example, antimicrobial powder may be poured into a wound or mixed with pork fat and rubbed on lacerations (Haak 1988).

Antimicrobials are often used in lower doses and quantities than recommended, and reasons for that are often understandable. A survey of pharmacy customers in the Philippines concluded that insufficient stock in drugstores is an important reason for buying other quantities than recommended. Some customers stated that they had done so earlier, without problems, and others stated that they wanted to test the drug first for undesirable side effects (Lansang et al. 1990). Antimicrobial purchasers in India believed in the balance of strength and weakness of the body, and preferred to buy a small test dose of the antimicrobial first (Dua et al. 1994). Vietnamese antimicrobial purchasers stated that they preferred not to take a full course of antimicrobials because of fears of harmful effects and doubts on actual necessity (Duong et al. 1997).

16.3.1.2 Powerful Medicine

Antimicrobials are often perceived as 'strong' medicine, capable of curing almost any kind of disease. In their classic study in West Africa, Bledsoe and Goubaud (1985) reported that people had specific criteria for selecting medicines. Colour is an important factor that determines perceived efficacy. Multicoloured capsules are believed to be particularly effective because the different colours imply that several kinds of medicine have been combined to make a very powerful drug. A secondary school graduate reported that he took a 'red and black' capsule, the antimicrobial ampicillin, after a hard day's work on the farm, to treat the sore body and to wake up refreshed for another day (Bledsoe and Goubaud 1985).

16.3.1.3 Antimicrobials as Preventive Agents

In many cultures there is the belief that antimicrobials can prevent disease. For example, mothers in Ghana believe that antimicrobials can be used to prevent cough and fever (Denno et al. 1994). In the Philippines, taking an antimicrobial is a common practice to prevent diarrhoea – especially after eating foods of doubtful hygienic status. In Zimbabwe (Nyazema 1992) and in the Philippines (Abellanosa and Nichter 1996), STDs are believed to be preventable by taking an antimicrobial immediately after visiting a prostitute. Rural populations in Brazil reported that they considered Ambra-Sinto® (Tetracycline-HCl) the medicine of choice against measles (Haak 1988).

16.3.2 Antimicrobial Dispensing and Sales

Drug dispensers have a great deal of influence on community drug use. People often prefer to purchase drugs directly from pharmacies for various reasons, including that more value is placed on drugs than on the medical consultation. Their position is often to mediate between health professionals and the popular sector (Van der Geest 1982). Purchasing drugs directly from pharmacies is easy because of the omni-presence of pharmacies in most regions; generally lower costs; often closer social and cultural ties to pharmacy attendants; and the fact that a pharmacy visit is generally less time-consuming than visiting a health facility (Van der Geest 1982).

Day-to-day activities in pharmacies tend to be handled by pharmacy attendants or clerks, and supervision by qualified pharmacists is often minimal. These clerks or attendants have various educational backgrounds. Igun (1991, "Curative pharmaceutical action versus ORT: A dilemma". Unpublished paper. *International Conference on Social and Cultural Aspects of Pharmaceuticals, Woudschoten, Netherlands*) found that most Nigerians do not differentiate between untrained street vendors and pharmacists, and that all are regarded as knowledgeable. In Somalia, nurses and pharmacists were present in private pharmacies, but that a variety of lay people (family members and children) attended customers (Serkkola 1990). In some countries, anybody working in a pharmacy may be considered to be a 'pharmacist' (Duong et al. 1997).

Available studies give the impression that antimicrobial dispensing has taken serious forms. For example, a Bolivian study reported that antimicrobials were dispensed to 92% of adults and 40% of children with watery diarrhoea (Bartoloni et al. 1998). Drug store personnel in Thailand dispensed antimicrobials in various dosing schedules, regardless of the diagnosis, and most dispensed antimicrobials for 2 days or fewer (Thamlikitkul 1988). Sri Lankan pharmacies mostly dispensed two capsules of tetracycline when asked (Wolffers 1987), and another Bolivian study reported that typically four antimicrobial tablets were dispensed to customers (Bartoloni et al. 1998).

16.3.2.1 Knowledge of Correct Dispensing

Knowledge of dispensers of antimicrobial dispensing has not been well researched. One client simulation study showed that pharmacy attendants were not aware of which tetracycline preparations they had in stock. The study quoted a pharmacy attendant when he explained that knowledge on drugs was not even that important, and that it was sufficient to know the price, as the patient generally knows well what he wants (Wolffers 1987).

16.3.2.2 Client Demand

Client demand is an important determinant for dispensers' practices. Dispensers frequently defer to clients' ideas on appropriate care and necessary medicines, and may combine their advice with popular treatment strategies, such as avoidance of certain foods, drinks and behaviours. For example, chemists in Nairobi, Kenya, were willing to sell smaller doses of antimicrobials at the request of patients (Indalo 1997), and antimicrobials were provided in India at the presentation of prescriptions from non-allopathic doctors or even without a prescription (Dua et al. 1994). Pharmacy staff in Nigeria stated that they believed that parents want medicines that 'stop' their children's diarrhoea, and feared that parents would go to another pharmacy if those expectations were not met (Igun 1994).

16.3.2.3 Lack of Regulation and Enforcement

Dispensers often ignore national legislation and dispense drugs routinely without prescriptions. Dispensers in India stated that they knew that law enforcement was impossible because of the large number of small drug stores (Dua et al. 1994). In Kenya, chemists sold antimicrobials under the name 'Septrin,' as requested by customers, although another antimicrobial was actually provided (Indalo 1997). A particularly important factor is the influence of private physicians as models for dispensers' practices. Dispensers frequently say to customers that a given drug is widely used by medical practitioners, and Sri Lankan pharmacy attendants carefully studied physicians' prescriptions to ensure that their advice was in line with that of physicians (Wolffers 1987).

16.3.3 Antimicrobial Prescription by Qualified Prescribers

16.3.3.1 Knowledge of Correct Prescribing

Knowledge of diagnostics and therapeutics is believed to determine the appropriateness of antimicrobial prescribing (Kunin et al. 1987). Following this line of thinking, poor prescribing can be improved by updating knowledge. However, studies have demonstrated that practice is more complex. In a Peruvian study most physicians (36 out of 40) knew when antimicrobials were needed to treat diarrhoea, but a practice assessment demonstrated that almost all prescribed an antimicrobial to one or more surrogate patients (Paredes et al. 1996). Indonesian private prescribers were well aware of the national guidelines on oral rehydration therapy, but found it difficult to send patients away without an antimicrobial agent. They stated that they worried that patients might have other infections, that they expected to receive other types of treatments from private than from public physicians, and that prescribing an antimicrobial immediately would avoid the illness to become more serious. Only the increased cost was seen as a minor disadvantage for patients (Ismail et al. 1991).

Various authors tried to show that differences in prescribing practices relate to differences in levels of knowledge and training (Stein et al. 1984; Bosu and Ofori-Adjei 1997; Okeke et al. 1999). Evidence to support this assumption is limited. For example, in Bangladesh, prescribing rates for metronidazole (not recommended) were the same for 'doctors' and 'medical assistants' (Guyon et al. 1994). Public and private sector physicians in an Indonesian study agreed that viruses caused most diarrhoeal disease, and that antimicrobials were not effective. However, over half of physicians in both groups prescribed antimicrobials for the treatment of diarrhoea (Gani et al. 1991).

16.3.3.2 Fear of Bad Outcomes

Prescribers often fear that disease outcomes may be poor without antimicrobial treatment, especially in developing countries. Whereas prescribers in industrialised countries may fear legal action for not practising evidence-based medicine, their colleagues in developing countries may fear the risk of losing clientele when they do not deliver a fast cure, or when they prescribe a medication with unpleasant side effects. Peruvian physicians mentioned a need to prevent possible 'complications' from diarrhoea. Not prescribing an antimicrobial was seen as very risky, and they emphasised that proper follow-up of children with diarrhoea is not ensured. To address that risk, antimicrobials were mostly prescribed during first visits (Paredes et al. 1996). As antimicrobial use is often considered risk free

(Van Staa 1993), and side effects or toxicity thought to be minimal (Ismail et al. 1991), there appear to be few reasons for not prescribing an antimicrobial. Nurses in primary care clinics in Zimbabwe cited a fear of bad outcomes when not prescribing antimicrobials, even for simple health problems where they were not indicated. Consequences of not treating a potential case of pneumonia were felt to be far greater than unjustified use of an antimicrobial (Stein et al. 1984).

16.3.3.3 Perceived Patient Demand

Patient demand is often believed to be a strong influence on physicians' prescribing decisions. Physicians in the Philippines stated that if they do not prescribe, patients may shop around for another doctor, an undesirable course of events (Van Staa 1993). An Indonesian study reported on parents' desire for potent drugs, leading both public and private doctors to prescribe antimicrobials to their children (Gani et al. 1991). Paediatricians in Trinidad attributed over-prescribing of antimicrobials to parents' demands and their own concern for secondary bacterial infections (Mohan et al. 2004). Nevertheless, the true influence of patient demand in prescribing is not clear. Most physicians who participated in a Peruvian study believed that mothers expected a prescription for their children, but observations of patient–doctor interactions showed that mothers' behaviour (passive or demanding) did not affect whether drugs were prescribed, which ones, and how many. Physicians rarely changed their prescribing habits on the basis of mothers' opinions of treatment options (Paredes et al. 1996). This would imply that doctors make their choices largely independent of patient desires or demands. More so, Peruvian mothers visiting physicians who correctly treated children with watery diarrhoea (without antimicrobials), often left the consultation unhappy. Paradoxically, they even called these physicians charlatans (Paredes et al. 1996).

16.3.3.4 Peer Norms and Local Medical Cultures

Peer norms and standards of senior clinicians appear to influence antimicrobial prescribing. For example, in Indonesia in the early 1960s, an influential senior paediatrician used streptomycin and phenobarbital to treat diarrhoea, and apparently some nurses continue to use this formula until today (Ismail et al. 1991). In some countries even unique local medical culture may develop. For example, in Peru the medical profession developed a concept of 'diarrea parenteral' which is different from western knowledge about disease causation. 'Diarrea parenteral' may occur when an infection is present that affects a system other than the gastrointestinal tract (e.g. ARI) and that may cause diarrhoea to develop. Antimicrobials are felt to be necessary to treat the primary infection, which would concurrently result in a cure of the diarrhoea episode (Paredes et al. 1996).

16.3.3.5 Availability of Laboratory Services

Lack of access to quality laboratory services is often regarded as a deterrent to rational use of antimicrobials. In Bangladesh, more than 90% of antimicrobials were reported to be used without establishing a laboratory diagnosis. In Trinidad, physicians explained that they do not normally request laboratory investigations, as they consider them unnecessary and the waiting time for results is felt to be too long (Mohan et al. 2004). Similarly, lack of laboratory facilities, or the inability of patients to pay for such services, was blamed for over-prescribing of antimicrobials in a Pakistani study (Nizami et al. 1996). An attractive alternative to using laboratory services may be to use antimicrobials as a diagnostic tool. When a patient does not recover after an initial antimicrobial treatment, further diagnostic activities may be employed (Hardon 1991).

Interestingly, the availability and accessibility of laboratory facilities and personnel does not necessarily motivate physicians to use them. All hospitals in a Malaysian study had facilities for microbiological culture, but only 20% of antimicrobial prescriptions were made on the basis of microbiological reports (Lim and Cheong 1993). Javato-Laxer and colleagues (Javato-Laxer et al. 1989) explained that because of their failure to determine aetiologies of infections, physicians often prefer to use broad-spectrum antibiotics, believing that this will cover all possible aetiologies and unusual pathogens.

16.3.3.6 Unstable Antimicrobial Supply

The ability of prescribers to provide appropriate antimicrobial therapy may be limited by the unavailability of indicated antimicrobials. From Bangladesh and India, it was reported that primary care facilities prescribed antimicrobials according to availability patterns, and less according to specific needs of the patient (Uppal et al. 1993; Guyon et al. 1994). A Tanzanian study concluded that overuse of antimicrobials in hospitals was caused by the wider range of antimicrobials available as compared to health centres (Massele and Mwaluko 1994).

16.3.3.7 Pressure of Pharmaceutical Promotion

Company sales representatives and commercially oriented drug publications are known to be an important source of information for prescribers (Bosu and Ofori-Adjei 1997). The few studies that carefully followed and interviewed pharmaceutical representatives (Kamat and Nichter 1987; Wolffers 1987) concluded that promotional activities increased the haphazard supply of antimicrobials in some societies. In Indonesia, prescribers receive payment for issuing certain drugs during promotional events. In the Philippines, pharmaceutical companies reinforce the notion of risk-free medicines and promote a 'why worry' attitude among doctors (Van Staa 1993). Nevertheless, reports on commercial pressures in drug prescribing in developing countries tend to be anecdotal, and understanding of this influence is limited.

16.3.4 Economic Determinants

Economic motives are present in many aspects of antimicrobial prescribing, dispensing and consumer use. Where spending resources and making profits are present, they may overrule the importance of other determinants. In Zimbabwe dispensing doctors prescribed significantly more antimicrobials per encounter than non-dispensing doctors. Dispensing doctors also spent less time on each patient encounter, and prescribed clinically and economically less appropriately (Trap et al. 2002). In rural China, health system financing influenced antimicrobial prescription, in both frequency and type (Dong et al. 1999a.). Physicians tended to prescribe more expensive antimicrobials for insured patients, resulting in higher profits for themselves (Dong et al. 1999b). Interestingly, also patients requested more expensive drugs as they did not pay the full cost of prescriptions (Dong et al. 1999a, 1999b). This economic rationale is especially strong in private settings where patients pay for services (van Staa 1993). Prescribing a commercially interesting drug may have the dual financial incentive of earning on the patient by selling the specific drug, and a bonus from the industry for prescribing it. In Vietnam, dispensing doctors were reported to refuse to disclose the name of prescribed drugs, to ensure that patients return (Duong et al. 1997). Whereas the economic incentives in prescribing are often mentioned, very few papers carefully investigated this influence.

Contrary to economic motives in antimicrobial prescribing, economic motives are frequently mentioned when dispensing practices are concerned. Being entrepreneurs, dispensers may be prepared to negotiate the type and quantity of drugs to be procured. Quantities of antimicrobials dispensed by pharmacies in Bolivia were reported to vary according to clients' ability to pay (Bartoloni et al. 1998). In India, pharmacies changed prescriptions so that they would suit the financial means of customers (Dua et al. 1994). In Nigeria, pharmacists were convinced that all pharmacies prescribe drugs for watery diarrhoea, instead of only ORT and hence they would do so too (Igun 1994).

Economic factors in consumers' use of drugs are little explored, and more work appears to be needed to understand these influences. An often forgotten notion is that spending much money on medications may sometimes be interpreted by patients as obtaining high quality of care, and therefore be part of the healing process. Low-cost solutions (without antimicrobial therapy) are sometimes understood to be inferior care!

16.3.5 Pharmaceutical Industry Influence

Influence of the pharmaceutical industry is often mentioned in papers dealing with medicine use. Reports from Brazil mentioned pharmacies trying to sell more of certain drugs because of incentives provided by suppliers (Haak 1988), and a report from the Philippines (Van Staa 1993) describes how pharmaceutical representatives visit pharmacies and physicians to boost consumption of defined drugs. In Sri Lanka, pharmacy attendants admitted that sales representatives were their major source of information on drugs (Wolffers 1987).

Apparently the pharmaceutical industry is able to change drug consumption patterns, but how this influence works seems to be less well investigated and known. Attention has been called for this lack of information on industry marketing practices in developing countries (Van der Geest et al. 1996), and applying 'marketing' methods may offer options that have been overlooked so far.

16.4 Discussion

Antimicrobials have been part of daily life in developing countries for a long time, and local 'cultures' have been built around their indications, efficacy, dosing and duration of use. And there are more determinants of antimicrobial use, some of them well explored, and others less. This realisation is important in efforts to improve antimicrobial prescribing, dispensing and consumer use. Thorough information on what determines antimicrobial use by all actors is critically important. Unfortunately there is a serious lack of such determinant research. Antimicrobial prescribing has received more research attention than antimicrobial dispensing, and antimicrobial use by consumers, possibly because of an assumption that most antimicrobial use is based on doctors' prescription. Practices and perceptions of dispensers are relatively little studied (Radyowijati and Haak 2001; Radyowijati and Haak 2003).

Whereas the majority of interventions to improve antibiotic use focus on strengthening knowledge through training and education, the assumption that this will lead to improved practices remains to be proven. As demonstrated in a few studies, physicians with adequate knowledge may nonetheless practice 'incorrectly.' Moreover, in developing countries there may be only a thin line between perceptions in antimicrobial use between prescribers, dispensers and their patients. The three categories often belong to the same or similar ethnic or geographical group(s), and they may share cultural perceptions on illness and health, and of the place of antimicrobials in these. Behaviours may even be copied between each of them. Hence, distinguishing determinants of antimicrobial use for prescribers, dispensers and consumers may be partly artificial. Nonetheless, available research data have shown some interesting aspects, which will be discussed below.

16.4.1 Consumers

Community use of antimicrobials is shaped by various sources of advice, ranging from lay networks to dispensers and health-care providers. People like to pass on successful experiences of antimicrobial use to others in their network. How prescribers' and dispensers' practices influence this knowledge exchange, or how pharmaceutical propaganda affects it, remains largely unknown.

Antimicrobial agents have a powerful image, and are often seen as the ultimate choice for a variety of complaints, a perception that is not necessarily confined to community members. Research does not explain whether this powerful image comes from prescribers, patients or both. The powerful image of antimicrobials may cause antimicrobials to cost more than other drugs (Duong et al. 1997). This, in turn, may lead to procurement of partial therapies or even of procuring other drugs than the prescribed antimicrobial agent. A better understanding of the powerful image of antimicrobials is needed to understand the complexity of care-seeking patterns in the use of antimicrobials, and to design and implement effective programmes to improve antimicrobial use.

16.4.2 Dispensers

Relatively little is known about the influence of dispensers in antimicrobial use. Little research has been carried out on what dispensers actually know on correct antimicrobial use, how they acquire their knowledge and how they apply it. Moreover, most of literature on dispensing activities focuses on quantitative aspects, and little is known about the characteristics of dispensers themselves. Although some incidental observations and remarks have been made on modelling influences by physicians' practices, little is also known on the role that dispensers actually see for themselves; what kind of diseases they feel they can handle themselves; which diseases they refer to physicians. Research is needed to understand dispensers' cultural ideas on antimicrobials, their social position in the community, and their knowledge about this class of drugs.

16.4.3 Prescribers

Physicians' knowledge of correct prescribing is a fascinating subject. A number of papers have shown that prescribers may demonstrate correct knowledge, but that they may still practice differently. This suggests that other determinants are stronger. Some of these determinants have been explored in the literature, but much remains to be explored. The often heard assumption that lack of quality diagnostic services drives poor antimicrobial use remains to be proven, as even when laboratory facilities are available, prescribers may not use them.

Prescribers often reported a fear of poor disease outcomes without using antimicrobial agents. This fear may link to the powerful image of antimicrobials and the notion that without prescribing them, not all clinical options have been tried out. Reports give examples of how doctors deal with insecurity on diagnosis and treatment, but it continues to be unknown whether these fears can be addressed in other ways than applying antimicrobial agents. On the other hand, as antimicrobials are often perceived as risk-free agents, doctors may feel that there are few arguments not to prescribe an antimicrobial agent. Risk perceptions of prescribers and how these can be used to improve practices of antimicrobial use are largely unexplored.

Peer norms and local medical cultures impress as important determinants of prescribing by doctors, pharmacist dispensing and community use. Interestingly, few studies have investigated the role of peer influence in inappropriate use of antimicrobials.

Determinants of antimicrobial use are complex, involving a variety of motivations on the part of prescribers, dispensers and consumers. Systematically expanding our knowledge about factors that determine antimicrobial use, their responsiveness to change, and cost-effective strategies for achieving these changes, will be indispensable to effectively counteract the rise of resistance.

16.5 Conclusion

Solutions to the problem of inappropriate use have often been sought in seeking to 'regulate' or 'reform' antimicrobial use. These proposals include the development of prescribing guidelines, strengthening medical and public education (Okeke et al. 1999), and the development of programmes to 'audit' antimicrobial use and resistance, and improving the availability of laboratory services (Hadi et al. unpublished). Some called for a 'change of mentality' of all actors in the chain of antimicrobial use, without explaining how mentalities can be changed. While there is certainly much value in these proposals, it is a question whether at present there is enough information to develop the envisaged 'reforms' and 'auditing systems', whether prescribing guidelines will be used if knowledge only partially determines prescribing or if economic motives are stronger determinants in prescribing and dispensing. All these efforts may remain 'rowing against the tide', unless determinants of antimicrobial use in a given health system are fully understood and efforts taken to address them.

In its report on a Global Strategy for the Containment of Antimicrobial Resistance (WHO 2001), the World Health Organization emphasised that resistance to first-line drugs by common pathogens 'costs money, livelihoods, and lives and threatens to undermine the effectiveness of health delivery programs.' Resistance was even considered 'a threat to global stability and national security.' Developing countries were said to have an important role in the emergence of resistance: 'antimicrobial use is the key driver of resistance', as stated by the Strategy. Improving the behaviours of prescribers, dispensers and consumers must be the major focus of interventions. Sadly, the world's literature on determinants of antimicrobial use in developing countries is very limited. System improvements and regulation are pillars in WHO's Strategy, but little is known about the effectiveness of these approaches in developing countries, where health systems and regulatory structures suffer serious limitations. Hence, if we know so little about why and how people use antimicrobials, how can we hope to change their behaviour? WHO appears to recognise this, and admits that 'despite the mass of literature on antimicrobial resistance, there is depressingly little on the effectiveness of interventions.'

Recommendations for future research directions based on the lessons in the current review have been presented elsewhere (Radyowijati and Haak 2001). Improved knowledge of determinants of antimicrobial use must be regarded as the missing link in efforts to improve antimicrobial use. In the past this may have been hard to accomplish. However, recently a variety of user-friendly and rapid research methods have become available and the use of such methods is now easily applicable (WHO 1993; Arhinful et al. 1996). Combining quantitative and qualitative methods, and 'triangulating' findings, may yield valuable information.

Acknowledgments The research behind this presentation was supported by a grant from the United States Agency for International Development through the Applied Research on Child Health Project.

References

Abellanosa, I., and Nichter, M. 1996. Antibiotic prophylaxis among commercial sex workers in Cebu City, Philippines. Patterns of use and perception of efficacy. Sex. Transm. Dis. 23:407–412.

Arhinful, D. K., Das, A. M., and Hadiyono, J. P. 1996. How to use qualitative methods to design interventions. Washington DC: INRUD.

Bartoloni, A., Cutts, F., Leoni, S., Austin, C. C., Mantella, A., Guglielmetti, P., Roselli, M., Salazar, E., and Paradisi, F. 1998. Patterns of antimicrobial use and antimicrobial resistance among healthy children in Bolivia. Trop. Med. Int. Health 3:116–123.

Bledsoe, H. C., and Goubaud, F. M. 1985. The reinterpretation of western pharmaceuticals among the Mende of Sierra Leone. Soc. Sci. Med. 21:275–282.

Bojalil, R., and Calva, J. J. 1994. Antibiotic misuse in diarrhea. A household survey in a Mexican community. J. Clin. Epidemiol. 47:147–156.

Bosu, W. K., and Ofori-Adjei, D. 1997. Survey of antibiotic prescribing pattern in government health facilities of the Wasa-west district of Ghana. East Afr. Med. J. 74:138–142.

Burghart, R. 1988. Penicillin: An ancient ayurvedic medicine. In The context of medicines in developing countries: Studies in pharmaceutical anthropology, eds. S. van der Geest and S.R. Whyte pp. 289–297. Dordrecht: Kluwer.

Calva, J. 1996. Antibiotic use in a peri-urban community in Mexico: A household and drugstore survey. Soc. Sci. Med. 42:1121–1128.

Denno, M.D., Bentsi-Enchill, A., Mock, N. C., and Adelson, J. W. 1994. Maternal knowledge, attitude and practice regarding childhood acute respiratory infections in Kumasi. Ann. Trop. Paediatr. 14:293–301.

Dong, H., Bogg, L., Rehnberg, C., and Diwan, V. 1999a. Association between health insurance and antibiotics prescribing in four counties in rural China. Health Policy 48:29–45.

Dong, H., Bogg, L., Rehnberg, C., and Diwan, V. 1999b. Health financing policies: Providers' opinions and prescribing behaviour in rural China. Int. J. Technol. Assess. Health Care 15:686–698.

Dua, V., Kunin, M., and White, L. 1994. The use of antimicrobial drugs in Nagpur, India. A window on medical care in a developing country. Soc. Sci. Med. 38:717–724.

Duong, V. D., Binns, W. B., and Le, V. T. 1997. Availability of antibiotics as over-the-counter drugs in pharmacies: A threat to public health in Vietnam. Trop. Med. Int. Health 2:1133–1139.

Etkin, N. I., Ross, P. L., and Muazzamu, I. 1990. The indigenization of pharmaceuticals: therapeutic transitions in rural Hausaland. Soc. Sci. Med. 30:919–928.

Gani, L., Arif, H., Widjaja, K., Adi, R., Prasadja, H., Tampubolon, L. H., Lukito, E., and Jauri, R. 1991. Physicians prescribing practice for treatment of acute diarrhoea in young children in Jakarta. J. Diarrhoeal Dis. Res. 9:194–199.

Guyon, A. B., Barman, A., Ahmed, J. U., Ahmed, A. U., and Alam, M. S. 1994. A baseline survey on use of drugs at the primary health care level in Bangladesh. Bull. World Health Organ. 72:265–271.

Haak, H. 1988. Pharmaceuticals in two Brazilian villages: lay practices and perceptions. Soc. Sci. Med. 27:1415–1427.

Haak, H., and Hardon, A. P. 1988. Indigenized pharmaceuticals in developing countries: widely used, widely neglected. Lancet 2:620–621.

Hardon, A. P. 1987. The use of modern pharmaceuticals in a Filipino village: Doctors' prescription and self-medication. Soc. Sci. Med. 25:277–292.

Hardon, A. P. 1991. Confronting ill health: medicines, self care and the poor in Manila. Quezon City: Health Action Information Network.

Hui, L., Li, X. S., Zeng, X. J., Dai, Y. H., and Foy, H. M. 1997. Patterns and determinants of use of antibiotics for acute respiratory tract infection in children in China. Pediatr. Infect. Dis. J. 16:560–564.

Igun, U. A. 1987. Why we seek treatment here: Retail pharmacy and clinical practice in Maiduguru, Nigeria. Soc. Sci. Med. 24:689–695.

Igun, U. A. 1994. Reported and actual prescription of oral rehydration therapy for childhood diarrheas by retail pharmacists in Nigeria. Soc. Sci. Med. 39:797–806.

Indalo, A. A. 1997. Antibiotic sale behaviour in Nairobi: A contributing factor to antimicrobial drug resistance. East Afr. Med. J. 74:171–173.

Ismail, R., Bakri, A., Nazir, M., and Pardede, N. 1991. The behaviour of healthcare providers in managing diarrhoeal disease in Palembang City, south Sumatera, Indonesia. Paediatr. Indones. 31:123–135.

Javato-Laxer, M., Navarro, E., and Littana, R. 1989. Antimicrobial patterns in hospitals: determinants and proposed interventions. Philipp. J. Microbiol. Infect. Dis. 18:41–46.

Kamat, V. R., and Nichter, N. 1987. Monitoring product movement: An ethnographic study of pharmaceutical sales representatives in Bombay, India. In Private health providers in developing countries: Serving the public interest? eds S. Bennett, B. McPake, and A. Mills. London: Zed Books.

Khallaf, N., Wahba, S., Herman, E., and Black, R. 1991. Recommendation from Egyptian pharmacies for children with acute respiratory illnesses. Lancet, 338:248.

Kunin, C. M., Lipton, H. L., Tupasi, T. E., Sacks, T., Scheckler, W. E., Jivani, A., Goic, A., Martin, R. R., Guerrant, R. L., and Thamlikitkul, V. 1987. Social, behavioural, and practical factors affecting antibiotic use worldwide. Report of Task Force 4. Rev. Infect. Dis. 9:S270–S285.

Lansang, M. A., Lucas Aquino, R., Tupasi, T. E., Mina, V. S., Salazar, L. S., Juban, N., Limjoco, T. T., Nisperos, L. E., and Kunin, C. M. 1990. Purchase of antibiotics without prescription in Manila, the Philippines: inappropriate choices and doses. J. Clin. Epidemiol. 43:61–67.

Lim, V. K. E., and Cheong, Y. M. 1993. Patterns of antibiotic usage in hospitals in Malaysia. Singapore Med. J. 34:525–528.

Massele, A. Y., and Mwaluko, G. M. 1994. A study of prescribing patterns at different health care facilities in Dar Es Salaam, Tanzania. East Afr. Med. J. 71:314–316.

Mohan, S., Dharamraj, K., Dindial, R., Mathur, D., Parmasad, V., Ramdhanie, J., Matthew, J., and Pinto Pereira, L.M. 2004. Physician behaviour for antimicrobial prescribing for paediatric upper respiratory tract infections: a survey in general practice in Trinidad, West Indies. Ann. Clin. Microbiol. Antimicrob. 3:11.

Nizami, S. Q., Khan, I. A., and Bhutta, Z. A. 1996. Drug prescribing practices of general practitioners and paediatricians for childhood diarrhoea in Karachi, Pakistan. Soc. Sci. Med. 42:1133–1139.

Nyazema, N. Z. 1992. Layman's perception of antimicrobial agents: A challenge to health education strategy in Zimbabwe. East Afr. Med. J. 69:126–129.

Obaseiki-Ebor, E. E., Akerele, J. O., and Ebea, P. O. 1987. A survey of antibiotic outpatients prescribing and antibiotic self-medication. J. Antimicrob. Chemother. 20:759–763.

Okeke, N. I., Lamikanra, A., and Edelman, R. 1999. Socioeconomic and behavioural factors leading to acquired bacterial resistance to antibiotics in developing countries. Emerg. Infect. Dis. 5:18–27.

Paredes, P., de la Peña, M., Flores-Guerra, E., Diaz, J., and Trostle, J. 1996. Factors influencing physicians' prescribing behaviour in the treatment of childhood diarrhoea: Knowledge may not be the clue. Soc. Sci. Med. 42:1141–1153.

Radyowijati, A., and Haak, H. 2001. Determinants of antimicrobial use in the developing world. Child health research project special report, 4(1). www.childhealthresearch.org/doc/AMRvol4.pdf.

Radyowijati, A., and Haak, H. 2003. Improving antibiotic use in low-income countries: An overview of evidence on determinants. Soc Sci. Med. 57:733–744.

Schorling, J. B., de Souza, M. A., and Guerrant, R. L. 1991. Patterns of antibiotic use among children in an urban Brazilian slum. Int. J. Epidemiol. 20:293–299.

Serkkola, A. 1990. Medicines, pharmacy and family: triplicity of self-medication in Mogadishu, Somalia. Occasional papers #11. Helsinki: University of Helsinki, Institute of Development Studies.

Singh, J., and Raje, N. 1996. The rise of western medicine in India. Lancet 348:1598.

Soroffman, B. 1992. Drug promotion in self-care and self-medication. J. Drug Issues 22:377–388.

Stein, C. M., Todd, W. T., Parirenyatwa, D., Chakonda, J., and Dizwani, A. G. 1984. A survey of antibiotic use in Harare primary care clinics. J. Antimicrob. Chemother. 14:149–156.

Sturm, A. W., van der Pol, R., Smits, A. J., van Hellemondt, F. M., Mouton, S. W., Jamil, B., Minai, A. M., and Sampers, G. H. 1997. Over-the-counter availability of antimicrobial agents: Self-medication and patterns of resistance in Karachi, Pakistan. J. Antimicrob. Chemother. 39:543–547.

Thamlikitkul, V. 1988. Antibiotic dispensing by drug store personnel in Bangkok, Thailand. J. Antimicrob. Chemother. 21:125–131.

Trap, B., Hansen, E. H., and Hogerzeil, H. V. 2002. Prescription habits of dispensing and non-dispensing doctors in Zimbabwe. Health Policy Plan. 17:288–295.

Uppal, R., Sarkar, U., Giriyappanavar, C. S., and Kacker, V. 1993. Antimicrobial drug use in primary health care. J. Clin. Epidemiol. 46:671–673.

Van der Geest, S. 1982. The illegal distribution of western medicines in developing countries: Pharmacists, drug pedlars, injection doctors and others. A bibliographic exploration. Med. Anthropol. 6(4):197–219.

Van der Geest, S. 1984. Anthropology and pharmaceuticals in developing countries. Med. Anthropol. Q. I + II 15:59–60 and 87–90.

Van der Geest, S., Whyte, S. R., and Hardon, A. P. 1996. The anthropology of Pharmaceuticals: A biographical approach. Ann. Rev. Anthropol. 25:153–178.

Van Staa, A. 1993. Myth, and metronidazole in Manila: the popularity of drugs among prescribers and dispensers in the treatment of diarrhoea. Master thesis in Medicine and Cultural Anthropology, University of Amsterdam.

Wolff, M. J. 1993. Use and misuse of antibiotics in Latin America. Clin. Infect. Dis. 17:S346–S351.

Wolffers, I. 1987. Drug information and sales practices in some pharmacies of Colombo, Sri Lanka. Soc. Sci. Med. 25:319–321.

WHO. 1993. How to investigate drug use in health facilities – selected drug use indicators. Geneva: World Health Organization, WHO/DAP/93.1.

WHO. 1998. The Role of the pharmacist in self care and self medication. Geneva: World Health Organization, WHO/DAP/98.13.

WHO. 1999. Development of a global strategy for the Phase I. Draft document compiled from presentations by groups of experts at a workshop held at WHO 4–5 February 1999.

WHO. 2001. WHO Global Strategy for Containment of Antimicrobial Resistance. Geneva: World Health Organization, WHO/CDS/CSR/DRS/2001.2

Chapter 17
Antimicrobial Use and Resistance in Africa

Iruka N. Okeke and Kayode K. Ojo

Abstract Antimicrobial resistance is becoming increasingly important in high-burden infectious diseases in Africa. Resistance to affordable antimalarials and antibacterials has worsened patient prognosis, increased health-care costs, and is a barrier to effective health care. This chapter outlines the African situation for a few infectious diseases and discusses mitigating factors, particularly selective pressure from antimicrobial use. Overuse of antimicrobials for prevention and treatment of real or supposed infections, unregulated antimicrobial sales as well as poor antimicrobial quality assurance may be important contributors to the problem by providing selective pressure for the emergence and spread of resistant strains. In addition to addressing these problems, containment of resistant strains and improvement of diagnostic infrastructure are necessary interventions for resistance control.

17.1 Scale and Status of the Antimicrobial Resistance in Africa

Africa bears the greatest infectious disease burden, and with it considerable burden from antimicrobial resistance. The consequences for patients infected with resistant pathogens are dire because many endemic pathogens cause life-threatening disease, because second-line drugs are more expensive than first-line ones, and because many newer, patent-protected drugs are not available in much of Africa. Young children, the malnourished, the aged, and the immunocompromised are less able to clear infection without appropriate chemotherapeutic intervention, so that the most vulnerable sub-populations bear the brunt of resistance. Resistance is a problem with all infectious diseases, and notoriously with HIV, malaria, tuberculosis, enteric, and respiratory pathogens as well as sexually transmitted bacteria. Many of these infections are the focus of specific chapters in this volume.

I.N. Okeke (✉)
Department of Biology, Haverford College, Haverford, PA, USA
e-mail: iokeke@haverford.edu

A. de J. Sosa et al. (eds.), *Antimicrobial Resistance in Developing Countries*,
DOI 10.1007/978-0-387-89370-9_17, © Springer Science+Business Media, LLC 2010

17.1.1 Malaria

Chloroquine, the principal antimalarial used in Africa until the mid-1980s, has been one of the most cost-effective drugs of all time. The demise of chloroquine began with the emergence of resistance to this safe and effective drug in South America and South Asia in the late 1950s, and subsequent dissemination of resistant Plasmodia from Asia to East Africa in the late 1970s (Kean 1979). Once chloroquine resistance had appeared in Africa, it spread throughout endemic regions over a period of about 10 years. Antifolate drugs, which work by inhibiting parasite dihydropterase synthetase (DHPS, inhibited by sulfonamides) and dihydrofolate reductase (DHFR, inhibited by pyrimethamine), provided a more expensive, typically affordable, but far too temporary solution to the problem of chloroquine resistance. Parasites with point mutations in *dhps, dhfr* genes, or both, emerge rarely but because antimalarial drugs typically have long in vivo half-lives and parasite genomes recombine, resistance rapidly became commonplace in much of Africa. This again necessitated the introduction of newer antimalarials and significantly raised the cost of antimalarial chemotherapy (Hastings et al. 2002).

Malaria is perhaps the quintessential example of how a dry antimicrobial pipeline, drug misuse, a high disease burden, intensive transmission, and microbial evolution have combined to yield a resistance problem of phenomenal proportions. The consequences of this resistance are being felt acutely in all malaria-endemic areas but particularly in Africa where falciparum malaria, the most deadly form of the disease, prevails and very high entomologic inoculation rates facilitate pathogen transmission (Arrow et al. 2004). The only respite in recent years has been the incorporation of an ancient ethnomedicine, the active principle from *Artemisia annua*, into allopathic medicine, rather than from de novo drug development. Artemisinin is an extremely efficacious medicine but its current supply falls far short of its demand. Although the availability of artemisinin combination therapies is showing gradual improvement, these drugs remain out of the reach of many people who are at risk of dying from this curable disease (Laxminarayan 2004).

In spite of the current dire situation, older antimalarials have had a 'good run' (Hastings et al. 2002; Phillips 2001). Resistance to quinine, one of the oldest chemotherapeutic agents is still uncommon. Resistance to cheap synthetic antimalarials appeared much more rapidly, beginning with chloroquine, but still took hold less rapidly than antibacterial drug resistance. Malaria parasites have much longer generation times and a lower capacity for horizontal gene transfer. These factors may account for the decades in which chloroquine and folate inhibitors were highly effective. Nonetheless, resistance has emerged. Antifolates enjoyed a shorter term of use than their predecessor, chloroquine, and active steps must be taken to conserve the dwindling repertoire of currently available antimalarials until recently initiated drug development efforts start to yield fruit. The example of malaria illustrates that when drug development rates are raised, infectious disease control and antimicrobial stewardship must continue to be part of global anti-infective strategy.

17.1.2 Cholera in the Time of Resistance

Cholera is a deadly dehydrating diarrheal disease caused by cholera toxin-producing *Vibrio cholerae* strains. Although referred to as 'Asiatic' cholera because an endemic focus exists in the Indian sub-continent, cholera has caused at least seven pandemics in recent history and the most recent focus of the on-going pandemic is Africa where over two-thirds of outbreaks in the last decade have occurred (Griffith et al. 2006). This present pandemic has been characterized by appearance of new *V. cholerae* lineages, including strains belonging to serotypes other than O1 and, critically, antimicrobial-resistant isolates.

Until the mid-1900s, there was little alternative to letting cholera epidemics 'run their course'. The discovery that cholera is waterborne with a marine reservoir, coupled with the development of rehydration technology and antimicrobials, has provided many ways to intervene in the life cycle of cholera bacteria. Most patients who are promptly and adequately rehydrated can overcome cholera without antibiotics. Antimicrobials, however, shorten the course of infection and halt the shedding and transmission of infectious organisms, which, it has been proposed, may be more virulent than vibrios ingested from the wild (Merrell et al. 2002). Antimicrobials may also be lifesaving for high-risk patients, such as those who arrive late to treatment centers, by shortening the course of illness and reducing stool volume. Tetracycline remained the empiric drug of choice for this purpose for many years. It was replaced by trimethoprim–sulfamethoxazole in the 1980s which most recently has given way to the quinolones as drugs of choice. Shifts in recommended therapies have, in every case, been precipitated by resistance, which in turn can in many cases be loosely linked to cholera-specific use.

In 1977, tetracycline-resistant *V. cholerae* emerged in Tanzania, following a cholera antimicrobial prophylaxis campaign using at least 1788 kg of the drug, aimed at curbing spread of the disease (Mhalu et al. 1979; Towner et al. 1980). In 1980, there was a significant cholera outbreak in Kenya, after a hiatus of at least 5 years. Tetracycline was again used in mass prophylaxis campaigns and by 1982, tetracycline-resistant *V. cholerae* had emerged in Kenya. Shortly after, strains stably harboring plasmids, usually belonging to incompatibility group C, were reported from Tanzania and Kenya (Glass et al. 1983; Finch et al. 1988). Although they all appeared and were described around the same time, resistance plasmids from Tanzania and Kenya encoded different resistance profiles and had different restriction patterns from each other and from

Bangladesh and Nigerian resistance plasmids (Towner et al. 1980; Finch et al. 1988). Molecular evaluation of subsequent resistant *V. cholerae* isolates typically found regionally conserved plasmids, some of which carried class 1 integrons (Ceccarelli et al. 2006; Mwansa et al., 2006).

Evidence suggests that the emergence of resistance in *V. cholerae* has increased mortality in recent African outbreaks. A devastating outbreak in the Goma refugee camp in July 1994 resulted in the death of about 12,000 refugees from multiresistant *Vibrio cholerae* O1 (Siddique et al. 1995). In Guinea Bissau, Dalsgaard et al. (2000) attributed the appearance of resistance during the second wave of a 1996–1998 epidemic in part to acquisition of a 150 kb resistance plasmid bearing a class 1 integron and genes encoding resistance to all antimicrobials commonly used in empiric treatment of cholera. The emergency nature of cholera epidemics coupled with poor diagnostic infrastructure at many locations where these epidemics erupt means that empiric treatment increasingly employs ineffective drugs, epidemics are more extensive than they might otherwise be, and more patients are at risk of dying.

17.1.3 *Multidrug-Resistant* Salmonella

An estimated 16 million cases of typhoid fever occur annually worldwide and about 3–4% of these patients die (Ivanoff et al. 1994). These figures do not reflect under-reporting, particularly from Africa where typhoid cases have long been misdiagnosed as malaria and vice versa (Amexo et al. 2004; Kariuki et al. 2004; Nsutebu et al. 2003; Reyburn et al. 2004; Schram 1971). Pre-antibiotic era records suggest that mortality rates for untreated typhoid fever approach 13% and therefore resistance crucially impacts mortality rates for this disease (Gupta 1994). Resistance of *Salmonella enterica* serovar Typhi to chloramphenicol was first recorded in 1950 (Colquhoun and Weetch 1950). By the 1980s resistance to early first-line drugs had spread across Asia, where mortality rates approaching pre-antibiotic era were recorded (Gupta 1994). This rapid dissemination is largely linked to the spread of transmissible incompatibility group H1 plasmids bearing multiple resistance genes (Wain and Kidgell 2004).

Salmonella Typhi resistance in Africa has been less well studied and documented. Multidrug-resistant *S.* Typhi were reported from South Africa in 1992 (Coovadia et al. 1992). Subsequently, reports of resistant typhoid bacilli have come from Ghana, Kenya, and Nigeria (Reviewed by Cooke and Wain 2004). A Kenyan report observed that at least 70% of *S.* Typhi isolates were multidrug resistant, carried incH1 plasmids, and that multiresistant strains were clonal (Kariuki et al. 2004). Surveillance is compromised by diagnostic deficiencies and the true burden from the disease in Africa, as well as the prevalence of resistance, is not known.

Recent reports suggest that multidrug-resistant non-typhoidal *Salmonella* (NTS) are an important cause of invasive illness and death among young children and HIV-positive adults (Kariuki et al. 2005, 2006). It is thought that hyper-endemnicity, malnutrition, and HIV may combine to contribute to the alarming incidence of life-threatening community-acquired NTS infections in parts of Kenya and Malawi (Berkley et al. 2005; Gordon et al. 2002; Graham et al. 2000; Kankwatira et al. 2004; Kariuki, et al. 2006; Wolday and Erge 1998). Additionally, it is possible that clonal expansion of hypervirulent strains may account for a substantial proportion of these infections (Kariuki et al. 2006).

17.1.4 Neisseria gonorrhoeae

In Africa, sexually transmitted infections (STIs) have a major impact on health including adverse consequences that are exacerbated by structurally deficient health-care systems (Buve et al. 2001; Gerbase et al. 1998; Mullick et al. 2005). Before the era of antibiotics, invasive STIs invariably progressed to long-term complications, a situation that remains for sexually transmitted diseases of viral etiology. Introduction of cheap but effective antimicrobials substantially reduced mortality and spread of sexually transmitted diseases like chlamydial infections, syphilis, and gonorrhea. For many bacterial STIs, penicillins and tetracycline were commonly used for treatment in developing countries until they were challenged by the appearance of plasmid-mediated resistance in *N. gonorrhoeae* strains from geographically diverse areas of Tanzania (West et al. 1995), Zimbabwe (Mason and Gwanzura 1988), Ghana (Addy 1994), Ethiopia (Habte-Gabr et al. 1983) and Nigeria (Obaseiki-Ebor et al. 1985). Limited data obtainable from Africa suggest that chlamydial infections and syphilis are still responding to these treatments (Tapsall 2005).

Increasing resistance to penicillins and tetracycline required changes in STI treatment protocols, introducing drugs like trimethoprim–sulfamethoxazole, gentamicin, kanamycin, and thiamphenicol in many parts of Africa. In the last two decades, newer, more expensive therapies, including ceftriaxone, ciprofloxacin, and spectinomycin, were also recommended for the treatment of gonorrhea (Tapsall 2001). In spite of their apparent, if temporary, success it is pertinent to note that most of these drugs are too expensive for patients in growing economies where a substantial part of the problem resides. Strains of *N. gonorrhoeae* that are less susceptible to even these newer antimicrobials have already been reported in South Africa (Moodley et al. 2004), Djibouti (Haberberger et al. 1990), and Tanzania (West et al. 1995). These observations indicate that *N. gonorrhoeae* is becoming resistant to increasing number of antibiotics and strongly suggest that even expensive antibiotics may soon cease to be sufficiently effective in the treatment of gonococcal disease (Tapsall 2005).

17.1.5 Resistance Concerns in Other Pathogens

Resistance is a concern in virtually every organism pathogenic for humans. Recent concerns include the appearance of extensively drug-resistant tuberculosis (XDR-TB) in South Africa, and the high prevalence of multiply resistant diarrheagenic *E. coli* in many parts (Gassama et al. 2001; Okeke et al. 2000; Senerwa, et al., 1989; Singh ct al. 2007; Vila et al. 1999). Many recently reported resistant pathogens were identified by clinical alerts after clonal expansion, that is, after prospects for containment had substantially declined. Inadequate resistance surveillance for all these high-burden illness accounts for late detection and heavy consequences.

17.2 Antimicrobial Use and Potential Contributions to Resistance

17.2.1 Curative Antimicrobial Use by Health Professionals

Antimicrobials are the most needed drugs in sub-Saharan Africa. The current demand for them exceeds supply and the number of professionals trained to disburse antimicrobials does not meet the current need. A weak and unregulated supply chain promotes imprudent drug use and drug distribution from unsanctioned providers. There are several documented examples of antimicrobial overuse or misuse by health professionals, unsanctioned providers as well as patients. One Nigerian study recorded 96.7% of inpatients and 50.3% of outpatients as receiving at least one antibacterial drug (Chukwuani et al. 2002). The average number of antibacterials per inpatient and outpatient in that study were 2.4 and 1.1, respectively. Only 4.2% of this antimicrobial prescribing was supported by susceptibility testing.

Health-care professionals often face inadequate diagnostic support and drug supply shortages, which make precise diagnoses of life-threatening infections difficult, and also limit chemotherapeutic choices. They typically work without information about local susceptibility patterns that could come from improved surveillance. Education, particularly continuing education programs, may not be effective or frequent enough to promote rational drug use. Health professionals also face competition because the health system is pluralistic and, more critically, because non-professionals can disburse allopathic medicines.

Due to the high infectious disease burden in sub-Saharan Africa, if appropriate antimicrobial use is to meet current demand, justifiable selective pressure will grow. Therefore, in addition to discouraging unjustified use, resistance control must also include interventions to lower the prevalence of preventable diseases.

17.2.2 Prophylactic Use and the Expansion of Indications for Antimicrobials Threatened by Resistance

Prophylactic antimicrobial use in humans and animals has been an important focus of the resistance discourse in Europe and America and curbing this practice is a necessary for containment. There is an equally long if less appreciated history of prophylactic use in Africa, and not just in visitors. For example, sulfonamides were intensively used during an outbreak of meningitis in Nigeria in 1949–1950 and for more than a decade afterwards, sulfonamides were used as prophylactics in outbreaks. Sulfonamides were available from hundreds of treatment centers and a black-market for these drugs also thrived (Schram 1971). Similarly, penicillin replaced arsenic as the drug of choice during the WHO's yaws control campaign (Zahra 1956). In areas where the yaws infection rate exceeded 5%, all children were given penicillin as 'juvenile mass treatment'. During the campaign, officials had to account for empty penicillin vials, to prevent them being stolen and refilled with counterfeit product. At least 1.5 million Nigerians received penicillin as a result and the campaign is credited as being one of the reasons why many people developed a fondness for injections (Schram 1971). While prophylaxis may have been the only recourse in the yaws and meningitis cases, antimicrobial prophylaxis has also featured in outbreaks of feco-orally transmitted disease. In 1977, the use of tetracycline for chemoprophylaxis preceded the emergence and spread of plasmid-borne resistance in *V. cholerae*. Towner et al. (1980), documented this occurrence, and stressed the need to accompany antimicrobial use policies with appropriate surveillance.

By the 1980s, resistance rates among common pathogens and commensals approached 100% for many drugs introduced very early for infectious disease prophylaxis and treatment (Paul et al. 1982). Like the aforementioned examples, most prophylactic use is no longer supported today, but the legacy of this use persists in the form of endemic resistant strains and genes. Additionally, undocumented prophylactic antimicrobial abuse continues in the informal sectors (Dalsgaard et al. 2000; Whyte et al. 2002). More recently, justifiable antimicrobial prophylaxis reemerged with HIV epidemic, particularly as trimethoprim–sulfamethoxazole was recommended as a drug of choice, to prevent opportunistic infections, in AIDS patients in low resource areas. Until antiretrovirals are appropriately rolled out and HIV prevention begins to show a marked effect, the selective pressure afforded by preventive and curative use of antimicrobials in the immunocompromised will be unavoidably high and will compromise the effectiveness of these drugs in the infections for which they were originally indicated. More recently, antimalarial activity of the valuable trimethoprim–sulfamethoxazole combination has also been observed and documented and, if acted upon, will expand the indications for this overstretched drug combination (Thera, et al.

2005). The changing patterns of trimethoprim–sulfamethoxazole use that have occurred in the last 15 years have not been accompanied by systematic surveillance and hence the need to revisit policy on the use of this drug has continued to be postponed.

17.2.3 Unregulated Antimicrobial Use

Unsanctioned distributors are repeatedly cited for the harm they cause overall and for their specific contribution to the resistance problem. Unsanctioned practitioners often dispense medicines that are not indicated, in insufficient regimen, and typically do not store medicines appropriately. Undocumented medicine outlets are susceptible to infiltration by counterfeiters. However, it must be acknowledged that while these problems are associated with unsanctioned providers they are, on occasion, documented from sanctioned providers (Babalola and Lamikanra 2002; Okeke and Lamikanra 1995; Taylor et al. 2001). Thus the problem of unregulated drug use is exacerbated by unsanctioned providers but it does not end there.

Unsanctioned providers operate outside the formal health system and therefore cannot access educational programs for resistance control. They, however, are accessible to more patients due to their broader distribution as well as their willingness to negotiate treatment protocols with clients. Patients are major drivers of antimicrobial use and self-medication is very common in Africa. Different generations prefer different drugs so that self-medication choices are part of a dynamic popular culture (Okeke and Lamikanra 2003; Whyte et al. 2002). Because the patient, rather than a primary care health professional, manages care, pluralism is common and medical histories are typically undocumented and access to effective care may be delayed (Needham et al. 2001). It is unlikely that imprudent use can be contained if activities of unsanctioned providers are ignored or heavily proscribed. However, a lack of quality training and facilities in many cases presents formidable challenges for devising ways of regulating their activities and improving-on or replacing the services they render.

17.2.4 Poor-Quality Antimicrobials

Many antimicrobials dispensed in Africa and elsewhere do not conform to pharmacopeial specifications (Amin et al. 2005; Basco 2004; Okeke and Lamikanra 1995, 2001; Taylor et al. 2001; Winstanley et al. 2004). Active constituents in some of these medicines have been degraded by the high ambient temperature and humidity. Climate conditions and challenges with storage have even altered the physicochemical properties of excipients so that the dosage form is no longer competent to protect and deliver the active compound

(Kayumba et al. 2004; Okeke and Lamikanra 2001; Risha et al. 2002). Degraded medicines contain less than stated dose, therefore, patients presumed to be consuming adequate regimen may not be doing so. Some degradation products are even toxic (Okeke and Lamikanra 1995). Worse than the problem of degraded drugs is that of outright counterfeits, which often contain little or no antimicrobial or the wrong antimicrobial (Basco 2004; Cockburn et al. 2005). In addition to exposing the sick to possible toxic side effects, these drugs may increase the chance of selecting resistance (when they contain less than stated dose or a substitute medicine), and allow the patient to continue to transmit the pathogen.

African countries typically have appropriate legislation proscribing the manufacture and distribution of sub-standard drugs of any kind. As with unsanctioned drug distribution, they often lack resources to enforce these laws and to impose strict penalties on counterfeiters and their agents. There is also a deficit of resources needed to identify counterfeits or ensure the quality of locally manufactured and imported medicines. These deficits combine with the high infectious burden to make African, and similar under-resourced countries very attractive to counterfeiters.

17.3 Conclusion: Prospects for Controlling Antimicrobial Resistance by Improving Drug Use

Drug use by health professionals can be improved through educational interventions and drug use guides such as essential drug lists and standard treatment guidelines (Okeke et al. 2005). Educational interventions should begin with curricular modules on drug resistance and rational antimicrobial choice but must be supplemented with appropriate and regular continuing education (Agyepong et al. 2002; Bexell et al. 1996; WHO 2001). Where untrained practitioners are in a position to influence drug use, unless they receive information about drug use, the impact of educational interventions may be limited. For example, Agyepong et al. (2002) deliberately trained housekeeping staff in patient counseling because they occasionally substitute for dispensers. Patients also need to be informed, to increase adherence with prescribed regimen, to reduce self-medication – particularly when antimicrobial supply is unregulated – and to reduce the pressure on prescribers to recommend unjustified antimicrobials.

Drug misuse is exacerbated by poor supply and access. When the optimal antimicrobials are unavailable, sub-optimal ones will be used, medicines will be hoarded and the duration of some regimen may be shortened. All of these practices impinge upon the resistance problem. Furthermore, shortages of quality medicines increase the likelihood that expired or counterfeit drugs will enter the supply chain.

There is evidence to suggest that carefully regulated national drug use programs can slow the rate of resistance emergence and spread or even reverse it. Mathematical models demonstrate that increasing drug diversity may be the most effective strategy (Bergstrom et al. 2004). Unfortunately, antimicrobial choices in Africa are becoming less diverse because many new, patent-protected drugs are unaffordable or unavailable while use of previously effective older drugs must be withdrawn because of resistance. Narrow-spectrum drugs, which could reduce unnecessary pressure substantially, are difficult to introduce without adequate diagnostic support.

There are challenges associated with increasing drug diversity for Africa but other drug strategies with documented impact have been inadequately exploited. In Denmark banning glycopeptides and other agricultural antimicrobial growth promoters led to a significant reduction in resistance among enteroccocal isolates (Aarestrup et al. 2001). In Malawi, it was demonstrated that complete and sustained withdrawal of chloroquine resulted in the re-emergence of susceptible *Plasmodium falciparum* (Kublin et al. 2003). Unfortunately, efforts to effect a similar intervention in Kenya were unsuccessful because chloroquine could not be withdrawn from the informal sector whereas the mere suggestion that chloroquine would be withdrawn from Nigeria brought about an angry response from health professionals who were justifiably concerned about antimalarial availability and affordability (Okeke 2007).

Active component substitution by counterfeiters can also compromise drug use strategies. Drug use programs depend on a reliable supply of quality assured medicines and a sufficient and knowledgeable staff base for their distribution. Both factors are linked to political support for health programs and resources, without which drug use programs are susceptible to compromise or sabotage that could impinge upon resistance control.

Acknowledgment INO is a Branco Weiss Fellow of the Society-in-Science, ETHZ, Zürich, Switzerland.

References

Aarestrup, F.M., Seyfarth, A.M., Emborg, H.D., Pedersen, K., Hendriksen, R.S., and Bager, F. (2001). Effect of abolishment of the use of antimicrobial agents for growth promotion on occurrence of antimicrobial resistance in fecal enterococci from food animals in Denmark. *Antimicrob Agents Chemother*, 45, 2054–2059.

Addy, P.A. (1994). Susceptibility pattern of Neisseria gonorrhoeae isolated at the Komfo, Anokye Teaching Hospital, Ghana to commonly prescribed antimicrobial agents. *East Afr Med J*, 71, 368–372.

Agyepong, I.A., Ansah, E., Gyapong, M., Adjei, S., Barnish, G., and Evans, D. (2002). Strategies to improve adherence to recommended chloroquine treatment regimes: a quasi-experiment in the context of integrated primary health care delivery in Ghana. *Soc Sci Med*, 55, 2215–2226.

Amexo, M., Tolhurst, R., Barnish, G., and Bates, I. (2004). Malaria misdiagnosis: effects on the poor and vulnerable. *Lancet*, 364, 1896–1898.

Amin, A.A., Snow, R.W., and Kokwaro, G.O. (2005). The quality of sulphadoxine-pyrimethamine and amodiaquine products in the Kenyan retail sector. *J Clin Pharm Ther*, 30, 559–565.

Arrow, K., Panosian, C., and Gelband, H. (2004). Saving lives, buying time. Economics of malaria drugs in an age of resistance. The National Academies Press, Washington, DC.

Babalola, O., and Lamikanra, A. (2002). Pattern of antibiotic purchases in community pharmacies in South Western Nigeria. *J Soc Admin Pharm*, 19, 33–38.

Basco, L.K. (2004). Molecular epidemiology of malaria in Cameroon. XIX. Quality of antimalarial drugs used for self-medication. *Am J Trop Med Hyg*, 70, 245–250.

Bergstrom, C.T., Lo, M., and Lipsitch, M. (2004). Ecological theory suggests that antimicrobial cycling will not reduce antimicrobial resistance in hospitals. *Proc Natl Acad Sci U S A*, 101, 13285–13290.

Berkley, J.A., Lowe, B.S., Mwangi, I., Williams, T., Bauni, E., Mwarumba, S., Ngetsa, C., Slack, M.P., Njenga, S., Hart, C.A., Maitland, K., English, M., Marsh, K., and Scott, J.A. (2005). Bacteremia among children admitted to a rural hospital in Kenya. *N Engl J Med*, 352, 39–47.

Bexell, A., Lwando, E., von Hofsten, B., Tembo, S., Eriksson, B., and Diwan, V.K. (1996). Improving drug use through continuing education: a randomized controlled trial in Zambia. *J Clin Epidemiol*, 49, 355–357.

Buve, A., Weiss, H.A., Laga, M., Van Dyck, E., Musonda, R., Zekeng, L., Kahindo, M., Anagonou, S., Morison, L., Robinson, N.J., and Hayes, R.J. (2001). The epidemiology of gonorrhoea, chlamydial infection and syphilis in four African cities. *AIDS*, 15, S79–88.

Ceccarelli, D., Salvia, A.M., Sami, J., Cappuccinelli, P., and Colombo, M.M. (2006). New cluster of plasmid-located class 1 integrons in *Vibrio cholerae* O1 and a *dfrA15* cassette-containing integron in *Vibrio parahaemolyticus* isolated in Angola. *Antimicrob. Agents Chemother.*, 50, 2493–2499.

Chukwuani, C.M., Onifade, M., and Sumonu, K. (2002). Survey of drug use practices and antibiotic prescribing pattern at a general hospital in Nigeria. *Pharm World Sci*, 24, 188–195.

Cockburn, R., Newton, P.N., Agyarko, E.K., Akunyili, D., and White, N.J. (2005). The global threat of counterfeit drugs: why industry and governments must communicate the dangers. *PLoS Med*, 2, e100.

Colquhoun, J., and Weetch, R.S. (1950). Resistance to chloramphenicol developing during treatment of typhoid fever. *Lancet*, 2, 621–623.

Cooke, F.J., and Wain, J. (2004). The emergence of antibiotic resistance in typhoid fever. *Travel Med Infect Dis*, 2, 67–74.

Coovadia, Y.M., Gathiram, V., Bhamjee, A., Mlisana, K., Pillay, N., Garratt, R.M., Madlalose, T., and Short, M. (1992). The emergence of multi-antibiotic-resistant strains of Salmonella typhi in northern Natal-KwaZulu. *S Afr Med J*, 81, 280–281.

Dalsgaard, A., Forslund, A., Petersen, A., Brown, D.J., Dias, F., Monteiro, S., Molbak, K., Aaby, P., Rodrigues, A., and Sandstrom, A. (2000). Class 1 integron-borne, multiple-antibiotic resistance encoded by a 150-kilobase conjugative plasmid in epidemic *Vibrio cholerae* O1 strains isolated in Guinea-Bissau. *J Clin Microbiol*, 38, 3774–3779.

Finch, M.J., Morris, J.G., Jr., Kaviti, J., Kagwanja, W., and Levine, M.M. (1988). Epidemiology of antimicrobial resistant cholera in Kenya and East Africa. *Am J Trop Med Hyg*, 39, 484–490.

Gassama, A., Sow, P.S., Fall, F., Camara, P., Gueye-N'diaye, A., Seng, R., Samb, B., M'Boup, S., and Aidara-Kane, A. (2001). Ordinary and opportunistic enteropathogens associated with diarrhea in Senegalese adults in relation to human immunodeficiency virus serostatus. *Int J Infect Dis*, 5, 192–198.

Gerbase, A.C., Rowley, J.T., and Mertens, T.E. (1998). Global epidemiology of sexually transmitted diseases. *Lancet*, 351 Suppl 3, 2–4.

Glass, R.I., Huq, M.I., Lee, J.V., Threlfall, E.J., Khan, M.R., Alim, A.R., Rowe, B., and Gross, R.J. (1983). Plasmid-borne multiple drug resistance in *Vibrio cholerae* serogroup O1, biotype El Tor: evidence for a point-source outbreak in Bangladesh. *J Infect Dis*, 147, 204–209.

Gordon, M.A., Banda, H.T., Gondwe, M., Gordon, S.B., Boeree, M.J., Walsh, A.L., Corkill, J.E., Hart, C.A., Gilks, C.F., and Molyneux, M.E. (2002). Non-typhoidal salmonella bacteraemia among HIV-infected Malawian adults: high mortality and frequent recrudescence. *AIDS*, 16, 1633–1641.

Graham, S.M., Walsh, A.L., Molyneux, E.M., Phiri, A.J., and Molyneux, M.E. (2000). Clinical presentation of non-typhoidal *Salmonella* bacteraemia in Malawian children. *Trans R Soc Trop Med Hyg*, 94, 310–314.

Griffith, D.C., Kelly-Hope, L.A., and Miller, M.A. (2006). Review of reported cholera outbreaks worldwide, 1995–2005. *Am J Trop Med Hyg*, 75, 973–977.

Gupta, A. (1994). Multidrug-resistant typhoid fever in children: epidemiology and therapeutic approach. *Pediatr Infect Dis J*, 13, 134–140.

Haberberger, R.L., Jr., Fox, E., Polycarpe, D., Abbatte, E.A., and Said, S. (1990). Antibiotic susceptibility patterns of Neisseria gonorrhoeae in Djibouti during June 1988. *Trans R Soc Trop Med Hyg*, 84, 738.

Habte-Gabr, E., Geyid, A., and Serdo, D. (1983). Beta-lactamase-producing Neisseria gonorrhoeae in Addis Ababa. *Ethiop Med J*, 21, 199.

Hastings, I.M., Bray, P.G., and Ward, S.A. (2002). Parasitology. A requiem for chloroquine. *Science*, 298, 74–75.

Hastings, I.M., Watkins, W.M., and White, N.J. (2002). The evolution of drug-resistant malaria: the role of drug elimination half-life. *Philos Trans R Soc Lond B Biol Sci*, 357, 505–519.

Ivanoff, B., Levine, M.M., and Lambert, P.H. (1994). Vaccination against typhoid fever: present status. *Bull World Health Organ*, 72, 957–971.

Kankwatira, A.M., Mwafulirwa, G.A., and Gordon, M.A. (2004). Non-typhoidal salmonella bacteraemia – an under-recognized feature of AIDS in African adults. *Trop Doct*, 34, 198–200.

Kariuki, S., Mwituria, J., Munyalo, A., Revathi, G., and Onsongo, J. (2004). Typhoid is over-reported in Embu and Nairobi, Kenya. *Afr J Health Sci*, 11, 103–110.

Kariuki, S., Revathi, G., Muyodi, J., Mwituria, J., Munyalo, A., Mirza, S., and Hart, C.A.(2004). Characterization of multidrug-resistant typhoid outbreaks in Kenya. *J Clin Microbiol*, 42, 1477–1482.

Kariuki, S., Revathi, G., Kariuki, N., Muyodi, J., Mwituria, J., Munyalo, A., Kagendo, D., Murungi, L., and Anthony Hart, C. (2005). Increasing prevalence of multidrug-resistant non-typhoidal salmonellae, Kenya, 1994–2003. *Int J Antimicrob Agents*, 25, 38–43.

Kariuki, S., Revathi, G., Kariuki, N., Kiiru, J., Mwituria, J., and Hart, C.A. (2006). Characterisation of community acquired non-typhoidal *Salmonella* from bacteraemia and diarrhoeal infections in children admitted to hospital in Nairobi, Kenya. *BMC Microbiol*, 6, 101.

Kayumba, P.C., Risha, P.G., Shewiyo, D., Msami, A., Masuki, G., Ameye, D., Vergote, G., Ntawukuliryayo, J.D., Remon, J.P., and Vervaet, C. (2004). The quality of essential antimicrobial and antimalarial drugs marketed in Rwanda and Tanzania: influence of tropical storage conditions on in vitro dissolution. *J Clin Pharm Ther*, 29, 331–338.

Kean, B.H. (1979). Chloroquine-resistant falciparum malaria from Africa. *JAMA*, 241(4), 395.

Kublin, J.G., Cortese, J.F., Njunju, E.M., Mukadam, R.A., Wirima, J.J., Kazembe, P.N., Djimde, A.A., Kouriba, B., Taylor, T.E., and Plowe, C.V. (2003). Reemergence of chloroquine-sensitive *Plasmodium falciparum* malaria after cessation of chloroquine use in Malawi. *J Infect Dis*, 187, 1870–1875.

Laxminarayan, R. (2004). Act now or later? Economics of malaria resistance. *Am J Trop Med Hyg*, 71, S187–195.

Mason, P.R., and Gwanzura, L. (1988). Characterisation by plasmid profiles, serogroups, and auxotypes of *Neisseria gonorrhoeae* from Harare, Zimbabwe. *Genitourin Med*, 64, 303–307.

Merrell, D.S., Butler, S.M., Qadri, F., Dolganov, N.A., Alam, A., Cohen, M.B., Calderwood, S.B., Schoolnik, G.K., and Camilli, A. (2002). Host-induced epidemic spread of the cholera bacterium. *Nature*, 417, 642–645.

Mhalu, F.S., Mmari, P.W., and Ijumba, J. (1979). Rapid emergence of El Tor *Vibrio cholerae* resistant to antimicrobial agents during first six months of fourth cholera epidemic in Tanzania. *Lancet*, 1, 345–347.

Moodley, P., Moodley, D., and Willem Sturm, A. (2004). Ciprofloxacin resistant *Neisseria gonorrhoeae* in South Africa. *Int J Antimicrob Agents*, 24, 192–193.

Mullick, S., Watson-Jones, D., Beksinska, M., and Mabey, D. (2005). Sexually transmitted infections in pregnancy: prevalence, impact on pregnancy outcomes, and approach to treatment in developing countries. *Sex Transm Infect*, 81, 294–302.

Mwansa, J.C., Mwaba, J., Lukwesa, C., Bhuiyan, N.A., Ansaruzamman, M., Ramamurthy, T., Alam, M., and Balakrish Nair, G. (2006). Multiply antibiotic-resistant Vibrio cholerae O1 biotype El Tor strains emerge during cholera outbreaks in Zambia. *Epidemiol Infect*, 135, 847–853.

Needham, D.M., Foster, S.D., Tomlinson, G., and Godfrey-Faussett, P. (2001). Socio-economic, gender and health services factors affecting diagnostic delay for tuberculosis patients in urban Zambia. *Trop Med Int Health*, 6, 256–259.

Nsutebu, E.F., Martins, P., and Adiogo, D. (2003). Prevalence of typhoid fever in febrile patients with symptoms clinically compatible with typhoid fever in Cameroon. *Trop Med Int Health*, 8, 575–578.

Obaseiki-Ebor, E.E., Oyaide, S.M., and Okpere, E.E. (1985). Incidence of penicillinase producing *Neisseria gonorrhoeae* (PPNG) strains and susceptibility of gonococcal isolates to antibiotics in Benin City, Nigeria. *Genitourin Med*, 61, 367–370.

Okeke, I.N, Lamikanra, A., Czeczulin, J., Dubovsky, F., Kaper, J.B, and Nataro, J.P. (2000). Heterogeneous virulence of enteroaggregative *Escherichia coli* strains isolated from children in Southwest Nigeria. *J Infect Dis*, 181, 252–260.

Okeke, I. (2007). The microbial rebellion: trends and containment of antimicrobial resistance in Africa. In T. Falola, and M.M. Heaton (Eds.), *HIV/AIDS, illness, and African well-being*. University of Rochester Press. Rochester, NY. pp. 155–181.

Okeke, I.N., and Lamikanra, A. (1995). Quality and bioavailability of tetracycline capsules in a Nigerian semi-urban community. *Int J Antimicrob Agents*, 5, 245–250.

Okeke, I.N., and Lamikanra, A. (2001). Quality and bioavailability of ampicillin capsules dispensed in a Nigerian semi-urban community. *Afr J Med Med Sci*, 30, 47–51.

Okeke, I.N., and Lamikanra, A. (2003). Export of antimicrobial drugs by West African travelers. *J Travel Med*, 10, 133–135.

Okeke, I.N., Klugman, K.P., Bhutta, Z.A., Duse, A.G., Jenkins, P., O'Brien, T.F., Pablos-Mendez, A., and Laxminarayan, R. (2005). Antimicrobial resistance in developing countries. Part II: strategies for containment. *Lancet Infect Dis*, 5, 568–580.

Paul, M.O., Aderibigbe, D.A., Sule, C.Z., and Lamikanra, A. (1982). Antimicrobial sensitivity patterns of hospital and non-hospital strains of *Staphylococcus aureus* isolated from nasal carriers. *J Hyg (Cambridge)*, 89, 253–260.

Phillips, R.S. (2001). Current status of malaria and potential for control. *Clin Microbiol Rev*, 14(1), 208–226.

Reyburn, H., Mbatia, R., Drakeley, C., Carneiro, I., Mwakasungula, E., Mwerinde, O., Saganda, K., Shao, J., Kitua, A., Olomi, R., Greenwood, B.M., and Whitty, C.J. (2004). Overdiagnosis of malaria in patients with severe febrile illness in Tanzania: a prospective study. *BMJ*, 329, 1212.

Risha, P.G., Shewiyo, D., Msami, A., Masuki, G., Vergote, G., Vervaet, C., and Remon, J.P. (2002). In vitro evaluation of the quality of essential drugs on the Tanzanian market. *Trop Med Int Health*, 7, 701–707.

Schram, R. (1971). *A history of the Nigerian health services*. Ibadan University Press, Ibadan.

Senerwa, D., Olsvik, O., Mutanda, L.N., Lindqvist, K.J., Gathuma, J.M., Fossum, K., and Wachsmuth, K. (1989). Enteropathogenic Escherichia coli serotype O111:HNT isolated from preterm neonates in Nairobi, Kenya. *J Clin Microbiol*, 27, 1307–1311.

Siddique, A.K., Salam, A., Islam, M.S., Akram, K., Majumdar, R.N., Zaman, K., Fronczak, N., and Laston, S. (1995). Why treatment centres failed to prevent cholera deaths among Rwandan refugees in Goma, Zaire. *Lancet*, 345, 359–361.

Singh, J.A., Upshur, R., and Padayatchi, N. (2007). XDR-TB in South Africa: no time for denial or complacency. *PLoS Med*, 4, e50.

Tapsall, J. (2001). Antimicrobial resistance in Neisseria gonorrheae. Geneva, Switzerland: Department of Communicable Disease Surveillance and Response, World Health Organization.

Tapsall, J.W. (2005). Antibiotic resistance in *Neisseria gonorrhoeae*. *Clin Infect Dis*, 41 Suppl 4, S263–268.

Taylor, R.B., Shakoor, O., Behrens, R.H., Everard, M., Low, A.S., Wangboonskul, J., Reid, R.G., and Kolawole, J.A. (2001). Pharmacopoeial quality of drugs supplied by Nigerian pharmacies. *Lancet*, 357, 1933–1936.

Thera, M.A., Sehdev, P.S., Coulibaly, D., Traore, K., Garba, M.N., Cissoko, Y., Kone, A., Guindo, A., Dicko, A., Beavogui, A.H., Djimde, A.A., Lyke, K.E., Diallo, D.A., Doumbo, O.K., and Plowe, C.V. (2005). Impact of trimethoprim-sulfamethoxazole prophylaxis on falciparum malaria infection and disease. *J Infect Dis*, 192, 1823–1829.

Towner, K.J., Pearson, N.J., Mhalu, F.S., and O'Grady, F. (1980). Resistance to antimicrobial agents of *Vibrio cholerae* El Tor strains isolated during the fourth cholera epidemic in the United Republic of Tanzania. *Bull World Health Organ*, 58, 747–751.

Vila, J., Vargas, M., Casals, C., Urassa, H., Mshinda, H., Schellemberg, D., and Gascon, J. (1999). Antimicrobial resistance of diarrheagenic *Escherichia coli* isolated from children under the age of 5 years from Ifakara, Tanzania. *Antimicrob Agents Chemother*, 43, 3022–3024.

Wain, J., and Kidgell, C. (2004). The emergence of multidrug resistance to antimicrobial agents for the treatment of typhoid fever. *Trans R Soc Trop Med Hyg*, 98, 423–430.

West, B., Changalucha, J., Grosskurth, H., Mayaud, P., Gabone, R.M., Ka-Gina, G., and Mabey, D. (1995). Antimicrobial susceptibility, auxotype and plasmid content of *Neisseria gonorrhoeae* in northern Tanzania: emergence of high level plasmid mediated tetracycline resistance. *Genitourin Med*, 71, 9–12.

WHO (2001). Interventions and strategies to improve the use of antimicrobials in developing countries. World Health Organization, Geneva.

Whyte, S.R., Geest, S.v.d., and Hardon, A. (2002). *Social lives of medicines* Cambridge University Press, Cambridge, UK; New York.

Winstanley, P., Ward, S., Snow, R., and Breckenridge, A. (2004). Therapy of falciparum malaria in sub-saharan Africa: from molecule to policy. *Clin Microbiol Rev*, 17, 612–637

Wolday, D., and Erge, W. (1998). Antimicrobial sensitivity pattern of Salmonella: comparison of isolates from HIV-infected and HIV-uninfected patients. *Trop Doct*, 28(3), 139–141.

Zahra, A. (1956). Yaws eradication campaign in Nsukka division, Eastern Nigeria. A preliminary review. *Bull World Health Organ*, 15, 911–915.

Chapter 18
Antimicrobial Drug Resistance in Asia

Yu-Tsung Huang and Po-Ren Hsueh

Abstract Antimicrobial resistance has become a major health problem world-wide, but marked variations in resistance profiles of bacterial pathogens are found among countries and in different patient settings, especially in Asia. These differences represent the effects of appropriate and inappropriate use of antibiotics in humans and animals. In this chapter, bacterial as well as fungal drug resistance is discussed.

18.1 Drug Resistance in Bacterial Infection

18.1.1 Community-Acquired Infections

18.1.1.1 Respiratory Tract Infections

1. *Streptococcus pneumoniae*. Antimicrobial resistance in *S. pneumoniae* has been an important problem in the treatment of infections for decades especially in the west Pacific area which reported the highest resistance rate in the world (Felmingham et al. 2005). In a multicenter surveillance involving 11 Asian countries, 52.4% isolates were not susceptible to penicillin and 29.4% of the isolates had penicillin minimum inhibitory concentrations (MICs) ≥ 2 mg/L (Song et al. 2004a). The erythromycin resistance rate varied greatly in different geographic areas; it was more than 70% in China, Hong Kong, Korea, Taiwan, and Vietnam but only 1.3% in India (Song et al. 2004a). Multilocus sequence typing revealed that two major clones (Taiwan 19^F and Spain 23^F clone) were spreading in Asia and could be a major reason for the increasing antimicrobial resistance of pneumococcus (Song et al. 2004a). The fluoroquinolone resistance was highest in isolates from Hong Kong, with a levofloxacin resistance rate (MIC of ≥ 4 mg/L) of 13.3%, and

P.-R. Hsueh (✉)
Divisions of Clinical Microbiology and Infectious Diseases, Departments of
Laboratory Medicine and Internal Medicine, National Taiwan University Hospital,
National Taiwan University College of Medicine, Taipei, Taiwan
e-mail: hsporen@ntuedu.tw

A. de J. Sosa et al. (eds.), *Antimicrobial Resistance in Developing Countries*,
DOI 10.1007/978-0-387-89370-9_18, © Springer Science+Business Media, LLC 2010

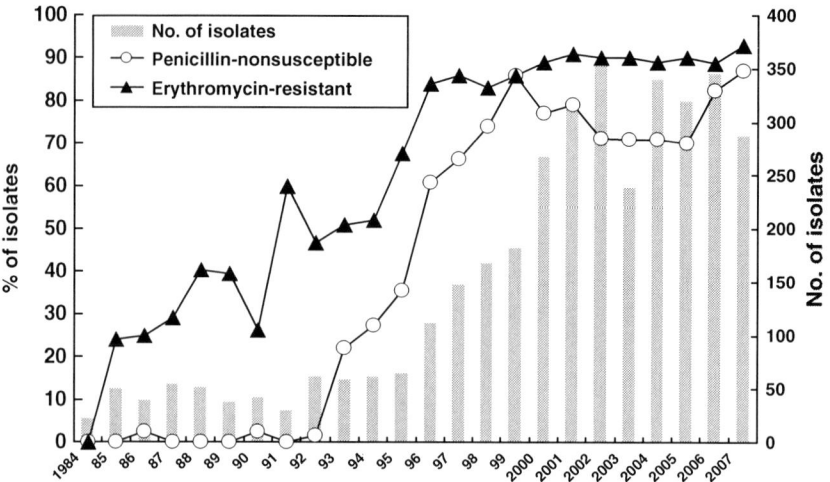

Fig. 18.1 Trends of nonsusceptibility of *S. pneumoniae* (by the disk diffusion method) to penicillin and erythromycin from National Taiwan University Hospital from 1984 to 2007

these isolates were related to the Spanish 23^F clone (Ho et al. 2001). In Taiwan, a hospital wide surveillance from National Taiwan University Hospital from 1984 to 2007 disclosed an increasing trend of nonsusceptibility (by the disk diffusion method) to erythromycin (93% in 2007) and penicillin (87% in 2007) (Fig. 18.1).

2. *Haemophilus influenzae*. Except for Israel, Malaysia, and Saudi Arabia, immunization against *H. influenzae* type b (Hib) vaccination has not been widely used in the vast majority of Asian areas including China, Hong Kong, India, Japan, Korea, Southeast Asia, and Taiwan (Morris et al. 2008). The antimicrobial resistance of *H. influenzae* has remained a concern in treating infected patients in this area (Invasive Bacterial Infections Surveillance [IBIS] Group of the International Clinical Epidemiology Network, 2002). In addition, two types of beta-lactam resistance in *H. influenza* have become clinically important: (1) acquired beta-lactamase production (plasmid mediated bla_{TEM} or bla_{ROB-1}) and (2) mutations in the penicillin-binding proteins, also known as beta-lactamase-negative ampicillin resistance (BLNAR) (Tristram et al. 2007). In a study conducted during 1999–2000, the beta-lactamase-producing percentage of *H. influenza* isolates was 8.5% in Japan, 17.1% in Hong Kong, and 64.7% in South Korea and only one BLNAR isolate was detected in Japan (Inoue et al. 2004). In a subsequent study in Japan, although the prevalence of beta-lactamase (mostly bla_{TEM}) producing strains remained stable (from 6.8% in 1998 to 8.0% in 2003), a remarkable increase in BLNAR genotypes caused by clonal dissemination resulted in a decline of beta-lactam susceptibilities (Sanbongi et al. 2006). In Korea, the prevalence of BLNAR ranged from 24.4 to 33.3% in a recent study (Kim et al. 2007b). Resistance to fluoroquinolones was noted in Japan and Hong Kong (both <3%) and none of

these isolates revealed clonal dissemination (Ho et al. 2004; Yokota et al. 2008). In Taiwan, high nonsusceptibility rates of *H. influenzae* to amoxicillin (58%), TMP-SMX (52%), and azithromycin (31%) were found (Hsueh et al. 2000). Fifty-six percent of these isolates produced beta-lactamases and only five isolates (1.7%) exhibited BLNAR genotypes (Hsueh et al. 2000). As vaccination in infants is not widely used in Western Pacific area and South Asia, monitoring of antibiotics susceptibilities is mandatory for *H. influenzae*.

3. *Moraxella catarrhalis*. A high percentage (>90%) of *M. catarrhalis* isolates worldwide have been reported to carry bla_{BRO-1} and bla_{BRO-2} and are susceptible to beta-lactam and beta-lactamase inhibitors with low fluoroquinolone MICs (Hoban and Felmingham 2002; Hsueh et al. 2000; Inoue et al. 2004).

4. Macrolide-resistant *Streptococcus pyogenes*. The prevalence of macrolide-resistant *S. pyogenes* was higher in the Asia/Pacific area than in Northern America or Western Europe (Beekmann et al. 2005). Resistance to macrolides is mediated by the production of modifying enzymes by *erm*A, *erm*B genes, or efflux pump by *mef*A gene (Leclercq 2002). The prevalence of erythromycin resistance in *S. pyogenes* in Taiwan declined to 17% in 2002 compared with more than 40% in earlier studies and this decline was correlated with reduced erythromycin prescription after applying restrictions on antibiotic use since 2001 (Hsueh et al. 2005a). High non-susceptible rates of *S. pyogenes* to the new ketolide, telithromycin, and quinupristin/dalfopristin were reported even before their marketing in Taiwan (17 and 8%, respectively) (Hsueh et al. 2003). Fortunately, beta-lactams and the new fluoroquinolones have remained potent against *S. pyogenes* in Taiwan (Hsueh et al. 2003). In Japan, resistance to macrolides was as high as 62.3% in the 1970s (Nakae et al. 1977) but had declined to less then 10% in recent surveillance studies after macrolides restriction (Ikebe et al. 2005). Telithromycin remained potent against *S. pyogenes* (1.4% resistant) in Japan but up to 10.9% of isolates remained non-susceptible to ciprofloxacin and 4.8% remained non-susceptible to levofloxacin in recent studies (Ikebe et al. 2005). In Korea, the resistance rates of *S. pyogenes* to erythromycin and clindamycin in 1995 were 29.4 and 10.1%, respectively, and had increased to 51.0 and 33.7%, respectively, in 2002 (Kim and Lee 2004). In Iran, resistance of *S. pyogenes* to erythromycin remained uncommon (0.6%) but resistance to tetracycline was high as 42% (Jasir et al. 2000). *S. pyogenes* resistance to telithromycin in Europe increased in the past decade, and the rate of this increase was correlated with the prescription of macrolides in individual countries. Restriction in macrolides usage may therefore be important in controlling macrolides resistance.

18.1.1.2 Diarrheal Diseases and Bacteremia

1. *Salmonella typhi* and *S. paratyphi*. Multidrug-resistant *S. enterica* serotype Typhi, which is resistant to first-line antimicrobial agents such as ampicillin,

trimethroprim/sulfamethoxazole, and chloramphenical, has caused several outbreaks in Southeast Asia and the Indian subcontinent (Parry et al. 2002). Strains with reduced susceptibilities to fluoroquinolones emerged as a major problem in the area mentioned above due to their poorer response to treatment (Parry et al. 2002). Recently, nalidixic acid has been recommended for better prediction of fluoroquinolones treatment response (CLSI). Chau et al. (2007) reported that *S. typhi* and *S. paratyphi* had various rates of multidrug resistance (0–37.5%) and nalidixic acid resistance (0–51%) in Asian countries (Fig. 18.2). MDR strains carrying bld_{TEM} gene have emerged in Korea and were indistinguishable from MDR isolates in India and Indonesia (Lee et al. 2004). Development of vaccination becomes crucial due to the increasingly limited treatment choices against the emerging MDR strains.

2. Non-typhoid *Salmonella* species (NTS). *S. enterica* serotypes Cholerasuis and Typhimurium are known for their drug resistance and for causing invasive diseases (Chiu et al. 2002, 2004a). A nationwide study in Taiwan found the overall ciprofloxacin resistance was 1.4% in *S. enterica* serotype Typhimurium and 7.5% in *S. enterica* serotype Cholerasuis (Hsueh et al. 2004a). A marked increase in the fluoroquinolone resistance rate of *S. enterica* serotype Cholerasuis was reported in Taiwan in 2003 (70%) (Chiu et al. 2004b). Ciprofloxacin-resistant *S. enterica* serotype Typhimurium has been reported in China, Japan, Korea, and Thailand (Cui et al. 2008; Izumiya et al. 2005; Choi et al. 2005; Kulwichit et al. 2007). In Korea, *S. enterica* serotype Enteriditis isolates carried a higher rate of resistance to nalidixic acid than serotype Typhimurium isolates (21.6% versus 12.1%) (Choi et al. 2005). Swine have been noted as the origin and as a reservoir of these fluoroquinolone-resistant NTS (Chiu et al. 2002; Hsueh et al. 2004a). Ceftriaxone-resistant *S. enterica* serotype Typhimurium carrying an AmpC beta-lactamase (CMY-2) was isolated in Taiwan (Yan et al. 2005). Antibiotics usage, especially quinolones in the livestock industry, should be regulated to prevent worsening of the current situation. Continual susceptibility monitoring is necessary to ensure that effective treatment options remain available.

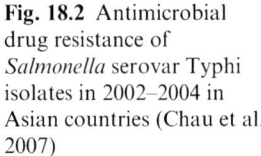

Fig. 18.2 Antimicrobial drug resistance of *Salmonella* serovar Typhi isolates in 2002–2004 in Asian countries (Chau et al. 2007)

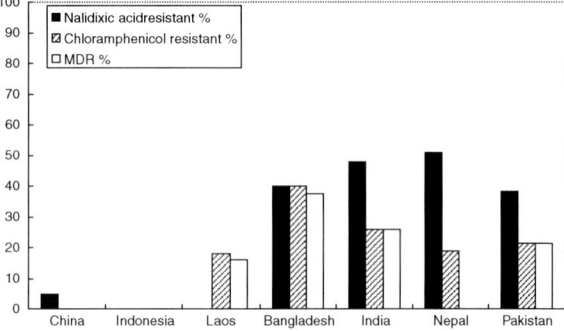

18.1.1.3 Community-Acquired Methicillin-Resistant *Staphylococcus aureus* (CAMRSA)

Community-acquired MRSA infection (CAMRSA) emerged during the last decade and causes considerable morbidity and mortality. Like nosocomial MRSA infection, clonal spreading of CAMRSA between community and hospital has been noted (Boucher and Corey 2008). The isolates of sequence type (ST59) identified by the multilocus sequence typing (MLST) method appeared to be the major clone of CAMRSA in northern Taiwan, accounting for more than 90% of CAMRSA (Boyle-Vavra et al. 2005). In Hong Kong, three MLST types of CAMRSA have been identified: ST30-IV (the more common), ST59-V, and ST8-IVA, exhibiting a diverse genetic distribution (Ho et al. 2007). In Japan, CAMRSA was found in 10–17% of *S. aureus* isolates from bullous impetigo and could be divided into three sequence types: ST89 (66.7%), ST8 (20%), and ST91 (13.1%) (Takizawa et al. 2005). An increasing incidence of CAMRSA-related soft tissue infection in Singapore has been reported with a predominant ST30 strain (Hsu et al. 2006). In Korea, CAMRSA accounted for only 5.9% of all MRSA isolates and ST72 was the predominant strain (Kim et al. 2007a). With the flourish of global travel, monitoring of CAMRSA might be necessary even when its initial prevalence is low.

18.1.1.4 Sexually Transmitted Diseases

1. *Neisseria gonorrhoeae. N. gonorrhoeae* has developed several mechanisms against antibiotics in recent decades including penicillinase-producing *N. gonorrhoeae* (PPNG), chromosomally mediated resistant *N. gonorrhoeae* (CMRNG), tetracycline-resistant *N. gonorrhoeae* (TRNG), and quinolone-resistant *N. gonorrhoeae* (QRGN) (Matsumoto 2008). PPNG and CMRNG were noted in various parts of the world and QRNG was mainly distributed in Asian countries (Matsumoto 2008). In 2006, QRNG emerged in the western Pacific and Southeast Asia as illustrated in Fig. 18.3 (data from the WHO Western Pacific Gonococcccal Antimicrobial Surveillance Programme 2008). In Taiwan, Hsueh et al. (2005b) reported the emergence of QRNG and the identification of several predominant clones, with the incidence of QRNG rising from 25% in 1999–2000 to 93.1% in 2003. The penicillin non-susceptible rate was 96% and all the isolates were non-susceptible to tetracycline (96% resistant) (Hsueh et al. 2005a). In China, Su et al. (2007) reported that the prevalence of PPNG, TRNG, and QRNG rose in a recent 8-year period (1999–2006). In India, a recent survey reported that penicillin resistance varied from 20 to 79%, 0 to 45.6% for TRNG, and 10.6 to 100% for QRNG (Ray et al. 2005). In Sri Lanka, the prevalence of penicillin resistance was 96.8% and ciprofloxacin resistance was 8.2%. In Bangladesh, the prevalence of TRNG was 76%, QRNG was 50%, the penicillin-resistant rate was 33%, and 1.5% of isolates had decreased susceptibility to ceftriaxone (Bala et al. 2008). In Pakistan, QRNG was first reported in 1999 and its proportion

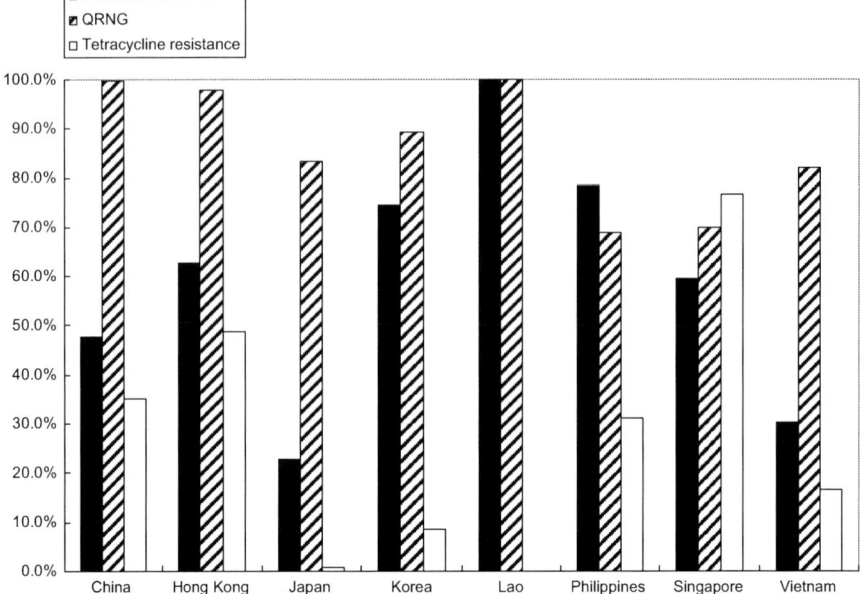

Fig. 18.3 Prevalence of antibiotics resistance in *Neisseria gonorrhoeae* from the west Pacific area and Southeast Asia (QRNG, quinolone-resistant *N. gonorrhoeae*) (data from the WHO Western Pacific Gonococccal Antimicrobial Surveillance Programme, 2008)

increased to 42% in 2002 (Jabeen et al. 2006). Current data indicate that almost all oral regimens for gonorrhea treatment may not be appropriate in Asia. The emergence of decreased ceftriaxone susceptibility in India and China should be closely monitored.

18.1.2 Nosocomial Infections

18.1.2.1 Gram-Negative Pathogens

1. Extended-spectrum beta-lactamses (ESBLs)-producing Enterobacteriaceae. Since the first identification of *K. pneumoniae* isolates carrying SHV-2-type ESBL in China, high levels of ESBL phenotype were noted in the Asia–Pacific region (Paterson et al. 2005; Hawkey 2008). In a recent surveillance from 1998 to 2002, the highest rates of confirmed ESBL-producing isolates were found in *K. pneumoniae* strains from Singapore (35.6%), followed by those from China (30.7%), the Philippines (21.9%), Taiwan (13.5%), Hong Kong (11.6%), and Japan (10%) (Hirakata et al. 2005). The percentage of *E. coli* isolates carrying ESBLs in these Asian countries were as follows: China (24.5%), Hong Kong (14.3%), Singapore (11.3%), Taiwan (5.6%),

Philippines (5.0%), and Japan (2.4%) (Hirakata et al. 2005). The predominant ESBL types varied greatly among different pathogens in these regions (Paterson and Bonomo 2005; Hawkey 2008). In Taiwan, TEM-type ESBLs were rare and data suggested that the most predominant ESBL types could be SHV-5, SHV-12, CTX-M-3, or CTX-M-14 in *K. pneumoniae* and *E. coli* isolates even among different institutes, whereas SHV-12 and CTX-M-3 were the more common ESBL types in *E. cloacae* and *S. marcescens*, respectively (Yu et al. 2006). In China, *E. coli* and *K. pneumoniae* producing CTX-M-type ESBL (CTX-M-14 and CTX-M-3) outnumbered those producing SHV-12 (Yu et al. 2007). In Korea, previous studies showed that while TEM-type ESBLs were more common, CTX-M groups were emerging in recent years (Hawkey 2008). In Japan, the early emergence of CTX-M group (CTX-M 2, TOHO-1, and TOHO-2) was unique in the Asia–Pacific region (Hawkey 2008). In India, CTX-M-15 was the most common ESBL type (Hawkey 2008). The emergence of new non-TEM, non-SHV ESBL (VEB-1) has been reported from Thailand and Vietnam (Paterson and Bonomo 2005).

2. *Pseudomonas aeruginosa*. The percentage of clinical isolates of *P. aeruginosa* with aminoglycoside resistance in Asia ranged from 4.2 to 10% for amikacin and 10.1 to 25% for tobramycin, which were lower than for isolates from Latin America (Gales et al. 2001; Jones et al. 2002). A global study found that the prevalence of *P. aeruginosa* isolates susceptible to levofloxacin in the Asia–Pacific area was 63.6%, which was similar to those from North America and Europe (63.3 and 67%, respectively) and better than those from Latin America (45.3%) (Reinert et al. 2007). In Japan, the percentages of ciprofloxacin-resistant *P. aeruginosa* isolates from urinary tract and respiratory tract were 40–60% and 15–25%, respectively (Yamaguchi et al. 2005). In Korea, the ciprofloxacin susceptibilities of *P. aeruginosa* isolates from non-tertiary hospitals ranged from 60.1% in 2002 to 59.1% in 2006 (Yoo et al. 2008). In Taiwan, the prevalence of ciprofloxacin-resistant *P. aeruginosa* ranged from 15 to 36% in 2002 (Hsueh et al. 2002). Asia-wide, the prevalence of carbapenem-susceptible *P. aeruginosa* ranged from 70.8 to 81.6% (Hadadi et al. 2008). The mechanisms of carbapenem resistance in *P. aeruginosa* are associated with the loss of an outer membrane protein (OprD), multidrug efflux pump, interplay between impermeability and certain β-lactamases, and production of carbapenemases including metallo-beta-lactamases (MBLs) (Walsh et al. 2005). VIM-type MBL genes were reported to be predominant in *P. aeruginosa* isolates from Europe, China, and Taiwan in comparison with the predominant IMP-type MBLs in Japan (Walsh et al. 2005). In Taiwan, the prevalence of the VIM-type MBL genes among CRPA ranged from 17.0 to 36.0% and was as high as 76.9% among multidrug-resistant *P. aeruginosa* (MDRPA), which are defined as those intermediately resistant or resistant to anti-pseudomonal penicillin/beta-lactamase inhibitor combinations and all antimicrobial agents available for clinical use against *P. aeruginosa* (cephalosporins, aztreonam, carbapenems, aminoglycosides, and ciprofloxacin) (Yan et al. 2001). These prevalence rates of MBL in CRPA were markedly higher in

Taiwan than in China (10%), Korea (11.4%), and Japan (12.5%) (Kimura et al. 2005; Yu et al. 2006). Clonal dissemination of MBL-producing CRPA among individual institutes was identified in several studies (Kimura et al. 2005; Yu et al. 2006).

3. *Acinetobacter baumannii*. The resistance of *A. baumannii* to antimicrobial agents is mediated by almost all of the known resistance mechanisms in bacteria including modification of target sites, enzymatic inactivation, active efflux, and decreased influx of drugs (Dijkshoorn et al. 2007). Like *P. aeruginosa*, MBLs in *A. baumannii* could result in carbapenem resistance. However, the Ambler class D oxacillinases (e.g., OXA-23, OXA-24, and OXA-58) are the more widespread carbapenemases seen in *A. baumannii* and have led to more limited treatment options (Dijkshoorn et al. 2007; Maragakis and Perl 2008). In addition, the mechanisms responsible for the development of multidrug resistance in *A. baumannii* remained largely unresolved. Recent global surveillance results showed that susceptibilities of *A. baumannii* to imipemen and amikacin in the Asia/Pacific area (69.2 and 59.6%, respectively) fell into the range between those of Northern America (88.6 and 85.0%, respectively) and Latin America (60.6 and 23.4%, respectively) (Reinert et al. 2007). In early reports from Taiwan, the prevalence of imipenem-resistant *A. baumannii* was less than 5% but this had increased to 32.1% of respiratory and 50% of urine isolates in a recent study (Kuo et al. 2008). In Korea, the resistance rate of *A. baumannii* was 20% in a recent study (Lee et al. 2005) The susceptibility rate of *A. baumannii* to meropenem or imipenem was >96% in Japan (Ishii et al. 2008). In a recent Iranian study, 23% of *Acinetobacter* spp. were resistant to ciprofloxacin and 79% were resistant to imipenem (Hadadi et al. 2008). By contrast, only 10.4% of *A. baumannii* isolates from China were resistant to imipenem (Xiao et al. 2008). Further surveillance of the trend of cabapenem resistance in Asia is needed.

18.1.2.2 Gram-Positive Pathogens

1. Methicillin-resistant and vancomycin-intermediate *S. aureus (MRSA and VISA)*. A remarkable increase in the prevalence of MRSA in nosocomial infections has been observed since MRSA was first identified in Taiwan (from 26.7% in 1990 to 75–84% in 1998–2000) (Hsueh et al. 2004b). The percentage of MRSA among *S. aureus* isolates was 62.9% in China and 64.0% in Korea (Xiao et al. 2008, Kim et al. 2004). Although quinupristin/dalfopristin revealed favorable activity against MRSA isolates in an Asia/Pacific surveillance (susceptibilities >99%), high rates of nonsusceptibility (MICs, ≥ 2 mg/L) in MRSA isolates up to 31% were reported from Taiwan (Luh et al. 2000; Biedenbach et al. 2007). Since its identification in 1997 in Japan, MRSA with reduced susceptibility to vancomycin (VISA)

has been reported in Korea, Thailand, and Taiwan (Hiramatsu et al. 1997; Kim et al. 2000; Trakulsomboon et al. 2001; Wang et al. 2004). In a surveillance study of VISA, 4.3% of MRSA isolates were heteroresistant VISA (hVISA) which were defined as strains of *S. aureus* that contain subpopulations of vancomycin-intermediate daughter cells but vancomycin MICs of their parent strain are only 1–4 mg/L. Previous studies found hVISA strains among MRSA isolates from Japan (8.2%), India (6.3%), South Korea (6.1%), the Philippines (3.6%), Vietnam (2.4%), Singapore (2.3%), and Thailand (2.1%), but not found among those from China, Indonesia, Saudi Arabia, Sri Lanka, or Taiwan (Song et al. 2004b). Although VISA remains sporadic in Asia, the emergence of VISA could be anticipated.
2. Vancomycin-resistant enterococci (VRE). The first human case of VRE (*E. faecalis*) infection was reported in 1996 in Japan and was also isolated from chickens given avoparcin for growth promotion (Fujita et al. 1998). VRE has caused several outbreaks in Korea and 11% of *E. faecium* isolates were VRE (Chong and Lee 2000). An increase in VRE isolation associated with the continuous widespread use of glycopeptides in a Taiwanese university hospital was observed (Hsueh et al. 2002). Furthermore, interhospital and nosocomial spread of some VRE clones, particularly one vanB2 *E. faecium* clone, or long-term persistence of multiple clones in hospitalized patients still exists (Lu et al. 2001).

18.2 Drug Resistance of *Mycobacterium tuberculosis* (TB)

M. tuberculosis strains that are resistant to an increasing number of second-line drugs used to treat multidrug-resistant tuberculosis (MDR TB), and resistant to at least rifampin and isoniazid, have become a global threat to public health (Shah et al. 2007). The so-called extensively drug-resistant TB (XDR TB) is defined as MDR TB that is resistant to any fluoroquinolone and at least one of three injectable second-line drugs (WHO 2006). In Taiwan, the incidence of primary resistance ranged from 4.7 to 12% for isoniazid, 0.7 to 5.9% for rifampin, 1 to 6% for ethambutol, and 4 to 11% for streptomycin (Hsueh et al. 2006). The overall rates of MDR TB among new cases and previously treated cases were 1–3% and 15–46%, respectively, which are similar to previously reported rates (Hsueh et al. 2006). However, the prevalence of MDR TB cases is on the rise in India, with reported proportions ranging from 1.1 to 5.3% in newly diagnosed TB and higher proportions (varied from 8 to 67%) in previously treated patients (Prasad 2005). In a recent study from India, a total of 5 (7.4%) of 68 MDR TB strains were XDR TB (Mondal and Jain 2007). Continuous surveillance of clinical isolates of *M. tuberculosis* is needed to identify XDR TB, especially in patients who have a history of TB and have received prior anti-TB treatment.

18.3 Drug Resistance in Fungi

Data from a 8.5-year global antifungal surveillance revealed no consistent trend toward increasing resistance to fluconazole and voriconazole among the common species (*C. albicans*, *C. glabrata*, or *C. tropicalis*) (Pfaller et al. 2007). The resistance rates to both fluconazole and voriconazole among isolates of *C. glabrata* were the lowest rates to both agents seen in the Asia–Pacific region and the highest in North America (Pfaller et al. 2007). A national surveillance of candidosis from Japan found fluconazole resistance in 1.8% of *C. albicans*, 0.8% of *C. parapsilosis*, 5.2% of *C. glabrata* and 3.2% of *C. tropicalis* isolates, respectively (Takakura et al. 2004). Thirty-five percent of isolates less susceptible ($\geq$16 mg/L) to fluconazole showed resistance to voriconazole ($\geq$2 mg/L). In Taiwan, Hsueh et al. (2005c) reported that all isolates of *Candida* species other than *C. glabrata* and *C. krusei* were susceptible to fluconazole, 27% of *C. glabrata* isolates were not susceptible to fluconazole and all of them were dose-dependently susceptible or resistant to itraconazole. All other *Candida* spp. isolates had voriconazole MICs $\leq$0.5 mg/L except for 5.1% isolates of *C. glabrata* which had MICs between 2 and 4 mg/L. Ruan et al. (2008) reported the fluconazole susceptibilities of *C. glabrata* isolates shifted rapidly from susceptible (64% in 1999–2001 to 19% in 2007) to susceptible dose dependent during a recent 7-year period (27% in 1999–2001 and 75% in 2007). *A. flavus* was less susceptible to amphotericin B, with the MICs of 1 mg/L in 50% and 2 mg/L in 90%. All *Aspergillus* isolates were inhibited by $\leq$1 mg/L of voriconazole (Hsueh et al. 2005c).

References

Bala M., Jain R.K., and Ray K. 2008. Antimicrobial susceptibility profile of resistance phenotypes of *Neisseria gonorrheae* in India. Sex. Transm. Dis. 35:588–91.

Beekmann S.E., Heilmann K.P., Richter S.S., García-de-Lomas J., and Doern G.V. 2005. The GRASP Study Group. Antimicrobial resistance in *Streptococcus pneumoniae*, *Haemophilus influenzae*, *Moraxella catarrhalis* and group A beta-haemolytic streptococci in 2002–2003. Results of the multinational GRASP Surveillance Program. Int. J. Antimicrob. Agents. 25:148–56.

Biedenbach D.J., Bell J.M., Sader H.S., Fritsche T.R., Jones R.N., and Turnidge J.D. 2007. Antimicrobial susceptibility of Gram-positive bacterial isolates from the Asia–Pacific region and an in vitro evaluation of the bactericidal activity of daptomycin, vancomycin, and teicoplanin: a SENTRY Program Report (2003–2004). Int. J. Antimicrob. Agents 30:143–9.

Boucher H.W., and Corey G.R. 2008. Epidemiology of methicillin-resistant *Staphylococcus aureus*. Clin. Infect. Dis. 46 Suppl 5:S344–9.

Boyle-Vavra S., Ereshefsky B., Wang C.C., and Daum R.S. 2005. Successful multiresistant community-associated methicillin-resistant *Staphylococcus aureus* lineage from Taipei, Taiwan, that carries either the novel *Staphylococcal* chromosome cassette mec (SCCmec) type VT or SCCmec type IV. J. Clin. Microbiol. 43:4719–30.

Chau T.T., Campbell J.I., Galindo C.M., Van Minh Hoang N., Diep T.S., Nga T.T., Van Vinh Chau N., Tuan P.Q., Page A.L., Ochiai R.L., Schultsz C., Wain J., Bhutta Z.A., Parry C.M., Bhattacharya S.K., Dutta S., Agtini M., Dong B., Honghui Y., Anh D.D., Canh do G., Naheed A., Albert M.J., Phetsouvanh R., Newton P.N., Basnyat B., Arjyal A., La T.T., Rang N.N., Phuong le T., Van Be Bay P., von Seidlein L., Dougan G., Clemens J.D., Vinh H., Hien T.T., Chinh N.T., Acosta C.J., Farrar J., and Dolecek C. 2007. Antimicrobial drug resistance of *Salmonella* enterica serovar *typhi* in Asia and molecular mechanism of reduced susceptibility to the fluoroquinolones. Antimicrob. Agents Chemother. 51:4315–23.

Chiu C.H., Wu T.L., Su L.H., Chu C., Chia J.H., Kuo A.J., Chien M.S., and Lin T.Y. 2002. The emergence in Taiwan of fluoroquinolone resistance in *Salmonella enterica* serotype choleraesuis. N. Engl. J. Med. 346:413–9.

Chiu C.H., Su L.H., and Chu C. 2004a *Salmonella enterica* serotype Choleraesuis: epidemiology, pathogenesis, clinical disease, and treatment. Clin. Microbiol. Rev. 17:311–22

Chiu C.H., Wu T.L., Su L.H., Liu J.W., and Chu C. 2004b. Fluoroquinolone resistance in *Salmonella enterica* serotype Choleraesuis, Taiwan, 2000–2003. Emerg. Infect. Dis. 10:1674–6.

Choi S.H., Woo J.H., Lee J.E., Park S.J., Choo E.J., Kwak Y.G., Kim M.N., Choi M.S., Lee N.Y., Lee B.K., Kim N.J., Jeong J.Y., Ryu J., and Kim Y.S. 2005. Increasing incidence of quinolone resistance in human non-typhoid *Salmonella enterica* isolates in Korea and mechanisms involved in quinolone resistance. J. Antimicrob. Chemother. 56:1111–4.

Chong Y., and Lee K. 2000. Present situation of antimicrobial resistance in Korea. J. Infect. Chemother. 6:189–95.

Cui S., Li J., Sun Z., Hu C., Jin S., Guo Y., Ran L., and Ma Y. 2008. Ciprofloxacin-resistant *Salmonella enterica* serotype Typhimurium, China. Emerg. Infect. Dis. 14:493–5.

Dijkshoorn L., Nemec A., and Seifert H. 2007. An increasing threat in hospitals: multidrug-resistant *Acinetobacter baumannii*. Nat. Rev. Microbiol. 5:939–51.

Felmingham D., White A.R., Jacobs M.R., Appelbaum P.C., Poupard J., Miller L.A., and Grüneberg R.N. 2005. The Alexander Project: the benefits from a decade of surveillance. J. Antimicrob. Chemother. 56 Suppl 2:ii3–ii21.

Fujita N., Yoshimura M., Komori T., Tanimoto K., and Ike Y. 1998. First report of the isolation of high-level vancomycin-resistant *Enterococcus faecium* from a patient in Japan. Antimicrob. Agents Chemother. 42:2150.

Gales A.C., Jones R.N., Turnidge J., Rennie R., and Ramphal R. 2001. Characterization of *Pseudomonas aeruginosa* isolates: occurrence rates, antimicrobial susceptibility patterns, and molecular typing in the global SENTRY Antimicrobial Surveillance Program, 1997–1999. Clin. Infect. Dis. 15;32 Suppl 2:S146–55.

Hadadi A., Rasoulinejad M., Maleki Z., Yonesian M., Shirani A., and Kourorian Z. 2008. Antimicrobial resistance pattern of Gram-negative bacilli of nosocomial origin at 2 university hospitals in Iran. Diagn. Microbiol. Infect Dis. 60:301–5.

Hawkey P.M. 2008. Prevalence and clonality of extended-spectrum beta-lactamases in Asia. Clin. Microbiol. Infect. Suppl 1:159–65..

Hirakata Y., Matsuda J., Miyazaki Y., Kamihira S., Kawakami S., Miyazawa Y., Ono Y., Nakazaki N., Hirata Y., Inoue M., Turnidge J.D., Bell J.M., Jones R.N., and Kohno S.; SENTRY Asia-Pacific Participants. 2005. Regional variation in the prevalence of extended-spectrum beta-lactamase-producing clinical isolates in the Asia-Pacific region (SENTRY 1998–2002). Diagn. Microbiol. Infect .Dis. 52:323–9.

Hiramatsu K., Aritaka N., Hanaki H., Kawasaki S., Hosoda Y., Hori S., Fukuchi Y., and Kobayashi I. 1997. Dissemination in Japanese hospitals of strains of Staphylococcus aureus heterogeneously resistant to vancomycin. Lancet. 350:1670–3.

Ho P.L., Yam W.C., Cheung T.K., Ng W.W., Que T.L., Tsang D.N., Ng T.K., and Seto W.H. 2001. Fluoroquinolone resistance among *Streptococcus pneumoniae* in Hong Kong linked to the Spanish 23F clone. Emerg. Infect. Dis. 7:906–8.

Ho P.L., Chow K.H., Mak G.C., Tsang K.W., Lau Y.L., Ho A.Y., Lai E.L., and Chiu S.S. 2004. Decreased levofloxacin susceptibility in *Haemophilus influenzae* in children, Hong Kong. Emerg. Infect. Dis. 10:1960–2

Ho P.L., Cheung C., Mak G.C., Tse C.W., Ng T.K., Cheung C.H., Que T.L., Lam R., Lai R.W., Yung R.W., and Yuen K.Y. 2007. Molecular epidemiology and household transmission of community-associated methicillin-resistant *Staphylococcus aureus* in Hong Kong. Diagn. Microbiol. Infect. Dis. 57:145–51.

Hoban D., and Felmingham D. 2002. The PROTEKT surveillance study: antimicrobial susceptibility of *Haemophilus influenzae* and *Moraxella catarrhalis* from community-acquired respiratory tract infections J. Antimicrob. Chemother. 50 Suppl S1:49–59.

Hsu L.Y., Koh Y.L., Chlebicka N.L., Tan T.Y., Krishnan P., Lin R.T., Tee N., Barkham T., and Koh T.H. 2006. Establishment of ST30 as the predominant clonal type among community-associated methicillin-resistant *Staphylococcus aureus* isolates in Singapore. J. Clin. Microbiol. 44:1090–3.

Hsueh P.R., Liu Y.C., Shyr J.M., Wu T.L., Yan J.J., Wu J.J., Leu HS, Chuang Y.C., Lau Y. J, and Luh K.T. 2000. Multicenter surveillance of antimicrobial resistance of *Streptococcus pneumoniae*, *Haemophilus influenzae*, and *Moraxella catarrhalis* in Taiwan during the 1998–1999 respiratory season. Antimicrob. Agents Chemother. 44:1342–5.

Hsueh P.R., Liu C.Y., and Luh K.T. 2002. Current status of antimicrobial resistance in Taiwan. Emerg. Infect. Dis. 8:132–7.

Hsueh P.R., Teng L.J., Lee C.M., Huang W.K., Wu T.L., Wan J.H., Yang D., Shyr J.M., Chuang Y.C., Yan J.J., Lu J.J., Wu J.J., Ko W.C., Chang F.Y., Yang Y.C., Lau Y.J., Liu Y.C., Leu H.S., Liu C.Y., and Luh K.T.; SMART Program 2001. Data. 2003. Telithromycin and quinupristin-dalfopristin resistance in clinical isolates of *Streptococcus pyogenes*: SMART Program 2001 Data. Antimicrob. Agents Chemother. 47:2152–7.

Hsueh P.R., Teng L.J., Tseng S.P., Chang C.F., Wan J.H., Yan J.J., Lee C.M., Chuang Y.C., Huang W.K., Yang D., Shyr J.M., Yu K.W., Wang L.S., Lu J.J., Ko W.C., Wu J.J., Chang F.Y., Yang Y.C., Lau Y.J., Liu Y.C., Liu C.Y., Ho S.W., and Luh K.T. 2004a. Ciprofloxacin-resistant *Salmonella enterica* Typhimurium and Choleraesuis from pigs to humans, Taiwan. Emerg. Infect. Dis. 10:60–8.

Hsueh P.R., Teng L.J., Chen W.H., Pan H.J., Chen M.L., Chang S.C., Luh K.T., and Lin F.Y. 2004b. Increasing prevalence of methicillin-resistant *Staphylococcus aureus* causing nosocomial infections at a university hospital in Taiwan from 1986 to 2001. Antimicrob. Agents Chemother. 48:1361–4.

Hsueh P.R., Shyr J.M., and Wu J.J. 2005a. Decreased erythromycin use after antimicrobial reimbursement restriction for undocumented bacterial upper respiratory tract infections significantly reduced erythromycin resistance in *Streptococcus pyogenes* in Taiwan. Clin. Infect. Dis. 40:903–5.

Hsueh P.R., Tseng S.P., Teng L.J., and Ho S.W. 2005b High prevalence of ciprofloxacin-resistant *Neisseria gonorrhoeae* in Northern Taiwan. Clin. Infect. Dis. 40:188–92.

Hsueh P.R., Lau Y.J., Chuang Y.C., Wan J.H., Huang W.K., Shyr J.M., Yan J.J., Yu K.W., Wu J.J., Ko W.C., Yang Y.C., Liu Y.C., Teng L.J., Liu C.Y., and Luh K.T. 2005c. Antifungal susceptibilities of clinical isolates of *Candida* species, *Cryptococcus neoformans*, and *Aspergillus* species from Taiwan: surveillance of multicenter antimicrobial resistance in Taiwan program data from 2003. Antimicrob. Agents Chemother. 49:512–7.

Hsueh P.R., Liu Y.C., So J., Liu C.Y., Yang P.C., and Luh K.T. 2006. *Mycobacterium tuberculosis* in Taiwan. J. Infect. 52:77–85

Ikebe T., Hirasawa K., Suzuki R., Isobe J., Tanaka D., Katsukawa C., Kawahara R., Tomita M., Ogata K., Endoh M., Okuno R., and Watanabe H. 2005. Antimicrobial susceptibility survey of *Streptococcus pyogenes* isolated in Japan from patients with severe invasive group A streptococcal infections. Antimicrob. Agents Chemother. 49:788–90.

Invasive Bacterial Infections Surveillance (IBIS) Group of the International Clinical Epidemiology Network. 2002. Are *Haemophilus influenzae* infections a significant problem in India? A prospective study and review. Clin. Infect. Dis. 34:949–57.

Inoue M., Lee N.Y., Hong S.W., Lee K., and Felmingham D. 2004. PROTEKT 1999–2000: a multicentre study of the antibiotic susceptibility of respiratory tract pathogens in Hong Kong, Japan and South Korea. Int. J. Antimicrob. Agents 23:44–51.

Ishii Y., Yamaguchi K., and ; Meropenem Surveillance Group. 2008. Evaluation of the susceptibility trends to meropenem in a nationwide collection of clinical isolates in Japan: a longitudinal analysis from 2002 to 2006. Diagn. Microbiol. Infect. Dis. 61:346–50.

Izumiya H., Mori K., Kurazono T., Yamaguchi M., Higashide M., Konishi N., Kai A., Morita K., Terajima J., and Watanabe H. 2005. Characterization of isolates of *Salmonella enterica* serovar typhimurium displaying high-level fluoroquinolone resistance in Japan. J Clin Microbiol. 43:5074–9.

Jabeen K., Khan E., and Hasan R. 2006. Emergence of quinolone-resistant *Neisseria gonorrhoeae* in Pakistan. Int. J. STD. AIDS 17:30–3.

Jasir A., Tanna A., Noorani A., Mirsalehian A., Efstratiou A., and Schalen C. 2000. High rate of tetracycline resistance in *Streptococcus pyogenes* in Iran: an epidemiological study. J. Clin. Microbiol. 38:2103–7.

Jones R.N., Kirby J.T., Beach M.L., Biedenbach D.J., and Pfaller M.A. 2002. Geographic variations in activity of broad-spectrum beta-lactams against *Pseudomonas aeruginosa*: summary of the worldwide SENTRY Antimicrobial Surveillance Program (1997–2000). Diagn. Microbiol. Infect. Dis. 43:239–43.

Kim E.S., Song J.S., Lee H.J., Choe P.G., Park K.H., Cho J.H., Park W.B., Kim S.H., Bang J.H., Kim D.M., Park K.U., Shin S., Lee M.S., Choi H.J., Kim N.J., Kim E.C., Oh M.D., Kim H.B., and Choe K.W. 2007a. A survey of community-associated methicillin-resistant *Staphylococcus aureus* in Korea. J. Antimicrob. Chemother. 60:1108–14.

Kim H.B., Jang H.C., Nam H.J., Lee Y.S., Kim B.S., Park W.B., Lee K.D., Choi Y.J., Park S.W., Oh M.D., Kim E.C., and Choe K.W. 2004. In vitro activities of 28 antimicrobial agents against *Staphylococcus aureus* isolates from tertiary-care hospitals in Korea: a nationwide survey. Antimicrob. Agents Chemother. 48:1124–7.

Kim I.S., Ki C.S., Kim S., Oh W.S., Peck K.R., Song J.H., Lee K., and Lee N.Y. 2007b. Diversity of ampicillin resistance genes and antimicrobial susceptibility patterns in *Haemophilus influenzae* strains isolated in Korea. Antimicrob. Agents Chemother. 51:453–60.

Kim S., and Lee N.Y. 2004. Epidemiology and antibiotic resistance of group A streptococci isolated from healthy schoolchildren in Korea. J. Antimicrob. Chemother. 54:447–50.

Kim M.N., Pai C.H., Woo J.H., Ryu J.S., and Hiramatsu K. 2000. Vancomycin-intermediate *Staphylococcus aureus* in Korea. J Clin Microbiol. 38:3879–81.

Kimura S., Alba J., Shiroto K., Sano R., Niki Y., Maesaki S., Akizawa K., Kaku M., Watanuki Y., Ishii Y., and Yamaguchi K. 2005. Clonal diversity of metallo-beta-lactamase-possessing *Pseudomonas aeruginosa* in geographically diverse regions of Japan. J. Clin. Microbiol. 43:458–61.

Kulwichit W., Chatsuwan T., Unhasuta C., Pulsrikarn C., Bangtrakulnonth A., and Chongthaleong A. 2007. Drug-resistant nontyphoidal *Salmonella* bacteremia, Thailand. Emerg. Infect. Dis. 13:501–2

Kuo L.C., Yu C.J., Kuo M.L., Chen W.N., Chang C.K., Lin H.I., Chen C.C., Lu M.C., Lin C.H., Hsieh W.F., Chen L.W., Chou Y., Huang M.S., Lee C.H., Chen S.C., Thai S.L., Chen P.C., Chen C.H., Tseng C.C., Chen Y.S., Hsiue T.R., and Hsueh P.R. 2008. Antimicrobial resistance of bacterial isolates from respiratory care wards in Taiwan: a horizontal surveillance study. Int. J. Antimicrob. Agents. 31:420–6.

Leclercq R. 2002. Mechanisms of resistance to macrolides and lincosamides: nature of the resistance elements and their clinical implications. Clin. Infect. Dis. 34:482–92.

Lee K., Yong D., Yum J.H., Lim Y.S., Kim H.S., Lee B.K., and Chong Y. 2004. Emergence of multidrug-resistant *Salmonella* enterica serovar *typhi* in Korea. Antimicrob. Agents Chemother. 48:4130–5.

Lee K., Yum J.H., Yong D., Lee H.M., Kim H.D., Docquier J.D., Rossolini G.M., and Chong Y. 2005. Novel acquired metallo-beta-lactamase gene, *bla*(SIM-1), in a class 1 integron from *Acinetobacter baumannii* clinical isolates from Korea. Antimicrob. Agents. Chemother. 49:4485–4491.

Lu J.J., Perng C.L., Ho M.F., Chiueh T.S., and Lee W.H. 2001. High prevalence of *vanB2* vancomycin resistant *Enterococcus faecium* in Taiwan. J. Clin. Microbiol. 39:2140–5.

Luh K.T., Hsueh P.R., Teng L.J., Pan H.J., Chen Y.C., Lu J.J., Wu J.J., and Ho S.W. 2000. Quinupristin-dalfopristin resistance among gram-positive bacteria in Taiwan. Antimicrob. Agents Chemother. 44:3374–80.

Maragakis L.L., and Perl T.M. 2008. *Acinetobacter baumannii*: epidemiology, antimicrobial resistance, and treatment options. Clin. Infect. Dis. 46:1254–63.

Matsumoto T. 2008. Trends of sexually transmitted diseases and antimicrobial resistance in *Neisseria gonorrhoeae*. Int. J. Antimicrob. Agents. 31 Suppl 1:S35–9.

Mondal R., and Jain A. 2007. Extensively drug-resistant *Mycobacterium tuberculosis*, India. Emerg. Infect. Dis. 13:1429–31.

Morris S.K., Moss W.J., and Halsey N. 2008. *Haemophilus influenzae* type b conjugate vaccine use and effectiveness. Lancet Infect Dis. Jul;8(7):435–43.

Nakae M., Murai T., Kaneko Y., and Mitsuhashi S. 1977. Drug resistance in *Streptococcus pyogenes* isolated in Japan. Antimicrob. Agents Chemother. 12:427–8.

Paterson D.L., and Bonomo R.A. 2005. Extended-spectrum beta-lactamases: a clinical update Clin Microbiol Rev. Oct;18(4):657–86.

Parry C.M., Hien T.T., Dougan G., White N.J., and Farrar J.J. 2002. Typhoid fever. N. Engl. J. Med. 347:1770–82.

Pfaller M.A., Diekema D.J., Gibbs D.L., Newell V.A., Meis J.F., Gould I.M., Fu W., Colombo A.L., and Rodriguez-Noriega E.; Global Antifungal Surveillance Study. 2007. Results from the ARTEMIS DISK Global Antifungal Surveillance study, 1997 to 2005: an 8.5-year analysis of susceptibilities of Candida species and other yeast species to fluconazole and voriconazole determined by CLSI standardized disk diffusion testing. J. Clin. Microbiol. 45:1735–45.

Prasad R. 2005. Current MDR status. Indian. J. Tuberc. 52:121–31.

Ray K., Bala M., Kumari S., and Narain J.P. 2005. Antimicrobial resistance of *Neisseria gonorrhoeae* in selected World Health Organization Southeast Asia Region countries: an overview. Sex. Transm. Dis. 32:178–84.

Reinert R.R., Low D.E., Rossi F., Zhang X., Wattal C., and Dowzicky M.J. 2007. Antimicrobial susceptibility among organisms from the Asia/Pacific Rim, Europe and Latin and North America collected as part of TEST and the in vitro activity of tigecycline. J. Antimicrob. Chemother. 60:1018–29.

Ruan S.Y., Chu C.C., and Hsueh P.R. 2008. In vitro susceptibilities of invasive isolates of Candida species: rapid increase in rates of fluconazole susceptible-dose dependent *Candida glabrata* isolates. Antimicrob. Agents Chemother. 52:2919–22.

Sanbongi Y., Suzuki T., Osaki Y., Senju N., Ida T., and Ubukata K. 2006. Molecular evolution of beta-lactam-resistant *Haemophilus influenzae*: 9-year surveillance of penicillin-binding protein 3 mutations in isolates from Japan. Antimicrob. Agents Chemother. 50:2487–92.

Shah N.S., Wright A., Bai G.H., Barrera L., Boulahbal F., Martín-Casabona N., Drobniewski F., Gilpin C., Havelková M., Lepe R., Lumb R., Metchock B., Portaels F., Rodrigues M.F., Rüsch-Gerdes S., Van Deun A., Vincent V., Laserson K., Wells C., and Cegielski J.P. 2007. Worldwide emergence of extensively drug-resistant tuberculosis. Emerg. Infect. Dis. 13:380–7.

Song J.H., Jung S.I., Ko K.S., Kim N.Y., Son J.S., Chang H.H., Ki H.K., Oh W.S., Suh J.Y., Peck K.R., Lee N.Y., Yang Y., Lu Q., Chongthaleong A., Chiu C.H., Lalitha M.K., Perera J., Yee T.T., Kumarasinghe G., Jamal F., Kamarulzaman A., Parasakthi N., Van P.H., Carlos C., So T., Ng T.K., and Shibl A. 2004a. High prevalence of antimicrobial resistance among clinical *Streptococcus pneumoniae* isolates in Asia (an ANSORP study). Antimicrob. Agents Chemother. 48:2101–7.

Song J.H., Hiramatsu K., Suh J.Y., Ko K.S., Ito T., Kapi M., Kiem S., Kim Y.S., Oh W.S., Peck K.R., and Lee N.Y.; Asian Network for Surveillance of Resistant Pathogens Study Group. 2004b. Emergence in Asian countries of Staphylococcus aureus with reduced susceptibility to vancomycin. Antimicrob. Agents Chemother. 48:4926–8.

Su X., Jiang F., Qimuge, Dai X., Sun H., Ye S. 2007. Surveillance of antimicrobial susceptibilities in *Neisseria gonorrhoeae* in Nanjing, China, 1999–2006. Sex. Transm. Dis. 34:995–9.

Takakura S., Fujihara N., Saito T., Kudo T., Iinuma Y., and Ichiyama S. 2004. National surveillance of species distribution in blood isolates of Candida species in Japan and their susceptibility to six antifungal agents including voriconazole and micafungin. J. Antimicrob. Chemother. 53:283–9.

Takizawa Y., Taneike I., Nakagawa S., Oishi T., Nitahara Y., Iwakura N., Ozaki K., Takano M., Nakayama T., and Yamamoto T. 2005. A Panton-Valentine leucocidin (PVL)-positive community-acquired methicillin-resistant *Staphylococcus aureus* (MRSA) strain, another such strain carrying a multiple-drug resistance plasmid, and other more-typical PVL-negative MRSA strains found in Japan. J. Clin. Microbiol. 43:3356–63.

Trakulsomboon S., Danchaivijitr S., Rongrungruang Y., Dhiraputra C., Susaemgrat W., Ito T., Hiramatsu K. 2001. First report of methicillin-resistant *Staphylococcus aureus* with reduced susceptibility to vancomycin in Thailand. J. Clin. Microbiol. 39:591–5.

Tristram S., Jacobs M.R., and Appelbaum P.C. 2007. Antimicrobial resistance in *Haemophilus influenzae*. Clin Microbiol Rev. 20:368–89

Walsh T.R., Toleman M.A., Poirel L., and Nordmann P. 2005. Metallo-beta-lactamases: the quiet before the storm? Clin. Microbiol. Rev. 18:306–325

Wang J.L., Tseng S.P., Hsueh P.R., and Hiramatsu K. 2004. Vancomycin heteroresistance in methicillin-resistant *Staphylococcus aureus*, Taiwan. Emerg. Infect. Dis. 10:1702–4.

WHO Western Pacific Gonococccal Antimicrobial Surveillance Programme. 2008. Surveillance of antibiotic resistance in *Neisseria gonorrhoeae* in the WHO Western Pacific Region, 2006. Commun. Dis. Intell. 32:48–51.

World Health Organization. Global tuberculosis control: WHO report. Geneva: the Organization; 2006. WHO/HTM/TB/2006.362.

Xiao Y.H., Wang J., and Li Y; on behalf of the MOH National Antimicrobial Resistance Investigation Net. 2008. Bacterial resistance surveillance in China: a report from Mohnarin 2004–2005. Eur. J. Clin. Microbiol. Infect. Dis. 27:697–708.

Yamaguchi K., and Ohno A.; Levofloxacin Surveillance Group. 2005. Investigation of the susceptibility trends in Japan to fluoroquinolones and other antimicrobial agents in a nationwide collection of clinical isolates: a longitudinal analysis from 1994 to 2002. Diagn. Microbiol. Infect. Dis. 52:135–43.

Yan J.J., Hsueh P.R., Ko W.C., Luh K.T., Tsai S.H., Wu H.M., and Wu J.J. 2001. Metallo-beta-lactamases in clinical *Pseudomonas* isolates in Taiwan and identification of VIM-3, a novel variant of the VIM-2 enzyme. Antimicrob. Agents Chemother. 45:2224–8.

Yan J.J., Chiou C.S., Lauderdale T.L., Tsai S.H., and Wu J.J. 2005. Cephalosporin and ciprofloxacin resistance in *Salmonella*, Taiwan. Emerg. Infect. Dis. 11:947–50.

Yoo J., Sohn E.S., Chung G.T., Lee E.H., Lee K.R., Park Y.K., Lee Y.S. 2008. Five-year report of national surveillance of antimicrobial resistance in *Pseudomonas aeruginosa* isolated from non-tertiary care hospitals in Korea (2002–2006). Diagn. Microbiol. Infect. Dis. 60:291–4.

Yokota S., Ohkoshi Y., Sato K., and Fujii N. 2008. Emergence of fluoroquinolone-resistant Haemophilus influenzae strains among elderly patients but not among children. J. Clin. Microbiol. 46:361–5.

Yu W.L., Chuang Y.C., and Walther-Rasmussen J. 2006. Extended-spectrum beta-lactamases in Taiwan: epidemiology, detection, treatment and infection control J. Microbiol. Immunol. Infect. 39:264–77

Yu Y., Ji S., Chen Y., Zhou W., Wei Z., Li L., and Ma Y. 2007. Resistance of strains producing extended-spectrum beta-lactamases and genotype distribution in China. J. Infect. 54:53–7.

Chapter 19
Antimicrobial Drug Resistance in Latin America and the Caribbean

Manuel Guzmán-Blanco and Raul E. Istúriz

Abstract Antimicrobial resistance in Latin America and the Caribbean is widespread and is a limitation for proper treatment of both community- and hospital-acquired infections. Resistance to respiratory pathogens such as *Streptococcus pneumoniae* and *Escherichia coli* from urinary tract infections is prevalent throughout the region in ambulatory patients with infections. High rates of resistant to antibiotics used in hospital-acquired infections is a growing threat, including *MRSA*, *ESBL-producing* Enterobacteriaceae and carbapenem-resistant *Pseudomonas aeruginosa* and *Acinetobacter baumannii*.

19.1 Introduction

Antimicrobial resistance has been the most significant factor limiting the efficacy of antibiotics since the beginning of the antimicrobial era. Bacteria become resistant to antimicrobial drugs by one of several essential but complex mechanisms that could include endogenous chromosomal mutations; the acquisition of exogenous genes that encode a variety of resistance factors, either or both exerted in a way that results in the inability of the antibiotic to achieve sufficient concentrations at its target site or in the inactivation or modification of the target site in a manner that prevents the drug from binding to and/or interacting with its critical target area in the bacterial cell.

Initially, in the mid-20th century, the known mechanisms of resistance were relatively simple, commonly requiring a sole but important mutation, or the acquisition of single genes that encoded one inactivating enzyme, but the dawn of the 21st century has witnessed the appearance and dissemination of increasingly complex mechanisms and the simultaneous utilization by bacterial

R.E. Istúriz (✉)
Department of Medicine, Infectious Diseases, Centro Médico de Caracas,
San Bernardino, Caracas, Venezuela; Centro Medico Docente La Trinidad,
Caracas, Venezuela
e-mail: mgrijm@cantv.net

A. de J. Sosa et al. (eds.), *Antimicrobial Resistance in Developing Countries*,
DOI 10.1007/978-0-387-89370-9_19, © Springer Science+Business Media, LLC 2010

pathogens of multiple or sophisticated means that, in some cases, render these bacteria resistant to all available agents.

19.2 Surveillance of Antimicrobial Resistance, Infection Control in Latin America and the Caribbean

The emergence and spread of antimicrobial drug resistance in Latin American and Caribbean countries contribute to the worldwide threat to undermine the fight against the spread of bacterial infectious diseases. In these countries, ADR may have several distinct contributing causes, including inappropriate use of available antibiotics, environmental changes and rapid population growth, and in the area, the problem of ADR may be exacerbated by malnutrition, poor sanitation and overcrowding. Self-medication and over-the-counter sale of antibiotics have played a critical role in the community.

In the early years of the 21st century, resistance in Latin America and the Caribbean included multi-resistant organism in the highly sophisticated third-level medical centers coming from the community or the hospital itself. Although surveillance systems for detection of resistance are available, widespread knowledge and use of the data generated by them are not yet a reality in the area. The Pan American Health Organization has promoted surveillance of resistance among countries of the region and that program is active and includes monitoring of quality. Access is feasible in the PAHO web page (www.paho.org); also the Pan-American society of infectious diseases (www.api.org) and, most recently, some country-based programs may be accessed through Internet as the Venezuelan program of surveillance of resistance (www.provenra.org).

Surveillance of resistance is a first step for applying wise prescription policies, which has to be linked to education and infection control programs. In this chapter, we will review information obtained from such systems, including those sponsored by industry along with information from the Pan-American Health Organization (PAHO), the Pan-American Association of Infectious Diseases (API), local surveillance systems and data from individual investigators.

19.2.1 Antibiotic Resistance Among Gram-Positive Bacteria

Gram-positive bacteria are common causes of respiratory tract infections, skin and skin structure infections, bloodstream invasion and infections of all organ systems, as well as many other infectious diseases in humans all over the globe.

19.2.1.1 Penicillin and β-Lactam-Resistant Pneumococcus

Penicillin resistance as well as resistance to cephalosporins and other antibiotic classes has become increasingly common among strains of *Streptococcus pneumoniae* since the 1970s. Some of these resistant strains are capable of producing invasive disease and indeed have caused bacteremia, severe pneumonia, and meningitis and have been responsible for significant morbidity and mortality. Invasive pneumococcal disease (IPD) in the area occurs in all age groups, but more commonly in children less than 2 years of age. Fortunately, the introduction and generalized utilization of a 7-valent pneumococcal conjugate vaccine (PCV7) has been associated with declining rates of IPD (Whitney et al. 2003).

SIREVA is a PAHO-sponsored surveillance system on invasive infection caused by *S. pneumoniae* in children less than 5 years, which has been active since 1993 (Kertesz et al. 1998) and since 1999 the system processes data from Argentina, Brazil, CAREC (Caribbean Epidemiology Centre), Chile, Costa Rica, Colombia, Cuba, Dominican Republic, Ecuador, El Salvador, Guatemala, Honduras, México, Nicaragua, Panama, Paraguay, Peru, Uruguay, and Venezuela. The program has a well-established external quality assurance system, as recently demonstrated (Lovgren et al. 2007).

In their first publication of 1649 (Kertesz et al. 1998) invasive isolates, overall, 24.9% had diminished susceptibility to penicillin (16.7% intermediate and 2% high level). Five serotypes 14, 5, 1, 6B, and 23F accounted for 59% of the invasive infections. The rates of reduced susceptibility to penicillin in SPN varied by country in another publication from this group in the evaluation of 1578 children with pneumonia in five Latin American countries (Hortal et al. 2000). In this study, the rate of resistance to penicillin (MIC$\geq$2 µg/ml) was 20% in Argentina, 2% in Brazil, 11.6% in Colombia, 20.8% in Mexico, and 18.55% in Uruguay. Resistance to trimethoprin–sulfamethoxazole (TMP–SMX) was elevated and relatively lower than erythromycin. Additional data from Venezuela revealed 69.5% susceptibility to penicillin, 21.9% intermediate, and 8.5% resistant. The most prevalent serotypes were, in order, 14, 5, 19A, and 6B.

Surveillance on pneumococcal infections in older children has been evaluated by PAHO since 2000 and from industry supported studies. Data from 2004 (PAHO 2006) using the screening test with oxacillin disk revealed diminished susceptibility to penicillin in 17.5% of strains from Argentina, 21.5% from Brazil, 17.55% from Chile, 29% from Colombia, 5.5% from Guatemala, 4% from Paraguay, and 9.5% from Uruguay. In another publication from SENTRY in year 2000 (Guzmán-Blanco et al. 2000) on 1261 isolates of *S. pneumoniae* from community-acquired respiratory infections in the region, 68.3% were susceptible to penicillin, 98.2% to amoxicillin, 87.7% to clarithromycin, 59.6% to TMP/SXT, and 99.95% to levofloxacin. The highest rate of resistance to penicillin was from Mexico (I: 42.2% and R: 18.3%) and the lowest was from Brazil (I: 18.9 and R: 4%). In this study, resistance was defined as MIC $\geq$2 µg/ml. Importantly, in a subsequent report from this group (Castanheira et al. 2006) resistance to penicillin in SPN increased over 7 years from 2.9 to 11% in Brazil. In addition another study by PROTEK (Mendes et al. 2003), with data from Brazil, Argentina, and Mexico, showed 42.1% of SPN with diminished susceptibility to penicillin and 15.3% resistance to erythromycin. There was no resistance to linezolid, telitromycin, teicoplanin, or vancomycin.

19.2.1.2 Methicillin-Resistant *Staphylococcus aureus* (MRSA)

Reports of penicillin resistance in *S. aureus* began to appear within only a few years of the introduction of penicillin in the early 1940s, and now, 60 years later, virtually all staphylococcal strains have acquired genes that allow them to produce β-lactamases that make them resistant to the natural penicillins, aminopenicillins, and the so-called anti-pseudomonal penicillins (insert reference). Oxacillin and methicillin were developed and successfully used clinically in the early 1960s, but over relatively short periods of time, resistance to oxacillin and methicillin emerged. Similar to penicillin-resistant organisms, these strains, collectively called methicillin-resistant *S. aureus* or MRSA have appeared and spread, in a clonal fashion, initially within the hospital environment and more recently in the community, where new strains have been described.

19.2.1.3 Healthcare-Associated MRSA

Most nosocomial MRSA is multidrug resistant. They tend to colonize and infect patients during hospitalization or stays in long-term care facilities, after surgery or after contact with persons who had an MRSA infection or used illicit drugs. In the 1970s, periodic outbreaks were described in various parts of the world, in association with high levels of oxacillin or methicillin use and in intensive care environments, but since the 1980s, MRSA became a significant worldwide problem, first in large hospitals and later in smaller community hospitals.

Nosocomial MRSA is a growing problem in the region. Information gathered in the PAHO-sponsored program of nosocomial infections (PAHO 2006), reported for the year 2004, country MRSA prevalence as follows: Argentina 42.5% of 5851 isolates, Bolivia 36% of 1167, Chile 80% of 246, Colombia 47% of 4214, Costa Rica 58% of 674, Cuba 6% of 80, Ecuador 25% of 1363, Guatemala 64% of 1483, Honduras 125 of 393, Mexico 52% of 497, Nicaragua 20% of 296, Paraguay 44% of 980, Peru 80% of 1407, Uruguay 59% of 1431, and Venezuela 25% of 2114. Data reported to the Pan-American Association of Infectious Diseases for the year 2006 (Casellas and API Comité Resistencia Antibacterianos 2006) demonstrated the following rates of resistance of hospital-acquired MRSA: Argentina 51%, Bolivia 55%, Brazil 54%, Chile 29%, Ecuador 25%, Mexico 32%, Panama 28%, Paraguay 30%, Uruguay 24%, and Venezuela 27%.

Molecular epidemiology of the spread of MRSA in Latin America has been evaluated, and three predominant clones have been identified in the region: the Brazilian clone (Sader et al. 1994) that has spread to Argentina (Da Silva Coimbra et al. 2000), the Pediatric clone (Gomes et al. 2001), and a clone from Cordoba, Argentina (Sola Gribaudo Vindel et al 2002). Additional clones have circulated in the area, including the Chilean clone in Colombia and

Paraguay (Cruz et al. 2005) and the New York/Japanese clone in both Brazil and Mexico (Melo et al. 2004)

19.2.1.4 Community-Associated MRSA (CA-MRSA)

MRSA began to be detected as a common cause of infection in the community in persons who had not had contact with the health-care system as recent as the mid-1990s and prevalence is increasing rapidly (CDC 2003; Kazakova et al. 2005; Moran et al. 2006). Organisms responsible for these infections are new strains as defined by molecular typing and have spread throughout the globe. These MRSA isolates express the smaller and more basic type IV *SCCmec* and are less often multi-resistant. Strains of CA-MRSA show various degrees of susceptibility to clindamycin, TMP–SMX, and tetracyclines. They have a high prevalence of genes encoding the Panton–Valentine leukocidine, an exotoxin associated with necrotizing skin and pulmonary disease and abscess formation and have been associated with significant morbidity and mortality.

The first report of CA-MRSA from Latin America was from Brazil in 2003 (Ribeiro et al. 2005). Three well-characterized strains harbored the SCC mec type IV, Panton–Valentine leukocidin, enterotoxin, and α-hemolysin genes. A large outbreak of CA-MRSA affected inmates in jails and people from the community in Montevideo, Uruguay, since January 2002 (Ma et al. 2005). At the end of the outbreak, more than 1000 patients had been affected and more than 12 deaths had occurred. Skin and soft tissue infections accounted for more than 65% of the cases, but severe forms of pneumonia were reported, including four deaths. A new clone was identified, UR6, and its pathogenicity was related to the presence of Panton–Valentine leukocidin. TMP–SMX was very active in the treatment of infections of the skin (Moszkowicz et al. 2004). Later, two cases of severe skin infection caused by CA-MRSA were reported from Bogota, also encoding PV and SCCmec type IV (Alvarez et al. 2006). PAHO has performed surveillance of community-acquired MRSA infections since 2005. A total of 12.4% of 845 isolates of *S. aureus* from the community in Venezuela were resistant to oxacillin, while 36.4% of 730 isolates from hospitalized patients were resistant (PROVENRA 1988–2006). No clinical information was available.

Vancomycin-intermediate *S. aureus* (VISA) and vancomycin-resistant *S. aureus* with intermediate resistance (MIC, 8–16 mg/l) to vancomycin have been described since 1996 in strains with heavy and prolonged exposure to vancomycin that develop thickening of their cell wall which may cause vancomycin molecules to become entrapped in the outer portion of the wall and not reaching their target site in the cytoplasmic membrane. No VISA strains have been reported from the PAHO surveillance area. Fully vancomycin-resistant (MIC$\geq$64 mg/l) strains have been found in a few patients in the United States (Reference). In a report from Brazil, 9 of 41 isolates of MRSA from patients on treatment with vancomycin had criteria for heteroresistance to it. One of the isolates had an MIC of 8 μg/ml (Melo et al. 2004)

19.2.1.5 *Enterococcus*

Strains of *Enterococcus* spp. have become multidrug resistant in a somewhat slower manner than strains of *Staphylococcus* spp. It took more than 30 years of vancomycin use, for example, for vancomycin-resistant *Enterococcus* (VRE) to appear as clinically important pathogens. But from the mid-1980s, and especially from the 1990s, a dramatic rise in the incidence of infections produced by VRE has been observed in hospitals and especially ICUs in many countries including the United States (Reference). Vancomycin resistance results from closely related gene clusters termed VanA and VanB which encode resistance by changing the target for vancomycin action, from D-alanine–D-alanine to D-alanine–D-lactate. Deficiencies in infection control practices and selective antibiotic pressure play a role in VRE outbreaks. Vancomycin, extended-spectrum cephalosporins, and drugs with anti-anaerobic spectrum have been associated with colonization and infection with VRE.

Vancomycin-resistant enterococci are found in different countries of the region. In the annual report from PAHO surveillance system of the year 2004 (PAHO 2006), 0.3% of 1193 *Enterococcus faecalis* strains from Argentina were VRE, while 33% of 203 *E. faecium* were resistant. In Chile, 25.6% of 164 *E. faecalis* were resistant and 58.6% of 116 *E. faecium*. Lower figures are reported in *E. faecalis* from Colombia (4% in 1802 isolates), Costa Rica (0% in 99 isolates), Ecuador (0% in 435 isolates), Guatemala (0% in 399 isolates), Honduras (0% in 240 isolates), and Venezuela (3% in 324 isolates). Data from year 2006 (Casellas and API Comité Resistencia Antibacterianos 2006) reported lower numbers of VR *E. faecium* from Argentina, Colombia, Mexico, Panama, Paraguay, and Uruguay (less than 1%) and higher numbers from Ecuador 3–5%, Venezuela 11–15%, and Brazil 20%. Most of these isolates come from patients in intensive care units. A recent increase in vancomycin-resistant *E. faecium* was reported from a large hospital in Mexico City (Cuellar-Rodríguez et al. 2007). VRE was isolated from 27 patients, the MIC for vancomycin was >256 µg/ml and vanA genotype was identified.

19.2.2 Antibiotic Resistance Among Gram-Negatives Bacteria

The emergence and spread of resistance in Enterobacteriaceae and non-fermenting Gram-negative bacteria are facts that have complicated and at times made impossible effective treatment of a variety of infections from the community and within the hospital environment. Bacteria resistant to all anti-microbials currently available are not only a theoretical threat but also a reality around the globe. Enterobacteriaceae notably *E. coli*, *Klebsiella* spp., and *Enterobacter* spp. are common and important causes of urinary tract, blood-stream, pulmonary, and intra-abdominal infections; *Salmonella* at times invasive is an additional problem organism.

Resistance to beta-lactams in Gram-negative bacilli is widely spread in the region. These include bacteria responsible for common community-acquired infections such as intestinal and urinary tract infections, and also multi-resistant organisms responsible for nosocomial infections. Shortly after the reintroduction of cholera in Latin America in 1991, PAHO established a surveillance program for detection of resistance in enteric pathogens. The surveillance program report of 2004 (PAHO 2006) showed large differences in ampicillin resistance in *Salmonella* spp. and *Shigella* spp. *Vibrio cholerae* in the first epidemic in Peru in year 1991 was susceptible to antibiotics commonly used, but resistance was found to three of them in a second outbreak in 1998 (Ibarra and Alvarado 2007), demonstrating that continuing surveillance is important.

In Brazil, ampicillin resistance in *Salmonella enterica* serotype Enteritidis (*S.* Enteritidis) was as low as 3%, while in *Shigella flexneri* it was as high as 90%. In Chile, 10% of *Salmonella* strains were resistant to ampicillin versus 78% of *Shigella* spp. In Colombia, ampicillin resistance in *S.* Enteritidis was 11%, but in *S.* Typhimurium it was 74%. *Shigella flexneri* was resistant to ampicillin in 84% of the strains in the same year. Data from Costa Rica are related to *Salmonella typhimurium*, and the rate of resistance to ampicillin was 44%, resistance in *Shigella* was 76%. In Cuba, resistance in *Salmonella* was 8% and in *Shigella* was 50%. Data from Mexico are related to 7% resistance in *Salmonella* and 75% in *Shigella*. In Paraguay, 3% resistance in *Salmonella* and 65% in *Shigella* and in Venezuela, 14% resistance in *Salmonella* and 83% in *Shigella*. Resistance to TMP/SXT was very closely related to resistance to ampicillin. A recent publication from Cuba (Puig et al. 2007) showed 11% of resistance to ampicillin in 120 human isolates of *Salmonella*.

Resistance in diarrheogenic *E. coli* in children less than 5 years was demonstrated in Mexico with more than 70% of resistance to ampicillin and TMP–SMX in hospitalized patients (Estrada-García et al. 2005). *Escherichia coli*, *Klebsiella pneumoniae*, and *Proteus mirabilis* are the most common organisms causing urinary tract infections in ambulatory patients in Latin America (Andrade et al. 2006; Casellas and API Comité Resistencia Antibacterianos 2006). A serious problem is increasing resistance to commonly used oral antimicrobials, including fluoroquinolones. Data from the PAHO surveillance system for year 2004 revealed resistance to ampicillin from 53 to 79% in ambulatory isolates of *E. coli* from urine; resistance to ciprofloxacin varied from 7% in Uruguay to 32% in Venezuela and resistance to TMP/SXT was between 38 and 68%. The most active drug was nitrofurantoin. In a publication from the SENTRY study group (Andrade et al. 2006), results from Argentina, Brazil, Chile, Mexico, and Venezuela in community-acquired urinary tract infection for year 2003 were presented. *Escherichia coli* were responsible for 71.6% of infections in female and 48% in males. Overall, resistance to ampicillin was 53%, to ampicillin/sulbactam 23.3%, to ciprofloxacin 21.6%, to SXT 40.4%, and to NIT 6.9%. The highest resistance to ciprofloxacin was from Mexico, 72.2%, and the lowest from Brazil, 11.1%. A report from Cuba

Table 19.1 Resistance to antibiotics in community-acquired *Escherichia coli* in urinary tract infections, Latin America (PAHO Surveillance System 2004)

Country	Isolates #	AMP (%)	CIPRO (%)	SXT (%)	NIT (%)
Argentina	11715	55	9	35	2
Ecuador	2335	64	27	54	5
Guatemala	821	79	32	68	3
Paraguay	1875	59	10	38	7
Uruguay	1162	57	7	33	3
Venezuela	1151	53	32	55	7

Source: PAHO (2006).

(Alvarez et al. 2006) of 2401 community-acquired urinary tract infections caused by *E. coli* between 2001 and 2004, 74.75 were resistant to ampicillin, 18.6% to nitrofurantoin, 29.25 to ciprofloxacin, and 84.5% to TMP/SXT.

Gram-negative bacilli are an important cause of nosocomial infection in Latin America. In Table 19.1 most common organisms isolated from different sources in hospitalized patients are presented, according to the SENTRY program for the years 1997–2004 (Guzmán-Blanco et al. 2000). *Escherichia coli* and *K. pneumoniae* were the most frequent isolates from nosocomial bacteremia, while *P. aeruginosa* was the predominant organism in nosocomial pneumonia (Table 19.2).

Table 19.2 Distribution of pathogens identified in Latin America SENTRY* January 1997–December 2004 (33,613 isolates)

	Bacteremia (16,654)	Pneumonia (4770)	Wound (2930)
1	*S. aureus* (22%)	*P. aeruginosa* (27%)	*S. aureus* (33%)
2	*E. coli* (18%)	*S. aureus* (22%)	*E. coli* (14%)
3	CoNS (13%)	*Klebsiella* (11%)	*P. aeruginosa* (13%)
4	*Klebsiella* (11%)	*Acinetobacter* (9%)	*Enterococcus* (7%)
5	*P. aeruginosa* (7%)	*Enterobacter* (5%)	*Klebsiella* (7%)
6	*Enterobacter* (5%)	*S. pneumoniae* (5%)	*Enterobacter* (6%)
7	*Acinetobacter* (4%)	*E. coli* (4%)	CoNS (5%)
8	*S. pneumoniae* (4%)	*H. influenzae* (4%)	*Acinetobacter* (4%)
9	*Enterococcus* (4%)		

*A total of 33,163 isolates from Argentina, Brazil, Chile, Colombia, Mexico, Uruguay, and Venezuela collected in 1997–2004, and tested by broth microdilution (CLSI) at the University of Iowa, USA, Courtesy of Dr. Helio S. Sader.

19.2.2.1 Extended-Spectrum *β*-Lactamase (ESBL)-Producing Bacteria

A total of 8% of 2707 nosocomial strains of *E. coli* were ESBL-producing strains, with a range of 4.5% from Uruguay and 11.7% from Venezuela. In the same publication, 46.9% of 1225 strains of *K. pneumoniae* were ESBL producers, with a range from 26.2% from Venezuela and 52% from Mexico (Sader 2004). The first isolation of an ESBL-producing bacteria was from Argentina in 1981 (Casellas and Goldberg 1989), in a strain of *K. pneumoniae*. The enzyme was identified as SHV-5. Another strain of *K. pneumoniae*, isolated in Santiago, Chile, in 1985, was later proven to be SHV 5 ESBL (Gutmann et al. 1989). The CTX type of β-lactamase, currently the most prevalent in the southern part of Latin America, was first described in an outbreak of neonatal infection with *S. typhimurium* in late 1989 (Guzmán-Blanco et al. 2000).

Reports of ESBL-producing bacteria include isolation from nosocomial bloodstream infection from Mexico (Gonzalez-Vertiz et al. 2001), infections caused by *S. enteritidis* from Trinidad and Tobago (Cherian et al. 1999), infections in neonatal intensive care unit from Brazil (Otman et al. 2002) and from Dominican Republic (Sanchez et al. 2007). An outbreak with two clones, CTX-M2 and SHV-5, was recently reported from Rosario, Argentina (Casellas et al. 2005). A total of 24% of 498 Enterobacteriaceae analyzed in Curitiba, Brazil, in the period 2002–2003 were ESBL producers, including more than 50% of the *K. pneumoniae* (Da Silva Coimbra et al. 2000).

The proportion of Enterobacteriaceae-producing ESBL was higher in South America than in North America, according to the reports from the MYSTIC surveillance system, between 1997 and 2003 (Turner 2005). In that report, 18.1% of *E. coli* from South America were ESBL positive versus only 7.5% of isolates from the USA, 51.9% of *K. pneumoniae* from SA were ESBL versus 12.3% of the strains from NA, and 6.2% of *P. mirabilis* versus 3.9% in the USA.

Resistance rates to Cefotaxime and ciprofloxacin in nosocomial isolates of *E. coli* from some of the countries participants of the PAHO surveillance network are presented in Table 19.3 (PAHO 2006). Resistance to cefotaxime could be used as a marker for ESBL-producing organisms. Prevalence is higher in countries from Central America. Rates of resistance to ciprofloxacin are greater than the rate for cefotaxime, especially in Ecuador and Venezuela, where it does not appear linked to ESBL production. Data of *K. pneumoniae* from nosocomial isolates showed high prevalence of resistance to cefotaxime, most likely related to

Table 19.3 Resistance to ciprofloxacin and cefotaxime in nosocomial isolates of *Escherichia coli*, Latin America (PAHO Surveillance System 2004)

Country	# of isolates	% R CIPRO	% R CTX
Argentina	9276	15	9
Colombia	2056	30	ND
Costa Rica	1239	20	30
Ecuador	3105	42	6
Guatemala	1511	37	30
Paraguay	848	19	19
Venezuela	5768	34	8

Source: PAHO (2006).

the prevalence of ESBL-producing strains in Latin American countries. CTX-M are the most prevalent type of ESBL in Latin America, specially in the southern part (Quinteros et al. 2003). CTX-M 12 was first described in the region in Colombia, from a strain of *K. pneumoniae* (Villegas et al. 2004) (Table 19.3).

SHV type of ESBL was more prevalent (72%) than CTX-M type in 224 strains (Estrada-García et al. 2005) of Enterobacteriaceae isolated between 2001 and 2004 in eight hospitals in Caracas.

19.2.2.2 Fluoroquinolone Resistance

Resistance is primarily due to alterations in DNA gyrase and/or topoisomerase IV, to the activation of efflux pumps or to changes in porin expression. In clinical practice, this is most often exemplified by the need to advise antibiotic treatment for symptomatic urinary tract infections, likely due to *E. coli*, and having to resort to carbapenems – most commonly ertapenem – to treat those infections until the results of susceptibility studies permit de-escalation to oral treatment. This is especially common in patients who have received fluoroquinolones in the recent past. Non-fermenters also exhibit high levels of resistance to fluoroquinolones, and they are often multi-resistant. In *Pseudomonas*, for example, overexpression of efflux mechanisms seems to play a protagonic role, but these overly active pumps combine with other mechanisms such as DNA gyrase mutations and combined produce highly resistant strains.

19.2.2.3 Carbapenem Resistance

In South America, KPCs are the most frequent of the carbapenemases. One outbreak of *K. pneumoniae* producing KPC was reported from New York City (Bratu et al. 2005). Two isolates of *K. pneumoniae* identified in Medellin, Colombia, harbored KPC 2 and were highly resistant to the three carbapenems (Villegas et al. 2006). This was the first report of a KPC 2 from Latin America.

19.2.2.4 Enterobacter

Enterobacter cloacae and *E. aerogenes* are the most common *Enterobacter* species in clinical practice and in nosocomial infections. In humans, these micro-organisms cause health-care-related infections because of colonization in these settings, a pre-requisite for infection occurs more often than in the community. Emergence and spread of significant antibiotic resistance has been observed in all *Enterobacter* spp. The use of third-generation cephalosporins has resulted in the selection of strains with high-level AmpC phenotypes. Increasing resistant to fluoroquinolones is also a concern. Data from Table 19.5 reflect the high rate of resistance to antibiotics in nosocomial isolates of *E. cloacae*, reported in the PAHO surveillance system. The best option as of 2004 was carbapenems. The elevated proportion of resistance to cefepime could reflect the combination of the ampC β-lactamases with some of the classical

Table 19.4 Resistance in *Enterobacter cloacae*, Latin America, 2004

Country	No.	AMK (%)	CIP (%)	IPM (%)	FEP (%)	CAZ
Argentina	869	19	26	0.2	13	38%
Colombia	489	22	29	1	22	32%
Costa Rica	210	17	16	0	1	18%
Ecuador	352	25	29	1	19	35%
Guatemala	577	12	16	0	31	50%
Paraguay	151	13	41	0	39	43%
Venezuela	601	29	23	0	21	53%

AMK – amikacin; FEP – cefepime; CAZ – ceftazidime; CIP – ciprofloxacin; IPM – imipenem.
Source: Casellas and API Comité Resistencia Antibacterianos (2006).

ESBL. Data from the survey of API reflect that *more than one β-lactamase could be harbored in the same bacteria* (Casellas and API Comité Resistencia Antibacterianos 2006). Resistance to cefepime ranges between 6 and 32%, while resistance to ceftazidime and cefotaxime ranges between 15 and 67% (Table 19.4).

19.2.2.5 Acinetobacter

The majority of *Acinetobacter* infections is acquired in the hospital, primarily in intensive care settings and is caused by *A. baumannii*, a highly prevalent organism in the environment. Most strains are resistant to antimicrobial agents, and many to most agents.

19.2.2.6 Pseudomonas

Pseudomonas aeruginosa resistant to multiple drugs is a progressive growing problem. Most infections are acquired in the hospital and in intensive care units. Resistance is both intrinsic and acquired, the later escalating at alarming rates. Morbidity and mortality increase with resistance. Infections due to strains resistant to all commonly available antibiotics have been treated with more toxic drugs, but strains with decreased susceptibility even to polymyxins are appearing. *Pseudomonas aeruginosa* and *A. baumannii* are common agents in nosocomial infections in Latin America. The most recent problem has been increasing resistance to carbapenems, considered the last resort for many of these infections.

Data from Casellas in API survey number 7 are presented in Table 19.5. The first isolation of VIM2 was made from one *Pseudomonas fluorescens* from Chile and three strains of *P. aeruginosa* from Venezuela recovered in year 2002 as part of the SENTRY Antimicrobial Surveillance Program (Mendes et al. 2004).

Table 19.5 Resistance to carbapenems in *Pseudomonas aeruginosa* and *Acinetobacter baumannii*, Latin America, 2006

Country	PAE IMP (%)	PAE MERO (%)	ABA IMP (%)	ABA MERO (%)
Argentina	28	24	38	39
Ecuador	24	22	26	20
Paraguay	35	35	33	37
Venezuela	17	10	32	34

PAE – *P. aeruginosa*; ABA – *A. baumannii*; IMP – imipenem; MERO – meropenem.
Source: Casellas and API Comité Resistencia Antibacterianos (2006).

In a worldwide summary of metallo-β-lactamases identified in the Sentry program (Fritsche et al. 2005) reported IMP1 from *Acinetobacter* isolated in Argentina and Brazil, IMP 16 from Brazil, SPM 1 also from *P. aeruginosa* from Brazil and OXA 23 from *A. baumannii* isolated in Venezuela. A report from Colombia describe VIM.2 from *P. aeruginosa* from several cities and clonality was demonstrated in five of eight cities (Villegas et al. 2006). The same group described dissemination of *A. baumannii* clones with OXA 23 in Colombian hospitals) (Villegas et al. 2007).

19.3 Development and Use of New Antimicrobial Drugs in the Region

In the last decades, the Latin American region has been targeted by pharmaceutical industry as an area where good quality clinical research can be performed. Clinical investigators in the region have attained the highest standards of good clinical practices and research is less expensive. Therefore antibiotics such as linezolid and tigecycline have been developed with important participation of the region.

19.4 Summary

Resistance to antimicrobial agents is a growing problem in Latin America. Surveillance systems are in place, but need further development to base appropriate antibiotic selection on local epidemiological data.

References

Alvarez, E., M. Espino and R. Contreras (2006). "Determinacion de la susceptibilidad de *Escherichia coli* en aislamientos del tracto urinario por el sistema DIRAMIC." *Rev Panam infect* **8**: 10–15.

Andrade, S. S., H. S. Sader, R. N. Jones, A. S. Pereira, A. C. C. Pignatari and A. C. Gales (2006). "Increased resistance to first-line agents among bacterial pathogens isolated from urinary tract infections in Latin America: time for local guidelines?" *Memórias do Instituto Oswaldo Cruz* **101**: 741–748.

Bratu, S., D. Landman, R. Haag, R. Recco, A. Eramo, M. Alam and J. Quale (2005). "Rapid spread of carbapenem-resistant *Klebsiella pneumoniae* in New York City: a new threat to our antibiotic armamentarium." *Arch Intern Med* **165**(12): 1430–1435.

Casellas, J. and API Comité Resistencia Antibacterianos (2006). "Resultados Encuesta Número 7." *Revista Panamericana de Infectología* **8**(48–51).

Casellas, J., E. Nanini, M. Radice and E. Cocconi (2005). "Estudio de un brote debido a aislados de *Klebsiella pneumoniae* prodcutoras de betalactamasas de espectro expandido en un centro asistencial de Rosario, Argentina." *Rev Pan Infect* **7**: 21–27.

Casellas, J. M. and G. M. Goldberg (1989). "Incidence of strains producing extended spectrum beta-lactamases in Argentina." *Infection* **17**(6): 434–436.

Castanheira, M., A. Gales, A. Pignatari, R. Jones and H. Sader (2006). "Changing antimicrobial susceptibility patterns among *Streptococcus pneumoniae* and *Haemophilus influenzae* from Brazil: report from the SENTRY Antimicrobial Surveillance Program (1998–2004)." *Microbial Drug Resist* **12**(2): 91–98.

CDC (2003). "Methicillin-resistant *Staphylococcus aureus* infections among competitive sports participants – Colorado, Indiana, Pennsylvania, and Los Angeles County, 2000–2003." *MMWR Morb Mortal Wkly Rep* **52**(33): 793–795.

Cherian, B., N. Singh, W. Charles and P. Prabhakar (1999). "Extended-spectrum beta-lactamase-producing salmonella enteritidis in Trinidad and Tobago." *Emerg Infect Dis* **5**(1): 181–182.

Cruz, C., J. Moreno, A. Renzoni, M. Hidalgo, J. Reyes, J. Schrenzel, D. Lew, E. Castaneda and C. A. Arias (2005). "Tracking methicillin-resistant *Staphylococcus aureus* clones in Colombian hospitals over 7 years (1996–2003): emergence of a new dominant clone." *Int J Antimicrob Agents* **26**(6): 457–462.

Cuellar-Rodríguez, J., A. Galindo-Fraga, V. Guevara, z. C. Pérez-Jiméne, L. Espinosa-Aguilar, A. Rolón, A. Hernández-Cruz, E. López-Jácome, M. Bobadilla-del-Valle, A. Martínez-Gamboa, A. Ponce-de-León and S.-O. J. (2007). "Vancomycin-resistant enterococci, Mexico City." *Emerg Infect Dis* **13**(5): 798–799.

Da Silva Coimbra, M. V., L. A. Teixeira, R. L. Ramos, S. C. Predari, L. Castello, A. Famiglietti, C. Vay, L. Klan and A. M. Figueiredo (2000). "Spread of the Brazilian epidemic clone of a multiresistant MRSA in two cities in Argentina." *J Med Microbiol* **49**(2): 187–192.

Estrada-García, T., P.-G. Leova, J. F. Cerna, R. F. Velázquez, T. J. Ochoa, J. Torres and H. L. DuPont. (2005). "Drug-resistant diarrheogenic *Escherichia coli*, Mexico." *Emerg Infect Dis* **11**(8): 1306–1308.

Fritsche, T. R., H. S. Sader, M. A. Toleman, T. R. Walsh and R. N. Jones (2005). "Emerging metallo-β-lactamase-mediated resistances: a summary report from the worldwide SENTRY Antimicrobial Surveillance Program." *Clin Infect Dis* **41**(s4): S276–S278.

Gomes, A. R., I. S. Sanches, M. Aires de Sousa, E. Castaneda and H. de Lencastre (2001). "Molecular epidemiology of methicillin-resistant *Staphylococcus aureus* in Colombian hospitals: dominance of a single unique multidrug-resistant clone." *Microb Drug Resist* **7**(1): 23–32.

Gonzalez-Vertiz, A., D. Alcantar-Curiel, M. Cuauhtli, C. Daza, C. Gayosso, G. Solache, C. Horta, F. Mejia, J. I. Santos and C. Alpuche-Aranda (2001). "Multiresistant extended-spectrum β-lactamase-producing *Klebsiella pneumoniae* causing an outbreak of nosocomial bloodstream Infection." *Infect Control Hosp Epidemiol* **22**(11): 723–725.

Gutmann, L., B. Ferre, F. W. Goldstein, N. Rizk, E. Pinto-Schuster, J. F. Acar and E. Collatz (1989). "SHV-5, a novel SHV-type beta-lactamase that hydrolyzes broad-spectrum cephalosporins and monobactams." *Antimicrob Agents Chemother* **33**(6): 951–956.

Guzmán-Blanco, M., J. Casellas and H. Sader (2000). "Bacterial resistance to antimicrobial agents in Latin America. The giant is awakening." *Infect Dis Clin North Am* **14**(1): 67–81.

Hortal, M., R. Ruvinsky, A. Rossi, C. I. Agudelo, E. Castañeda, C. Brandileone, T. Camou, R. Palacio, G. Echaniz and J. L. D. Fabio (2000). "Impact of *Streptococcus pneumoniae* in pneumonias of Latin American children." *Revista Panamericana de Salud Pública/Pan Am J Public Health* **8**(3): 185–195.

Ibarra, J. O. and D. E. Alvarado (2007). "Antimicrobial resistance of clinical and environmental strains of *Vibrio cholerae* isolated in Lima-Peru during epidemics of 1991 and 1998." *Braz J Infect Dis* **11**: 100–105.

Kazakova, S. V., J. C. Hageman, M. Matava, A. Srinivasan, L. Phelan, B. Garfinkel, T. Boo, S. McAllister, J. Anderson, B. Jensen, D. Dodson, D. Lonsway, L. K. McDougal, M. Arduino, V. J. Fraser, G. Killgore, F. C. Tenover, S. Cody and D. B. Jernigan (2005). "A clone of methicillin-resistant *Staphylococcus aureus* among professional football players." *N Engl J Med* **352**(5): 468–475.

Kertesz, D., J. Di Fabio, M. Brandileone, E. Castaneda, G. Echaniz-Aviles, I. Heitmann, A. Homma, M. Hortal, M. Lovgren, R. Ruvinsky, J. Talbot, J. Weekes and J. Spika (1998). "Invasive *Streptococcus pneumoniae* Infection in Latin American Children: results of the Pan American Health Organization Surveillance Study." *Clin Infect Dis* **26**(6): 1355–1361.

Lovgren, M., J. A. Talbot, M. C. Brandileone, S. T. Casagrande, C. I. Agudelo, E. Castaneda, M. Regueira, A. Corso, I. Heitmann, A. Maldonado, G. Echaniz-Aviles, A. Soto-Nogueron, M. Hortal, T. Camou, J.-M. Gabastou, J. L. D. Fabio and the SIREVA Study Group (2007). "Evolution of an international external quality assurance model to support laboratory investigation of *Streptococcus pneumoniae*, developed for the SIREVA Project in Latin America, from 1993 to 2005." *J Clin Microbiol* **45**(10): 3184–3190.

Ma, X. X., A. Galiana, W. Pedreira, M. Mowszowicz, I. Christophersen, S. Machiavello, L. Lope, S. Benaderet, F. Buela, W. Vincentino, M. Albini, O. Bertaux, I. Constenla, H. Bagnulo, L. Liosa, T. Ito and K. Hiramatsu (2005). "Community-acquired methicillin-resistant *Staphylococcus aureus*, Uruguay." *Emerg Infect Dis* **11**(6): 973–976.

Melo, M., M. Silva-Carvalho, R. Ferreira, L. Coelho, R. Souza, C. Gobbi, R. Rozenbaum, C. Solari, B. Ferreira-Carvalho and A. Figueiredo (2004). "Detection and molecular characterization of a gentamicin-susceptible, methicillin-resistant *Staphylococcus aureus* (MRSA) clone in Rio de Janeiro that resembles the New York/Japanese clone." *J Hosp Infect* **58**(4): 276–285.

Mendes, C., M. E. Marin, F. Quiñones, J. Sifuentes-Osornio, C. Cuilty Siller, M. Castanheira, C. M. Zoccoli, H. López, A. Súcari, F. Rossi, G. Barriga Angulo, A. J. A. Segura, C. Starling, I. Mimica and D. Felmingham (2003). "Antibacterial resistance of community-acquired respiratory tract pathogens recovered from patients in Latin America: results from the PROTEKT surveillance study (1999–2000)." *Braz J Infect Dis* **7**: 44–61.

Mendes, R. E., M. Castanheira, P. Garcia, M. Guzman, M. A. Toleman, T. R. Walsh and R. N. Jones (2004). "First isolation of blaVIM-2 in Latin America: report from the SENTRY Antimicrobial Surveillance Program." *Antimicrob Agents Chemother* **48**(4): 1433–1434.

Moran, G. J., A. Krishnadasan, R. J. Gorwitz, G. E. Fosheim, L. K. McDougal, R. B. Carey, D. A. Talan and EMERGEncy ID Net Study Group (2006). "Methicillin-resistant S. aureus infections among patients in the emergency department." *N Engl J Med* **355**(7): 666–674.

Moszkowicz, M., W. Pedreira, A. Galiana and K. Hiramatsu (2004). "Efficacy of co-trimoxazole high dose short course in community-acquired methicillin resistant *Staphylococcus aureus* (CA-MRSA) outbreak in Uruguay jails 2003." *Int J Infect Dis* **8**(suppl): 185.

Otman, J., E. D. Cavassin, M. E. Perugini and M. C. Vidotto (2002). "An outbreak of extended-spectrum beta-lactamase-producing Klebsiella Species in a neonatal intensive care unit in Brazil." *Infect Control Hosp Epidemiol* **23**(1): 8–9.

PAHO (2006). Informe anual Red de Monitoreo/Vigilancia resistencia a los Antibióticos. Publicación Organización panamericana de la salud. . Washington, DC, Pan American Health Organization.

PROVENRA (1988–2006). Programa Venezolano de Vigilancia de Resistencia a los Antimicrobianos, Asociación Civil Programa Venezolano de Vigilancia de la Resistencia a los Antimicrobianos en Venezuela- 2006.

Puig, Y., M. Espino, T. Leyva, D. Martino, P. Mendez, P. Soto and Y. Ferrer (2007). "Susceptibilidad antimicrobiana en cepas de salmonella spp. de origen clinico y alimentario." *Rev Panam Infect* **9**(3): 12–16.

Quinteros, M., M. Radice, N. Gardella, M. M. Rodriguez, N. Costa, D. Korbenfeld, E. Couto and G. Gutkind (2003). "Extended-spectrum {beta}-lactamases in enterobacteriaceae in buenos aires, Argentina, Public Hospitals." *Antimicrob Agents Chemother* **47**(9): 2864–2867.

Ribeiro, A., C. Dias, M. C. Silva-Carvalho, L. Berquo, F. A. Ferreira, R. N. S. Santos, B. T. Ferreira-Carvalho and A. M. Figueiredo (2005). "First report of infection with community-acquired methicillin-resistant *Staphylococcus aureus* in South America." *J Clin Microbiol* **43**(4): 1985–1988.

Sader, H. S., A. C. Pignatari, R. J. Hollis and R. N. Jones (1994). "Evaluation of interhospital spread of methicillin-resistant *Staphylococcus aureus* in Sao Paulo, Brazil, using pulsed-field gel electrophoresis of chromosomal DNA." *Infect Control Hosp Epidemiol* **15**(5): 320–323.

Sader, H. and R. Jones (2004). "Gales A, Pignatary A and SENTRY participants group latin America: SENTRY antimicrobial surveilance program report. Latin America and brazilian results from 1997 through 2001." *Braz j Infect Dis* **8**:25–79.

Sanchez, J., J. Feris-Iglesias, J. Fernandez, E. Perez-Then, S. Ramirez, G. Ortega and L. Jimenez (2007). "Aislamiento de Klebsiella pneumnoiae productora de betalacatamsas de espectro extendido en recién nacidos en el hospital Roberto Reid Cabral de Santo Domingo, Republica Dominicana." *Rev Panam Infect* **7**(4): 15–20.

Sola, C., G. Gribaudo, A. Vindel, L. Patrito and J. L. Bocco (2002). "Identification of a novel methicillin-resistant Staphylococcus aureus epidemic clone in Cordoba, Argentina involved in nosocomial infections." *J Clin Microb* **40**:1427–1435.

Turner, P. J. (2005). "Extended-spectrum β-lactamases." *Clin Infect Dis* **41**(s4): S273–S275.

Villegas, M. V., A. Correa, F. Perez, T. Zuluaga, M. Radice, G. Gutkind, J. M. Casellas, J. Ayala, K. Lolans, J. P. Quinn and the Colombian Nosocomial Resistance Study Group (2004). "CTX-M-12 {beta}-lactamase in a *Klebsiella pneumoniae* clinical isolate in Colombia." *Antimicrob Agents Chemother* **48**(2): 629–631.

Villegas, M. V., J. N. Kattan, A. Correa, K. Lolans, A. M. Guzman, N. Woodford, D. Livermore, J. P. Quinn and the Colombian Nosocomial Bacterial Resistance Study Group (2007). "Dissemination of *Acinetobacter baumannii* clones with OXA-23 carbapenemase in Colombian Hospitals." *Antimicrob Agents Chemother* **51**(6): 2001–2004.

Villegas, M. V., K. Lolans, A. Correa, C. Jose Suarez, J. Lopez, M. Vallejo, J. Quinn and the Colombian Nosocomial Resistance Study Group (2006). "First detection of the plasmid-mediated class A carbapenemase KPC-2 in clinical isolates of *Klebsiella pneumoniae* from South America." *Antimicrob Agents Chemother* **50**(8): 2880–2882.

Whitney, C., M. Farley, J. Hadler, L. Harrison, N. Bennet, R. Lynfield, A. Reingold, D. Jackson, R. Faclan, J. Jorgensen and A. Schuchat (2003). "Decline in invasive pneumococcal diseases after the introduction of protein-polysaccharide conjugate vaccine." *N Eng J Med* **348**:1736–1746.

Chapter 20
Hospital Infections by Antimicrobial-Resistant Organisms in Developing Countries

Fatima Mir and Anita K.M. Zaidi

Abstract Antimicrobial resistance is a global public health concern. Hospitals in developing countries are fighting a progressively uphill battle against not only high burden of infectious diseases but also an ever-increasing proportion of multidrug-resistant pathogens. As many as 56% of *Staphylococcus aureus* isolates from hospital-acquired infections in South Asia have now become methicillin resistant. Vancomycin-resistant enterococci (VRE), with a reported prevalence of 6–12% in Asian hospital-based studies account for 10% of nosocomial urinary tract infections and a 31% mortality rate with VRE bacteremias. Third-generation cephalosporin-resistant *Escherichia coli* and *Klebsiella* account for case fatality rates between 12 and 52% in neonatal inpatients. Pan-resistance has been reported in 14–35.8% isolates of *Acinetobacter* spp. from developing country hospital inpatients over the last decade. *Pseudomonas* spp., an important neonatal killer in the Indo-Pak subcontinent, is ceftazidime resistant in 34–55% isolates and aminoglycoside resistant in 23–69% isolates. Alarmingly, these figures show a rising temporal trend and highlight antimicrobial-resistant organisms as major, untreatable threats in many resource-constrained countries. Antimicrobial resistance does not only mean increasing expense of treatment and poor clinical outcomes but also has a major impact on the way health systems are perceived and therefore accessed. There is an urgent need to devise comprehensive strategies and programs for improving hospital infection control and containing antimicrobial resistance in developing countries.

20.1 Introduction

Hospitals and health-care facilities, in developing countries, are hotbeds for evolution and transmission of multidrug-resistant (MDR) organisms. Reasons for easy DNA transfer in the nosocomial environment are manifold: high

F. Mir (✉)
Department of Paediatrics and Child Health, Aga Khan University, Karachi, Pakistan
e-mail: fatima.mir@aku.edu

A. de J. Sosa et al. (eds.), *Antimicrobial Resistance in Developing Countries*,
DOI 10.1007/978-0-387-89370-9_20, © Springer Science+Business Media, LLC 2010

antimicrobial selective pressure due to constant antibiotic use in a susceptible population cohort; spread from patient to patient and patient to personnel to patient when barrier precautions, hand washing, and equipment cleaning are not optimal (Harris et al. 2007; Kim et al. 2003); underdosing due to shortage of antibiotics (Safdar and Maki 2002); empiric prescribing due to lack of microbiological laboratory capacity; communication gap between laboratory personnel and clinicians (Tateda 2002; Archibald and Reller 2001; Urdea et al. 2006); and ultimately unaffordability of alternate agents. In resource-constrained settings, these risk factors are further compounded by deficient health systems, weak pharmaceutical drug management capacity, and scarce professional expertise (Boyce 2007; Nyamogoba and Obala 2002; Pearse 1997). Antimicrobial resistance advocacy and containment programs in developing country public hospitals are few and far between due to lack of visionaries and champions. In this scenario, poor legislation and quality check on counterfeit and substandard pharmaceutical products only accelerates the development of resistance (Gaudiano et al. 2007).

A hospital-acquired infection [nosokomeion (Greek): hospital] is an infection acquired during hospital stay which was not present or incubating at the time of the patient's admission. This includes infections acquired in the hospital but appearing after discharge, and also occupational infections among staff of the facility. The Center for Disease Control and Prevention (CDC) has attempted to define hospital-acquired infections for specific sites based on clinical and biological criteria and has provided standardized methods for collecting and comparing hospital infection rates (Jarvis 2003; Soleto et al. 2003).

20.2 Common Antimicrobial-Resistant Pathogens in Developing Countries

Hospitals especially intensive care areas tend to become colonized by highly resistant organisms capable of causing seriously untreatable infections. Antimicrobial-resistant organisms in hospital settings differ from those in the community. Agents causing nosocomial infections in developing world hospitals more or less follow developed world trends. Methicillin-resistant *S. aureus* (MRSA), a strain of *S. aureus* resistant to penicillinase-resistant penicillins like methicillin and nafcillin, is of primary importance. Resistance is conferred by acquisition of chromosomal DNA segment (mecA) which encodes a new penicillin-binding protein. MRSA infections are similar to those caused by methicillin-sensitive *S. aureus* (MSSA) affecting wound sites, lower respiratory and urinary tract, device-associated sites, septicemia, pressure sores, burns, and ulcers. Outbreaks have been reported commonly from high-risk units like ICU and burns units (Baddour et al. 2007; Ota et al. 2007; Perdelli et al. 2007; Szeto et al.2007; Oncul et al. 2002; Krishnan et al. 2002; Vidhani et al. 2001; Haddad et al. 1993). Risk factors are inadequate hand washing and isolation facilities

(Romero et al. 2006), patient overcrowding (Haddad et al. 1993), frequent patient and staff transfers, prolonged length of stay (LOS), debilitated patients, and multiple device sites. MRSA is reported now in various studies from Pakistan (Perwaiz et al. 2007), South Africa (Perovic et al. 2006), Malaysia (Norazah et al. 2003), Ethiopia (Geyid and Lemeneh 1991), Iraq, and Lagos (Kesah et al. 2003). Rates vary from less than 10% in Tunisia, Morocco, and Algeria to as high as 43% in Pakistan.

Coagulase-negative staphylococci (CONS), normal bacterial flora, are an important part of the nosocomial bug-brigade, mainly due to high antimicrobial selection. Infections caused by them are generally device associated and affect vulnerable populations like immunocompromised patients and recipients of prosthetic implants. Most nosocomial CONS are now only sensitive to vancomycin (de Allori et al. 2006). Africa and South Asia have high rates of *S. aureus* infections, whereas Latin America, Southeast Asia, and the Middle East have high reported rates of CONS infections. The preponderance of CONS might indicate a high rate of invasive device use.

Enterococcus spp. (normal bowel flora) have become important pathogens in the nosocomial environment due to acquisition of resistance to first- and second-line antibiotics. Vancomycin-resistant enterococci (VRE) are important causes of urinary tract, wound site, and endocardial (native and prosthetic valve) infections (Zarate et al. 2007; Yang et al. 2007; Zubaidah et al. 2006; Al-Otaibi et al. 2004). Resistance to vancomycin is acquired through a chromosomal DNA element (vanA) which is responsible for making the cell wall impermeable to vancomycin. Ever enterprising, *vanA* gene can also be transferred to *S. aureus* (d'Azevedo et al. 2006) thereby increasing MRSA rates too.

Enterobacteriaceae (*Klebsiella pneumoniae*, *E. coli*) are increasingly acquiring resistance against beta-lactam antibiotics (penicillins, cephalosporins) through production of extended-spectrum beta-lactamases (ESBLs) (Shehabi et al. 2004; Orrett 2005). Harris et al. (2007) contends that person to person spread is important in spread of ESBL-producing organisms. In Gujarat, India, 53% of *K. pneumoniae* and *E. coli* were found to be resistant to at least one of the three third-generation cephalosporins (Duttaroy and Mehta 2005). A comprehensive study from Delhi also correlated the resistance patterns of ESBL-producing *Klebsiella* spp. with increasing resistance to cefipime after its introduction into the Indian market (Grover et al. 2006). The main problem now in developing country hospitals is emerging resistance to carbapenems which will virtually make these pathogens untreatable (Mohanty et al. 2007).

Pseudomonas spp. continue to be important developing country nosocomial pathogens with widespread resistance to first-line ceftazidime (Kanafani et al. 2003; Qadri et al. 1994). An emerging nosocomial threat in the form of *Acinetobacter* spp. has been widely reported from developed countries in the 1990s and now in the 2000s is increasingly a developing world hospital pathogen (Alp et al. 2006; Prashanth and Badrinath 2005; Misbah et al. 2004; Jeena et al. 2001).

Our tertiary care hospital laboratory in Karachi maintains a database of antimicrobial-resistant isolates from inpatients as well as outpatients.

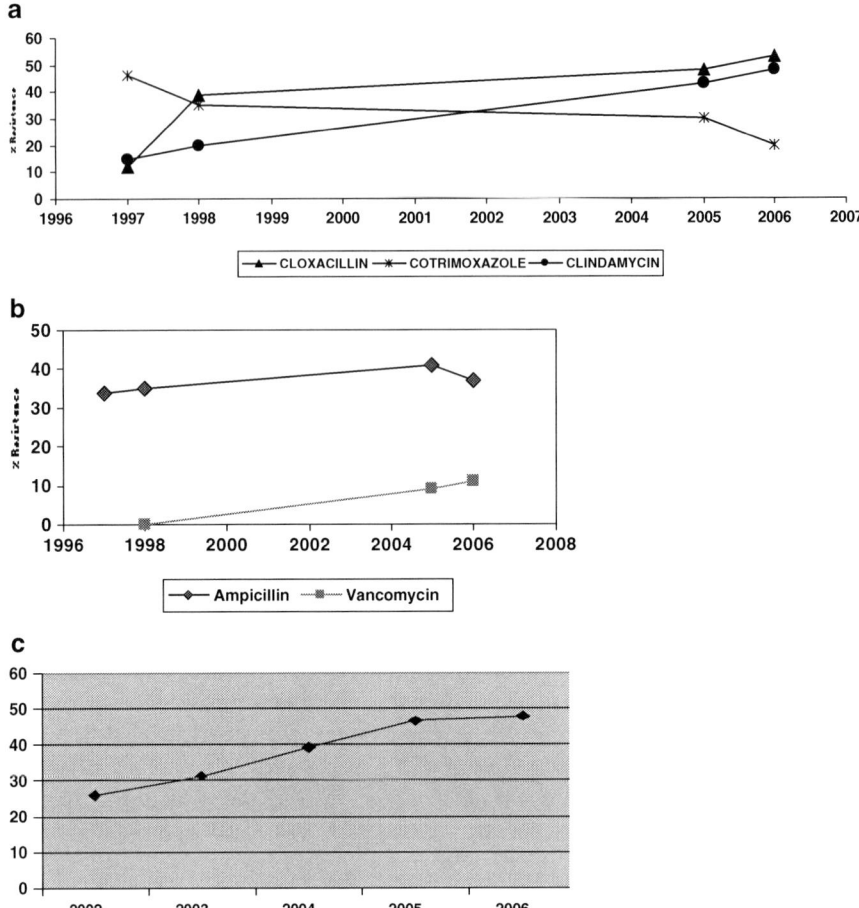

Fig. 20.1 Multidrug-resistant organisms: trends in inpatients 1996–2006, Aga Khan University Hospital, Karachi. (**a**) *Staphylococcus aureus* and methicillin-resistant *Staphylococcus aureus* (MRSA). (**b**)Antimicrobial-resistant enterococci. (**c**) Extended spectrum beta-lactamase producing Gram-negative organisms (Courtesy: Clinical Microbiology Laboratory, AKUH)

Figure 20.1(a–c) shows the temporal trend of MRSA, VRE, and ESBL-producing Gram-negative organisms as an indicator of the overall increase in antimicrobial resistance in the last decade.

20.3 Implications of Antimicrobial Resistance in Developing World Hospitals

In 1990, almost one-quarter of the 39.5 million deaths in developing countries were estimated to be due to infectious and parasitic diseases (Hart and Kariuki 1998). The developing world also contributed 98% to global child mortality, a major chunk due to infectious diseases. In this background of high infectious disease burden, rampant poverty and deprivation, the emergence of antimicrobial resistance in the community and hospitals of the developing world makes millennium development goals quite unattainable. The implications are manifold: for the individual, the hospital, and the public health system. For the patient, antimicrobial-resistant infections mean increased morbidity and mortality (vila-Figueroa et al. 1999; Murray 1992), financial drain (Yalcin 2003), and emotional trauma to self and dependents. For relatively well patients and hospital personnel, it means colonization and becoming unrecognized sources of spread within the hospital (Austin et al. 1999; Zavascki et al. 2006) and community. For the hospital administration, it is a clear indicator of poor administration and allocation of funds, poor infection control, poor quality assurance of antibiotics in use, and non-scientific approach to an essentially scientific problem (Ponce-de-Leon 1991). For the public health sector, it is an ethical and moral failure (Selgelid 2005; Dimond 2006). In developing countries where nosocomial infection rates are neither routinely assessed nor advertised, patients bear all or most hospital costs from their private funds and do not have a strong lobby to advocate their rights especially where nosocomial conditions prolong hospital stay and cost (Vincent et al. 2005). This lifts the check and balance of litigation from hospital administrations and orphans the cause of hospital infection control. Developing country hospital administrations must take primary responsibility and back their Infection Control Committees to curtail hospital infection rates and subsequent patient care charges. Cause champions in isolation cannot do much.

20.4 Antimicrobial-Resistant Infection Rates in Developing Country Hospitals

Most burden reviews from the developing world assess antimicrobial resistance in the community. Prevalence studies of drug-resistant nosocomial infections are generally single centre based. Many quality hospital-acquired infection burden estimates come from Latin America where nosocomial infection rates range from 10 to 26% with a severe impact on morbidity, mortality, and finances (Ribas et al. 2007; Rosenthal et al. 2003, 2006). Overall infection rates (number of nosocomial infections identified in a given time frame divided by total patients admitted or discharged during that time), however, are difficult to interpret! These rates treat each type of infection as equally important are not adjusted for length of stay, risk of infection, or predisposition to infection (device use, etc). They cannot be compared from one hospital to the other due to non-standardized health care in developing countries (Pottinger et al. 1998). Emphasis now is on adjusted rates for length of stay (per patient days), exposure to devices (catheter days or ventilator days), and severity of illness (risk stratification). A multicenter prospective cohort surveillance of device-associated infections which compared data from 55 ICUs of 46 hospitals in Argentina, Brazil, Colombia, India, Mexico, Morocco, Peru, and Turkey with the US National Nosocomial Infection

Surveillance (NNIS) report showed rates in developing countries to be at least threefold higher with high incidence of antimicrobial resistance (Rosenthal et al. 2006). A prevalence survey in 55 hospitals across four WHO regions showed the highest frequencies of nosocomial infections reported from hospitals in the Eastern Mediterranean and Southeast Asia regions (11.8 and 10% respectively) (Mayon-White et al. 1988). Studies from Morocco (Ramdani-Bouguessa et al. 2006; Jroundi et al. 2007), Saudi Arabia (Balkhy et al. 2006), Pakistan (Naeem et al. 2006), India, and Turkey (Meric et al. 2005) show a much higher hospital-acquired infection prevalence (17–25%). Commonest infection agents reported were *S. aureus* and Gram negatives and commonest sites were urinary tract and surgical sites. *Acinetobacter* spp. though an emerging ICU colonizer in the developing world was not very widely reported except from South Africa, Turkey, and Chile (Jeena et al. 2001; Misbah et al. 2004; Rosenthal et al. 2006). Bello et al. (2000) described high levels of carbapenem resistance and mortality.

20.5 Antimicrobial-Resistant Neonatal Infections in Developing Countries

In a comprehensive review of neonatal infections from 1990 to 2004, Zaidi et al. (2005) examined data from developing countries on rates of infections among hospital-born neonates, range of pathogens, antimicrobial resistance, and infection-control interventions. Reported rates of neonatal sepsis varied from 6.5 to 38 per 1000 live hospital-born babies (Kaushik et al. 1998; Nathoo et al. 1990). Rates of bloodstream infections ranged from 1.7 to 33 per 1000 live births. These bloodstream infection rates are 3–20 times higher than those reported for hospital-born babies in industrialized countries and are likely an underestimate of neonatal nosocomial infectious disease burden due to variable sensitivity and rigor of surveillance methods, lack of post discharge monitoring, and inadequate laboratory resources and capability.

About half of early-onset neonatal bloodstream infections in developing regions are due to *Klebsiella* spp., *E. coli*, *Pseudomonas* spp., and *Acinetobacter* spp. that cause common-source outbreaks because they thrive in multi-use containers of medications, liquid soaps and other solutions, including antiseptics, disinfectants, and inadequately reprocessed equipment (Rhinehart et al. 1991; Zaidi et al. 1995; Macias et al. 2004). Almost 70% of *K. pneumoniae*, other Gram-negative rods and *S. aureus*, the major pathogens in developing world neonates would not be covered by an empiric regimen of ampicillin and gentamicin (standard first-line cover in developing country nurseries and NICUs), and many might be untreatable in resource- constrained environments (Zaidi et al. 2005; WHO 2000). Hospital-acquired infections due to antimicrobial-resistant organisms are therefore sabotaging the expectation of improved neonatal outcomes (due to greater availability of specialized neonatal care) with their associated morbidity, mortality, and cost (Tables 20.1 and 20.2).

Table 20.1 Etiology of neonatal bloodstream infections reported from hospitals in developing countries

Pathogen	Africa		Southeast Asia		South Asia		Latin America and Caribbean		Middle East and Central Asia		All developing regions	
	n	%	n	%	n	%	n	%	n	%	N	%
All Gram positives	593	38.3	926	41.1	1857	31	333	38.3	87	32.8	3796	34.8
Staphylococcus aureus	218	14.1	181	8	1206	20.2	122	14.0	47	17.7	1774	16.2
Coagulase-negative *staphylococci*	117	7.6	621	27.5	356	5.9	152	17.5	33	12.5	1279	11.7
Group **B** Streptococci	133	8.6	43	1.9	31	0.5	28	3.2	4	1.5	239	2.2
All Gram negatives	935	60.4	1262	56	3793	63.4	524	60.3	178	67.2	6692	61.3
Klebsiella species	440	28.4	435	19.3	1450	24.2	163	18.8	68	25.7	2556	23.4
Escherichia coli	155	10	108	4.8	984	16.4	97	11.2	18	6.8	1362	12.5
Pseudomonas species	51	3.3	158	7.0	576	9.6	81	9.3	20	7.5	886	8.1
Acinetobacter species	4	0.3	290	12.9	251	4.2	15	1.7	5	1.9	565	5.2
Other Gram negatives	136	8.8	153	6.8	90	1.5	1	0.1	26	9.8	406	3.7
Candida species	5	0.3	33	1.5	170	2.8	12	1.4			220	2.0
Other pathogens	14	0.9	34	1.5	164	2.7					212	1.9
Total	1547	100	2255	100	5984	100	869	100	265	100	10920	100

Reprinted with permission from Elsevier (The Lancet, 2005, Vol. 365, 1175–1187)

Table 20.2 Antimicrobial resistance among *Escherichia coli*, *Klebsiella* and *Staphylococcus aureus* causing neonatal infections in hospitals of developing countries

Pathogen	Antibiotic	Africa		Latin America/Caribbean		Middle East/Central Asia		SouthAsia		SouthEast Asia		Total	
		Number tested	%	Number tested	%	Number tested	%	Number tested	%	Number tested	%	Number tested	%
Klebsiella species	Ampicillin	50	96			99	100	508	81	42	98	699	86
	Gentamicin	88	59	28	11	79	62	1013	74	42	86	1250	71
	Cefotaxime*	81	30	28	36	101	36	965	54	42	86	1217	52
	Amikacin	50	14	32	0	91	44	923	38	42	71	1138	37
Escherichia coli	Ampicillin	16	88	11	82	4	75	300	82	9	78	340	82
	Gentamicin	16	6	11	55	8	25	328	55	9	56	372	52
	Cefotaxime*	17	12	11	36	3	0	385	51	9	0	425	48
	Amikacin	10	0	12	0	2	0	391	27	9	56	424	26
Staphylococcus aureus	Methicillin†	67	27	183	16	27	4	445	56	119	28	841	39
	Cotrimoxazole	21	62	46	13			248	73			315	64

* Or other third-generation cephalosporins.

† Or first-generation cephalosporins, ceftriaxone, or cefotaxime.

Reprinted with permission from Elsevier (The Lancet, 2005, Vol. 365, 1175–1187.

20.6 Economic Implications of Antimicrobial Resistance

Cost analyses from many centers emphasize how nosocomial infections drain resources, public, and private. Rosenthal et al. (2006) analyzed ventilator-associated pneumonia (VAP) patients in Argentinian hospitals and reported mean extra length of stay (LOS) for cases (compared with controls) of 8.95 days, mean extra antibiotic cost $996 and mean extra total cost of $2255. Numbers like these in countries where per capita income is $500–1500, lead to a vicious circle of non-compliance, poor clinical outcomes, financial, and psychosocial ruin with propagation of antimicrobial resistance from the hospital into the community.

Bhutta et al. (2007, New born and Young Infant sepsis and antimicrobial resistance: burden and implications, unpublished) analyzed expected financial burden of nosocomial drug-resistant infections in neonates extrapolating that a quarter of the 1.44 million newborn deaths (36% of the global burden) and 600,000 young infant deaths each year in South Asia could be attributed to antimicrobial resistance with a cost estimate in excess of $120 million. Pearse (1997) makes a similar cost estimate exceeding US $110 million where hospital infection rates are as high as 15% and each patient requires seven extra days of hospital stay. Both figures make a strong case for hospital infection control in developing countries.

20.7 Antimicrobial Resistance Containment Strategies

A true state of public hospitals in the developing world can be garnered by an audit conducted by Baqi et al. (2007, Infection Control at a Public Sector Hospital in Karachi, Pakistan, unpublished) in 13 units of a large public sector hospital in Karachi. Factors promoting infection ranged from individual omissions like poor hand washing, minimal occupational safety practices, suboptimal disinfection of instruments between patients, and improper device care to more institutional lapses like absence of infection control policy on paper and in practice, and non-existent infection control training of clinical and non-clinical staff. Most of these lapses can be met with cost-effective measures but require championing by dedicated individuals and hospital administrations.

Hospital Infection Control Programs initiated in the 1950s in the developed world have evolved over the subsequent 50 odd years into multi-disciplinary forums. They require cohesion and co-operation among key players like clinicians, microbiologists, pharmacists, nursing staff, and non-clinical services like central sterilization, food, laundry, housekeeping, and maintenance. Team effort in a resource-constrained and politicized public sector hospital setting has to stem from consistent practical steps and the rationale of "think big, start small". Infection control in developing countries, therefore, must choose the common sense of the doable (hand hygiene, aseptic procedures, staff awareness messages, appropriate device care, and transmission-based precautions) over the intractable (hiring of more specialized manpower, pharmaceutical quality control and restriction, sophisticated ICUs, and greater fund allocation toward infection control). There are few reports of successful hospital-based infection control programs in developing countries. However, where the impact of some fledgling programs has been described (Hanssens et al. 2005; Ejilemele and Ojule 2005; Lopes et al. 1999), others are increasingly analytical and "can do" particularly in Latin America, Asia, and South Africa (Leblebicioglu et al. 2007; Rosenthal et al. 2006; Macias and Ponce-de-Leon 2005). Figure 20.2(a and b) shows the positive impact of antibiotic restriction in an intensive care (ICU) of a tertiary care hospital in Karachi, Pakistan as well as overall improvement in compliance with rational antibiotic use.

Ironically, the key to containing AMR is improving access to drugs – the right ones. To this end, WHO has developed a Global Strategy for Containment of Antimicrobial Resistance and an Essential Medicines List which tabulates critical drugs required to treat specific complaints. Analyses show that in countries with essential drug policies (Wertheimer and Santella 2007), individuals have greater access to the drugs they really need, yet resort to fewer injections and antimicrobial prescriptions when confronted with possible infections (Hogerzeil et al. 1993).

a

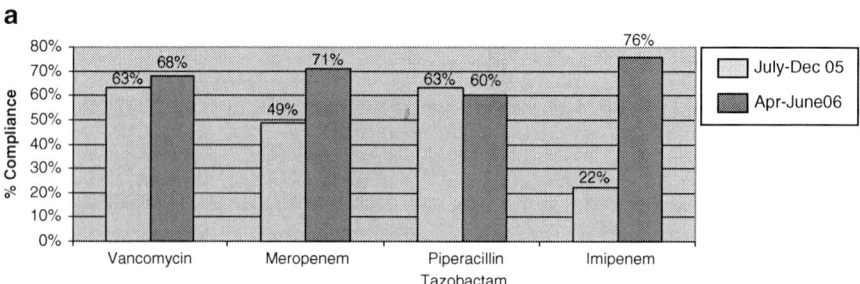

b

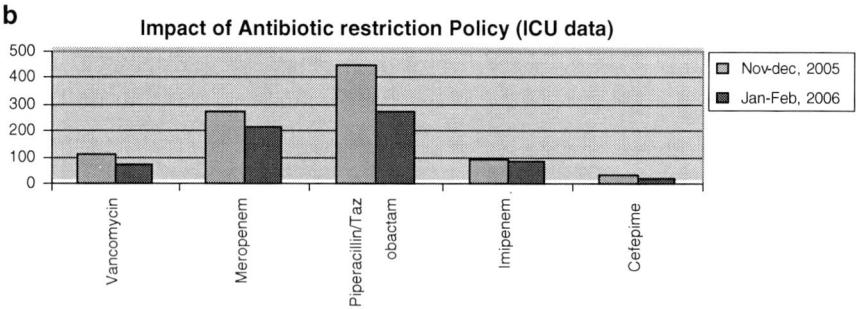

Fig. 20.2 Antibiotic Utilization Review: Nov–Dec, 2005 to Jan–Feb, 2006 at Aga Khan University Hospital (AKUH) Karachi. (**a**) Overall Compliance with antibiotic restriction policy. (**b**) Impact on Daily Refined Doses of Antibiotics per 1000 patient-days in ICU post-antibiotic restriction policy implementation (Courtesy: Salwa Ahsan, Senior Pharmacist, Aga Khan University Hospital Pharmacy Pilot project)

Studies from Ethiopia (Desta et al. 1997), Niger (Mallet et al. 2001), Bolivia (Bartoloni et al. 1998), Zambia (Bexell et al. 1996), and India (Dineshkumar et al. 1995) show increasing awareness and commitment of institutions toward controlling and quantifying their antimicrobial resistance burden. An important step toward uniting the antimicrobial taskforces in different countries is the Alliance for Prudent Use of Antibiotics (APUA) which aims at creating a common forum for infection control surveillance teams the world over and ensures that uniform messages of containment are disseminated to check spread of resistant organisms from the hospitals into communities.

Collaboration with global experts and funding through public–private sector partnerships has also empowered developing countries to launch programs which engender and strengthen capacity for surveillance, containment, and diagnosis of antimicrobial resistance. Country-based networks like the "Pakistan Antimicrobial Resistance Network" (PARN) www.parn.org.pk are results of such collaborations and create platforms for clinicians and technicians to gather and disseminate information on infection control and antimicrobial resistance burden in hospitals. Health professionals can improve empiric prescriptions through accessing local hospital antibiograms and learn about standard operating procedures for performing antimicrobial sensitivities.

Periodic audits are the key in developing countries to assess where we are going wrong (Harvey et al. 2004). Simple interventional studies will help in finding customized solutions in resource-poor settings; case in point, a study in Panama showed lower resistance rates and 50% reduction in antimicrobial expenditures after initiation of a systematic program to discontinue empiric antimicrobial treatment for early-onset infection after 3 days if infants were doing well, cultures were negative, and laboratory markers of infection were normal (Saez-Llorens et al. 2000). Similarly, Brusaferro et al. (2006) found 100% sensitivity of capture systems based on biweekly prevalence studies where patients were selected on basis of fever, antibiotic use and devices provided 62% population was covered. Each developing country hospital must therefore analyze the indicators which give maximum information for reform and customize their program to local conditions. Tools to assist in design and implementation of "bundled" infection-control interventions in resource-limited settings and guidelines for prevention of healthcare-associated infections are available in printed manuals and online (Panknin 2006; Coen 2007; Torres 2006).

In summary, hospitals in the developing world, need to air and share their experiences and resources in order to maintain a consistent, rational, evidence-based, and cost-effective offensive in the fight against increasingly antimicrobial-resistant pathogens.

Acknowledgment Dr. Fatima Mir received research training support from the National Institute of Health's Fogarty International Center (1 D43 TW007585-01) which made this work possible.

References

Al-Otaibi, F.E., Kambal, A.M., and Baabbad, R.A. 2004. Enterococcal bacteremia in a teaching hospital in the Central region of Saudi Arabia. Saudi. Med. J. 25: 21–25.

Alp, E., Esel, D., Yildiz, O., Voss, A., Melchers, W., and Doganay, M. 2006. Genotypic analysis of Acinetobacter bloodstream infection isolates in a Turkish university hospital. Scand. J. Infect. Dis. 38: 335–340.

Archibald, L.K. and Reller, L.B. 2001. Clinical microbiology in developing countries. Emerg. Infect. Dis. 7: 302–305.

Austin, D.J., Bonten, M.J., Weinstein, R.A., Slaughter, S., and Anderson, R.M. 1999. Vancomycin-resistant enterococci in intensive-care hospital settings: transmission dynamics, persistence, and the impact of infection control programs. Proc. Natl. Acad. Sci. U.S.A 96: 6908–6913.

Baddour, M.M., Abuelkheir, M.M., Fatani, A.J., Bohol, M.F., and Al-Ahdal, M.N. 2007. Molecular epidemiology of methicillin-resistant *Staphylococcus aureus* (MRSA) isolates from major hospitals in Riyadh, Saudi Arabia. Can. J. Microbiol. 53: 931–936.

Balkhy, H.H., Cunningham, G., Chew, F.K., Francis, C., Al Nakhli, D.J., Almuneef, M.A., and Memish, Z.A. 2006. Hospital- and community-acquired infections: a point prevalence and risk factors survey in a tertiary care center in Saudi Arabia. Int. J. Infect. Dis. 10: 326–333.

Bartoloni, A., Cutts, F., Leoni, S., Austin, C.C., Mantella, A., Guglielmetti, P., Roselli, M., Salazar, E., and Paradisi, F. 1998. Patterns of antimicrobial use and antimicrobial resistance among healthy children in Bolivia. Trop. Med. Int. Health 3: 116–123.

Bello, H., Dominguez, M., Gonzalez, G., Zemelman, R., Mella, S., Young, H.K., and Amyes, S.G. 2000. In vitro activities of ampicillin, sulbactam and a combination of ampicillin and sulbactam against isolates of Acinetobacter calcoaceticus-Acinetobacter baumannii complex isolated in Chile between 1990 and 1998. J. Antimicrob. Chemother. 45: 712–713.

Bexell, A., Lwando, E., von, H.B., Tembo, S., Eriksson, B., and Diwan, V.K. 1996. Improving drug use through continuing education: a randomized controlled trial in Zambia. J. Clin. Epidemiol. 49: 355–357.

Boyce, J.M. 2007. Environmental contamination makes an important contribution to hospital infection. J. Hosp. Infect. 65 Suppl 2: 50–54.

Brusaferro, S., Regattin, L., Faruzzo, A., Grasso, A., Basile, M., Calligaris, L., Scudeller, L., and Viale, P. 2006. Surveillance of hospital-acquired infections: a model for settings with resource constraints. Am. J. Infect. Control 34: 362–366.

Coen, P.G. 2007. How mathematical models have helped to improve understanding the epidemiology of infection. Early Hum. Dev. 83: 141–148.

d'Azevedo, P.A., Dias, C.A., and Teixeira, L.M. 2006. Genetic diversity and antimicrobial resistance of enterococcal isolates from Southern region of Brazil. Rev. Inst. Med. Trop. Sao Paulo 48: 11–16.

de Allori, M.C., Jure, M.A., Romero, C., and de Castillo, M.E. 2006. Antimicrobial resistance and production of biofilms in clinical isolates of coagulase-negative Staphylococcus strains. Biol. Pharm. Bull. 29: 1592–1596.

Desta, Z., Abula, T., Beyene, L., Fantahun, M., Yohannes, A.G., and Ayalew, S. 1997. Assessment of rational drug use and prescribing in primary health care facilities in north west Ethiopia. East Afr. Med. J. 74: 758–763.

Dimond, B. 2006. Infection control: the rights of the patient. Br. J. Nurs. 15: 670–671.

Dineshkumar, B., Raghuram, T.C., Radhaiah, G., and Krishnaswamy, K. 1995. Profile of drug use in urban and rural India. Pharmacoeconomics 7: 332–346.

Duttaroy, B. and Mehta, S. 2005. Extended spectrum b lactamases (ESBL) in clinical isolates of Klebsiella pneumoniae and Escherichia coli. Indian J. Pathol. Microbiol. 48: 45–48.

Ejilemele, A.A. and Ojule, A.C. 2005. Knowledge, attitude and practice of aspects of laboratory safety in Pathology Laboratories at the University of Port Harcourt Teaching Hospital, Nigeria. Niger. J. Clin. Pract. 8: 102–106.

Gaudiano, M.C., Di, M.A., Cocchieri, E., Antoniella, E., Bertocchi, P., Alimonti, S., and Valvo, L. 2007. Medicines informal market in Congo, Burundi and Angola: counterfeit and sub-standard antimalarials. Malar. J. 6: 22.

Geyid, A. and Lemeneh, Y. 1991. The incidence of methicillin resistant *S. aureus* strains in clinical specimens in relation to their beta-lactamase producing and multiple-drug resistance properties in Addis Abeba. Ethiop. Med. J. 29: 149–161.

Grover, S.S., Sharma, M., Chattopadhya, D., Kapoor, H., Pasha, S.T., and Singh, G. 2006. Phenotypic and genotypic detection of ESBL mediated cephalosporin resistance in Klebsiella pneumoniae: emergence of high resistance against cefepime, the fourth generation cephalosporin. J. Infect. 53: 279–288.

Haddad, Q., Sobayo, E.I., Basit, O.B., and Rotimi, V.O. 1993. Outbreak of methicillin-resistant *Staphylococcus aureus* in a neonatal intensive care unit. J. Hosp. Infect. 23: 211–222.

Hanssens, Y., Ismaeil, B.B., Kamha, A.A., Elshafie, S.S., Adheir, F.S., Saleh, T.M., and Deleu, D. 2005. Antibiotic prescribing pattern in a medical intensive care unit in Qatar. Saudi. Med. J. 26: 1269–1276.

Harris, A.D., Perencevich, E.N., Johnson, J.K., Paterson, D.L., Morris, J.G., Strauss, S.M., and Johnson, J.A. 2007. Patient-to-patient transmission is important in extended-spectrum beta-lactamase-producing Klebsiella pneumoniae acquisition. Clin. Infect. Dis. 45: 1347–1350.

Hart, C.A. and Kariuki, S. 1998. Antimicrobial resistance in developing countries. BMJ 317: 647–650.

Harvey, K., Kalanj, K., and Stevanovic, R. 2004. Croatian pharmaceutical sector reform project: rational drug use. Croat. Med. J. 45: 611–619.

Hogerzeil, H.V., Bimo, Ross-Degnan, D., Laing, R.O., Ofori-Adjei, D., Santoso, B., zad Chowdhury, A.K., Das, A.M., Kafle, K.K., Mabadeje, A.F., et al.. 1993. Field tests for rational drug use in twelve developing countries. Lancet 342: 1408–1410.

Jarvis, W.R. 2003. Benchmarking for prevention: the Centers for Disease Control and Prevention's National Nosocomial Infections Surveillance (NNIS) system experience. Infection 31 Suppl 2: 44–48.

Jeena, P., Thompson, E., Nchabeleng, M., and Sturm, A. 2001. Emergence of multi-drug-resistant Acinetobacter anitratus species in neonatal and paediatric intensive care units in a developing country: concern about antimicrobial policies. Ann. Trop. Paediatr. 21: 245–251.

Jroundi, I., Khoudri, I., Azzouzi, A., Zeggwagh, A.A., Benbrahim, N.F., Hassouni, F., Oualine, M., and Abouqal, R. 2007. Prevalence of hospital-acquired infection in a Moroccan university hospital. Am. J. Infect. Control 35: 412–416.

Kanafani, Z.A., Kara, L., Hayek, S., and Kanj, S.S. 2003. Ventilator-associated pneumonia at a tertiary-care center in a developing country: incidence, microbiology, and susceptibility patterns of isolated microorganisms. Infect. Control Hosp. Epidemiol. 24: 864–869.

Kaushik, S.L., Parmar, V.R., Grover, N., Grover, P.S., and Kaushik, R. 1998. Neonatal sepsis in hospital born babies. J. Commun. Dis. 30: 147–152.

Kesah, C., Ben, R.S., Odugbemi, T.O., Boye, C.S., Dosso, M., Ndinya Achola, J.O., Koulla-Shiro, S., Benbachir, M., Rahal, K., and Borg, M. 2003. Prevalence of methicillin-resistant *Staphylococcus aureus* in eight African hospitals and Malta. Clin. Microbiol. Infect. 9: 153–156.

Kim, P.W., Roghmann, M.C., Perencevich, E.N., and Harris, A.D. 2003. Rates of hand disinfection associated with glove use, patient isolation, and changes between exposure to various body sites. Am. J. Infect. Control 31: 97–103.

Krishnan, P.U., Mahalakshmi, P., and Shetty, N. 2002. Strain relatedness of endemic MRSA isolates in a burns unit in South India–a five year study. J.Hosp.Infect. 52: 181–184.

Leblebicioglu, H., Rosenthal, V.D., Arikan, O.A., Ozgultekin, A., Yalcin, A.N., Koksal, I., Usluer, G., Sardan, Y.C., and Ulusoy, S. 2007. Device-associated hospital-acquired infection rates in Turkish intensive care units. Findings of the International Nosocomial Infection Control Consortium (INICC). J. Hosp. Infect. 65: 251–257.

Lopes, J.M., Starling, C.E., Lessa, C., and Couto, B.R. 1999. Joint effort to improve quality in a Brazilian pediatric public hospital through cross-infection control. J. Pediatr. (Rio J.) 75: 361–366.

Macias, A.E., Munoz, J.M., Herrera, L.E., Medina, H., Hernandez, I., Alcantar, D., and Ponce de, L.S. 2004. Nosocomial pediatric bacteremia: the role of intravenous set contamination in developing countries. Infect. Control Hosp. Epidemiol. 25: 226–230.

Macias, A.E. and Ponce-de-Leon, S. 2005. Infection control: old problems and new challenges. Arch. Med. Res. 36: 637–645.

Mallet, H.P., Njikam, A., and Scouflaire, S.M. 2001. Evaluation of prescription practices and of the rational use of medicines in Niger. Sante 11: 185–193.

Mayon-White, R.T., Ducel, G., Kereselidze, T., and Tikomirov, E. 1988. An international survey of the prevalence of hospital-acquired infection. J. Hosp. Infect. 11 Suppl A: 43–48.

Meric, M., Willke, A., Caglayan, C., and Toker, K. 2005. Intensive care unit-acquired infections: incidence, risk factors and associated mortality in a Turkish university hospital. Jpn. J. Infect. Dis. 58: 297–302.

Misbah, S., AbuBakar, S., Hassan, H., Hanifah, Y.A., and Yusof, M.Y. 2004. Antibiotic susceptibility and REP-PCR fingerprints of Acinetobacter spp. isolated from a hospital ten years apart. J. Hosp. Infect. 58: 254–261.

Mohanty, S., Singhal, R., Sood, S., Dhawan, B., Kapil, A., and Das, B.K. 2007. Citrobacter infections in a tertiary care hospital in Northern India. J. Infect. 54: 58–64.

Murray, B.E. 1992. Problems and dilemmas of antimicrobial resistance. Pharmacotherapy 12: 86S–93S.

Naeem, I., Naqvi, B.S., Hashmi, K., and Gauhar, S. 2006. Paediatric nosocomial infections: resistance pattern of clinical isolates. Pak. J. Pharm. Sci. 19: 52–57.

Nathoo, K.J., Mason, P.R., and Chimbira, T.H. 1990. Neonatal septicaemia in Harare Hospital: aetiology and risk factors. The Puerperal Sepsis Study Group. Cent. Afr. J. Med. 36: 150–156.

Norazah, A., Lim, V.K., Rohani, M.Y., Alfizah, H., Koh, Y.T., and Kamel, A.G. 2003. A major methicillin-resistant *Staphylococcus aureus* clone predominates in Malaysian hospitals. Epidemiol. Infect. 130: 407–411.

Nyamogoba, H. and Obala, A.A. 2002. Nosocomial infections in developing countries: cost effective control and prevention. East Afr. Med. J. 79: 435–441.

Oncul, O., Yuksel, F., Altunay, H., Acikel, C., Celikoz, B., and Cavuslu, S. 2002. The evaluation of nosocomial infection during 1-year-period in the burn unit of a training hospital in Istanbul, Turkey. Burns 28: 738–744.

Orrett, F.A. 2005. Clinical sources and prevalence of resistance to antimicrobials of Klebsiella pneumoniae strains in Trinidad. Saudi. Med. J. 26: 1766–1770.

Ota, R., Takahashi, C., Shiraishi, T., and Tominaga, M. 2007. Genesis of methicillin-resistant *Staphylococcus aureus* and selective antibacterial injection pressure. Kansenshogaku Zasshi 81: 370–378.

Panknin, H.T. 2006. Prevention of ventilator-associated pneumonia: review of national and international guidelines. Pflege. Z. 59: suppl-8.

Pearse, J. 1997. Infection control in Africa. Nosocomial infection. Afr. Health 19: 10–11.

Perdelli, F., Dallera, M., Cristina, M.L., Sartini, M., Ottria, G., Spagnolo, A.M., and Orlando, P. 2007. A new microbiological problem in intensive care units: environmental contamination by MRSA with reduced susceptibility to glycopeptides. Int. J. Hyg. Environ. Health 211(1–2):213–218.

Perovic, O., Koornhof, H., Black, V., Moodley, I., Duse, A., and Galpin, J. 2006. *Staphylococcus aureus* bacteraemia at two academic hospitals in Johannesburg. S. Afr. Med. J. 96: 714–717.

Perwaiz, S., Barakzi, Q., Farooqi, B.J., Khursheed, N., and Sabir, N. 2007. Antimicrobial susceptibility pattern of clinical isolates of methicillin resistant *Staphylococcus aureus*. J. Pak. Med. Assoc. 57: 2–4.

Ponce-de-Leon, S. 1991. The needs of developing countries and the resources required. J. Hosp. Infect. 18 Suppl A: 376–381.

Pottinger J.M., Herwaldt L.A., and Perl T.M. 1998. Basics of Surveillance-an overview. In A Practical Handbook for Hospital Epidemiologists. The Society for Healthcare Epidemiology of America. SLACK Incorporated and SHEA.

Prashanth, K. and Badrinath, S. 2005. Epidemiological investigation of nosocomial Acinetobacter infections using arbitrarily primed PCR & pulse field gel electrophoresis. Indian J. Med. Res. 122: 408–418.

Qadri, S.M., Huber, T.W., Lee, G.C., and al-Hajjar, S. 1994. Antimicrobial resistance of bacterial pathogens at two tertiary-care centers, in Riyadh and Texas. Tex. Med. 90: 59–62.

Ramdani-Bouguessa, N., Bes, M., Meugnier, H., Forey, F., Reverdy, M.E., Lina, G., Vandenesch, F., Tazir, M., and Etienne, J. 2006. Detection of methicillin-resistant *Staphylococcus aureus* strains resistant to multiple antibiotics and carrying the Panton-Valentine leukocidin genes in an Algiers hospital. Antimicrob. Agents Chemother. 50: 1083–1085.

Rhinehart, E., Goldmann, D.A., and O'Rourke, E.J. 1991. Adaptation of the Centers for Disease Control guidelines for the prevention of nosocomial infection in a pediatric intensive care unit in Jakarta, Indonesia. Am. J. Med. 91: 213S–220S.

Ribas, R.M., Freitas, C., and Gontijo Filho, P.P. 2007. Nosocomial bloodstream infections: organisms, risk factors and resistant phenotypes in the Brazilian University Hospital. Braz. J. Infect. Dis. 11: 351–354.

Romero, D.V., Treston, J., and O'Sullivan, A.L. 2006. Hand-to-hand combat: preventing MRSA infection. Adv. Skin Wound Care 19: 328–33, quiz.

Rosenthal, V.D., Guzman, S., and Orellano, P.W. 2003. Nosocomial infections in medical-surgical intensive care units in Argentina: attributable mortality and length of stay. Am. J. Infect. Control 31: 291–295.

Rosenthal, V.D., Maki, D.G., Salomao, R., Moreno, C.A., Mehta, Y., Higuera, F., Cuellar, L. E., Arikan, O.A., Abouqal, R., and Leblebicioglu, H. 2006. Device-associated nosocomial infections in 55 intensive care units of 8 developing countries. Ann. Intern. Med. 145: 582–591.

Saez-Llorens, X., Castrejon de Wong, M.M., Castano, E., De, S.O., De, M.D., and De, A., I. 2000. Impact of an antibiotic restriction policy on hospital expenditures and bacterial susceptibilities: a lesson from a pediatric institution in a developing country. Pediatr. Infect. Dis. J. 19: 200–206.

Safdar, N. and Maki, D.G. 2002. The commonality of risk factors for nosocomial colonization and infection with antimicrobial-resistant *Staphylococcus aureus*, enterococcus, gram-negative bacilli, Clostridium difficile, and Candida. Ann. Intern. Med. 136: 834–844.

Selgelid, M.J. 2005. Ethics and infectious disease. Bioethics 19: 272–289.

Shehabi, A.A., Mahafzah, A.M., and Al-Khalili, K.Z. 2004. Antimicrobial resistance and plasmid profiles of urinary Escherichia coli isolates from Jordanian patients. East Mediterr. Health J. 10: 322–328.

Soleto, L., Pirard, M., Boelaert, M., Peredo, R., Vargas, R., Gianella, A., and Van der, S.P. 2003. Incidence of surgical-site infections and the validity of the National Nosocomial Infections Surveillance System risk index in a general surgical ward in Santa Cruz, Bolivia. Infect. Control Hosp. Epidemiol. 24: 26–30.

Szeto, C.C., Chow, K.M., Kwan, B.C., Law, M.C., Chung, K.Y., Yu, S., Leung, C.B., and Li, P.K. 2007. *Staphylococcus aureus* peritonitis complicates peritoneal dialysis: review of 245 consecutive cases. Clin. J. Am. Soc. Nephrol. 2: 245–251.

Tateda, K. 2002. From the perspective of clinical laboratory physicians. Rinsho Byori 50: 672–677.

Torres, A. 2006. The new American Thoracic Society/Infectious Disease Society of North America guidelines for the management of hospital-acquired, ventilator-associated and healthcare-associated pneumonia: a current view and new complementary information. Curr. Opin. Crit. Care 12(5):444–445.

Urdea, M., Penny, L.A., Olmsted, S.S., Giovanni, M.Y., Kaspar, P., Shepherd, A., Wilson, P., Dahl, C.A., Buchsbaum, S., Moeller, G., and Hay, B. 2006. Requirements for high impact diagnostics in the developing world. Nature 444 Suppl 1: 73–79.

Vidhani, S., Mehndiratta, P.L., and Mathur, M.D. 2001. Study of methicillin resistant *S. aureus* (MRSA) isolates from high risk patients. Indian J. Med. Microbiol. 19: 13–16.

vila-Figueroa, C., Cashat-Cruz, M., randa-Patron, E., Leon, A.R., Justiniani, N., Perez-Ricardez, L., vila-Cortes, F., Castelan, M., Becerril, R., and Herrera, E.L. 1999. Prevalence of nosocomial infections in children: survey of 21 hospitals in Mexico. Salud Publica Mex. 41 Suppl 1: S18–S25.

Vincent, J.L., Brun-Buisson, C., Niederman, M., Haenni, C., Harbarth, S., Sprumont, D., Valencia, M., and Torres, A. 2005. Ethics roundtable debate: a patient dies from an ICU-acquired infection related to methicillin-resistant *Staphylococcus aureus* – how do you defend your case and your team? Crit Care 9: 5–9.

Wertheimer, A.I. and Santella, T.M. 2007. Innovation and the WHO's essential medicines list: giving credit where credit is due. Res. Social. Adm Pharm. 3: 137–144.

WHO and UNICEF. Management of the child with a serious illness or severe malnutrition. Guidelines for care at the first referral level in developing countries. WHO/FCH/CAH/00. 1.2000. Geneva: World Health Organization, 2000.

Yalcin, A.N. 2003. Socioeconomic burden of nosocomial infections. Indian J. Med. Sci. 57: 450–456.

Yang, K.S., Fong, Y.T., Lee, H.Y., Kurup, A., Koh, T.H., Koh, D., and Lim, M.K. 2007. Predictors of vancomycin-resistant enterococcus (VRE) carriage in the first major VRE outbreak in Singapore. Ann. Acad. Med. Singapore 36: 379–383.

Zaidi, A.K., Huskins, W.C., Thaver, D., Bhutta, Z.A., Abbas, Z., and Goldmann, D.A. 2005. Hospital-acquired neonatal infections in developing countries. Lancet 365: 1175–1188.

Zaidi, M., Angulo, M., and Sifuentes-Osornio, J. 1995. Disinfection and sterilization practices in Mexico. J. Hosp. Infect. 31: 25–32.

Zarate, M.S., Gales, A., Jorda-Vargas, L., Yahni, D., Relloso, S., Bonvehi, P., Monteiro, J., Campos-Pignatari, A., and Smayevsky, J. 2007. Environmental contamination during a vancomycin-resistant Enterococci outbreak at a hospital in Argentina. Enferm. Infecc. Microbiol. Clin. 25: 508–512.

Zavascki, A.P., Barth, A.L., Gaspareto, P.B., Goncalves, A.L., Moro, A.L., Fernandes, J.F., and Goldani, L.Z. 2006. Risk factors for nosocomial infections due to Pseudomonas aeruginosa producing metallo-beta-lactamase in two tertiary-care teaching hospitals. J. Antimicrob. Chemother. 58: 882–885.

Zubaidah, A.W., Ariza, A., and Azmi, S. 2006. Hospital-acquired vancomycin-resistant enterococci: now appearing in Kuala Lumpur Hospital. Med. J. Malaysia 61: 487–489.

Part IV
Cost, Policy, and Regulation
of Antimicrobials

Chapter 21
The Economic Burden of Antimicrobial Resistance in the Developing World

S.D. Foster

Abstract Antimicrobial drugs are a scarce resource whose misuse, in both industrialized and developing countries, has contributed to an increased economic burden on the health systems of developing countries. The price differential between amoxicillin and the combination of amoxicillin and clavulanic acid, for example, is on the order of a factor of 20; the change in standard therapy for malaria from chloroquine (CQ) and sulfadoxine/pyrimethamine (SP) to artemisinin-containing therapy (ACT) has increased the cost of treating a case of malaria by a factor of 10 or more. Looked at another way, the same malaria drugs budget will now treat only one-tenth the number of patients as before. These cost increases are forcing health staff and policy makers to confront terrible choices, between using a drug which they know to be ineffective in many cases and which may lead to increased morbidity or mortality, or spending ever-increasing amounts on higher cost antibiotics and antimicrobials, often at the expense of buying enough to meet their needs. Improving laboratory capacity to detect and monitor antibiotic resistance can be a cost-effective strategy in many cases, especially as resistance forces us to use more expensive and scarce antimicrobials.

The specter of the permanent loss of many classes of antimicrobials due to resistance – and the failure of the pharmaceutical industry to engage enthusiastically in the search for new ones – means that the developing world, which still bears the vast majority of the world's infectious disease burden, will soon find itself with fewer and fewer options for treatment. "Resistance without borders" will disproportionately affect the developing world. It is in everyone's interest to use these scarce and dwindling resources as efficiently and as carefully as possible.

S.D. Foster (✉)
Alliance for the Prudent Use of Antibiotics (APUA); Boston University,
School of Public Health, Boston, MA, USA
e-mail: susan.foster@tufts.edu

A. de J. Sosa et al. (eds.), *Antimicrobial Resistance in Developing Countries,*
DOI 10.1007/978-0-387-89370-9_21, © Springer Science+Business Media, LLC 2010

21.1 Introduction

Antimicrobial use in the developing world has undoubtedly saved millions of lives. But this achievement is now threatened by the development of antimicrobial resistance. Whereas the main cause of antimicrobial resistance in the developed world is the overuse of antimicrobials, paradoxically in the developing world much of the problem stems from underuse. Antimicrobials have long been available without prescription from a variety of both formal and informal sector outlets, and patients are used to take the amount of antimicrobial they can afford, rather than the amount they need. Even injectable antimicrobials and syringes are widely available from traditional healers, small informal outlets, and shops (Reeler 1990). In some settings, the fraction of the population getting antibiotics (mostly broad spectrum) from what have been termed "medically unsupervised components of the health care system" – the informal sector – may exceed the fraction of those using formal clinics by a factor of 40 (Becker et al. 2002). The result is high – and growing – levels of antimicrobial resistance for nearly all of the diseases of public health importance. Policy makers and health authorities face a stark tradeoff – find the funds to pay the higher cost for newer drugs and diagnostics, or witness the loss of healthy years of life through avoidable death and disability.

This chapter will review the main types of economic burden and "cost," examine the drug supply chain for opportunities to intervene to reduce resistance, present data on treatment of meningitis from Papua New Guinea to examine the issues around switching from an older to a newer antibiotic and the role of diagnostics, review some of the unintended consequences of antimicrobial use, and finally, using the example of the recent change of standard malaria therapy, identify the issues around the funding of changes to therapy caused by resistance.

21.2 What is Meant by "Costs"?

Many people think of costs first in financial terms – the amount of money actually paid for something. That notion is the building block for other types of analysis. Economists most often think in terms of "opportunity cost" – a reflection of the cost of the alternative which is given up when a decision is made to use resources in a particular way. Dealing first with the financial costs, Table 21.1 shows the wholesale international costs of a typical adult dose of some of the oral antimicrobials most commonly used in primary health care. The variation in costs is substantial – the cost difference between amoxicillin and the combination of amoxicillin plus clavulanic acid, for example, is on the order of a factor of 20 – which illustrates the implications of the loss of a cheaper antimicrobial and the need to replace it with a more expensive one. If for example, what has been termed "resistance-induced substitution"

Table 21.1 Costs of typical adult dose of commonly used oral antibiotics

Drug	Price/ unit ($)	Usual adult dose	No. units	Price/ dose ($)
Amoxicillin 250 mg tab/cap	0.02	Bid × 14 days	28	0.43
Amoxicillin/ clavulanic acid	0.30	Bid × 14 days	28	8.40
Ampicillin 250 mg tab/cap	0.01	250–500 mg PO qid × 5–7 days	28	0.41
Azithromycin 250 mg tab/cap	0.22	500 mg first day, then 250 mg once daily × 4 days*	6	1.31
Chloramphenicol 250 mg tab/cap	0.01	250–500 mg PO q6 h × 5–7 days	28	0.39
Ciprofloxacin 250 mg tab/cap	0.02	250–500 mg PO q6 h × 7–14 days**	56	1.09
Cotrimoxazole (b) 400 + 80 mg tab/cap	0.02	800/160 mg q12 h × 7–10 days	28	0.47
Doxycycline 100 mg tab/cap	0.01	200 mg on the first day of treatment (administered 100 mg every 12 h) followed by a maintenance dose of 100 mg/day	11	0.11
Erythromycin 250 mg tab/cap	0.03	250–500 mg PO q6–8 h × 5–7 days	40	1.05
Metronidazole 400–500 mg tab/cap	0.01	250–500 mg PO q8 h × 7–10	21	0.14
Nalidixic acid 500 mg tab/cap	0.04	1 g every 6 h × 7–10 days	56	2.09
Nitrofurantoin 100 mg tab/cap	0.01	50–100 mg every 6 h × 7–10 days	28	0.24
Penicillin V 250 mg tab/cap	0.01	500–750 mg q6 h × 7 days	56	0.71
Tetracycline 250 mg tab/cap	0.01	1.5 g, followed by 0.5 g every 6 h for 4 days, to a total dosage of 9 g#	36	0.27

*CAP = community-acquired pneumonia
**Lower respiratory tract, skin, bone, and joint infections
#Gonorrhea

(Howard 2004) prompts a prescriber to use amoxicillin plus clavulanic acid, instead of amoxicillin alone, the opportunity cost of the drug cost alone is on the order of $8 per patient – without taking into account the additional longer term opportunity cost of increasing resistance to the combination therapy.

When an ineffective antimicrobial is used, funds are wasted, without effectively treating the patient. And where a newer, more expensive drug is used when an older one would still work, the extra cost of that new drug is also wasted. Where diagnostics are little used or not relied upon, as is the case in much of the developing world (particularly at lower levels of the health service and in the informal sector), much of the expenditure on drugs is wasted because they are not appropriate or indicated for the patient's condition. Antibiotics may be given for acute respiratory infections which often are viral in origin, or antimalarials given for pneumonia; work by Kallander et al. has shown that young febrile children are often treated empirically for malaria, when in fact they have pneumonia (Kallander et al. 2004). Another type of economic cost attributable to drug resistance – losses in terms of premature mortality and avoidable morbidity – is where a disease remains untreated because resistance is present, but the appropriate drugs are not available. A study in Ghana found that the recommended treatment for pelvic inflammatory disease, ceftriaxone and ciprofloxacin, was not available at hospital, health centers, or in the private sector in a region of Ghana (Bosu and Mabey 1998).

Improved diagnostics can play a critical role in reducing losses and the economic burden of resistance. One of the most important recent developments is the development of a rapid point of care test for malaria, but there is evidence that clinicians who are used to making their decisions solely on clinical grounds are disregarding the results of the test and prescribing antimalarials despite a negative test (Lubell et al. 2008). Where such a diagnostic test is available and is used, but the prescriber discounts the results when making his or her clinical decision, the cost of that test is part of the economic cost of the waste. And now that expensive artemisinin-containing treatments (ACT) are in wide use, this means the waste of both the diagnostic test and the expensive antimalarial – and still the patient has not been appropriately treated for the cause of fever and exposed to unnecessary side effects. And patients who do have malaria may find the drug is out of stock as a result of having been used inappropriately. The problem is not limited to malaria or to lower levels of the health-care system. In a teaching hospital in Thailand, antibiotics were administered to a sample of 307 patients in a study, but only in 9% (27) of the patients were the use of antibiotics entirely appropriate; in 36% of the patients, antibiotics were used without evidence of infection (Aswapokee et al. 1990).

There is another important problem caused by the inappropriate use of antimicrobials without monitoring the development of resistance – that of thinking a drug is still working when in fact it has lost its potency due to resistance (Hastings et al. 2007). Using data from Africa, Hastings et al. (2007) show that antimalarial drugs lose their efficacy long before the failure is perceived by the population; "the most insidious consequence of presumptive treatment [for malaria] may be that perceived drug efficacy remains high even for a drug that is failing badly, leading to its continued use. . . ." They advocate use of monitoring of resistance to "minimize any future delays in implementing antimalarial drug policy change, so that the long and fatal delays associated with the replacement of chloroquine are not repeated" (Hastings et al. 2007). Researchers in South Africa have made a similar discovery with regard to extensively drug-resistant tuberculosis (XDR-TB). In KwaZulu-Natal, funding for susceptibility testing was cut back, while a DOTS approach based on four drugs continued to be used – even though resistance was developing to all four: "in the absence of susceptibility test results at the commencement of treatment, patients have been treated unintentionally with 1 or 2 active drugs only … blinded, standardized treatment allowed for selection of increasingly resistant organisms" (Pillay and Sturm 2007). The cost of treating resistant tuberculosis with second-line drugs and the associated excess mortality is far higher than that of first-line drugs, so the economies made by not continuing susceptibility testing are far outweighed by the excess cost of second-line therapy and spread of resistant tuberculosis. The resulting 53 XDR-TB cases, 52 of whom died within a short period of being diagnosed (Gandhi et al. 2006) set off a world-wide alarm about the rapid emergence of this highly lethal strain of TB. There may be a crucial and deadly time lag between perceiving drug resistance and taking action to move to a new drug.

21.3 The Drug Supply System: Identifying Opportunities to Intervene

Providing a specific drug to a patient involves many steps. The selection, procurement, distribution, prescription, dispensing, and finally, the consumption of the drugs by the patient occur in a sequential flow (Foster 1991; Quick et al. 1997). Opportunities to intervene to reduce drug resistance occur along the flow; missed opportunities often cause significant economic losses to the health system and to the society as well as additional unnecessary morbidity and mortality for patients. The nature of the delivery program is also important; it is easier (although not easy) to make a change of drug therapy in a vertical, highly directed program such as a tuberculosis control program and many HIV/AIDS treatment programs which use highly active antiviral therapy (HAART) when compared with a disease such as diarrheal disease or acute respiratory infections (ARI), which are largely treated through self-treatment or lower level health facilities, either public or private. Diffusion of new treatment protocols is somewhat easier in a system where there is a regular program of information dissemination, training workshops, and production of guidelines and manuals. Nonetheless, introducing a new therapy can take years and is costly in both time and resources. Identification of where these opportunities exist in different settings and for different diseases can therefore help in slowing the spread of antimicrobial resistance. Some of the main steps in the sequence of drug supply and issues raised by resistance for several diseases of public health importance are set out in Table 21.2.

Syndromic management can be exceptionally cost-effective in high-prevalence settings (Grosskurth et al. 1995) but will increasingly need to be informed by both prevalence data and antimicrobial resistance data. However, reliance on syndromic approaches in low-prevalence settings can result in waste of drugs without providing effective treatment of the underlying condition, and as more pathogens

Table 21.2 The sequence of drug supply and areas for intervention in diseases of public health importance

	Selection of drugs	Procurement and distribution	Diagnosis and prescription	Dispensing (both public and private)	Patient adherence and use
HIV (HAART)	At individual and program level, depending on individual resistance	Usually a public sector program, guided by funders	According to established protocols, both first- and second-line therapies; depends on lab tests available (viral load, CD4 ct)	Within control program; some private sector distribution but not large in most settings (unaffordable for most)	Key issue is patient adherence – lifelong therapy
Malaria	Various – national level to ART, private sector various; need for surveillance of ART effectiveness	Public and private sector – depending on funding source. Quality control important	Rapid tests available, but not always used or trusted – mostly syndromic/empiric prescription even with rapid test available	Self-treatment from private sector drug sellers, lower level health facilities: improve advice	Poor adherence due to failure to resolve symptoms quickly. Dosing in young children poses problems
Tuberculosis	Protocols need to be based on surveillance data and adjust for resistance	Usually a public sector program (some exceptions, e.g.,India)	Diagnosis insensitive, depends on 150-year-old technology (sputum smear); patient returns 4–5 times before diagnosis is made	Usually within a controlled environment (some exceptions, e.g., India)	Key issue is patient adherence after initiation phase – long therapy, costly, and onerous for patient
ARI*	Protocols need to be based on surveillance data and adjusted for resistance	Private sector and lower level public health facilities. Quality control important	Often confused with malaria, not treated; resistance to commonly used AB; sensitivity testing not often used	Self-treatment from private sector drug sellers, lower level health facilities: improve advice	Administration and dosing of AB for very young children may pose problems

Table 21.2 (continued)

	Selection of drugs	Procurement and distribution	Diagnosis and prescription	Dispensing (both public and private)	Patient adherence and use
STI**	Syndromic management needs to be based on surveillance data for each pathogen	Some public clinics, much private sector supply. Quality control of market/street drugs for STIs an issue	Syndromic management in high-prevalence settings, but asymptomatic infections in women remain untreated; presumptive treatment in, e.g., MCH or FP settings?	Self-treatment from private sector drug sellers; stigma reduces use of formal facilities; partner(s) often not treated	Self-treatment means often incomplete doses and ineffective drugs; partners are seldom treated so infection recurs
Enteric pathogens	Protocols need to be based on surveillance data and adjusted for resistance	Mostly through public and private sector, not vertical program (except cholera)	Variable depending on availability of lab; usually symptom-based; often ORS would be more appropriate	Often self-treatment or treatment at primary level or informal sector	Antibiotics often used when oral rehydration would be more appropriate

* ARI = Acute respiratory infection.
** STI = Sexually transmitted infection.

become resistant, a syndromic approach will require more drugs to address all the possibilities. And as the cost of a treatment rises, the cost-effectiveness of syndromic management in low-prevalence areas becomes even less attractive. In Bangladesh, the treatment of women for sexually transmitted infections (STIs) using a WHO-recommended syndromic management approach resulted in over-treatment in about 87% of cases – very few cases were true STIs. Furthermore, the syndromic approach missed all of the true cervical infections, resulting in a small fraction of the women coming forward being treated appropriately. Most of the antimicrobial expenditure was wasted (Hawkes et al. 1999).

21.4 The Economics of Switching Therapy

Dealing with drug resistance gives rise to perhaps the most important economic decision facing health authorities in the developing world: the decision about if and when to switch therapy for a disease of public health importance. In the United States, it is "common knowledge" among medical residents that a threshold of 25% is the point at which a switch to a new antibiotic is made. For community-acquired pneumonia (CAP), the Infectious Disease Society of America and American Thoracic Society (IDSA/ATS) produced guidelines in 2007, also suggesting a 25% threshold for discontinuing macrolide use for empirical therapy of CAP when there is high-level resistance; the threshold was reached by expert consensus, "without explicit consideration of the implications of this threshold for patient outcomes or for tradeoffs between effectiveness, cost and future resistance..." (Daneman et al. 2008). In Europe and the United States, the data exist but are not always explicitly considered; for most of the developing world, however, there are few data on which to base decisions to switch therapy. Policy makers face a tradeoff – finding funds for increased expenditures on drugs and diagnostics, or accepting the cost in life years lost or lived with preventable disability. The switching may occur within a vertically organized control program, e.g., those for tuberculosis or malaria, or at primary care level, both public and private.

The economics of switching therapy for malaria was described by Phillips and Phillips-Howard (1996), who developed a formula to help identify the threshold at which a switch should be made. A switch was deemed appropriate when the cost of the old drug plus the "cost of failure," times its rate of failure, was greater than the cost of the new drug plus its associated cost of failure (Phillips and Phillips-Howard 1996). It is fairly straightforward to calculate the cost of a drug – but what elements should be included in the "cost of failure"? Determinants of the cost of failure fall into two main categories – those which are borne by the patients themselves and others which are borne by society more widely. For example, patient-level factors would include whether an illness is rapidly fatal or likely to produce important sequelae; whether a second-line

Table 21.3 Factors in "cost of failure" for selected diseases of public health importance

	Malaria	ARI*	Meningitis	Tuberculosis	Gonorrhea	Cholera	HIV
Rapidly fatal?	Yes, especially in children	Can be in young children	Yes, and may lead to neurological sequelae	No	No	Yes	No
Affordable rapid diagnostic test available?	Yes	No	No	No	No	NA	Yes
Diagnosis costly and invasive?	No	No	Yes	No	No	No	No
Second-line treatment cost increase over first-line	10× more (author's calculations, see text)	5–20× more (author's calculations, see text)	8–10× more (Duke et al. 2003	12–17× more (Resch et al. 2006)	Varies greatly – 3–20× more (author's calculations)	Varies depending on location	5–10× more
Transmission an important factor?	Some	Yes	Yes	Yes	Yes	Yes	Yes
Implications of loss of drug and pipeline prospects	Nothing currently in pipeline after ACT	All more expensive	Few options, more expensive	Second- and third-line toxic, expensive, limited	Few options remaining (only cephalosporins)	Few options remaining in some settings	New options more costly, unavailable

*Acute respiratory infection
Source: Author's estimates and sources cited.

treatment is readily available and how much it costs, and whether patients are likely to reach, and receive, second-line treatment in time; and whether a diagnostic test is available and whether it is invasive or costly to use. Public health factors would include the degree of transmissibility of the disease if treatment failure occurs; how likely it is that treatment failure will be noticed and acted upon in time; and the extent to which use of the drug will reduce its efficacy in the future and the wider implications of the potential loss of the drug. Clearly these factors are different for different diseases and treatments. These "cost of failure" factors for several different diseases of public health importance are illustrated in Table 21.3.

The practical difficulties and costs involved in implementing such a switch are considerable. In Malawi, the switch from chloroquine to sulfadoxine/ pyrimethamine (SP, brand name Fansidar®) as first-line antimalarial took more than a year, even though SP had been used as a second-line therapy for some time already. Kazembe et al. (2002. "The process of changing first-line malaria treatment from chloroquine to sulphadoxine pyrimethamine: the experience from Malawi", unpublished manuscript) describe the process as follows:

> During this lag period, health workers were trained about the 'new treatment', the pharmaceutical industry was informed, IEC materials were produced for both health workers and the general public, the central medical stores procured enough SP for all the health centers for a period of about 6 months, and a contract was signed with a local retail chain distributor to procure SP from the central medical stores to sell at a discounted 'affordable' price through its outlets in trading centers including the rural areas. The idea was to make sure that, by the time of the official launch, the local shops and health centers should have some stocks of SP available so as to minimize 'panic' since the main message of the official launch was to be that chloroquine is no longer useful and that SP should be used instead for the home treatment of malaria. There was a high profile launch by the Minister of Health with a press conference explaining the reason behind the change (Kazembe et al. 2002, unpublished manuscript).

Kazembe et al. also illustrate the importance of clear communications with patients. The radio and poster messages emphasized that SP is "stronger" and therefore only needs to be given in a single dose. Some mothers were afraid to give this "strong" drug to their younger children, and as a result children continued to be brought in quite late to the hospital with severe malaria because their mothers feared giving them SP (Kazembe et al. 2002, unpublished manuscript).

21.5 The Important Role for Diagnostics: A Case Study of Meningitis

Much of the cost of failure can be reduced by making sure the drug is the right one for the illness, and that the pathogen remains susceptible to the drug. Key to this process is the availability of diagnostic tests. Research in Papua New Guinea on treatment of meningitis in children (Duke et al. 2003) illustrates the difficult choices posed by antimicrobial resistance in a rapidly fatal disease, for which the drug price differential is substantial. They compared chloramphenicol with ceftriaxone for the treatment for meningitis, using data from three hospitals in Papua New Guinea. Although their setting was that of tertiary hospitals that were able to carry out diagnostic tests, the dilemma they faced is even more stark for staff at lower level health facilities where there is no diagnostic capacity. The study originally set out to compare chloramphenicol with ceftriaxone, but chloramphenicol resistance was occurring in 20% of *Haemophilus influenzae* strains; children treated with chloramphenicol had poor outcomes. The treatment was then changed to ceftriaxone, with a switch to chloramphenicol if the bacteria isolated were susceptible; in this setting diagnostic services were available and played a key role. Four treatment combinations are possible in this setting: two depend on diagnostics and two do not. Estimates of the costs of each sequence are presented below (Table 21.4).

The drug cost of resistance in this setting can be roughly calculated as the difference between chloramphenicol at $3.29 and ceftriaxone at $16.20, or about $13 per patient with meningitis (costs of mortality and sequelae are not included in this estimate). As mortality and sequelae from chloramphenicol treatment alone are considered unacceptably high, switching to ceftriaxone is a valid, but costly option. The sequence is crucial; using chloramphenicol first, then switching to ceftriaxone is somewhat cheaper than the sequence of ceftriaxone first, then switching to chloramphenicol ($832 compared with $1109) but the outcomes are much worse when chloramphenicol is used first: about 20% of children had resistant infections and died or had neurological sequelae. The cost of saving the 20% of children with resistant infections would be about $10 per child.

If there is a laboratory which can perform antibiotic sensitivity testing (AST), and there is time to perform the test, then the decision to switch therapy can be done on a case-by-case basis. In developing country settings the cost of performing AST has been estimated at approximately $2–3.[1] (Note that the cerebrospinal fluid (CSF) has already been obtained as part of routine diagnostic procedures and the cost of this procedure is not included.) We can estimate the value of having susceptibility testing available by comparing the two therapies which use ceftriaxone first and have acceptable results in terms of mortality and neurological sequelae, namely therapy 2 (ceftriaxone alone, $1620 per 100 children) with therapy 3 (ceftriaxone, then a switch to chloramphenicol if resistance is not present, at $1109 per 100 children). Using susceptibility testing for 100 children, here estimated at $300 ($3 per child × 100 children), is about $211 less than the extra cost of using ceftriaxone for all treatments ($1620–$1109 = $511) when the switch is made to chloramphenicol in 80% of cases when no resistance is found. In other words, using susceptibility testing at a cost of $3 per test is cost saving, and produces a savings of $211 per 100 children in this setting. At the current rate of resistance of 20%, it makes

---

[1] Dr. T.O'Brien, personal communication, 2007 and author's calculations.

Table 21.4 Comparison of four options for treatment of meningitis*

Treatment sequence/option	Cost/day**	14 days total	Total per 100 cases
A. No antibiotic susceptibility testing available			
1. Chloramphenicol (CP) alone	0.24	3.29	329
2. Ceftriaxone (CEF) alone	1.16	16.20	1,620
B. With antibiotic susceptibility testing			
3. Ceftriaxone then chloramphenicol	3 days	11 days	14 d total
susceptible (80% of cases): 3 d CEF, 11 d CP	3.47	2.59	6.06
resistant (20% of cases): 14 d CEF	3.47	12.73	16.20
combined (average for all 100)			809
susceptibility testing (all cases) ***			300
Total, ceftriaxone then chloramphenicol per 100			1,109
4. Chloramphenicol then ceftriaxone	3 days	11 days	14 d total
susceptible (80% of cases): 14 d CP	0.71	2.59	3.29
resistant (20% of cases): 3 d CP, 11 d CEF	0.71	12.73	13.43
combined (average for all 100)			532
susceptibility testing (all cases) ***			300
Total, chloramphenicol then ceftriaxone per 100			832

* Data from Duke et al. (2003).

**Costs estimated on the basis of a child weighing 10 kg. Note that this may slightly overstate the cost of drugs in the Duke et al. study as the children in their study were 6–7 months old on average and thus might weigh less than 10 kg on average. Drug costs from International Drug Price Indicator Guide, FOB European port. To this cost would be added the costs of transport, insurance, and taxes. The transport and insurance are considered to be approximately equal for both drugs (1 g vials in both cases), although if import duties and taxes are levied on these drugs, the ceftriaxone would be slightly more expensive.

*** Estimated at $2–3 per case. Dr. T. O'Brien, personal communication and author's calculations.

sense to introduce susceptibility testing before making the switch to ceftriaxone for all children with meningitis. It is not until resistance levels hit about 60% that the cost of switching to ceftriaxone without testing predominates, if only the costs of testing and drugs are included.

In settings where susceptibility testing is not (yet) possible, the alternatives are more stark: to continue using a drug to which 20% of pathogens are resistant, with predictably bad outcomes for many children, or to step up to a new drug, ceftriaxone in this case, which will cure all the cases but at the cost of the waste of about 70–80% of drug costs. If ceftriaxone is available, switching to ceftriaxone when it is clear on clinical grounds that chloramphenicol is failing for 20% of children costs about $5 per child, but the ceftriaxone arrives too late for many and outcomes are poor.

A similar dilemma is posed by the treatment of Gram-negative sepsis in older children. A commonly used therapy in the developing world[2] is a combination of chloramphenicol and gentamicin (despite the possibility of ototoxicity with gentamicin (Govaerts et al. 1990) and of bone marrow aplasia with chloramphenicol); the alternatives include more effective monotherapy with ciprofloxacin or cefotaxime. Use of more effective antimicrobials would increase per-patient costs for a 15 kg child treated parenterally for 7 days by a factor of nearly 30 or 40, from $2.68 ($0.38 for gentamicin and $2.30 for chloramphenicol) to $108 or $73 for cefotaxime or ciprofloxacin, respectively (using 2 days parenteral, then oral therapy) (Bejon et al. 2005). Although these figures do not include any estimate of the cost of hearing loss in children, switching to more effective antimicrobials without antibiotic susceptibility testing is unfeasible in most low-income settings.

This example raises an interesting question: How does the threshold for switching drugs differ between diseases? The switch to artemisinin-containing treatments (ACT) for malaria has happened in many places where earlier therapies still had some efficacy (Kouyate 2007). Factors such as the lethality of the disease, the speed with which death or serious complications can occur, the likelihood of the patient returning to the health facility for second-line treatment in case of failure of the first-line treatment, and so on – the "cost of failure" – affect the threshold at which the decision to switch is made. There is also a need to differentiate between the decision making of a clinician faced with an individual patient and a public health official trying to maximize the number of people who can be treated within a constrained budget. In the developing world, unfortunately, the most important consideration is often the cost. The switch from the less effective and more toxic thiacetazone to rifampicin as a component of tuberculosis therapy was delayed by attempts to preserve it because of its lower procurement cost, even though it is not cost-effective (Kelly et al. 1994; Elliott and Foster 1996). The option of strengthening the laboratory to enable

---

[2] Virtually every child with a fever who is admitted to hospital in Burundi is administered gentamicin and chloramphenicol, regardless of diagnosis. Personal communication of Burundian hospital administrator to S. Foster, July 2006.

more rapid and cheap diagnosis of antibiotic-resistant disease can, in many settings, produce important savings and improve treatment outcomes as well – if clinicians are willing to trust the results from the lab and prescribe accordingly. Introduction of new laboratory tests should be accompanied by a concerted effort to gain the trust and adherence of prescribers.

21.6 Wider and Unintended Consequences of Antimicrobial Use

The production and use of antimicrobials and consequent resistance can have many wider ramifications – what economists term "externalities," unintended effects, either positive or negative. It may be important to consider the "incidence" of the effects – who benefits and who is harmed by the use of antibiotics, and by antimicrobial resistance and the mitigating actions we must take. One example of such an externality is that decisions made with a focus on one particular disease may have consequences for other diseases or for health in general, which are unforeseen, or perhaps judged to pose an acceptable level of risk for the wider community. Cotrimoxazole is a cheap, broad-spectrum antibiotic which has long been used in primary health care. But in 2000 WHO made the recommendation that children born of HIV-infected women should receive presumptive prophylaxis with cotrimoxazole (Gill et al. 2004). In 2005 a WHO Expert Committee made the recommendation that all people diagnosed with HIV should receive cotrimoxazole prophylaxis (WHO 2006). The Expert Committee considered the issue of resistance, but left unanswered how to deal with the issue:

> The panel discussed the need for sentinel surveillance to monitor the incidence of resistance among invasive and carried organisms as an element of any CTX [cotrimoxazole] prophylaxis programme. However, it remains unclear as to what methods should be used to best monitor such resistance (WHO 2006).

Unfortunately resistance to cotrimoxazole in *Streptococcus pneumoniae* is growing rapidly, and resistance to a number of other pathogens may develop as well. According to Gill et al. (2004), the list of pathogens whose epidemiology could be altered by cotrimoxazole exposure includes *S. pneumoniae*, *S. aureus*, *H. influenzae*, *Neisseria* spp., and enteropathogens (*Shigella* spp., *Salmonella* spp., *Isospora belli*)., and *Neisseria* spp. (Gill et al.) The wider costs of the potential loss of this cheap and well-tolerated antibiotic may prove considerable.

The veterinary use of antimicrobials in the developing world is another potential source of unintended consequences. While the non-therapeutic veterinary use of antibiotics as growth promoters is not widespread as in the industrialized world, where as much as half of the antibiotics used in the United States are used in animal feed (Levy 2002), antibiotics are being used for veterinary therapy in the developing world, and they are available without prescription from some of the same informal sector shopkeepers and market sellers who sell human antibiotics. This use should be researched and monitored for the impact on human health.

21.7 Counterfeit of Antimicrobial Drugs

The strong demand for antibiotics creates a market for unscrupulous persons and counterfeiters. Tremendous amounts of money can be made by counterfeiting of antimicrobial drugs, and they are among the preferred targets of counterfeiters (Kelesidis et al. 2007). But counterfeiting of antibiotics – many of which contain too little of the active ingredient – also contributes to the creation of resistant strains. Almost every antimicrobial has been found in substandard or counterfeit form.

21.8 Environmental Cost of Antimicrobial Production and Use

Negative externalities are often borne by the poorest segments of society, those who are economically least able to mitigate their effects. The production of antimicrobials creates toxic waste which must be processed and disposed of appropriately. A study of the production of antimicrobials in India found that pharmaceutical production companies which produce generic drugs for the world market were responsible for discharging antimicrobial- and drug-laden effluent into streams and rivers (Larsson et al. 2007). Ciprofloxacin was the drug found in highest concentrations; "the concentration of the most abundant drug, ciprofloxacin (up to 31,000 μg/L) exceeds levels toxic to some bacteria by over 1000-fold." The amount of ciprofloxacin active ingredient discharged daily is equivalent to 5 days' consumption by Sweden's population of 9 million (Larsson et al. 2007). Treating this effluent appropriately would require investment in processing equipment which would raise the cost of the drugs produced, and in the current world competitive market, this cost would make some of these companies uncompetitive. Thus an economic incentive to improve the situation is missing.

21.9 Malaria: Who Pays for the Cost of Resistance?

Any discussion of the economic burden of resistance has to include a discussion of how the consequences of resistance, and in particular the switch to a newer and more expensive drug, will be funded. The recent experience with the treatment of malaria is a good case study of the economic impact of drug resistance. The fact that malaria is so widespread, and its treatment has become much more expensive, means that the economic choices surrounding drug therapy are of enormous importance to the health budgets of affected countries. As chloroquine resistance became more widespread, pressure grew to change therapy to a more effective drug; an eminent group of malariologists from around the world wrote a piece in *The Lancet* in 1999 (White et al. 1999) calling for a switch to combinations containing artemisinin. They were aware of the cost implications but promoted the widespread use of the new drugs without really addressing the issue of how they would be paid for in the short term:

> Cost is usually the major factor that determines the use of antimalarial drugs. Combinations with artemisinin derivatives would, in general, be expected to double the treatment cost for individual patients. But increased short-term costs should result in overall savings in the longer term. [...] Since chloroquine and PSD [SP] are already failing in many areas, combination treatment would be expected to improve cure rates with a reduction in the morbidity (and therefore costs) associated with treatment failure. [...] In parts of Vietnam and Thailand, where these drugs have been used systematically, there has been a reduction in the incidence of falciparum malaria, thus saving lives and money (White et al. 1999).

A WHO consultation on antimalarial combination therapies in 2001 considered the issues around switching to artemisinin-containing therapies (ACT), and whether it would be worth switching temporarily to sulfadoxine/pyrimethamine (SP, brand name Fansidar[R]) even though evidence was available that indicated that the effectiveness of SP was likely to be short lived (on the order of 3–4 years), and a further switch to ACT would soon be necessary. Awareness of the difficulties of switching therapy and the time and resources required (Kazembe et al. 2002, unpublished manuscript) yielded a consensus that given the transaction costs of making the switch, the intermediate step of SP would be more costly than it was worth. Pressure grew to switch to ACT which are on the order of 10 times more expensive than SP and chloroquine, and some termed it "unethical" to use alternative older drugs (Attaran et al. 2004). The Global Fund for AIDS, Tuberculosis and Malaria (GFATM) was meant to be a main financier of these drugs, but GFATM funds only about 41% of the malaria proposals it has received; [3] many countries have submitted multiple unfunded proposals, and have found it an unreliable source of funding for ACT. Burkina Faso, for example, has submitted six applications for malaria funding but only two have been funded (33%). So it is not clear how this desirable new therapy is to be paid for, and some are also asking the question of whether the switch was made too early, and too widely. The impact of the new policy at district level in Burkina Faso, where a combination of amodiaquine and SP was still partially effective, was nothing less than disastrous (Kouyate et al. 2007). ACT was available only in private pharmacies, at a very high cost (for Burkina citizens) of $6.50 per treatment. Burkina Faso's application to the GFATM for funding for ACT had been rejected in 2005, so chloroquine remains the de facto first-line treatment, as described below:

> Unfortunately there is no ideal world. As sufficient funds for high coverage provision of ACT are currently not available, an appropriate interim solution would be to use a pragmatic combination of two affordable drugs. The obvious choice would be the combination of pyrimethamine–sulfadoxine and amodiaquine, which has been shown

---

[3] Global Fund Observer, Issue 81: 29 November 2007.

> to be as effective as ACT in a number of SSA [Sub-Saharan African] countries, including Burkina Faso ... However, after it became clear that Burkina Faso would not receive GFATM funds for the purchase of ACT, the NMCP [National Malaria Control Program] of Burkina Faso asked the World Bank to use a portion of an existing US $12 million loan ... to purchase pyrimethamine–sulfadoxine and amodiaquine as an interim solution. This request was rejected with the argument that WHO recommends only ACT. As a result, chloroquine remains factually the first-line malaria treatment in Burkina Faso. These observations support the view that SSA countries continue to be victims of ignorance and lack of coordination between external donors and international organisations... (Kouyate et al. 2007).

Global policy, uninformed by local data on resistance patterns, can be devastating in practice, however, well intentioned. It is of no help to recommend an unaffordable strategy, however, "optimal" it might seem in theory. Burkina's experience with the GFATM is not unique. The experience of the GFATM's first round of awards and the results for malaria, particularly in Africa, were not good:

> Over the lifetime of these malaria awards, African countries, which experience more than 90% of the global malaria burden, will have access to only 56% of the total global malaria resources made available by the GFATM. Four countries outside Africa (China, Indonesia, Laos, and Sri Lanka), which bear only 3% of the world's malaria burden, would have access to 44% of the malaria funds awarded in this first cycle of funding. Support per person for antimalaria actions approved by GFATM remains paltry given the needs of Africa. Many national proposals for prevention and control received no support [...] Proposals endorsed for funding by the Board will provide on average US $0.17 per person at risk during the first year. At this level, the targets agreed at Abuja have no chance of being met (Teklehaimanot and Snow 2002).

And as noted above, the GFATM's subsequent rounds did not solve the problem of funds for ACT. How much is actually needed to implement this strategy of a more effective treatment for malaria? The amount is staggering:

> Depending upon which combinations are selected and how they are deployed, we estimate that between US $0.32 billion and US $3.41 billion each year over the next five years is required to meet the demand for ACT through the formal, regulated health sector in Africa. Our estimates of the immediate national costs needed to support ACT drugs such as Coartem® [ACT drug] amount to between 0.1 and 5.8% of GDP for countries outside of southern Africa and exceed the resources available to the GFATM (Snow et al. 2003).

Attaran et al. (2004), who were instrumental in the call for the switch to ACT, now ask "Where did it all go wrong?" and blame the donors – bilateral donors such as USAID, the World Bank, the GFATM, and above all the World Health Organization – for failing to secure the funds necessary (Attaran 2004). Yet at a time when the public health world was urgently scaling up antiretroviral treatment for HIV, coming to terms with the specter XDR-TB, and contending with the redirection of much of the public health resource base to the perceived threats of bioterrorism and pandemic flu, it is not altogether surprising that these funds were promised, but not secured.

21.10 Conclusion

Antibiotics are a scarce resource whose misuse, both in industrialized and developing countries, has contributed to an increased economic burden on the health systems of developing countries. Several examples have been presented in this chapter – the price differential between amoxicillin and the combination of amoxicillin and clavulanic acid, for example, is on the order of a factor of 20; the change in standard therapy for malaria from chloroquine (CQ) and sulfadoxine/pyrimethamine (SP) to artemisinin-containing therapy (ACT) has increased the cost of treating a case of malaria by a factor of 10 or more. Looked at another way, the same malaria drugs budget will now treat only one-tenth the number of patients as before. These cost increases are forcing health staff and policy makers to confront terrible choices, between using a drug which they know to be ineffective in many cases and which may lead to increased morbidity or mortality, or spending ever-increasing amounts on higher cost antibiotics and antimicrobials, often at the expense of buying enough to meet their needs. Improving laboratory capacity to detect and monitor antibiotic resistance can be a cost-effective strategy in many cases, especially as resistance forces us to use more expensive and scarce antimicrobials.

The specter of the permanent loss of many classes of antibiotics due to resistance – and the failure of the pharmaceutical industry to engage enthusiastically in the search for new antibiotics (Nathan 2004) – means that the developing world, which still bears the vast majority of the world's infectious disease burden, may find itself with fewer and fewer options for treatment. "Resistance without borders" will disproportionately affect the developing world. It is in everyone's interest to use these scarce and dwindling resources as efficiently and as carefully as possible.

References

Aswapokee, N., S. Vaithayapichet, et al. (1990). "Pattern of antibiotic use in medical wards of a university hospital, Bangkok, Thailand." *Rev Infect Dis* **12**(1): 136–41.

Attaran, A. (2004). "Where did it all go wrong?" *Nature* **430**(7002): 932–3.

Attaran, A., K. I. Barnes, et al. (2004). "WHO, the Global Fund, and medical malpractice in malaria treatment." *Lancet* **363**(9404): 237–40.

Becker, J., E. Drucker, et al. (2002). "Availability of injectable antibiotics in a town market in southwest Cameroon." *Lancet Infect Dis* **2**(6): 325–6.

Bejon, P., I. Mwangi, et al. (2005). "Invasive Gram-negative bacilli are frequently resistant to standard antibiotics for children admitted to hospital in Kilifi, Kenya." *J Antimicrob Chemother* **56**(1): 232–5.

Bosu, W. K. and D. Mabey (1998). "The availability and cost of antibiotics for treating PID in the Central Region of Ghana and implications for compliance with national treatment guidelines." *Int J STD AIDS* **9**(9): 551–3.

Daneman, N., D. E. Low, et al. (2008). "At the threshold: defining clinically meaningful resistance thresholds for antibiotic choice in community-acquired pneumonia." *Clin Infect Dis* **46**(8): 1131–8.

Duke, T., A. Michael, et al. (2003). "Chloramphenicol or ceftriaxone, or both, as treatment for meningitis in developing countries?" *Arch Dis Child* **88**(6): 536–9.

Elliott, A. M. and S. D. Foster (1996). "Thiacetazone: time to call a halt? Considerations on the use of thiacetazone in African populations with a high prevalence of human immunodeficiency virus infection." *Tuber Lung Dis* **77**(1): 27–9.

Foster, S. (1991). "Supply and use of essential drugs in sub-Saharan Africa: some issues and possible solutions." *Soc Sci Med* **32**(11): 1201–18.

Gandhi, N. R., A. Moll, et al. (2006). "Extensively drug-resistant tuberculosis as a cause of death in patients co-infected with tuberculosis and HIV in a rural area of South Africa." *Lancet* **368**(9547): 1575–80.

Gill, C. J., L. L. Sabin, et al. (2004). "Reconsidering empirical cotrimoxazole prophylaxis for infants exposed to HIV infection." *Bull World Health Organ* **82**(4): 290–7.

Govaerts, P. J., J. Claes, et al. (1990). "Aminoglycoside-induced ototoxicity." *Toxicol Lett* **52**(3): 227–251.

Grosskurth, H., F. Mosha, et al. (1995). "Impact of improved treatment of sexually transmitted diseases on HIV infection in rural Tanzania: randomised controlled trial." *Lancet* **346**(8974): 530–6.

Hastings, I. M., E. L. Korenromp, et al. (2007). "The anatomy of a malaria disaster: drug policy choice and mortality in African children." *Lancet Infect Dis* **7**(11): 739–48.

Hawkes, S., L. Morison, et al. (1999). "Reproductive-tract infections in women in low-income, low-prevalence situations: assessment of syndromic management in Matlab, Bangladesh." *Lancet* **354**(9192): 1776–81.

Howard, D. H. (2004). "Resistance-induced antibiotic substitution." *Health Econ* **13**(6): 585–95.

Kallander, K., J. Nsungwa-Sabiiti, et al. (2004). "Symptom overlap for malaria and pneumonia – policy implications for home management strategies." *Acta Trop* **90**(2): 211–4.

Kelesidis, T., I. Kelesidis, et al. (2007). "Counterfeit or substandard antimicrobial drugs: a review of the scientific evidence." *J Antimicrob Chemother* **60**(2): 214–36.

Kelly, P., A. Buve, et al. (1994). "Cutaneous reactions to thiacetazone in Zambia – implications for tuberculosis treatment strategies." *Trans R Soc Trop Med Hyg* **88**(1): 113–115.

Kouyate, B., A. Sie, et al. (2007). "The great failure of malaria control in Africa: a district perspective from Burkina Faso." *PLoS Med* **4**(6): e127.

Larsson, D. G. J., C. de Pedro, et al. (2007). "Effluent from drug manufactures contains extremely high levels of pharmaceuticals." *J Hazard Mater* **148**(3): 751–5.

Levy, S. B. (2002). *The antibiotic paradox : how the misuse of antibiotics destroys their curative power*. Cambridge, MA, Perseus Pub.

Lubell, Y., H. Reyburn, et al. (2008). "The impact of response to the results of diagnostic tests for malaria: cost-benefit analysis." *BMJ* **336**(7637): 202–5.

Nathan, C. (2004). "Antibiotics at the crossroads." *Nature* **431**(7011): 899–902.

Phillips, M. and P. A. Phillips-Howard (1996). "Economic implications of resistance to antimalarial drugs." *Pharmacoeconomics* **10**(3): 225–38.

Pillay, M. and A. W. Sturm (2007). "Evolution of the extensively drug-resistant F15/LAM4/KZN strain of Mycobacterium tuberculosis in KwaZulu-Natal, South Africa." *Clin Infect Dis* **45**(11): 1409–14.

Quick, J. D., J. R. Rankin, et al. (1997). "Managing drug supply." *Management Sciences for Health in collaboration with the World Health Organization*. West Hartford, CT: Kumarian Press.

Reeler, A. V. (1990). "Injections: a fatal attraction?" *Soc Sci Med* **31**(10): 1119–25.

Snow, R. W., E. Eckert, et al. (2003). "Estimating the needs for artesunate-based combination therapy for malaria case-management in Africa." *Trends Parasitol* **19**(8): 363–9.

Teklehaimanot, A. and R. W. Snow (2002). "Will the Global Fund help roll back malaria in Africa?" *Lancet* **360**(9337): 888–9.

White, N. J., F. Nosten, et al. (1999). "Averting a malaria disaster." *Lancet* **353**(9168): 1965–7.

WHO (2006). "WHO Expert Consultation on Cotrimoxazole Prophylaxis in HIV Infection." TRS WHO/HIV/2006.01.

Chapter 22
Strengthening Health Systems to Improve Access to Antimicrobials and the Containment of Resistance

Maria A. Miralles

Abstract This chapter provides an overview of contemporary health systems issues related to improving access to essential medicines, which includes many antimicrobials. Key conceptual frameworks are presented that guide the exploration of challenges to improving sustainable access, including appropriate use of medicines, as well the design and evaluation of interventions. Critical topical areas that have emerged as a result of the push to rapidly implement global programs that have a strong focus on improving access to antimicrobials are highlighted. Finally, a case study is presented to illustrate how a health systems approach was applied to guide the design of an intervention to improve access to and use of medicines, including antimicrobials, in Tanzania.

22.1 Introduction

The current global health environment is characterized by unprecedented levels of investments to increase access to medicines to treat infectious diseases that particularly afflict developing countries. Antimicrobials are the cornerstone of effective treatment for global programs, including the Stop TB Partnership's Global Drug Facility (GDF), and initiatives such as the President's Emergency Program for HIV/AIDS Relief (PEPFAR) and the President's Malaria Initiative (PMI), among others. Since its founding in January 2002, nearly half of the value of grants awarded by the Global Fund to Fight HIV/AIDS, Tuberculosis, and Malaria (GFATM) to 136 countries around the world has been directed to the procurement of medicines and related commodities alone, currently representing approximately US $4.7 billion. As of January 2008, more than 3.3 million tuberculosis patients have enrolled in DOTS programs and 1.4 million now have access to ARV therapy thanks to GFATM grants alone (GFATM, 2008).

M.A. Miralles (✉)
Center for Pharmaceutical Management, Management Sciences for Health, Arlington, VA 22203, USA
e-mail: mmiralles@msh.org

A. de J. Sosa et al. (eds.), *Antimicrobial Resistance in Developing Countries*,
DOI 10.1007/978-0-387-89370-9_22, © Springer Science+Business Media, LLC 2010

Unfortunately, the increased access to medicines and related technologies has not progressed at the same pace as the capacity of country health systems to effectively manage them and ensure their appropriate use, critical for the containment of antimicrobial resistance (AMR) (Trachtenberg and Sande, 2002; Nugent et al., 2008; Ruxin et al., 2005). The inability to effectively manage these medicines contributes to the erosion of the usefulness of these medicines as resistance emerges. As resistance emerges, health systems are faced with new and increasing burdens associated with more costly and prolonged treatments. The problem is immediate and the clock is ticking.

How to harness existing health system capacity and to build new capacity to management medicines with both the short and the long view in mind is the challenge. Much is known about factors that influence the use of antimicrobials in developing countries (Radyowijati and Haak, 2002) and what interventions can improve the use of medicines (WHO, 2001), but much less is known about how to take proven interventions to national scale (ICIUM, 2004). Barriers to scalability and sustainability of otherwise successful interventions are complex, often interrelated health systems issues such as lack of human resources, inadequate financing mechanism, and weak institutional capacity to carry out their mandates to protect the public interest, among others (Caines and Buse, 2004; UNHMP, 2005). Among the lessons learned from the health sector reforms of the 1980s and 1990s is that successful health system strengthening initiatives anticipate system-wide requirements and are sensitive to the potential for negative unintended impacts, including system failure (Roberts et al., 2004). The pace and magnitude of opportunities to positively impact on health systems as a result of global program inputs to improve access to medicines has given rise to a cry to not forget such lessons (AHSPR, 2004).

This chapter presents some key conceptual frameworks to guide the thinking of the types of challenges health systems face in improving regular, sustainable access to and the appropriate use of medicines, including antimicrobials. . While a comprehensive review of research findings to date is beyond the scope of this chapter, critical topical areas that have emerged as a result of the push to rapidly implement global programs that have a strong focus on improving access to antimicrobials are highlighted. Finally, a case study is presented to illustrate how a health systems approach was applied to guide the design of an intervention to improve access to and use of medicines, including antimicrobials, in Tanzania.

22.2 Medicines from the Health Systems Perspective

The WHO defines a health system as the constellation of organizations, people, and actions involved in promoting, maintaining, or restoring health (WHO, 2002). Influenced by the historical, social, political, and economic contexts in which they operate, health systems are dynamic. Health systems include the structures, organizations, and individuals that take actions that impact on health at all levels, central to household, and in the public and the private sectors. For these reasons, understanding how health systems work requires a multidisciplinary perspective.

There are different ways to conceptualize how health systems work but most models will agree that they are organized around the need to carry out key interrelated functions that address pharmaceuticals (i.e., medical products, vaccines and technologies), human resources, financing, information systems, stewardship, and governance. To a large extent, these functions involve the creation and management of critical resources needed to produce health. Figure 22.1 is an example of how these health systems functions relate to each other to achieve specific health impacts.

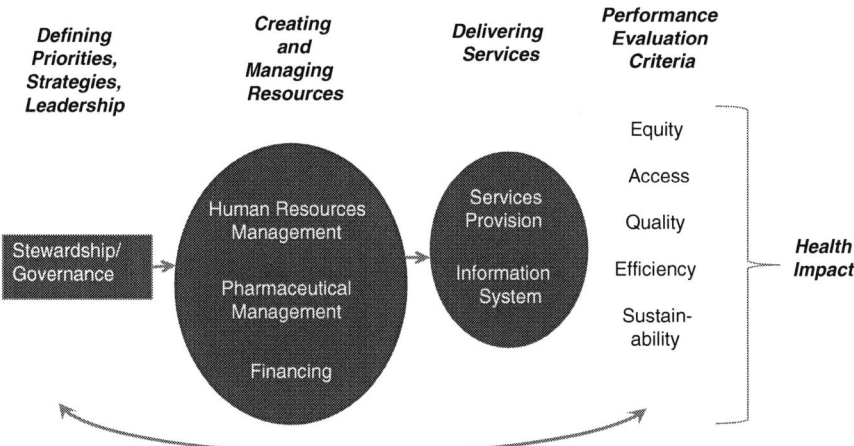

Fig. 22.1 Health system functions and performance criteria (adapted from Islam, 2007)

22.2.1 Pharmaceutical Management

Ensuring the timely and equitable access to, and appropriate use of, safe, effective, quality medicines and related products and services in any health-care setting (MSH n.d.) is the objective of pharmaceutical management, a key function of health systems, and the starting point for this exploration of how health systems, and their subsystems, work with respect to issues related to access and use of medicines (Fig. 22.2).

The practices that define pharmaceutical management within the context of a given health system can be categorized according to the following types of activities[1]:

Selection: the determination of which medicines and related health commodities are to be prioritized for use within the health system.

---

[1] For detailed discussion of best practices for each set of activities, see MSH n.d.

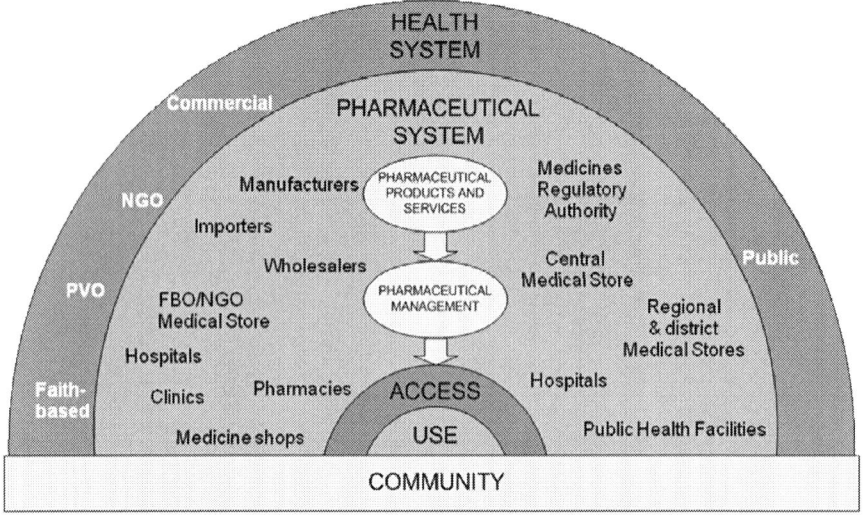

Fig. 22.2 Pharmaceutical systems and access to and use of medicines (adapted from CPM, 2003)

Procurement: the determination of the quantities of medicines needed at different levels of care and types of facilities, how many can be afforded, and purchasing them at competitive prices, of acceptable quality from reliable suppliers.

Distribution: the storage and transportation of medicines and commodities so that they are available where they need to be when needed with minimal losses due to damage, theft, and expiry.

Use: the prescription, dispensing, and consumption of medicines so that desired therapeutic outcomes are achieved.

Within a given health system, these activities are organized as interrelated core functions of system. The ability to effectively and efficiently carry out the activities of pharmaceutical management in a given system is constrained by the ability to take advantage of the existing human, financial and information systems capacity, and by the overarching policies and legislation that influence each of these as well as those specifically related to pharmaceutical products. In the end, a critical criterion for health services quality and a predictor of health services utilization is the availability of medicines.

22.2.2 Human Resources Management

The World Health Organization has estimated that more than four million health workers are needed across the world to fill the gap in the health workforce crisis (WHO, 2006). The specific lack of qualified people to carry out pharmaceutical management activities in developing countries cannot be overstated with limited human resources along the entire medicines chain, from research, manufacture, regulation, inspection, supervision, inventory management, diagnosis, prescription, dispensing. Challenges are encountered with training, recruitment, and retention. The most commonly cited causes of shortages occur from weak educational systems, the impact of diseases such as HIV/AIDS, and the "brain drain" of qualified personnel. Ironically, as global health initiatives seek to expand their programs to increase access, a competition for scarce qualified personnel has served to further aggravate the situation by drawing staff away from the very systems they aim to support.

Interventions to increase human resources capacity for pharmaceutical management aim at improving or expanding the skills sets of existing health facility personnel (Matowe, 2006), developing of cadres of health worker such as community health workers to perform additional specific vital tasks such as prescribing basic medicines (e.g., Senauer et al., 2007; Tawfik et al., 2002), and improving skills of private sector drug vendors (e.g., Rutta et al., 2007). Other options include partnering with or contracting private sector entities (e.g., Siddiqi et al., 2006). With a longer view to sustainability, a complementary approach aims at building local institutional capacity to develop human resources for pharmaceutical management is another (Matowe, 2007).

22.2.3 Information Systems Management

The ability to generate and use data on health system performance is support function of health systems which remains weak in most developing countries and has a profound impact on all aspects of pharmaceutical management. Making appropriate decisions about drug selection requires epidemiologic data (morbidity, mortality and other relevant health outcomes); effective procurement requires data for quantification of needs (historical consumption, morbidity), tendering and contracting (e.g., supplier past performance, price comparators), distribution (e.g., stock movement, stock outs, loss due to expiry, damage, etc.), and use (e.g., patient medication records, laboratory results, dispensing records). Information systems are also needed to track the availability and distribution of pharmaceutical services (e.g., registration, inspection and supervision records).

Some commonly encountered constraints that countries face include an unrealistic or inappropriate information system design (e.g., not taking into consideration technological, geographic or human resources constraints, insufficient consideration to needs of all stakeholders at each level of care) and plan (training plans, financing for recurrent costs). Most recently, ironically, some health systems are becoming overburdened by the various reporting requirements of multiple donor funded programs which do not contribute to the overall functioning of the health system but rather serve to inform donors or program progress.

One area that has been receiving increasing attention from the perspective of information systems is pharmacovigilance. The push to rapidly introduce new medicines into countries without the benefits of previous testing in their populations has highlighted the weaknesses of existing systems to collect and use data on medicines use problems, such as adverse reactions and non-responsiveness to treatment (Pirmohamed et al., 2007). The two major branches of pharmacovigilance programs are one related to clinical trials and another to post-marketing surveillance. Each one implies data collections efforts and analyses that imply potentially substantial inputs in terms of financial and human resources. A framework is needed to support strategic decision making and planning with respect to building viable pharmacovigilance systems within the context of larger health systems of varying strengths and weaknesses.

22.2.4 Finance

The health finance function of health systems is concerned with the way financial resources are generated and used to achieve health impact. With respect to improving access to medicines, this necessarily includes addressing financing for existing appropriate and efficacious medicines (and the services to support their appropriate use) as well as those that do not yet exist (UNHMP, 2005). Financing options related to use of existing resources are generally limited to reallocation to more cost-effective alternatives (for example, selection of less costly treatments) and improving system efficiencies (e.g., improved procurement practices, contracting services, minimizing waste due to theft). There is growing evidence of the impact of innovative approaches to increasing access to medicines such as regional pooled procurement (e.g., Fitzgerald and Gomes, 2003) and contacting out various pharmaceutical management functions has demonstrated some success in improving efficiencies while improving access (MSH n.d.; Liu et al., 2008).

To address issues of equity and sustainability of financing, options may include cost sharing and pooling financial risks through mechanisms such as insurance programs and community health funds. There is considerable evidence that schemes such as user fees and revolving medicines funds have the potential to create perverse incentives that may lead to over-prescribing as well as create additional barriers to access medicines (Haak, 2003). The management of pharmacy benefits as part of a health insurance scheme has been meeting with some success in transition countries and is an area receiving increasing interest in developing countries for the potential to contribute to both increased access to essential medicines and incentives to promote their appropriate use, but significant questions remain about how to implement national health insurance schemes in developing countries (Basaza et al., 2007).

Donor funding is another source of funding which is currently driving much of the concern about the recent increased access to medicines, sustainability, and AMR. Donor funding has also allowed for research in areas considered to be high-risk and low-return investments from the for profit industry perspective such as vaccine research for HIV and malaria vaccines.

Health financing, as other functions of the health system, should reflect established priority health goals and be supported by appropriate legislation and regulation (e.g., Basaza et al., 2007). And, as with the other health system functions, the ability of finance options and implementing mechanisms to achieve desired results is also facilitated or constrained by the limitations of other dependent health systems functions, including the ability to enforce corresponding legislation and regulation, the topic in the following section.

22.2.5 Stewardship and Governance

Stewardship refers to the role of government to safeguard the public interest by defining and acting on priorities and setting standards. Governance refers to how decisions are made and actions carried out in support of stated goals. The UNDP defines good governance by how well decisions and actions correspond to criteria such as transparency, accountability, participatory approaches, inclusiveness and equity, and adherence to the rule of law (UNDP, 1997). In terms of public health, the stewardship role and style of governance are expressed through the national health and medicines policies, the legal and regulatory framework regarding the manufacture, sale and use of medicines, and the ability to support and enforce these through appropriate financing and human resources (Gray, 2004). Together, the stewardship and governance functions of a health system define what pharmaceutical management activities will be undertaken by which entity and how these activities should be carried out.

The large amount of funding available for countries to increase access to medicines and services came with the concern for ensuring that funds are used appropriately. Many include systems of checks and balances, audits, and progress monitoring. For example, eligibility for GFATM grants requires the involvement of a multi-stakeholder Country Coordinating Mechanism (CCM) to ensure that the priority issues are addressed and that grant requests are appropriate. Unfortunately, there is evidence to suggest that these mechanisms often face great challenges to performing their responsibilities, including lack of adequate preparation and supporting systems (e.g., Shretta and Thumm, 2007), leaving them vulnerable to accusations of incompetence and corruption. The recognition of the impact of corruption on decreasing access to essential medicines systems has generated interest in finding ways to combat corruption (e.g., WHO, 2007, 2008). Much of this rests on the ability to enforce the implementation of the tools for governance and the recognized value of the contributions that civil society can make in this regard. Understanding the nature of governance for medicines, defining good governance for medicines, translating it throughout the pharmaceutical management system, and measuring it within an improvement framework is a relatively new and extremely relevant research topic for improving access to medicines.

22.3 Measuring Access

The health systems model presented in Fig. 22.1 identifies access as one of five performance evaluation criteria, the others being equity, quality, efficiency, and sustainability. The criteria of equity, efficiency, and sustainability refer to health systems resources allocation concerns. In terms of access to medicines, in general, and antimicrobials, in particular, must incorporate the qualifiers of quality, safe and effective products and services required to ensure their appropriate use. From a medicines management and research perspective, the unspecified use of the construct *access* can be problematic. This section addresses the operationalization of access.

22.3.1 Access Framework

One framework that has been used to focus on the specific question of access to medicines and has been used with success in Brazil, Cambodia, El Salvador, Ghana, India, and Tanzania is graphically presented in Fig. 22.3 (CPM, 2003). According to this framework, access is understood as a multidimensional construct defined by four key objectively measureable characteristics: geographic accessibility, affordability, availability, and acceptability. Each dimension is further defined as dynamic relationship between characteristics of the consumer or patient and the supplier or provider. For example, the relationship between consumer location and location where products are dispensed captures the dimension of geographic accessibility. Similarly, the relationship between the supply and the demand for medicines reflects the dimension of availability. These characteristics apply throughout the health system, including the public and private sectors. Another defining characteristic of this framework is that the notion of quality product and service is as the object of access is inherent or "built in". With this qualifier, the framework does not allow for consideration of expired nonessential products or inadequate dispensing services to be included in the measure of access. This conceptual framework was then further operationalized through the development of a set of indicators that would lend themselves to generating

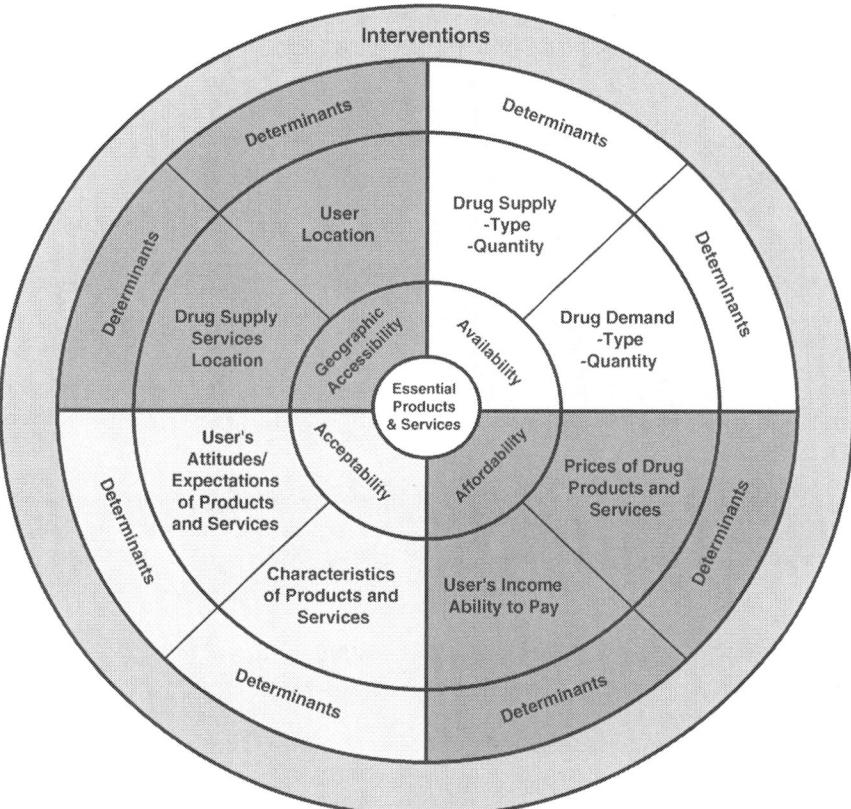

Fig. 22.3 Conceptualizing access to quality medicines and services (SEAM, 2000)

comparable results across sectors, over time, and to some extent, also allow for country comparisons. The framework further allows for tracking the impact of interventions to improve one or more dimensions of access to medicines on other dimensions of access. For example, it is plausible to expect that increasing the availability of quality medicines may result in higher prices, and therefore potentially less affordable.

The framework also differentiates between access to and use of medicines, with access as the necessary but not sufficient condition for use. Medicines use may be understood in terms of three types of behaviors: prescribing, dispensing, and patient adherence. Figure 22.4 depicts the most common types of observable problematic behaviors associated with inappropriate use of antimicrobials (Radyowijati and Haak, 2002) and how these problems are impacted by access issues that are systemic in nature. For this reason, the design of a successful intervention must take into consideration how people behave at the local level with respect to medicines use, as well as how the health system behave at higher levels to support access to medicines.

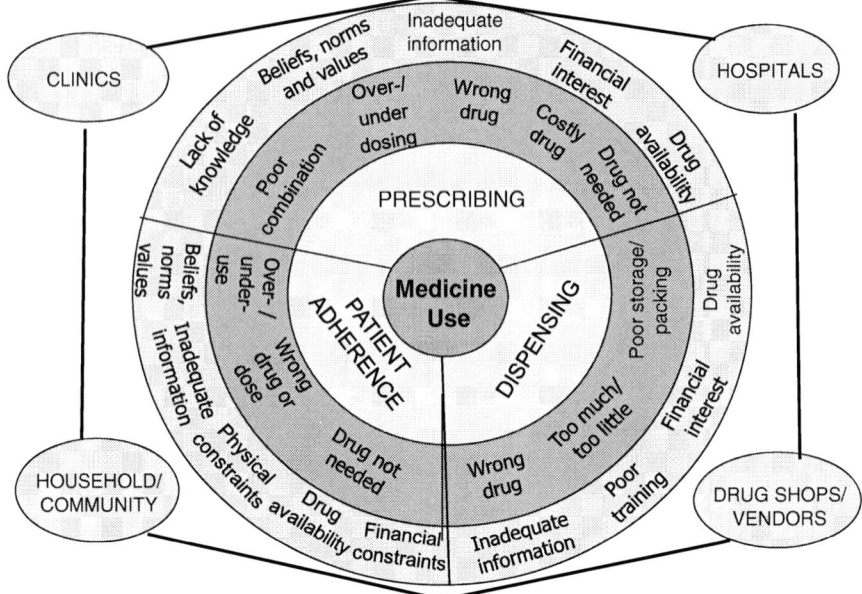

Fig. 22.4 Medicines use framework (adapted from MSH, n.d.)

22.4 Case Study – Improving Access to and Use of Medicines in Tanzania

In 2001, the Strategies for Enhancing Access to Medicines (SEAM) program received funding from the Bull and Melinda Gates Foundation to improve access to medicines important for public health through private sector initiatives. The access to medicines framework described above guided the research to define the nature of the problem. Health systems considerations guided the design of a feasible and viable intervention. The case study outlines how a health systems perspective was applied to address concerns about access to medicines, including appropriate use, from the Tanzania country program (SEAM, 2007).

Problem: In 2001, an extensive indicator-based assessment was conducted of the pharmaceutical sector and revealed a unique opportunity to improve access to and use of essential medicines through the private sector. Duka la dawa baridi (DLDB) are licensed private sector shops authorized by the Tanzania Food and Drugs Authority (TFDA) to provide nonprescription medicines. With an estimated 4600 plus shops nationwide, DLDB provided a relatively broad coverage for purchasing medicines in Tanzania. Although prescription medicines were prohibited for sale in DLDB, they were widely available illegally and the quality of medicines sold at the DLDB was found to be a significant problem given the difficulty in finding reliable and legal sources of medicines and other health-care commodities to sell, indicating the inability of the TFDA to enforce existing regulations. The survey also found that despite relatively high prices or medicines sold at the DLDB due to lack of alternatives, there was a demand for products and services from DLDB. In assessing the services offered by them, however, the assessment determined that dispensing staff lacked basic qualifications and training, and shop owners lacked business skills which potentially limited coverage capacity.

Strategy for Change: The long-term strategy to improve access to quality pharmaceutical products and services was to upgrade the basic level of services offered by DLDB through the creation of a system of accreditation. The explicit goal of the accredited drug dispensing outlet (ADDO) project was to improve access to affordable, quality medicines, and pharmaceutical services in retail drug outlets in rural or peri-urban areas where there are few or no registered pharmacies. Accreditation was built into the model to allow for the eventual participation of ADDOs in the future national health insurance scheme, contributing to the potential long-term financial sustainability of the ADDOs as well as a mechanism to ensure quality standards. The approach combined changing the behavior and expectations of those who use, own, regulate, or work in retail drug shops as well strengthening the capacity of the regulatory authorities to enforce the program. The success of the program rested on taking advantage of a favorable stakeholder environment to make rapid policy and regulatory changes.

Major program activities that contributed to this strategy and the ultimate creation of Duka la Dawa Muhimu (Swahili for essential drug shops) included the development of an accreditation program based on Ministry of Health/TFDA-instituted standards and regulations. Revision of existing legislation was required to improve legal access to a limited list of basic, high-quality prescription, and nonprescription essential medicines (including some key antimicrobials); and to ensure enforcement, inspection and local regulatory capacity had to be strengthened through provision of standard operating procedures and training. Meeting accreditation requirement required ADDO owners to have better business and supervisory skills, and dispensing staff to improve their basic knowledge and practices, so training and supervisory programs had to be designed and an implementation plan developed. Changing the behaviors of ADDO owners also required providing commercial incentives (e.g., access to loans, authorization to sell some prescription medicines) paired with increasing awareness of customers regarding quality and the importance of treatment compliance through marketing and public education.

Results: Required changes in legislation and regulations were achieved relatively quickly by ensuring alignment of all stakeholders early on in the process. The first ADDO shops received accreditation by the TFDA in August 2003. As of August 2005, more than 150 shops were accredited across the Ruvuma region. The SEAM Program conducted an evaluation of the ADDO shops in October and November 2004, comparing them with a control group of DLDB in the Singida region. Significant results include the following.

Quality of Medicines: The proportion of unregistered medicines in intervention region was reduced by a factor of 13, from 26 to 2%. In Singida, the proportion of unregistered medicines was also reduced from 29 to 10%, showing the effect of the broader work of the TFDA to improve registration, but the effect was not as great as in Ruvuma (Fig. 22.5). As a result of this improvement, people in intervention now have a 1 in 50 chance of buying an unapproved medicine, compared to a 1 in 10 chance for the people in the control region.

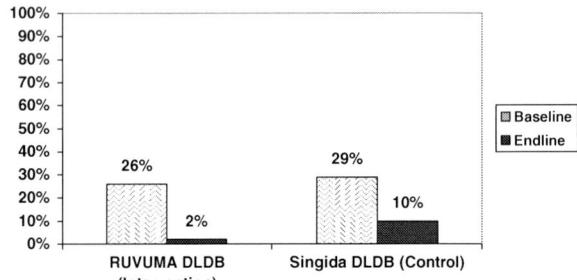

Fig. 22.5 Percentage of unregistered medicines for sale in study regions at baseline (SEAM, 2007)

Quality of Dispensing Services: End line data showed that fewer ADDO attendants (14%) sold or recommended antibiotics for upper respiratory tract infections (URTI) in Ruvuma than at baseline (39%) in 2001 or in Singida at end line (25%) (Fig. 22.6). ADDOs in Ruvuma now have a legal right to sell selected antibiotics and are selling them more responsibly than in 2001, while DLDB are still legally restricted from selling prescription medicines.

Thirty-two percent of malaria treatment encounters at ADDOs included the sale of an appropriate first-line antimalarial (sulfadoxine-pyrimethamine), compared with only 16% at baseline. In 24% of encounters, medicines were

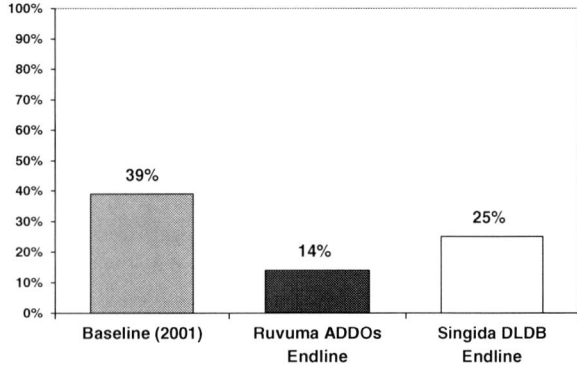

Fig. 22.6 Percentage of simulated upper respiratory tract infection clients dispensed or recommended antibiotics (SEAM, 2007)

dispensed exactly according to standard treatment guidelines, compared with 6% at baseline. There was a significant increase in the percentage of referrals for malaria symptoms in ADDOs in Ruvuma, from 42 to 62%. Medical personnel interpreted this finding as evidence of the efforts of ADDO dispensers to conduct business in an ethical and professional manner; however, future training will focus on increasing dispensers' knowledge and confidence regarding their ability to provide appropriate malaria treatment without referral.

Availability of Essential Medicines: The average availability of prescription medicines in ADDOs was 50% better than the average availability in DLDB. In Ruvuma, the average availability for antimalarials increased from 74 to 90%. Singida also experienced an increase in the availability of antimalarials (60–71%).

Affordability of Medicines and Pharmaceutical Services: The average median price for a set of tracer items was the same at both Ruvuma ADDOs and Singida DLDB at end line, but the median cost of a course of treatment for malaria and URTI was lower in Ruvuma: 60% less for malaria and 10% less for URTI. Analysis of patient registers shows that the customer base has remained stable after the conversion from DLDB, suggesting that prices have not affected sales at ADDOs. The consensus of regional stakeholders is that the increase in prices has not affected access in Ruvuma.

Evidence of Financial Sustainability of ADDO Shops: ADDOs are able to gauge profitability; the majority earns a profit each month. Out of 23 business owners interviewed, all but one believed that it was worth the investment to become an ADDO. New applications for ADDO shops continue to come in, indicating that the business is perceived by the private sector as financially successful.

In early 2006, Tanzania's National Malaria Control Program included ADDOs as part of its national strategy to provide access to malaria treatment. Based on this decision and with encouragement from the Global Fund to Fight AIDS, Tuberculosis, and Malaria, the Government of Tanzania is applying for funding to roll out ADDOs in high-impact regions. The aim is to improve access to artemisinin-based combination therapies (ACTs) to rural children under five. The President's Malaria Initiative is already using ADDOs as a platform for increasing access to malaria drugs and commodities, including subsidized ACTs.

In 2007, the accreditation status allowed for ADDOs to be reimbursed for medicines dispensed to government employees and for the eventual inclusion in the national roll out of the national health insurance program to all citizens. This phase of the program is currently being piloted in a few districts.

Improvements in the Regulatory System: The TFDA implemented a system of regulation in which the local government, acting on behalf of the TFDA, is responsible for regulating ADDOs, including licensing and inspection. Regional and district technical committees have been formed in all Ruvuma districts to carry out basic regulatory functions. At end line, only three ADDOs had been closed because of ethical or regulatory infractions, indicating that general adherence to standards is good. The closures are noteworthy since they indicate that violations are not tolerated and that regulations are being enforced. The TFDA also established a unit to supervise, coordinate, and support all ADDO-related regulatory activities.

Key Lessons Learned: There were several lessons learned from this experience. The most important was the demonstration that quality of pharmaceutical services in the private sector in developing countries can be substantially improved. Even prescription medicines can be rationally dispensed at grassroots drug outlets, but supervision and monitoring are needed to support improvements in rational use, and regulatory inspections are essential. Owner, dispenser, and local regulatory/inspector training (both initial and ongoing) was necessary, costly, and time-consuming. Although the decentralization of regulatory authority responsibilities showed some promise, the eventual achievement of financial sustainability remains to be seen. While building on existing systems and creating and aligning appropriate incentives to support commercial success were clearly important for sustainability, it is also not yet clear if there have been or will be any unintended impact on other health services. These are questions that require continued investigation.

In terms of approach, the broad-based support obtained from stakeholders from the public and private sectors, including consumers, were considered to be critical for success by ensuring complete transparency and modes for accountability, thereby supporting the credibility of the program. The success of the ADDO initiative has led to plans for the Tanzanian government to roll out the concept nationwide (over 20 additional regions) and to use ADDOs as a base for additional public health services (e.g., child health, HIV/AIDS support, malaria control).

22.5 Conclusion

Ensuring regular access to safe, affordable medicines of assured quality is arguably one of the most critical functions of a health system. It stands to reason that increasing access to medicines is meaningless if access is not regular and sustainable and use of those medicines is not appropriate. In the case of antimicrobials in particular, the increased threat of AMR as a negative effect of the inability of weak health systems to manage the rapid influx of medicines into countries is a reality that requires immediate and sustained attention. In particular, with respect to these major infectious diseases, the immediate challenge to health policy makers, planners and providers, and consumers and other stakeholders is to "get it right" before available treatments are no longer effective and treatment alternatives are not available. Getting it right requires appreciating the potential system-wide implications of having to manage these important medicines, as much as the implications of health system failures on assuring the equitable and sustainable access to medicines. For this, the value of health systems perspective is the application strong conceptual and analytical frameworks to guide the design of appropriate interventions and rigorous methodologies for evaluation and comparative analyses so that our understanding of how health systems work and can be improved and applied judiciously.

Acknowledgments The Strengthening Pharmaceutical Systems (SPS) Program is supported by the US Agency for International Development under the terms of cooperative agreement number GHN-A-00-07-00002-00 with Management Sciences for Health. The opinions expressed herein are those of the author and do not necessarily reflect the views of the US Agency for International Development.

References

Alliance for Health Policy and Systems Research (AHPSR). 2004. Strengthening Health Systems: The Role and Promise of Policy and Systems Research. Geneva: Global Forum for Health Research. Source: http://www.who.int/alliance-hpsr.

Basaza, R., Criel, B., and Van der Stuyft, P. 2007. Low enrollment in Ugandan community health insurance schemes: underlying causes and policy implications. BMC Health Services Research 7: 105. Source: http://www.pubmedcentral.nih.gov/articlerender.fcgi?artid = 1940250.

Caines, K. and Buse, K. 2004. Assessing the Impact of Global Health Partnerships: Synthesis of Findings from the 2004 DFID Studies on Global Health Partnerships: Assessing the Impact. London: DFID Health Resources Center.

Center for Pharmaceutical Management (CPM). 2003. Defining and Measuring Access to Essential Drugs, Vaccines, and Health Commodities: Report of the WHO-MSH Consultative Meeting, Ferney-Voltaire, France, December 11–13, 2000. Prepared for the Strategies for Enhancing Access to Medicines Program, Arlington, VA: Management Sciences for Health. Source: http://www.msh.org/seam/reports/measuring_access_Dec2000.pdf.

Fitzgerald, J. and Gomes B. 2003. An open competition model for regional price negotiations yields lowest ARV prices in the Americas. Presented to the 8th World STI/AIDS Conference, Punta del Este, Uruguay, December 2–5, 2003.

The Global Fund to Fight HIV/AIDS, Tuberculosis and Malaria (GFATM). 2008. Source: http://www.theglobalfund.org/en/performance/results/.

Gray, A. 2004. Access to Medicines and Drug Regulations in Developing Countries: A Resource Guide for DFID. London: DFID Health Systems Resource Centre.

Haak, H. 2003. Access to Antimalarial Medicines: Improving the Affordability and Financing of Artemisinin-Based Combination Therapies. WHO: Geneva. Source: http://www.emro.who.int/rbm/background%20documents/egy04/afford.pdf.

International Conference on Improving Use of Medicines (ICIUM). 2004. Source: http://mednet3.who.int/icium/icium2004/recommendations.asp

Islam, M., ed. 2007. Health Systems Assessment Approach: A How-To Manual. Arlington. Arlington, VA: Management Sciences for Health.

Liu, X. Hotchkiss, D. R., and Bose, S. 2008. The effectiveness of contracting-out primary care services in developing countries. Health Policy and Planning 23:1–13.

Management Sciences for Health (MSH). In press. *MDS-3: Managing Access to Medicines and Essential Health Commodities*. Arlington, VA: MSH.

Matowe, L. 2006. Using the Monitoring-Training-Planning Approach to Build Problem-Solving Skills for Health Care Staff to Support Pharmaceutical Management during ART Scale Up. Presented at the XVI International Aids Conference, Toronto, 13–18 August. Source: http://www.aids2006.org/PAG/PSession.aspx?s = 473.

Matowe, L. 2007. Building a regional network of academic experts to increase capacity in pharmaceutical management in East Africa. Presented at the Annual meeting of the American Public Health Association, November 7, 2007. Source: http://apha.confex.com/apha/135am/techprogram/paper_160781.htm.

Nugent, R., Pickett, J., and Back, E. 2008. Drug Resistance as a Global Health Policy Priority. Drug Resistance Working Group Background Paper. Center for Global Development. Source: http://www.cgdev.org/doc/ghprn/Concept%20Paper.pdf.

Pirmohamed, M., Atuah, K. N., Dodoo, A. N., and Winstanley, P. 2007. Pharmacovigilance in developing countries. BMJ 335:462.

Radyowijati, A. and Haak, H. 2002. Determinants of Antimicrobial Use in the Developing World. Child Health Research Project, Special Report 4:1.

Roberts, M, Hsiao, W., Berman P., and Reich, M. 2004. Getting Health Sector Reform Right: A Guide to Improving Performance and Equity. New York: Oxford University Press.

Rutta, E., Liana, J., and Mbwasi R. 2007. Building capacity for pharmaceutical services where pharmacies do not exist: The ADDO experience in Tanzania. Presented at the Annual Meeting of the American Public Health Association, November 7, 2007. Source: http://apha.confex.com/apha/135am/techprogram/paper_160700.htm.

Ruxin, J, Paluzzi, J. E., Wilson, P. A., Tozan, Y., Kruk, M., and Teklehaimanot, A. 2005. Emerging consensus in HIV/AIDS, malaria, tuberculosis, and access to essential medicines. Lancet 365: 618–21.

Senauer, K., Briggs, J., Saleeb, S., and Adeya, G. 2007. Improving child health through informed policy decisions and targeted interventions to strengthen medicine management in the community: The example of Senegal. Presented at the Annual Meeting of the American Public Health Association, November 7, 2007. Source: http://apha.confex.com/recording/apha/135am/pdf/free/4db77adf5df9fff0d3caf5cafe28f496/paper153789_1.pdf .

Shretta, R. and Thumm, M. 2007. Global Fund Grants for Malaria: Lessons learned in the implementation of ACT policies in Ghana, Nigeria and Guinea-Bissau. Rational Pharmaceutical Management Plus Program. Arlington, VA: Management Sciences for Health.

Siddiqi, S., Masud, T. I., and Sabri, B. 2006. Contracting but not without caution: experience with out-sourcing of health services in countries of the Eastern Mediterranean Region. Bulletin of the World Health Organization 84(11): 867–875.

Strategies for Enhancing Access to Medicines Program (SEAM). 2007. Tanzania: Accredited Drug Dispensing Outlets—*Duka la Dawa Muhimu* Summary Report. Arlington, VA: Management Sciences for Health. Source: http://www.msh.org/seam/country_programs/3.1.4.htm

Tawfik, J., Kinoti, S., and Blain, G. 2002. Introducing Antiretroviral Therapy (ART) on a Large Scale: Hope and Caution. AED Global Health, Population and Nutrition Group. Washington, D.C.: Academy for Educational Development.

Trachtenberg, J. and Sande, M. 2002. Emerging resistance to nonnucleoside reverse transcriptase inhibitors: A warning and a challenge. JAMA 288 (2):239–241. Source: http://jama.ama-assn.org/issues/v288n2/ffull/jed20035.html.

UNDP (United National Development Programme). 1997. Governance for Sustainable Human Development: A UN Policy Document. Source: http://mirror.undp.org/magnet/policy/.

UNHMP (United Nations Health Millennium Project). 2005. Prescription for Healthy Development: Increasing Access to Medicines. Report of the Task Force on HIV/AIDS, Malaria, TB and Access to Medicines, Working Group on Access to Essential Medicines.

UNHMP (UN Millennium Project). 2005. Prescription for Health: Increasing Access to Medicines. Report of the Task Force on HIV/AIDS, Malaria and Tuberculosis Access to Essential Medicines Working Group on Access to Essential Medicines. Sterling, VA: Earthscan.

WHO (World Health Organization). 2001. Interventions and strategies to improve the use of antimicrobials in developing countries. Geneva: WHO.

WHO (World Health Organization). 2002. World Health Report 2000: Health Systems: Improving Performance. Geneva: WHO.

WHO (World Health Organization). 2006. Human Resources for Health (HRH). Geneva: WHO. Source: http://www.who.int/whr/2006/en/.

WHO (World Health Organization). 2007. Everybody's Business: Strengthening Health Systems to Improve Health Outcomes. WHO's Framework for Action. Geneva: WHO.

WHO (World Health Organization). 2008. Good Governance for Medicines Programme. Geneva: WHO. Source: http://www.who.int/medicines/ggm/en/index.html.

Chapter 23
The Role of Unregulated Sale and Dispensing of Antimicrobial Agents on the Development of Antimicrobial Resistance in Developing Countries

Eric S. Mitema

Abstract Antimicrobial resistance has become a major medical and public health problem worldwide. This has been brought about by overuse and/or misuse of these drugs especially in developing countries. The role of unregulated sale and dispensing of antimicrobial agents on the development of resistance is presented in this chapter.

Developing countries use both branded and generic antimicrobial agents which include major classes like beta-lactams, tetracyclines, aminoglycosides, fluoroquinolones, sulfonamides, chloramphenicol, macrolides, lincosamides, polymixins, bacitracin, vancomycin, vovobiocin, and rifamycins among others. Despite existing legislative framework to control antibiotics in most developing countries, enforcement of these regulatory processes is still a problem. Some of the factors that have led to unregulated sale and dispensing of antibiotics in developing countries include inadequate legislative framework, lack of financial resources for implementation, inadequate education on the awareness of public health risks of antimicrobial resistance, social–cultural attitudes on antimicrobial agents, easy access to antimicrobial agents due to high prevalence of HIV/AIDS, and lack of surveillance and monitoring programs on antibiotic usage in medical and veterinary use among others. Intervention measures required to contain the upsurge of antimicrobial resistance in developing countries should include the following among others: increased public education and creation of awareness to the public and pharmacists, enactment of laws where none exists and their proper enforcement and regular surveillance, and monitoring of antimicrobial usage and emergence of resistance. Mitigation measures mentioned above will go along way in the containment of antimicrobial resistance and hence protection of public health from emerging antimicrobial resistance.

E.S. Mitema (✉)
Department of Public Health, Pharmacology and Toxicology, Faculty of Veterinary Medicine, University of Nairobi, Nairobi, Kenya
e-mail: esmitema@uonbi.ac.ke

A. de J. Sosa et al. (eds.), *Antimicrobial Resistance in Developing Countries*,
DOI 10.1007/978-0-387-89370-9_23, © Springer Science+Business Media, LLC 2010

23.1 Introduction

Antimicrobial agents are pharmacotherapeutic armaments that have been referred to as "magic bullets" capable of selectively killing bacterial pathogens and sparing the body cells. However, these once wonderful bullets over 60 years ago can no longer be effective due to upsurge of bacterial resistance associated with overuse or misuse. Antimicrobial resistance has therefore become a major medical and public health problem. The main factor responsible for the development and spread of bacterial resistance is injudicious use of antimicrobial agents, which has resulted in most gram-positive and gram-negative bacteria continuously developing resistance to the antimicrobials in regular use at different time periods (Urassa et al. 1997). Resistant bacteria can diminish the effectiveness of antibiotics and demand the use of more expensive or less safe alternatives (Tollefson et al. 1999).

23.2 Classes of Antimicrobial Agents

Developing countries use both branded and generic antimicrobial preparations for the management of bacterial infections, even though generics exceed in usage due to their low costs, which can be afforded by many people. Some of the newer classes of antimicrobial agents like third- and fourth-generation cephalosporins and fluoroquinolones, respectively, are rarely used due to their high costs.

Major classes of antibiotics which are used include beta-lactams (penicillin and cephalosporins), tetracyclines, aminoglycosides, fluoroquinolones, sulfonamides, chloramphenicol, macrolides, lincosamides, polymixins, bacitracins, vancomycin, novobiocin, and rifamycins (Kahn and Line 2005).

23.3 Regulation of Antimicrobial Agents

Most developing countries have statutory regulatory authorities or agencies established to control drugs including antibiotics. These regulatory authorities are usually established as acts of legislative framework and are responsible for the registration review process, issuance of importation or exportation licenses, distribution and approval for market authorization. What lacks in most developing countries is the enforcement of regulatory processes, which may be attributed to inadequate staff, political interference, or insufficient financial resources.

The role of regulatory agencies is to protect public health by ensuring that antimicrobial agents to be authorized are of good quality, safe, and efficacious. The review process by regulatory agencies in most developing countries is not vigilant enough due to either non-competent authorities or inadequate resources. In some cases, proper legislation for the review process is inadequate. Establishment of proper regulation in various developing countries to oversee approval process and distribution and to ensure that all antibiotics undergo prescriptions is fundamentally important to contain any upsurge of antimicrobial resistance.

23.3.1 Factors Contributing to Unregulated Sale and Dispensing

More than 65 years ago antimicrobial agents killed all bacteria in the tissues; however, these compounds no longer can kill many strains of bacteria which are now capable of making proteins that nullify their effects. Each such protein is expressed by a gene that the susceptible ancestors lacked. The problem is therefore the rise of prevalence of such resistant genes in bacterial populations. In developing countries it is estimated that 35% of the total health budget is spent on antimicrobials versus 11% in developed nations. It is also reported that in the developing world communicable diseases are thought to be responsible for 58.6% of deaths when compared to 34.2% in the developed world (Gwatkin and Guillot 2000). Control of bacterial diseases is therefore under threat in most developing countries where unregulated sale and distribution of antimicrobial agents is not properly enforced. It is now known that increased use of antimicrobial agents leads to emergence of resistant strains. Even in the developed countries where proper control of antibiotics is implemented, cases of antimicrobial resistances are still on the rise; the situation therefore would be more compromised in developing countries.

Unregulated antimicrobial agents can predispose to overuse and misuse of these products in the health-care system. Consequently, emerging resistant infectious agents can adversely affect mortality, treatment costs, disease spread, and duration of illness (Laxminarayan 2003). Multiple drug-resistant organisms have been prevalent in hospital in the past, where antimicrobial agents are commonly prescribed. However, recently community-acquired infections especially gastrointestinal and respiratory infections have become the leading causes of death in developing countries (Wenzel and Edmond 2000).

Data on antimicrobial resistance, unregulated sale, and prescription are very scarce in developing countries. It is now thought that unregulated sale of antibiotics could immensely increase resistant bacterial organisms and genes. Some of the factors that can lead to unregulated sale and dispensing in the developing world include the following:

a) Lack of proper legislative framework
b) Lack of financial resources for implementation of legislation
c) Lack or inadequate education on the awareness of public health risk of antimicrobial resistance
d) Socio-cultural attitudes on antimicrobial agents

e) Easy access to antimicrobial agents from some non-governmental organizations to poor people in rural areas
f) Lack of surveillance and monitoring programs on antibiotic usage in human and veterinary medicine
g) High prevalence of HIV/AIDS especially in Africa prompting easy access to lifesaving antibiotics like third- and fourth-generation cephalosporins and quinolones, respectively.

23.4 Prescription Patterns and Effect on Development of Resistance

Data on antimicrobial drug use and sale in most developing countries are scarce. Most countries do not publish national statistics on drug use and pharmaceutical companies keep information private for trade reasons. However, estimates on consumption complied from importation and companies from respective ministries of health and trade indicate bacterial resistance is low, but increasing in countries that are less- heavy consumers of antibiotics (Isturiz and Carbon 2000). In Kenya, data on prescription pattern for food animal and human use have been reported (Mitema et al. 2001 and Mitema and Kikuvi, 2004). In food animals, oxytetracyclines are the most prescribed in Kenya, whereas the penicillins (extended type) are the most prescribed in human medicine. Oxytetracyclines are broad spectrum and relatively cheap and this may influence their increased usage. Extended penicillins are also relatively affordable and have broad spectrum of activity on both gram-positive and gram-negative bacteria. In Pakistan, ampicillin/amoxicillin (Pakistan Pharmaceutical Index 2005) tops the list of prescription and this is similar to the Kenyan scenario.

Even though antimicrobial agents are regulated in most African countries, enforcement of prescriptions is minimal since most patients can still have access to antibiotics in some countries (Indalo 1997 and Mbori-Ngacha 1997). In Thailand, the haphazard use of antimicrobial agents has accelerated dissemination of antimicrobial resistance (Sirinavia and Dowell 2004). In other developing countries such as the Philippines (Lansang et al. 1990) and Vietnam (Larsson et al. 2000) the sale of antibiotics is largely unregulated and this has predisposed to antimicrobial resistance development.

Accessibility to these drugs can lead to enhanced selective pressure and development of resistant pathogens. Such an easy availability has provoked the development of resistance among infectious agents of pneumococcal meningitis, tuberculosis, typhoid fever, and nosocomial infections (Hart and Kariuki 1998; Okeke et al. 2005a, b).

The situation in veterinary practice in most developing countries is even worse, where lay users have access to antibiotics and can administer them in food animals. Some zoonotic and indicator organisms can develop resistance to particular antimicrobial agents and these can be transferred via the food chain to man (Mitema et al. 2001; Kikuvi et al. 2007). In developed nations like Denmark (DANMAP 2005), the use of antimicrobial agents in food animals from 2004 to 2005 remained unchanged with tetracyclines, beta-lactamases, and sulfonamides leading in that order in terms of usage. In humans, the beta-lactamases sensitive penicillins, extended-spectrum penicillins, and macrolides accounted for 70% of the usage in Denmark. Zoonotic bacteria like *Salmonella* Typhimurium, *S. enteritidis, and Campylobacter jejuni* isolates whose origin is outside Denmark showed an increase in resistance to commonly used antibiotics. In Sweden (SWEDRES/SVARM 2005) and Norway (NORM/NORM-VET 2005) the situation was generally similar to Denmark in which there was a decline in resistance associated with drug usage.

23.5 Some Selected Antibiotics and Impact on Bacterial Resistance Development

23.5.1 Penicillins

Penicillins contribute to over 60% of usage among antimicrobial agents in both developing (Mitema et al. 2001) and developed countries (DANMAP, 2005; SWEDRES/SVARM 2005; NORM/NORM-VET 2005). Penicillins have been among the most misused and overused antibiotics. The saturation of some environments, for instance, hospitals, poultry, and pig units with penicillin has produced selective pressure favoring penicillin-resistant organisms like staphylococci. Nearly 90% of all staphylococcal strains in hospital and in the environmental community including dairy farms are beta-lactamase producers. A similar trend is being observed for methicillin-resistant strains of *Staphylococcus aureus* (MRSA). Misuse of penicillin by either unregulated control or dispensing has also led to resistance development due to production of beta-lactamase by strains of *Haemophilus influenzae* and *Neisseria gonorrhoeae*, precluding use of penicillin as first-line agent for infections caused by these organisms.

23.5.2 Tetracycline

Tetracyclines have been extensively used in veterinary medicine in both developing (Mitema et al. 2001) and developed countries (DANMAP 2005; SWEDRES/SVARM 2005). Overusage due to either lack of regulation or indiscriminate dispensing has led to emergence of resistance even among once highly susceptible species, for instance, pneumococci, staphylococci, and group A streptococci (Chambers 2002). Tetracyclines have also been used in animal feeds to enhance growth and this practice has contributed to the steadily increasing spread of tetracycline resistance among enteric bacteria and plasmids that enable tetracycline resistance genes.

23.5.3 Newer Agents

Due to multidrug resistance of many bacteria like strains of MRSA, vancomycin-resistant strains of *Enterococcus faecium, Streptococcus pneumoniae, and Listeria monocytogenes*, newer and expensive streptogramins (quinupristin–dalfopristin, Synercid[R], and linezolid) have been approved for use especially against *E. faecium* which has developed resistant to vancomycin. These streptogramins are relatively expensive for users in developing countries; hence their use is still limited to a few hospital practices.

23.5.4 Fluoroquinolone

These are very useful, easily bioavailable agents and when given orally are effective for the treatment of a variety of serious bacterial infections such as those caused by *Pseudomonas aeruginosa* or strains of MRSA which formerly could only be treated by parenteral drugs. They have become drugs of choice for urinary tract infections superseding trimethoprim–sulfamethoxazole in *E. coli* and other gram-negative pathogens.

Their overuse has led to development of resistance by pneumococci and streptococci, which had been sensitive. Their use in strains of MRSA should not be undertaken until in vitro susceptibility tests are performed. WHO (2000) had recommended the use of fluoroquinolone in farm animals only after culture sensitivity tests in order to contain transfer of resistant determinants from animals to humans. Recently the United States Food and Drug Administration banned the use of enrofloxacin oral medication in flock treatment in poultry due to emergence of resistant *Campylobacter* in humans associated with its use in poultry. Other developing countries, for instance Kenya, have also followed suit banning the use of oral medication preparation of enrofloxacin.

23.6 Control of Antimicrobial Agents: Interventions

Containment of antimicrobial resistance is a multipronged approach, which includes among others, proper practices in regulation and dispensing of antimicrobial agents. Implementation of WHO (2001) strategy for the containment of antimicrobial resistance and WHO (2000) global principles has been undertaken in most developed countries, but some countries in the developing world have so far not implemented all interventions. In a recent study in Kenya, Ole-Mapenay (2007) reported an association between tetracycline and beta-lactam usage in food animals and occurrence of resistance among ecological *E. coli* isolates. In developing countries where regulation may be inadequate, the use of antibiotics will definitely increase and this may even enhance development of bacterial resistance. Some interventions for control and dispensing antibiotics will be discussed below to reduce the spread of antimicrobial resistance.

23.6.1 Public Education

Creation of awareness through public education as an intervention strategy in the control of unregulated antimicrobial agents is affordable since manpower is available. In the developed countries awareness of antimicrobial resistance is quite advanced through the media, literature reading, and other outlets. Educational interventions can change the behavior of people from using drugs without proper prescriptions and this can be reinforced through multiple interventions. Inclusion of pharmacotherapy modules into standard medical, veterinary, and allied health curricula has been shown to be cost-effective means for inculcating rational and evidence-based prescribing practice in medical training in both developing and developed countries (De Vries et al. 1995; Vollebbregt et al. 2006). Continuing education programs for medical and all health allied professionals are desirable, because in their absence, the practices invariably rely on information from pharmaceutical companies, which are not always consistent with rational antimicrobial use (Bhutta and Vitry 1997). Focused group workshops and large seminars should be initiated in various developing countries. The use of published pamphlets, newspapers, televisions, and radios should be encouraged to create awareness on the risks associated with using unregulated drugs. An important component of education intervention is long-term commitment and refresher courses for the public.

23.6.2 Pharmacy Education

Even though regulation may exist, most pharmacies in many developing countries will dispense antibiotics on demand or by suggestion after listening briefly to the buyer who may not be the patient in certain cases (Isturiz and Carbon, 2000 and Butta and Balchin 1996 and Indalo 1997). In Bangladesh, for instance, the situation has been reported to be even worse where up to 90% of drugs can be sold without prescription (Isturiz and Carbon, 2000). In a Vietnam study (Larsson et al. 2000), patients always seek treatment at drug retail outlets. Many pharmacists and drug sellers have not been formally trained in diagnosis. They will sell these products to make a sale and accommodate patients' ability to pay. For instance, it has been reported (Gachia and Holmes 2003) that drug sellers in Peru prefer small interactive group seminars, continued supportive contact, and small incentives like refreshments and client material. More educational programs should target pharmacists, pharmacy attendants, importers, wholesalers, and even the pharmaceutical agents to reduce the sale of non-prescribed antibiotics.

23.6.3 Establishment of Legislative Framework and Law Enforcement

Most developing countries have legislative acts to control antimicrobial agents just like any other ethical products and in a few war torn countries like Somali and Iraq may have none or not enforced at all. The problem in most developing countries is better regulation and enforcement of the laws (Sirinavia and Dowell 2004). Countries in the developing world should enact appropriate legislation and increase funding in enforcement of these laws.

23.6.4 Regular Surveillance and Monitoring

The WHO (2001) strategy for the containment of antimicrobial resistance and WHO (2000) global principles for the containment of antimicrobial resistance from food animals recommend regular surveillance and monitoring of antimicrobial use and resistance. Because unregulated antimicrobial agents can enhance emergence of antimicrobial resistance, it is also very important to investigate the extent of antimicrobial agents sold without prescriptions. Where a large proportion of antimicrobial agents are sold without prescriptions, proper legislative and enforcement should be enacted to minimize possible upsurge in antimicrobial resistance. An increase in antimicrobial resistance could lead to escalated costs, use of newer expensive antimicrobial agents, and longer convalescence period of sick patients.

References

Bhutta T.I. and Balchin C. (1996). Assessing the impact of a regulatory intervention in Pakistan. Soc Sci Med 42: 1195–202.
Bhutta T.I. and Vitry A. (1997). Treating dysentery with metronidazole in Pakistan. BMJ 314: 146–147.
Chambers H.F. (2002). Antimicrobial drugs. In: Basic and Clinical Pharmacology 8th ed. By B.G. Katzung, New York: Lange Medical Books/McGraw-Hill.
DANMAP (2005). Use of antimicrobial agents and occurrence of antimicrobial resistance in bacteria from food animals, foods and humans in Denmark. DANMAP 2005-July 2006. ISSN 1600–2032.
De Vries T.P., Henning R.H., Hogerzeil H.V., Bapna J.S., Bero L., Kafle K.K., Mabadeje A., Santos B. and Smith A.J. (1995). Impact of a short course in pharmacotherapy for undergraduate medical students: An international randomized study. Lancet 346: 1454–1457.
Gachia P.J. and Holmes K.K. (2003). STD trends and patterns of treatment for STD by physicians in private practice in Peru. Sex Transm Infect 79: 403–407.
Gwatkin, D.R. and Guillot M. (2000). The burden of disease among global poor. Current trends, future trends and implications for strategy. Washington DC: The World Bank. Health, Nutrition and population series.
Hart C.A. and Kariuki S. (1998). Antimicrobial resistance in developing countries. BMJ 317: 647–650.
Indalo A.A. (1997). Antibiotic sale behavior in Nairobi. A contributing factor to antimicrobial drug resistance. East Afr Med J 74: 171–173.

Isturiz R.E. and Carbon C. (2000). Antibiotic use in developing countries. Infect Control Hosp Epidemiol 21: 394–397.

Kahn C.M. and Line S. (2005). Antimicrobial agents. In: The Merk Veterinary Manual, 9th ed., Whitehouse, NJ: Merck **and Co** 2056–2098.

Kikuvi G.M., Schwarz S., Ombui J.N., Mitema E.S and Krehrenberg K. (2007). Streptomycin and chloramphenicol resistance genes in *Escherichia coli* from cattle, pigs and chicken form Kenya. Microbial Drug Res 13: 63–69.

Lansang M., Lucas-Aquino R., Tupasi T., Mina V.S., Salazar L.S., Juban N. Limjoco T.T., Nisperos L.E. and Kunin C.M. (1990). Purchase of antibiotics without prescription on Manila, the Philippines. J Clin Epidemiol 43: 61–67.

Larsson M., Kronvall G., Chuc N.T., Karlsson I., Lager F., Hanh H.D., Tomson G. and Falkenberg T. (2000). Antibiotic medication and bacterial resistance to antibiotics: A survey of children in a Vietnamese community. Trop Med Int Health 5: 711–721.

Laxminarayan R. (2003). Battling resistance to antibiotics and pesticides: An economic approach. Washington DC: Resources for future.

Mitema E.S., Kikuvi G.M., Stohr K. and Wegener H. (2001). An assessment of antimicrobial consumption in food producing animals in Kenya. J Vet Pharmacol Ther 24: 385–390.

Mitema E.S. and Kikuvi G. (2004). Surveillance of overall use of antimicrobial drugs in humans over a five-year period (1997–2001) in Kenya. J Antimicob Chemother 54: 966 67.

Mbori-Ngacha D.N. (1997). Rational approach to limiting emergence of antimicrobial drug resistance. East Afr Med J 74: 187–189.

NORM/NORM-VET (2005). Usage of antimicrobial agents and occurrence of antimicrobial resistance in Norway Tromso, Oslo 2006. ISSN: 1502–2307.

Okeke I.N., Laxminarayan R., Bhutta Z.A., Duse A.G., Jenkins P., O'Brien T.F., Pablos-Mendez A. and Klugman K.P. (2005a). Antimicrobial resistance in developing countries. Part I: Recent trends and current status. Lancet Infect Dis 5: 568–580.

Okeke I.N., Klugman K.P., Bhutta Z.A., Duse A.G., Jenkins, P., O'Brien, T.F., Pablos-Mendez A. and Laxminarayan R. (2005b). Antimicrobial resistance in developing countries. Part II: Recent trends and current status. Lancet Infect Dis 5: 481–493.

Ole-Mapenay I.M. (2007). An assessment of antimicrobial usage and occurrence of antimicrobial resistance in *E. coli* from food animals in Kenya. PhD Thesis. University of Nairobi, Kenya.

Pakistan Pharmaceutical Index. (2005). Selected Information on Drug Sales (1988–2002): Karachi, Pakistan: Information Medical Statistics. http://www.imshealth.com (Accessed June 30, 2005).

Sirinavia S. and Dowell S.F. (2004). Antimicrobial resistance in countries with limited resources. Unique challenges and limited alternatives. Semin Pediatr Infect Dis 15: 94–98.

SWEDRES/SVARM (2005). A report of Swedish antibiotic utilization and resistance in human medicine. SWEDRES 2005. ISSN 1400–3473.

Tollefson L, Fedorka-Cray P.J. and Angulo F.J. (1999). Public health aspects of antibiotic resistance monitoring in the USA. Acta Vet Scand Suppl 92: 67–75.

Urassa W., Lyamuya E. and Mhalu F. (1997). Recent trends on bacterial resistance to antibiotics. East Afr Med J 74: 129–133.

Vollebbregt J. A., van Oldenriik J., Kox D., van Galen S.R., Sturm B., Metz J.C., Richir M.C., de Haan M., Hugtenburg J.G. and de Vries T. P. (2006). Evaluation of a pharmacotherapy context-learning programme for preclinical medical students. Br J Clin Pharmacol 62: 666–672.

Wenzel R.P and Edmond M.B. (2000). Managing antibiotic resistance. New Engl J Med. 343: 1961–1963.

World Health Organization (2000). Global principles for the containment of antimicrobial resistance due to antimicrobial use in animals intended for food June 5–9, Geneva, Switzerland.

World Health Organization (2001). WHO global strategy for containment of antibiotic resistance. Geneva: 99.

Chapter 24
Counterfeit and Substandard Anti-infectives in Developing Countries

Paul N. Newton, Facundo M. Fernández, Michael D. Green,
Joyce Primo-Carpenter, and Nicholas J. White

Abstract There is considerable interest in optimizing the therapy for important infections in developing countries and in making the best treatments readily available and inexpensive. There is also great concern that resistance to anti-infective drugs is worsening, putting affordable treatments at risk. We argue that an important, but usually neglected aspect of these problems is drug quality. Drugs may be of poor quality if they are counterfeit, substandard or degraded. Few objective data on the prevalence of poor-quality drugs exist but surveys suggest that an alarming proportion of antimalarials and antibiotics in much of the developing world are of poor quality. For individual patients these will increase mortality and morbidity and lead to loss of faith in medicines and health systems. Counterfeit, substandard or degraded drugs with sub-therapeutic concentrations of the active ingredient or the wrong active ingredient are likely to engender the emergence and spread of resistance to these anti-infectives. Although modelling suggests that poor-quality drug should worsen drug resistance, there is sparse evidence from the field, as there has been little research. It will be very difficult to distinguish the effects of poor-drug quality and reduced patient adherence and incorrect health worker prescribing on the spread of resistance. Strengthening drug regulatory authorities, improving quality of drug production and facilitating the availability of relatively inexpensive, good-quality anti-infectives are likely to be key factors in improving drug quality.

24.1 Introduction

There is increasing concern that much of the anti-infective drug supply in the developing world is of poor quality. This problem is not new; counterfeit antimalarials were a severe problem in the 17th century when counterfeits of

P.N. Newton (✉)
Wellcome Trust–Mahosot Hospital–Oxford Tropical Medicine Research
Collaboration, Microbiology Laboratory, Mahosot Hospital, Vientiane, Lao PDR;
Centre for Tropical Medicine, Churchill Hospital, University of Oxford, UK
e-mail: paul@tropmedres.ac

A. de J. Sosa et al. (eds.), *Antimicrobial Resistance in Developing Countries*,
DOI 10.1007/978-0-387-89370-9_24, © Springer Science+Business Media, LLC 2010

the first potent antimalarial drug, cinchona bark (the source of quinine), were widely marketed in Europe (Newton et al. 2006a). Over the last decade, it has become increasingly apparent that counterfeit and substandard medicines are a very important but largely unrecognized public health problem. The use of poor-quality anti-infectives can undermine public confidence in health programs and waste scarce resources. More importantly, it can lead to serious health consequences such as treatment failure due to suboptimal dosing, adverse reactions, and increased morbidity and mortality (Newton et al. 2002; Primo-Carpenter 2004). To what extent the apparent increase is a real change or an artefact due to emerging interest remains uncertain, as both historical and contemporary data are sparse.

Poor-quality drugs can be classified as counterfeit, substandard or degraded. A counterfeit medicine is 'deliberately and fraudulently mislabelled with respect to identity and/or source. Counterfeiting can apply to both branded and generic products and counterfeits may include products with the correct ingredients or with the wrong ingredients, without active ingredients, with insufficient active ingredient or with fake packaging' (Wondemagegnehu 1999). In contrast, 'substandard drugs are genuine drug products which do not meet quality specifications set for them' (World Health Organisation 2005). Substandard products may arise as a result of lack of expertise or insufficient manufacturing infrastructure (Primo-Carpenter 2004). In addition, good-quality drugs can deteriorate after production due to inadequate storage but it can be difficult to distinguish such degraded drugs from those which left the factory as substandard (Keoluangkhot et al. 2008). The amount of active ingredient is not sufficient information to determine accurately whether a medicine is counterfeit and inspection of the packaging is required. In many reports it is unclear whether a poor-quality medicine is counterfeit, substandard or degraded. A recent paper referred to all poor-quality antimalarials and medicines clinically inappropriate as 'substandard' and did not attempt to distinguish substandard and counterfeit medicines (in the senses of the WHO definition) 'as neither is clinically suitable' (Bate et al. 2008). However, the distinctions are vital because the reasons for their dissemination and potential countermeasures differ considerably.

There are few published primary research reports (for reviews, see ten Ham 1992; Reidenberg and Conner 2001; Primo-Carpenter 2004; Deisingh 2005; Primo-Carpenter and McGinnis 2007; Harper 2006; Newton et al. 2006a) making it very difficult to assess objectively the prevalence and distribution of poor-quality anti-infectives or their consequences (IMPACT 2006). Only 5–15% of the 191 WHO member states report cases of counterfeit drugs (Rago 2002; World Health Organisation 2005). Counterfeit drug production is facilitated by the relatively high cost of genuine medicines, which gives the counterfeiters an economic incentive. However, it is not only expensive medicines that are counterfeited but relatively inexpensive anti-infectives, such as chloroquine, are also faked, presumably as the economy of scale still results in a significant profit. Inadequate law enforcement and light penalties, corruption, complex spaghetti-like trade arrangements and low awareness of counterfeits among the public and health workers exacerbate the problem. Many patients in developing countries obtain their medication from unlicensed vendors, who are usually not trained in pharmacy, do not observe Good Pharmacy Practice and provide inadequate courses without patient information (Foster 1995; Okeke et al. 1999; Rozendaal 2000). The high prevalence of unregistered drugs circulating in the tropics probably also contributes to a poor-quality drug supply. For example, in Burma and Vietnam only 43 and 40%, respectively, of medicines sampled were registered (Wondemagegnehu 1999), and in Kenya only half of sulphadoxine–pyrimethamine (SP) and amodiaquine preparations were registered (Amin and Snow 2005). In some surveys, registered medicines were associated with better drug quality than those unregistered, implying that action against unregistered medicines could improve drug quality (Phanouvong 2003). Only 20% of WHO member states are reported to have well-developed drug regulation and 30% have either no drug regulation or a capacity that hardly functions (Primo-Carpenter 2004; World Health Organisation 2005). Although it is unclear how these figures were derived, the lack of financial and human resources available to many drug regulatory authorities (DRAs) often makes investigation of poor-quality drugs and action impossible. The common absence of monitoring for therapeutic ineffectiveness, apart from relying on astute health-care workers, makes it very difficult to detect poor-quality medicines (Figueras et al. 2002).

24.2 The Prevalence of Poor-Quality Anti-Infectives

Counterfeits of the majority of commonly used anti-infective drugs have been described (see Table in Newton et al. 2006a). Herbal anti-infectives have a long history of being adulterated. Queen Hatshepsut of Egypt commissioned a plant hunting expedition in 1500 BC to obtain genuine herbs, as the markets were flooded with bogus botanicals (Kreig 1967). In the first century AD in Greece, Dioscorides first classified drugs by their therapeutic use, warned of the dangers of adulterated drugs and advised on their detection (World Health Organisation 1999). Congeners were used to adulterate *Valeriana officinalis* root used for treating cholera, and fakes of 'China Root' for the treatment of syphilis (Kreig 1967; Newton et al. 2006a). With the rise of industrialized scientific medicine in the mid-19th century, the widespread adulteration of medicine, especially quinine, prompted the first regulation of the trade in medicines, codes of practice for pharmacists and guides on the detection of counterfeit drugs in Europe and the USA (Heath 2004). In his screenplay 'The Third Man,' Graham Greene described the hunt for a counterfeit penicillin smuggler in post-war Vienna (Greene 1950). However, counterfeit drugs were first addressed at an international health meeting only 20 years ago and in 1988, the World Health

Assembly adopted a resolution against counterfeit and substandard pharmaceuticals (World Health Organisation 1999).

A recent search found references to 206 cases of counterfeit anti-infectives from 38 countries (Newton et al. 2006a). Additional reports on drug quality may be found in Primo-Carpenter and McGinnis (2007) and from Uganda (counterfeit quinine, Bogere and Nafula 2007), 'Congo' (counterfeit quinine, Gaudiano et al. 2007), sub-Saharan Africa (fake and substandard artemisinin derivatives, Atemnkeng et al. 2007; Anon. 2007b) and China (fake rabies vaccine, Anon. 2007a). Of 771 reports of counterfeit medicines received by WHO between 1982 and 1999, 48.4% were from the Western Pacific Region, with the majority (51.2%) labelled as anti-infectives (Wondemagegnehu 1999; World Health Organisation 2000). The International Medical Products Anti-Counterfeiting Taskforce (IMPACT 2006) recently cautioned against the off-quoted estimate of 10% of the global supply being counterfeit and suggested that 'many developing countries of Africa, parts of Asia, and parts of Latin America have areas where more than 30% of the medicines on sale can be counterfeit. Other developing markets, however, have less than 10%; overall, a reasonable estimate is between 10 and 30%'. The statement includes estimates that about 8% of over-the-counter drugs sold in China are counterfeit, that approximately 70% of medicines used by the Angolan population are forgeries and cites a random survey in Kenya which found that almost 30% of drugs in Kenya were counterfeit (IMPACT 2006).

Much of the data have been interpreted uncritically and some are not accurate. For example, we and others (Cockburn et al. 2005; Newton et al. 2006a; Primo-Carpenter and McGinnis 2007) have quoted a news story (Fackler 2002), which states that 192,000 people died in China in 2001 as a consequence of fake drugs, based on a story in the Shenzhen Evening News. However, this was a mistranslation and the original article states that 192,000 people died of drug-induced diseases (i.e. adverse effects) from the irrational use of drugs in 2001 and not because of counterfeiting (Newton et al. 2007). There is an urgent need for data, with sufficient sample size and random sampling design, that can estimate reliably the prevalence of counterfeit and substandard drugs within countries so that the extent of the problem can be gauged objectively, the effectiveness of interventions assessed and change through time (hopefully for the better) measured. However, considering the vast scale of the global pharmaceutical industry, even 1% of poor-quality medicines is unacceptable, implying millions of victims.

Antimalarials have been particularly targeted throughout history. However, as there are few data for other anti-infectives and as assays for antimalarials are relatively accessible, this may be an artefact of public health interest. Of those antimalarials used for chemoprophylaxis, mefloquine, chloroquine and doxycycline, and for those used for treatment sulphadoxine–pyrimethamine (SP), sulphalene–pyrimethamine, quinine, mefloquine, halofantrine, chloroquine, doxycycline, tetracycline, artesunate, artemether, dihydroartemisinin and dihydroartemisinin–piperaquine have been counterfeited. As far as we are aware, counterfeits of artemether–lumefantrine, parenteral artesunate and primaquine have not yet been reported. Counterfeit mefloquine and artesunate were first noticed in Cambodia in 1998 when suspiciously inexpensive tablets were found (Rozendaal 2000). 'Convenience' sampling in mainland SE Asia in 2000/2001 and 2002/2003 showed that 38 and 53%, respectively, of artesunate obtained from pharmacies and shops were counterfeit (Newton et al. 2001, 2003; Dondorp et al. 2004) (Fig. 24.1). A recent study in Cambodia demonstrated that 72% of quinine, 20% of artesunate, 27% of tetracycline, 9% of chloroquine and 8% of mefloquine samples failed drug quality tests. They described counterfeit quinine and tetracycline labelled as made by the 'Brainy Pharmaceutical Limited Partnership', which is not registered in either Thailand or Cambodia (Lon et al. 2006). A wide range of different counterfeit artesunate types, based on packaging and fake holograms, have been described in SE Asia (Newton et al. 2008, Fig 24.1) and counterfeit artesunate has recently been found in China (USP 2004). Most alarmingly, counterfeit dihydroartemisinin, dihydroartemisinin–piperaquine and artesunate have recently been discovered in Tanzania, Cameroon, Nigeria and Kenya (Newton et al. 2006b; Anon. 2007b; Fig 24.2). The high cost and desirability of artemisinin derivatives, the shift in antimalarial treatment policy to this class of drugs in Africa and the shortage of the raw material create a dire situation for individual patients, confidence in these medicines, and the spread of drug resistance. Indeed, most worryingly, there is evidence that *Plasmodium*

(A)

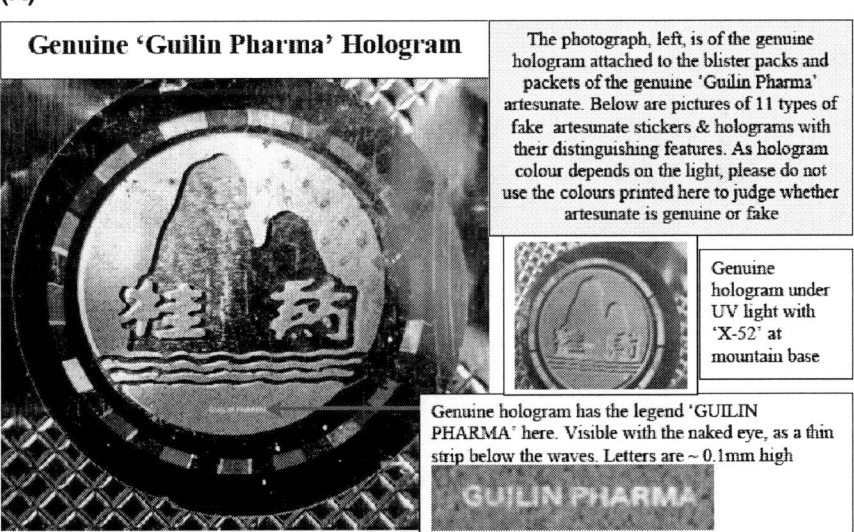

Fig. 24.1 Examples of genuine and counterfeit Guilin Pharmaceuticals Co., Ltd., holograms from Asia. Each hologram is ~13 mm in diameter. (**A**) Genuine Guilin Pharmaceuticals Co., Ltd., hologram; (**B**) Counterfeit Type 3 hologram; (**C**) Counterfeit Type 5 sticker; and (**D**) Counterfeit Type 10 hologram. Reproduced from Newton et al. (2008)

(B)

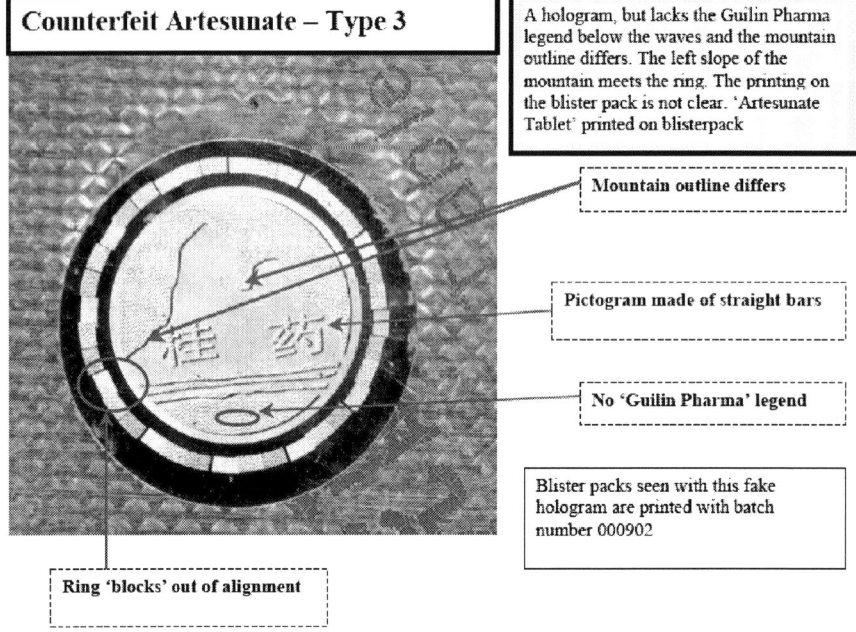

(C)

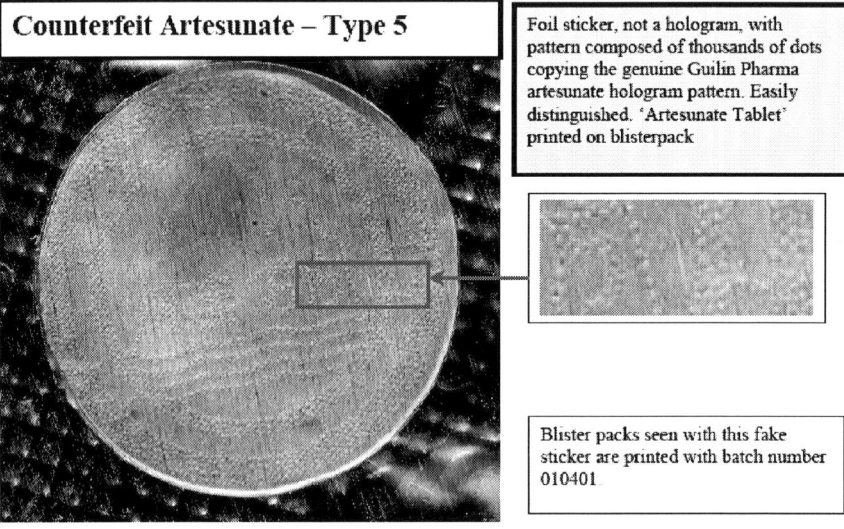

Fig. 24.1 (continued)

(D)

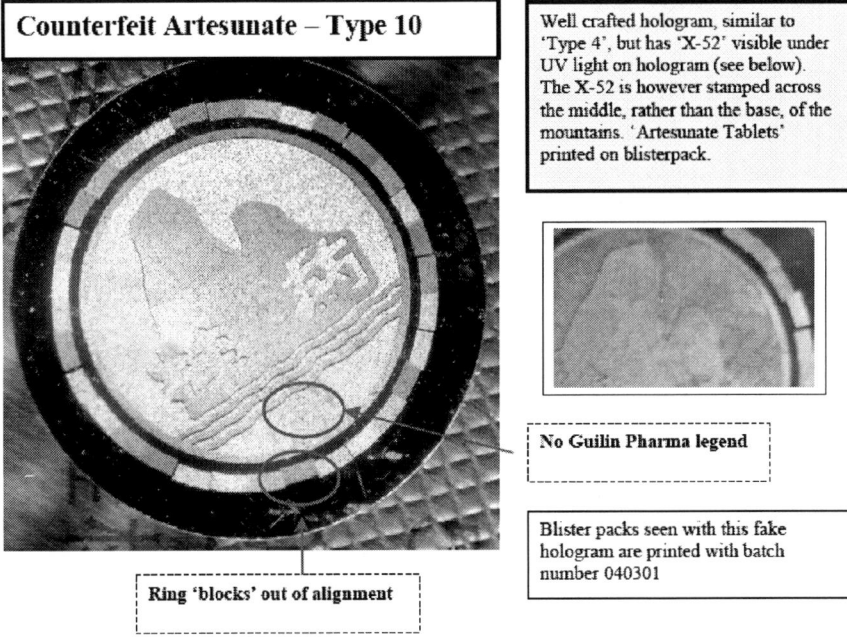

| Counterfeit Artesunate – Type 10 | Well crafted hologram, similar to 'Type 4', but has 'X-52' visible under UV light on hologram (see below). The X-52 is however stamped across the middle, rather than the base, of the mountains. 'Artesunate Tablets' printed on blisterpack. |

No Guilin Pharma legend

Blister packs seen with this fake hologram are printed with batch number 040301

Ring 'blocks' out of alignment

Fig. 24.1 (continued)

(A)

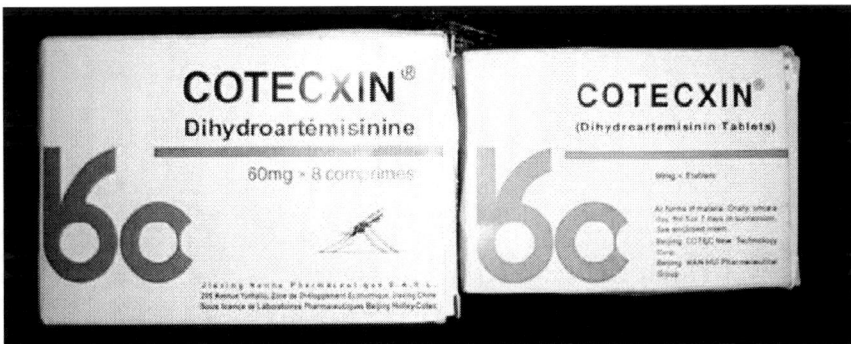

DOI: 10.1371/journal.pmed.0030197.g001

Fig. 24.2 Examples of genuine and counterfeit artemisinin derivatives from Africa. (**A**) Genuine (*left*) and counterfeit (*right*) Cotexcin™ (dihydroartemisinin) from Tanzania. Photograph by Manuela Sunjio. (**B**) Genuine (*right*) and counterfeit (*left*) Arsumax™ (artesunate) from Cameroon. Photograph by Manuela Sunjio. Reproduced from Newton et al. (2006b)

(B)

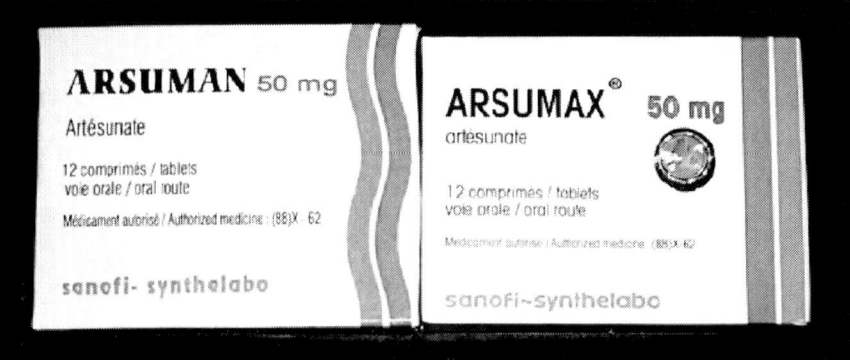

DOI: 10.1371/journal.pmed.0030197.g002

Fig. 24.2 (continued)

falciparum on the Thai/Cambodia border has developed resistance to the artemisinin derivatives (Anon. 2007c).

In Cameroon, of 284 antimalarial samples from 132 vendors, 32% of chloroquine, 10% of quinine and 13% of SP were probably counterfeit, although packaging was not compared with chemical composition (Basco 2004). Some of the 'quinine' contained chloroquine. Of drugs used for self-medication by 15 malaria-infected patients, six 'chloroquine' samples contained no active ingredient, one contained a lower than expected chloroquine content and two 'quinine' samples contained chloroquine (Basco 2004).

Substandard medicines have also been with us since they were first compounded and are an inevitable consequence of inadequate regulation and inspection of the pharmaceutical industry in much of the developing world. For example, substandard antimalarials have recently been recorded in Yemen, Cameroon, Nigeria, Kenya, Senegal, Uganda and Tanzania (Abdi et al. 1995; Taylor et al. 1995; Shakoor et al. 1997; Basco et al. 1997; Ogwal-Okeng et al. 1998; Smine et al. 2002; Risha et al. 2002; Amin et al. 2005; Hebron et al. 2005; Abdo-Rabbo et al. 2005). In Amazonian Venezuela, primaquine tablets were found to contain 19–168% of the stated content (Petralanda 1995) and a patient developed *P. vivax* malaria, after taking primaquine containing 46% of the stated content in Namibia (Kron 1996). A significant proportion of SP tablets available in Africa are substandard, and fail dissolution testing, presumably because of incorrect excipients and formulation, resulting in poor oral bioavailability and reduced efficacy (Risha et al. 2002; Amin et al. 2005). In Nigeria and Thailand, 31 and 0%, respectively, of sampled chloroquine preparations contained quantities of chloroquine outside pharmacopoeial limits (Shakoor et al. 1997). Samples of chloroquine, quinine, mefloquine and SP collected in Congo, Burundi and Angola demonstrated problems with packaging (loose tablets, no producer name nor expiration date), low content of active ingredient in one sample, different undeclared active ingredient in one sample and very poor dissolution profiles in about 50% of the samples (Gaudiano et al. 2007).

A large number of antibiotics have been shown to be counterfeit or substandard (Primo-Carpenter 2004). Of 10 brands of co-trimoxazole from drug shops in Bangladesh, seven were substandard (Chowdhury et al. 2004). Among 21 different samples of antibiotics used in a sexually transmitted disease program in Burma (benzathine benzylpenicillin, benzylpenicillin, ceftriaxone, chlortetracycline, ciprofloxacin, clotrimazole, co-trimoxazole, doxycycline, erythromycin) that were analysed, only three showed the official registration label. Three drugs were expired and six drugs did not have any expiration date. One product did not contain chlortetracycline, the active drug declared, and also did not show any in vitro activity against bacteria. Seven (33%) of the 21 products did not contain the stated dosage (Prazuck et al. 2002).

Considering the severe public health impact of antibiotic resistance in *Mycobacterium tuberculosis*, reports of substandard and counterfeit tuberculosis (TB) medications are of particular concern (Laing et al. 2004). In Nigeria, samples of isoniazid, rifampicin, pyrazinamide and streptomycin were substandard or fake (Taylor et al. 2001) and in a survey of six countries on three continents, 10% of TB medicines were substandard (Laserson et al. 2001). The existence of poor-quality TB drugs threatens the success of the directly observed treatment, short-course (DOTS) strategy for global control of TB. Clinically important variations in bioavailability of rifampicin and isoniazid in fixed dose combination products have been reported (Agrawal et al. 2004). Deteriorated TB drugs due to inappropriate drug management have been found in treatment facilities in southern Thailand, with grossly deteriorated ethambutol in 44% of the hospitals and institutes. Fourteen percent of ethambutol failed the content assay test and 62% of rifampicin and 26% of pyrazinamide failed the dissolution test (Rookkapan et al. 2005).

Antiviral therapy is also compromised by the availability of substandard and counterfeit antiretrovirals. Counterfeit HIV antiretrovirals (ARVs) have been described from 11 countries and fake oseltamivir has been intercepted (Newton et al. 2006a). As HIV spreads and concern about avian influenza increases, the prevalence of counterfeit and substandard ARVs and oseltamivir is likely to rise. An antidepressant (fluvoxamine) and a muscle relaxant (cyclobenzaprine hydrochloride) have been sold as antiretrovirals in DR Congo (Ahmad 2004). Counterfeits of the triple antiretroviral combination product (zidovudine 200 mg, lamivudine 150 mg, indinavir 40 mg) have been discovered in the Côte d'Ivoire, with samples containing zidovudine 201 mg, stavudine 40 mg and an unidentified substance and no lamivudine or indinavir (WHO 2003).

In a one published investigation of drug quality using random sampling, 48% of specimens from Nigerian pharmacies contained active ingredients outside pharmacopoeial limits (Taylor et al. 2001). No pyrazinamide or isoniazid tablets, chloroquine syrups or metronidazole suspensions were within acceptable quality limits. Samples of amoxicillin, ampicillin, benzylpenicillin, co-trimoxazole, cloxacillin, dapsone and metronidazole failed pharmacopoeial tests. As packaging was not assessed it is not possible to be certain whether these were counterfeit or substandard (Taylor et al. 2001). A recent survey of the packaging and drug quality of 293 oral anti-infective and paracetamol products in the Côte d'Ivoire suggested that 49 (17%) were counterfeit (Legris 2005).

There is surprisingly little information on degraded medicines and the consequences of hot and humid storage on anti-infective medicines in the tropics (Kayumba et al. 2004; Ashokraj et al. 2004, 2005). Hogerzeil et al. (1991) found that tetracycline, chloramphenicol and benzylpenicillin did not deteriorate over two summers in the Sudan. Similarly, Nazerali and Hogerzeil (1998) found that amoxicillin, ampicillin, benzylpenicillin, doxycycline, phenoxymethylpenicillin, procaine benzylpenicillin and tetracycline did not deteriorate over 2 years in Zimbabwe. In contrast, ampicillin, erythromycin, penicillin G, trimethoprim and chloroquine showed more than a 10% loss of their active ingredient during 2 years of storage in Burkina Faso. Loss tended to occur after 270 days, with the exception of penicillin G and trimethoprim, in which significant degradation occurred after 90 days. They recommended that these essential drugs should not be stored for more than 1 year in a tropical climate (Ballereau et al. 1997). Of antituberculosis medicines, the second-line cycloserine is particularly vulnerable to the effects of high humidity (Rao et al. 1968). There is an urgent need to quantify the relationships between storage temperature, humidity and time for the degradation of the quantity of active ingredient and bioavailability for key antimalarials and antibiotics.

There is almost no publicly available information on the quality of veterinary medicines and insecticides, which also has an important bearing on human infections. Fake insecticides in the EU (Sanderson 2006) have been reported as well as fake oxfendazole, a domestic mammal anthelminthic in Ireland (ten Ham 1992), and fake avian influenza vaccine in China (Anon. 2005); these veterinary drugs have been repackaged for human use (Saywell and McManus 2002). Other medicines with adjunctive roles in the treatment of infections are also counterfeited or are substandard, for example, ~500 children died after ingesting paracetamol containing the renal toxin diethylene glycol in south Asia, Africa and Central America (Hanif et al. 1995; Newton et al. 2006a). Contaminated substandard magnesium sulphate led to *Serratia* septicaemia (Sunenshine et al. 2007), and gentamicin-resistant *Pseudomonas aeruginosa* in gentamicin eye drops led to severe eye infections (Issack 2001), probably due to errors in manufacture rather than fraud. Similarly, it is uncertain whether the recent deaths due to heparin contaminated with oversulphated chondroitin sulphate (Kishimoto et al. 2008) were a consequence of counterfeiting or an error during manufacture. Also of serious and related concern is the recent description that instructions included with many artemisinin derivatives do not conform to international guidelines on dosing (Jackson et al. 2006; Kachur et al. 2006) – the instructions are, in effect, substandard and those following the package instructions risk underdosing. Diagnostic tests have also been copied with reports of counterfeit lactate test strips (Day et al. 2004) and HIV antibody kits (alleged to be made of pregnancy test kit materials) (Anon. 2006). Insecticide-treated bednets have been counterfeited, being sold without the insecticide treatment (Lengeler 2000). Recycled contaminated disposable syringes will also have an important bearing on infections. It has been estimated that 30–50% of used syringes in India find their way back into shops after washing and wrapping via waste scavengers (Baikait 2003). This must have important consequences for the dissemination of blood-borne viruses such as hepatitis B, C and HIV (Kermode 2004).

The currently available data do not allow any accurate generalizations about the prevalence of poor-quality drugs – existing information suggests that counterfeit and/or substandard anti-infectives are commonly sold in many sub-Saharan, south Asian, Southeast Asian and South American countries. There has been much debate and dispute as to the origin of counterfeit and substandard drugs. China and India are often stated to be the origin of much of the counterfeit drug supply (Raufu 2002, 2003; Mashelkar 2003) but the situation remains very unclear. A recent forensic investigation of the origins of fake artesunate suggested that at least some came from southern China, a hypothesis supported by subsequent arrests by the Chinese Ministry of Public Security of traders in these 'products' (Newton et al. 2008).

24.3 Strategies for the Laboratory Chemical Analysis of Counterfeit and Substandard Drugs in Developing Countries

The chemical characterization of anti-infective drugs suspected to be counterfeit is a crucial part of the overall strategy against this epidemic and should complement rapid field testing methods. The laboratory analytical methods used to detect counterfeit anti-infectives are not different from the tools commonly used in pharmaceutical laboratory analysis, but the scope of the intended work generally differs. Instrumental analysis methods vary in sample throughput, sensitivity, specificity, cost per sample and comprehensiveness of the information produced. Before deciding on the most appropriate technique, it is imperative to evaluate local technical expertise, training sources, servicing and replacement parts, the expected volume of samples, the information sought by the analyst and whether the technique should be fieldable and/or non-destructive.

Every instrumental pharmaceutical analysis technique has potential for the detection and 'fingerprinting' of counterfeit drugs. A detailed description of the most commonly used techniques exceeds the scope of this chapter, but excellent reviews exist (Ali 2000; Gilpin and Pachla 2005). Table 24.1 lists examples found in the recent peer-reviewed literature, with a summary of the advantages and disadvantages of each technique. The majority of these approaches rely on chemical analysis, but there are other approaches. For example, Iqbal et al.

Table 24.1 Summary of instrumental chemical analysis methods to assess medicine quality

Methodology	Application	References	Salient advantages	Salient disadvantages
Gas chromatography				
Gas chromatography (GC, with headspace sampling)	Forensic investigation of residual solvents and organic volatile impurities	Mulligan and Brueggemeyer (1996)	Specific for volatiles. Allows investigation of gases trapped in blisterpack. Can be fieldable	Analysis of non-volatiles requires derivatization
Gas chromatography-mass spectrometry (GC-MS)	Fingerprinting of counterfeit Captagon® (fenethylline) tablets	Alabdalla (2005)	Highly sensitive and specific. Databases can be searched to identify unknowns	Analysis of non-volatiles requires derivatization. MS detector requires extra maintenance, supplies and training
Pyrolysis GC-MS	Discrimination between wet granulation and direct compression tablet production methods	Shen et al. (2003)	Used to generate fingerprints from polymeric substances	Molecular weights are not directly determined
Liquid chromatography				
Liquid chromatography (spectrophotometric UV-Vis detection)	Multidimensional fingerprinting tool to detect counterfeit or substandard herbal medicines	Fan et al. (2006)	Ideal for non-volatiles, polar and labile substances. Offers comprehensive separations and a variety of column chemistries.	Instrumentation more complex than GC. Operates at high pressure. Slower throughput than GC. In general not fieldable. Requires high purity solvents
	Analysis of commonly counterfeited antimalarials (chloroquine, mefloquine, quinine, and artesunate) from informal markets in Congo, Burundi and Angola	Gaudiano et al. (2006)	Widely accepted technology	

Table 24.1 (continued)

Methodology	Application	References	Salient advantages	Salient disadvantages
	Gas chromatography			
	Artemisinin derivatives collected in Kenya and DR Congo	Atemnkeng et al. (2007)		
	Evaluation of potentially counterfeit oseltamivir capsules (Tamiflu®)	Lindegardh et al. (2006)		
Liquid chromatography-mass spectrometry (LC-MS)	Investigation of counterfeit Halfan® (halofantrine) antimalarials	Wolff et al. (2003)	Offers the most comprehensive separations. Can be used to identify unknowns via accurate mass measurements, and tandem MS (MS/MS) experiments. Can be used for selective and sensitive quantitation	More complex than LC with spectrophotometric detection. High maintenance technique. Requires intensive training and experience. Excipients may not be detected, depending on molecular weight and chemical properties
	Counterfeit Betnesol® (betamethasone sodium phosphate) tablets	Arthur et al. (2004)		
	Fingerprinting of counterfeit artesunate antimalarial tablets	Hall et al. (2006)		
Direct ionization mass spectrometry				
Direct analysis in real time (DART)	Counterfeit artesunate antimalarials. Several wrong active ingredients found including the combination drug pyrimethamine/ sulphadoxine, chloroquine, metamizole, artemisinin	Fernández et al. (2006)	Very high sample throughputs. Ideal for screening. Simple to interpret spectra, only protonated molecules are generated. Produces information on chemical composition of sample surface	Limited mass range to approximately 800-1000 Da. Most applicable to small, semi-volatile or volatile molecules. Contamination of sample surface may produce false positives
	Counterfeit antimalarial sample responsible for the death of a Burmese	Newton et al. (2006b)		

Table 24.1 (continued)

Methodology	Application	References	Salient advantages	Salient disadvantages
		Gas chromatography		
	patient. Sample contained acetaminophen and subtherapeutic artesunate, combined with LC			
Desorption electrospray ionization (DESI) MS	Latest generation counterfeit antimalarials containing low levels of artesunate on surface	Nyadong et al. (2007)	Very high sample throughputs. Ideal for screening. Generates multiply charged ions, very large mass range. Surface analysis technique that can be used for imaging. Offers moderate depth profiling capabilities. Specific reactions can be carried with spray solution. Instrumentation can be built in-house	Produces not only proton, but also alkali-metal adduct ions, complicating identification of unknowns and decreasing sensitivity. Contamination of sample surface may produce false positives. Liquid spray slightly damages sample
		Optical spectroscopies (IR, Raman, etc.)		
Near IR spectroscopy (NIR)	Characterization of counterfeit drugs seized in Brazil	Scafi and Pasquini (2001)	Non-invasive, portable	Lower specificity than MS. Relies on database searching and/or peak matching for identification of unknowns. Excipients may obscure signals from active ingredients

Table 24.1 (continued)

Methodology	Application	References	Salient advantages	Salient disadvantages
		Gas chromatography		
Fourier-transform mid-IR imaging spectroscopy	Spatially resolved information on the distribution of wrong active ingredients and wrong excipients present in counterfeit artesunate antimalarials, combined with DESI MS	Ricci, et al. (2007)	Spatially resolved distribution of active ingredients and excipients	Sophisticated instrumentation required. Concave or convex surfaces are difficult to image
Raman spectroscopy	Fingerprinting of a set of 50 different fake artesunate antimalarial samples	De Veij, et al. (2007)	Non-invasive, portable. Provides complementary information than IR	Lower specificity than MS. Signals are inherently weaker, and may be masked by fluorescence from excipients
Spatially offset Raman spectroscopy (SORS)	Authentication of pharmaceuticals through packaging layers	Eliasson and Matousek (2007)	Same advantages as above, plus the ability of probing samples through packaging	Only carried out in a few specialized laboratories at this point
		Other techniques		
X-ray powder diffraction (XRD)	Identification of counterfeit Viagra® tablets	Maurin et al. (2007)	Provides information on inorganic excipients, and crystalline forms	Expensive instrumentation required, generally only available at centralized laboratories or Universities. Not portable
Inductively coupled plasma (ICP) MS	Identification of counterfeit Viagra® tablets	Clough et al. (2006)	Information on inorganic sample constituents. Highly sensitive	Same as above. Not portable
Isotope ratio MS	Fingerprinting of packaging tapes	Carter et al. (2004)	Extremely useful in forensic investigations. Can provide information on geographical origin of inorganic excipients	Same as above

(2004) used an innovative antimicrobial activity assay of different ofloxacin preparations in Pakistan, against three ofloxacin-sensitive reference bacterial species and showed that three injectables and one tablet brand had reduced or no antimicrobial activity.

The chemical analysis of a counterfeit antimalarial sample has many components. First and foremost, the presence or absence of the expected active ingredient has to be investigated. The second task is to quantify the amount of active ingredient present, if any. These two tasks are most commonly accomplished via liquid and gas chromatography (LC and GC, respectively), with UV-Vis detection (fixed wavelength or diode array), mass spectrometry (MS) detection or both. Care should be taken in reporting the absence of a given active ingredient with respect to the methodological detection limit of the method of choice (i.e. 'lower than X mg/g'). Volatile residues, such as solvents used in the manufacturing process can be detected by sampling the gases entrapped in blister packs or other packaging prior to being opened. This is usually accomplished by headspace GC (Table 24.1). The next task in this type of analysis is to investigate the identity of any impurities, decomposition products or 'wrong' active ingredients that may have been added to the sample, unwittingly or not. This is a very complex task and only in rare cases will all components be identified with complete confidence. Mass spectrometry, online or offline, is a key technique. Identification with good certainty can also be accomplished by matching retention times in LC and GC with known standards, and by comparing the UV spectrum of unknowns with databases, but these procedures are time-consuming and require some a priori knowledge of the potential identity of the unknown. Following the identification of unknown active ingredients, the analysis of the identity, proportion and morphological properties of the excipients in the sample may be explored. This can be achieved with a combination of techniques, depending on local availability or existing international collaborations. Techniques useful for the examination of the identity and morphology of pharmaceutical excipients include Raman and infrared (IR) spectroscopy, X-ray diffraction, dissolution testing and LC. Where sample complexity warrants it, more advanced analyses can be further performed to investigate the homogeneity of the distribution of the active ingredient(s) with respect to the excipients and diluents. This can be a quite complex task, generally requiring Raman or IR-based spectroscopic imaging techniques.

Forensic investigation of fake drugs sometimes involves specialized 'fingerprinting analysis', which together with packaging inspection, is of crucial importance in determining the relationships between different fake 'recipes', and in providing evidence to the authorities as to how many sources of fake pharmaceuticals may be present (Flurer and Wolnik 1994; Mulligan et al. 1996; Drasar and Moravcova 2004; Hall et al. 2006; Fernández et al. 2006; Ricci et al. 2007; Newton et al. 2008). In fingerprinting analysis, the goal is to compare the pattern of spectral features from different samples produced by a given instrumental technique. For example, whole Raman or IR spectra of an unknown may be compared with previously collected spectra in a database to determine sample origin. The degree of similarities and differences points to the relationship between fake drug manufacturing sources, and to the evolution of the product composition over time. One of the most powerful approaches for such analysis is principal component analysis (PCA) (Wold et al. 1987), a variable projection method that graphically clusters samples in 2D or 3D space based on their multivariate fingerprints. In order to perform PCA on 'fingerprints' (spectra or chromatograms) collected on different instruments, analytical conditions should be kept strictly constant. Even so, instrument drift and/or differences in response of detectors, liquid pumps and radiation sources may cause significant fingerprint differences. To correct for these differences, several instrument standardization approaches have been devised, and implemented in commercial software packages (Eigenvector-Technologies 2007). Fingerprinting analysis can be extended to chemical data obtained from hyphenated methods of analysis, such as LC-MS and GC-MS. This is generally performed by first extracting a set of spectral peaks from the 2D data, followed by PCA analysis. An alternative approach involves the use of multi-way methods, such as parallel factor analysis (PARAFAC), which makes use of the full data set, without requiring feature extraction, but requiring more computer power and expertise (Bro 1997). These techniques have been used to analyse the great diversity of fake artesunate in SE Asia and, linked with packaging analysis, to suggest different trade routes (Newton et al. 2008). Pollen contamination of tablets may also give clues as to where the tablets were made (Newton et al. 2008). Forensic chemistry and packaging analysis have also recently been used to suggest that poor-quality intramuscular artemether was either substandard or degraded and not counterfeit (Keoluangkhot et al. 2008).

The capital equipment for MS-based techniques is expensive but new techniques such as DESI and DART do not require expensive consumables or sample preparation and hundreds of samples can be analysed per day. In contrast conventional HPLC-based analysis, which also requires expensive capital expenditure, requires expensive consumables, considerable sample preparation with a throughput of only tens of samples per day. Regional MS labs supporting countries may be a cost-effective solution to large-scale sampling of drug quality.

24.4 How to Identify Poor-Quality Medicines Rapidly and Inexpensively

It is important to be able to detect a counterfeit drug quickly using simple observational techniques so that this information can be acted upon to protect patients (Newton et al. 2006b). As counterfeits can contain the correct amount of the active ingredient, analysis of the packaging is crucial. A comparison of print quality on the package relative to the genuine product can usually provide initial clues to product authenticity. Inconsistencies in colouring, spelling, fonts and the location of words and symbols on the package create suspicions. Once these defects are identified, public alerts should be quickly disseminated. An example is the recent discovery of counterfeit dihydroartemisinin–piperaquine (Duo-cotecxin) and dihydroartemisinin (Cotecxin) in Kenya (Anon. 2007b). Examination of the counterfeit blister packs revealed obvious differences between the fake and genuine product (The Pharmacy and Poisons Board 2007). In the counterfeits, the name of the manufacturer was not indicated on the reverse of the blister pack and the writings run along the width of the strip. The print easily faded on rubbing with a wet finger. The lot number and expiry dates on the fakes were more difficult to see relative to the genuine. Also, extraneous writing such as 'fixed dose combination' and '8 tablets oral taken' appeared on the fake blister pack. This information was promptly disseminated via the Internet by the Pharmacy and Poisons Board of Kenya, providing the opportunity for health-care professionals to evaluate the quality of their stock rapidly. Keen observations using ones senses provide the most inexpensive way to detect a suspicious drug quickly, including observing the size and shape of the tablet and strange odours or tastes. Even the way a tablet crumbles can provide clues to its authenticity. However, in much of the developing world tablets are sold loose or not in the original packaging, making it very difficult to distinguish counterfeit medicines from substandard medicines. The United States Pharmacopoeia in partnership with the International Council of Nurses developed the *Tool for Visual Inspection of Medicines*. It is a useful checklist that serves to help identify suspicious medicinal products. It is available online at http://www.usp.org/pdf/EN/dqi/visualinspectionTool.pdf

The active ingredient is the principle component and in many cases the most expensive component of the drug formulation. Therefore, counterfeiters may attempt to eliminate or diminish the amount of the active ingredient in order to increase their profit margin. Counterfeit drugs commonly do not contain the specified active ingredient or they contain incorrect ingredients. Counterfeiters may take inexpensive paracetamol or aspirin and reformulate or repackage them as another product. Colorimetric tests provide rapid qualitative information on the presence of the specified active ingredient. These assays are designed to produce distinct colour changes when a particular active ingredient reacts or complexes with a specific reagent or dye. Colorimetric tests for many antimalarial drugs, such as artesunate, artemether, dihydroartemisinin (DHA), mefloquine, chloroquine and quinine, and oseltamivir, are available (Green et al. 2001, 2008; Newton 2006b). Thin-layer chromatography (TLC) also provides a sensitive inexpensive technique to detect specific active ingredients. This technique consists of placing a small volume of sample onto a thin layer of silica attached to a support such as glass, aluminium or plastic. The plate is inserted into a vessel containing a solvent mixture. By capillary action, the solvent mixture travels up the silica material dissolving and carrying the sample while separating the active ingredient from other components. Since the components of the sample mixture have characteristic affinities to the silica matrix, separation is achieved by migration through the matrix at different speeds. This property results in a characteristic retention factor (Rf) for each component. The active ingredients can be visualized by chemical treatment or UV absorbance. A TLC kit, the Minilab (German Pharma Health Fund, Oberursel, Germany), has been developed specifically to allow detection of essential medicines in the tropics (Jähnke et al. 1998; Jähnke 2004). TLC methods usually require very careful handling of flammable and/or toxic reagents and require a moderate amount of expertise for accurate interpretation. The assays may be modified to make them quantitative. A handheld filter photometer may be used to measure colour intensity as a function of active ingredient concentration for the colorimetric assays while the relative density of TLC sample spots may be compared with the corresponding reference. Measurements of characteristic physical or chemical properties of pharmaceuticals can be accomplished with inexpensive equipment. For example, tablet weight and dimensions, density of dissolved ingredients and pH, can be quickly and easily measured with a balance, caliper, refractometer and pH indicator paper, respectively (Green et al. 2003). In the specific case of artesunate tablets, the counterfeits tend to weigh more than the authentic ones. When a crushed tablet is mixed in alcohol, artesunate preferentially dissolves in the solvent while the insoluble excipients are filtered out. A good estimation of active ingredient concentration can be achieved by measuring the refractive index or density of the extract using a simple and inexpensive handheld refractometer (Green et al. 2007). A few drops of an alcoholic extract of the genuine product in water gives a pH of about 3.5, whereas the counterfeit yields a pH of about 6.5 (Green et al. 2003). Also, optical crystallographic characteristics of many drugs have been established, and can be used to confirm the presence of the active ingredient with a polarising microscope (Tillson and Eisenberg 1954; Jordan 1993).

24.5 The Consequences of Counterfeit and Substandard Anti-Infectives for Individual Patients

Individual patients suffer from the consequences of poor-quality anti-infective directly through increased mortality and morbidity and indirectly through wider consequences on public health with loss of confidence in health care, spread of drug-resistant pathogens, economic losses for producers and traders in the genuine articles and increased burden for health workers, DRAs, customs and police (Erhun et al. 2001). Although clinical trials may declare particular medicines to be highly efficacious in controlled conditions, their effectiveness in real life is probably usually much lower. Amin et al. (2004) demonstrated the importance of this distinction for antimalarial drugs in Kenya with effectiveness much lower than efficacy due to reduced adherence and reduced drug quality.

Mortality and morbidity will increase as a result of medicines having no or subtherapeutic amounts of active ingredient, such as artesunate for falciparum malaria or ofloxacin for typhoid. The potential dangers of subtherapeutic dosage, such as may occur with substandard drugs or counterfeits containing insufficient active ingredient, are illustrated by problems with primaquine therapy. When dosed without taking patient body weight into account, heavier Israeli tourists, and not their thinner co-travellers, developed *P. vivax* relapses (Schwartz et al. 2000). Patients may also suffer from surprising adverse effects of unexpected active ingredients. For example, counterfeit ampicillin-containing cassava and antimould powder, flour or turmeric, co-trimoxazole-containing diazepam, metronidazole made of chloroquine, oseltamivir-containing vitamin C and lactose, ceftazidime made of reused vials containing streptomycin, ofloxacin made of penicillin and glucose (Newton 2006a), halofantrine syrup containing a sulphonamide (Wolff et al. 2003), quinine-containing mepacrine (Primo-Carpenter and McGinnis 2007), quinine-containing chloroquine (Gaudiano et al. 2007) and some counterfeit artesunate tablets containing artemisinin, chloramphenicol, erythromycin, paracetamol, metronidazole and metamizole (Hall et al. 2006; Fernández et al. 2006; Ricci et al. 2007; Newton et al. 2008) have been described. Patients may be allergic to these pharmaceuticals, or may experience adverse events, which would be clinically confusing and time wasting as the physicians would be unaware of the true active ingredients. Some substandard drugs contain much more active ingredient than stated on the label. For example, in Nigeria 59% of antimalarials tested contained ≥110% of the stated active ingredient (Taylor et al. 2001). For anti-infectives with relatively narrow therapeutic ratios, such as quinine, this may increase the incidence of adverse effects. The substitution of aspirin for chloroquine may contribute to acidosis in children presenting with severe malaria (Newton et al. 2002). Harm to an individual due to the spread of resistant pathogens within that individual when he/she consumes a drug inappropriately is relatively uncommon. An important exception is *M. tuberculosis* in which clinically important antimicrobial resistance can emerge within a treated individual (Lipsitch and Samore 2002).

24.6 For the Community

Loss of faith in genuine medicines and the local health system and DRAs is inevitable in areas where drug quality is perceived as being poor. The switching of brands and of health-care systems are potential consequences. The use of counterfeit antimalarials, and the consequent failure of patients to improve, has led to false reports of *P. falciparum* and *P. vivax* drug resistance (Rozendaal 2000; Petralanda 1995; Basco et al. 1997). Health practitioners themselves also loose confidence in the medications that they rely upon.

Anti-infectives containing sub-therapeutic amounts of the active ingredient, whether counterfeit or substandard, may increase the risk of selection and spread of drug-resistant pathogens (Taylor et al. 1995; Okeke et al. 1999, 2005a; Green 2006). The selection of drug-resistant pathogens depends on a wide variety of factors – the biomass of infecting pathogens, host immunity, the relationship between the pharmacokinetic profile of the drug and pharmacodynamic effects on the pathogen, the antimicrobial susceptibility of the pathogen and the fitness of resistant mutants. The dose absorbed is a critical determinant of cure. If resistant pathogens infect or unusually arise de novo within a host and encounter sublethal concentrations of a slowly eliminating antimicrobial, they will have a survival advantage and multiply faster than sensitive pathogens (Price and Nosten 2001; White 2004). Counterfeits containing no active ingredient will not provide this selective advantage. Counterfeits containing subtherapeutic amounts of the stated ingredient, or wrong active ingredients (Basco 2004; Hall et al. 2006; Fernández et al. 2006; Gaudiano et al. 2007; Newton et al. 2008) with antimicrobial effects, may facilitate the emergence and spread of drug-resistant pathogens – consequences that will be very hard to understand if the wrong active ingredients are 'invisible'. For diseases treated with combination therapy, such as tuberculosis, HIV and falciparum malaria, poor-quality drugs risk the spread of resistance due to both poor-quality drug and 'unprotected' co-drugs (Price and Nosten 2001; White 2004).

Two important reasons for patients having subtherapeutic plasma drug concentrations are reduced adherence to optimum regimens (Yeung and White 2005) and regimens of inadequate dose and/or duration (Kachur et al. 2006). A third important reason is poor-quality medicines, containing either low concentrations of the active ingredient(s) or low bioavailability. Although it is a logical presumption, there is very limited field evidence that poor quality of the drug supply is a contributor to resistance because it is very difficult to tease apart the effects of the misuse of anti-infective by medical practitioners, patient adherence and poor-quality drugs. There is evidence from Europe that antimicrobial resistance is related to high antibiotic use (Goossens et al. 2005) but such data are sparse from the developing world (Okeke et al. 2005b). It remains unclear if excessive drug use per se is important or the inappropriate delivery of antimicrobials by health-care workers and/or inappropriate consumption by patients for many pathogen–antimicrobial combinations. Tan et al. (1993) found that use of antibiotics in the month prior to an infection was a risk factor for the development of a systemic penicillin-resistant *S. pneumoniae* infection, suggesting that poor-quality antibiotics would have a similar effect. The high prevalence of substandard chloramphenicol and co-trimoxazole in Burma (Wondemagegnehu 1999) may well have contributed to the high frequency of antibiotic resistance in *S.* Typhi there (Shwe et al. 2002). Poor-quality rifampicin and pyrazinamide are also likely to be fuelling the spread of multidrug-resistant *M. tuberculosis*. The likely importance of subtherapeutic medication in the emergence of drug resistance is illustrated by the addition of chloroquine to salt on the Thai/Cambodia border in the 1950s – which may have been crucial in the emergence of chloroquine-resistant *P. falciparum* parasites (Payne 1988) and the beginning of the loss of the utility of this important medicine in Asia and then Africa.

There have been no specific attempts to model the consequences of poor-drug quality on the spread of antimalarial or antibacterial resistance. Models of the emergence and spread of antimalarial drug resistance (White and Pongta-vornpinyo 2003; White 2004) suggest that poor-quality drugs are important. For malaria, White and Pongtavornpinyo (2003) discuss the 'window of opportunity' for the selection of de novo resistant parasites during which blood concentrations of antimalarial are below the minimum inhibitory concentration (MIC) for the new resistant mutants but above the MICs for the drug-sensitive parasites. In this window, the net growth of the single drug-resistant mutant outstrips the growth of the drug-sensitive organisms to reach the density required to produce transmissible gametocytes (Stepniewska and White 2008). Therefore, for slowly eliminated drugs, poor-quality medicines that produce blood concentrations within this selective window will be prone to select resistance. Very poor-quality drugs that produce blood concentrations below this 'window', which is specific for different drugs, host immunity and parasite sensitivity, are unlikely to select for resistant parasites. However, once resistance has arisen and been transmitted to other patients, different relationships will apply. Long elimination half-lives will strongly select for resistant parasites, amplifying resistance (White and Pongtavornpinyo 2003; White 2004), and poor-quality drugs with long half-lives may be important in the spread of resistance, irrespective of how poor quality they are. The relationship between the pharmacokinetic (PK) and pharmacodynamic (PD) profiles of a medicine for particular pathogens will be important in determining the effect of subtherapeutic blood concentrations on the development of resistance. If the slope of the antimicrobial concentration–effect relationship is steep, only a narrow range of drug concentrations provide partial inhibition/killing of the pathogens, whereas a flatter slope offers a wider range of concentrations for this to happen (White 1999). Among bacteria the maximum differential selection for resistance occurs at antimicrobial concentrations that exert between 20 and 80% of the maximum effect (Lipsitch and Levin 1997). The PK/PD parameters most often associated with the clinical efficacy of antimicrobial agents are the ratio of the maximum antimicrobial concentration (C_{max}) to MIC, ratio of the area under the curve (AUC) to MIC and the duration serum antimicrobial levels exceed the MIC. Antibiotics vary in the importance of these three parameters in determining in vivo activity, for example, for fluoroquinolones and aminoglycosides, C_{max}/MIC and AUC/MIC are more important whereas for B-lactams, it is the duration for which serum antimicrobial levels exceed the MIC (Craig 2001). These parameters will also be important in the risks of the emergence and spread of resistant bacteria. Combining modelling and clinical research, Thomas et al. (1998) demonstrated that an AUC/MIC ratio of <100 was associated with the appearance of resistance among a wide range of pathogens causing nosocomial lower respiratory tract infections on an intensive care unit. How these findings relate to poor drug quality waits to be determined, in part because there are few data on the prevalence of poor-quality drugs and few data describing or predicting their consequence in the emergence and spread of antimicrobial resistance.

The recent discovery of fake 'artesunate' containing small quantities of the active ingredient raises the likelihood that these counterfeits will facilitate the selection and spread of *P. falciparum* parasites resistant to artemisinin derivatives, risking the catastrophic loss of these vital medicines (Newton et al. 2006b). This will lead to therapeutic failure and the need for the development of new anti-infectives. Multiple reports of fake vaccines are of considerable concern because of their affects on the incidence of key preventable disease and public confidence in vaccines – which can be tenuous. Fake pneumococcal vaccine may affect the incidence of resistant pneumococcal disease in the whole population as vaccines against pneumococci in children also reduce disease in adults within the same family and community (Whitney et al. 2003).

24.7 Counterfeit Drugs and Policy

There is an urgent need to assess the prevalence of poor-quality anti-infective in the developing world, to examine their impact, both on the outcome of individual patients and in the emergence and spread of antimicrobial drug resistance and to investigate the most cost-effective solutions. The recent 'Declaration of Rome' has led to the formation of an International Medical Products Anti-Counterfeiting Taskforce by WHO. Increased political and financial support for police action, close police–DRA cooperation, the involvement of INTER-POL, border inspections in cooperation with DRAs would help to suppress the trade, but will not eliminate it. A wide range of sophisticated overt and covert quality assurance markers and electronic tracking systems have been developed but are unlikely to be effective in the poorest countries. As tragically demonstrated by the sophisticated fake artesunate holograms, criminal counterfeiters are able to make complicated copies of overt protective devices (Newton et al. 2006b). Good liaison between police forces is extremely important. In a successful cross-border operation in 2000, Belgian customs officers seized 57,000 packs of fake halofantrine capsules en route from China to Nigeria (Anon. 2003). Recently, a group of forensic scientists and health workers working with WHO in the Western Pacific Region, under the chairmanship of INTERPOL, were able to determine the source of at least part of the trade in fake artesunate and precipitate an effective Chinese Government criminal investigation (Newton et al. 2008). Investigative journalists in the developing world, rather than physicians or scientists, have been very important in documenting the public health problems of poor drug quality (Newton et al. 2006a). A recent documentary exposed the criminal export of fake medicines from India to Africa (du Tetre and Koutouzis 2006).

The interventions necessary to reduce the prevalence of substandard drugs are more straightforward, as criminal deception is not involved, but costly and labour intensive involving improvements in industrial quality control and Good Manufacturing Practice. Programs supporting under-resourced DRAs are likely to play a vital role in improving the quality of the drug supply whether counterfeit or substandard. The increased provision of free or inexpensive medicines for key diseases would undercut the counterfeiters and reduce the criminal financial incentive.

24.7.1 Conflict of Interest Statement

None of the authors have a conflict of interest.

Acknowledgments PNN and NJW are supported by the Wellcome Trust and thank the WHO-Western Pacific Region and the Government of the Lao PDR for help and advice. FMF thanks WHO-Western Pacific Region and the US National Science Foundation for support. The authors thank many people for their help and for discussions, especially those at INTERPOL, USP and WPRO.

References

Abdi, Y.A., Rimoy, G., Ericsson, O., Alm, C., and Massele, A.Y. 1995. Quality of chloroquine preparations marketed in Dar es Salaam, Tanzania. Lancet 346:1161.

Abdo-Rabbo, A., Bassili, A., and Atta, H. 2005. The quality of antimalarials available in Yemen. Malar. J. 4:28.

Alabdalla, M.A. 2005. Chemical characterization of counterfeit Captagon tablets seized in Jordan. Forensic Sci. Int. 152:185–8.

Ali, S.L. 2000. Counterfeit drugs and analytical tools for their discrimination: European perspectives. Pharm. Chem. J. 34:31–2.

Agrawal, S., Singh, I., Kaur, K.J., Bhade, S., Kaul, C.L., and Panchagnula, R. 2004. Bioequivalence trials of rifampicin containing formulations: extrinsic and intrinsic factors in the absorption of rifampicin. Pharmacol. Res. 50:317–27.

Ahmad, K. 2004. Antidepressants are sold as antiretrovirals in DR Congo. Lancet 363:713.

Amin, A.A., Hughes, D.A., Marsh, V., Abuya, T.O., Kokwaro, G.O., Winstanley, P.A., Ochola, S.A., and Snow, R.W. 2004. The difference between effectiveness and efficacy of antimalarial drugs in Kenya. Trop. Med. Int. Health 9:967–74.

Amin, A.A., and Snow, R.W. 2005. Brands, costs and registration status of antimalarial drugs in the Kenyan retail sector. Malar. J. 4:36.

Amin, A.A., Snow, R.W., and Kokwaro, G.O. 2005. The quality of sulphadoxine-pyrimethamine and amodiaquine products in the Kenyan retail sector. J. Clin. Pharm. Ther. 30:559–65.

Anon. 2003. GlaxoSmithKline case study (7th April 2003). Judicial Protection of IPR in China. Available at: http://www.chinaiprlaw.com/english/news/news14.htm. Accessed 22 February 2005.

Anon. 2005. Thirteen fake bird flu vaccine-makers punished. China View, 7th November 2005. http://news.xinhuanet.com/english/2005-11/07/content_3745572.htm. Accessed 10 January 2006.

Anon. 2006. Two accused over 'fake' HIV tests. BBC News 30th October 2006. Available at: http://news.bbc.co.uk/1/hi/world/south_asia/6099064.stm. Accessed 22 August 2007.

Anon. 2007a. 17 nabbed for fake rabies vaccine. Available at: http://www.china.org.cn/english/China/219629.htm. Accessed 13 August 2007.

Anon. 2007b. Malaria drugs recalled in Kenya. Available at: http://news.bbc.co.uk/1/hi/world/africa/6951586.stm. Accessed 20 August 2007.

Anon. 2007c. Resistance to artemisinin derivatives along the Thai-Cambodian border. Wkly. Epidemiol. Rec. 82: 360.

Arthur, K.E., Wolff, J.C., and Carrier, D.J. 2004. Analysis of betamethasone, dexamethasone and related compounds by liquid chromatography/electrospray mass spectrometry. Rapid Commun. Mass Spectrom. 18:678–84.

Ashokraj, Y., Agrawal, S., Varma, M.V., Singh, I., Gunjan, K., Kaur, K.J., Bhade, S.R., Kaul, C.L., Caudron, J.M., Pinel, J., and Panchagnula, R. 2004. Quality control of anti-tuberculosis fixed-dose combination formulations in the global market: an in vitro study. Int. J. Tuberc. Lung Dis. 8:1081–8.

Ashokraj, Y., Kohli, G., Kaul, C.L., and Panchagnula, R. 2005. Quality control of anti-tuberculosis FDC formulations in the global market: Part II-accelerated stability studies. Int. J. Tuberc. Lung Dis. 9:1266–72.

Atemnkeng, M.A., De Cock, K., and Plaizier-Vercammen, J. 2007. Quality control of active ingredients in artemisinin-derivative antimalarials within Kenya and DR Congo. Trop. Med. Int. Health 12:68–74.

Baikait R. 2003. Unsafe injections blamed for 1.3 million premature deaths annually. Pharmabiz (Mumbai) 21st May 2003.

Ballereau, F., Prazuck, T., Schrive, I., Lafleuriel, M.T., Rozec, D., Fisch, A., and Lafaix, C. 1997. Stability of essential drugs in the field: results of a study conducted over a two-year period in Burkina Faso. Am. J. Trop. Med. Hyg. 57:31–6.

Basco, L.K., Ringwald, P., Manene, A.B., and Chandenier, J. 1997. False chloroquine resistance in Africa. Lancet 350:224.

Basco, L.K. 2004. Molecular epidemiology of malaria in Cameroon. XIX. Quality of anti-malarial drugs used for self-medication. Am. J. Trop. Med. Hyg. 70:245–50.

Bate, R., Coticelli, P., Tren, R., and Attaran, A. 2008. Antimalarial drug quality in the most severely malarious parts of Africa – a six country study. PLoS ONE. 3:e2132.

Bogere, H., and Nafula, T. 2007. Fake quinine on the market. Daily News, 9th May 2007. Available at http://www.monitor.co.ug/news/news05102.php. Accessed 10 May 2007.

Bro, R. 1997. PARAFAC. Tutorial and applications. Chemom. Intell. Lab. Syst. 38:149–71.

Carter, J.F., Grundy, P.L., Hill, J.C., Ronan, N.C., Titterton, E.L., and Sleeman, R. 2004. Forensic isotope ratio mass spectrometry of packaging tapes. Analyst 129:1206–10.

Chowdhury, M.M.H., Rana, M.S., Amin, M.N., and Faruque, A. 2004. Quality assessment of some paediatric cotrimoxazole suspensions marketed in Bangladesh. Hamdard Med. XLVII(2):58–62.

Clough, R., Evans, P., Catterick, T., and Evans, E.H. 2006. D34s measurements of sulfur by multicollector inductively coupled plasma mass spectrometry. Anal. Chem. 78:6126–32.

Cockburn, R., Newton, P.N., Agyarko, E.K., Akunyili, D., and White, N.J. 2005. Global threat of counterfeit drugs: why industry and governments must communicate the dangers. PLoS Med. 2:e100.

Craig, W.A. 2001. Does the dose matter? Clin. Infect. Dis. 33 Suppl 3:S233–7.

Day, J.N., Hien, T.T., and Farrar, J. 2004. Expiry-date tampering. Lancet 363:172.

De Veij, M., Vandenabeele, P., Alter Hall, K., Fernandez, F.M., Green, M.D., White, N.J., Dondorp, A.M., Newton, P.N., and Moens, L. 2007. Fast screening and chemical finger-printing of counterfeit antimalarial tablets by raman spectroscopy. J. Raman Spectrosc. 38:181–7.

Deisingh, A.K. 2005. Pharmaceutical counterfeiting. Analyst 130:271–9.

Dondorp, A.M., Newton, P.N., Mayxay M., Van Damme, W., Smithuis, F.M., Yeung, S., Petit, A., Lynam, A.J., Johnson, A., Hien, T.T., McGready, R., Farrar, J.J., Looareesuwan, S., Day, N.P.J., Green, M.D., and White N.J. 2004. Fake antimalarials in Southeast Asia are a major impediment to malaria control: multinational cross-sectional survey on the prevalence of fake antimalarials. Trop. Med. Int. Hlth. 9:1241–6.

Drasar, P., and Moravcova, J. 2004. Recent advances in analysis of Chinese medical plants and traditional medicines. J. Chromatogr. B. 812:3–21.

du Tetre P., and Koutouzis, M. 2006. Trafic mortel. BFC productions. Available at http://www.dailymotion.com/video/xjkmd_trafic-mortel. Accessed 22 August 2007.

Eigenvector-Technologies. 2007. PLS Toolbox for Matlab. 4.1. Eigenvector Technologies: Manson.

Eliasson, C., and Matousek, P. 2007. Noninvasive authentication of pharmaceutical products through packaging using spatially offset Raman spectroscopy. Anal. Chem. 79:1696–701.

Erhun, W.O., Babalola, O.O., and Erhun, M.O. 2001. Drug regulation and control in Nigeria: the challenge of counterfeit drugs. J. Health Popul. Dev. Ctries. 4:23–34.

Fan, X.H., Cheng, Y.Y., Ye, Z.L., Lin, R.C., and Qian, Z.Z. 2006. Multiple chromatographic fingerprinting and its application to the quality control of herbal medicines. Anal. Chim. Acta 555:217–24.

Fackler, M. (2002) China's fake drugs kill thousands. San Francisco Examiner, July 29.

Fernández, F.M., Cody, R.B., Green, M., Hampton, C.Y., McGready, R., Sengaloundeth, S., White, N.J., and Newton, P.N. 2006. Characterization of solid counterfeit drug samples by desorption electrospray ionization and direct-analysis-in-real-time coupled to time-of-flight mass spectrometry. ChemMedChem 1:702–5.

Figueras, A., Pedrós, C., Valsecia, M., and Laporte, J.R. 2002. Therapeutic ineffectiveness: heads or tails? Drug Saf. 25:485–7.

Flurer, C.L., and Wolnik, K.A. 1994. Chemical profiling of pharmaceuticals by capillary electrophoresis in the determination of drug origin. J. Chromatogr. A 674:153–63.

Foster, S. 1995. Treatment of malaria outside the formal health services. Trop. Med. Hyg. 98:29–34.

Gaudiano, M.C., Antoniella, E., Bertocchi, P., and Valvo L. 2006. Development and validation of a reversed-phase LC method for analysing potentially counterfeit antimalarial medicines. J. Pharm. Biomed. Anal. 42:132–5.

Gaudiano, M.C., Di Maggio, A., Cocchieri, E., Antoniella, E., Bertocchi, P., Alimonti, S., and Valvo, L. 2007. Medicines informal market in Congo, Burundi and Angola: counterfeit and sub-standard antimalarials. Malar. J. 6:22.

Gilpin, R.K., and Pachla, L.A. 2005. Pharmaceuticals and related drugs. Anal. Chem. 77:3755–69.

Goossens, H., Ferech, M., Vander Stichele, R., Elseviers, M., and ESAC Project Group 2005. Outpatient antibiotic use in Europe and association with resistance: a cross-national database study. Lancet 365:579–87.

Green, M.D., Mount, D.L., and Wirtz, R.A. 2001. Authentication of artemether, artesunate and dihydroartemisinin antimalarial tablets using a simple colorimetric method. Trop. Med. Int. Health. 6:980–2.

Green, M.D., Newton, P.N., and Fernández, F. 2003. Simple low-cost strategies to rapidly identify counterfeit drugs in developing countries, in Combating Pharmaceutical Fraud and Counterfeiting, SMI Conference Documentation. London: SMI Publishing.

Green, M.D. 2006. Antimalarial drug resistance and the importance of drug quality monitoring. J. Postgrad. Med. 52:288–90.

Green, M.D., Nettey, H., Villalba-Rojas, O., Pamanivong, C., Khounsaknalath, L., Grande Ortiz, M., Newton, P.N., Fernández, FM, Vongsack, L., and Manolin, O. 2007. Use of refractometry and colorimetry as field methods to rapidly assess antimalarial drug quality. J. Pharm. Biomed. Anal. 43:105–10.

Green, M.D., Nettey, H., and Wirtz, R.A. 2008. Determination of oseltamivir quality by colorimetric and liquid chromatographic methods. Emerg. Infect. Dis. 14, 552–6.

Greene, G. 1950. The Third Man. London: Heinemann.

Hall, K., Newton, P.N., Green, M.D., De Veij, M., Vandenabeele, P., Pizzanelli, D., Mayxay, M., Dondorp, A., and Fernández, F. 2006. Characterization of counterfeit artesunate antimalarial tablets from SE Asia. Am. J. Trop. Med. Hyg. 75:804–11.

Hanif, M., Mobarak, M.R., Ronan, A., Rahman, D., Donovan, J., and Bennish, M.L. 1995. Fatal renal failure caused by diethylene glycol in paracetamol elixir: the Bangladesh epidemic. BMJ. 311:88–91.

Harper, J. 2006. Counterfeit Medicines Survey Report, Council for Europe. Strasbourg: Council of Europe Publishing.

Heath, W.J. 2004. America's first drug regulation regime: the rise and fall of the Import Drug Act of 1848. Food Drug. Law J. 59: 169–200.

Hebron, Y., Tettey, J.N., Pournamdari, M., and Watson, D.G. 2005. The chemical and pharmaceutical equivalence of sulphadoxine/pyrimethamine tablets sold on the Tanzanian market. J. Clin. Pharm. Ther. 30:575–81.

Hogerzeil, H.V., de Goeje, M.J., and Abu-Reid, I.O. 1991. Stability of essential drugs in Sudan. Lancet 338:754–5.

IMPACT (2006) Counterfeit medicines: an update on estimates. Available at: www.who.int/medicines/services/counterfeit/impact/TheNewEstimatesCounterfeit.pdf. Accessed 11 August 2007.

Iqbal, M., Hakimm, S.T., Hussain, A., Mirza, Z., Qureshi, F., and Abdulla, E.M.M. 2004. Ofloxacin: laboratory evaluation of the antibacterial activity of 34 brands representing 31 manufacturers available in Pakistan. Pak. J. Med. Sci. 20: 349–56.

Issack, M.I. 2001. Substandard drugs. Lancet 358: 1463.

Jackson, Y., Chappuis, F., Loutan, L., and Taylor, W. 2006. Malaria treatment failures after artemisinin-based therapy in three expatriates: could improved manufacturer information help to decrease the risk of treatment failure ? Mal. J. 5:81.

Jähnke, R.W.O., Pachaly, P., Gobina, N.P., Schuster A., Nigge, O. J., Dwornik, K., Rubeau, V., Smine, A., Phanouvong, S., Davydova, N., Bradby, S. and Hajjou, M. 1998. Concise quality control guide on essential drugs, Vol. II, Thin Layer Chromatography. Frankfurt: German Pharma Health Fund,. Supplement 1999, 2002, 2003, and 2004.

Jähnke, R.W.O. 2004. Counterfeit medicines and the GPHF-Minilab for rapid drug quality verification. Pharm. Ind. 66:1187–93.

Jordan, D.D. 1993. Optical crystallographic characteristics of some USP drugs. J. Pharm. Sci. 82: 1269–71.

Kachur, S.P., Black, C., Abdulla, S., and Goodman, C. 2006. Putting the genie back in the bottle? Availability and presentation of oral artemisinin compounds at retail pharmacies in urban Dar-es-Salaam. Mal. J. 5:25.

Kayumba, P.C., Risha, P.G., Shewiyo, D., Msami, A., Masuki, G., Ameye, D., Vergote, G., Ntawukuliryayo, J.D., Remon, J.P., and Vervaet, C. 2004. The quality of essential antimicrobial and antimalarial drugs marketed in Rwanda and Tanzania: influence of tropical storage conditions on in vitro dissolution. J. Clin. Pharm. Ther. 29:331–8.

Keoluangkhot, V., Green, M., Nyadong, L., Fernández, F., Mayxay, M., and Newton, P.N. 2008. Impaired clinical response in a patient with uncomplicated falciparum malaria who received poor quality and underdosed intramuscular artemether. Am. J. Trop. Med. Hyg. 78, 552–555.

Kermode, M. 2004. Unsafe injections in low-income country health settings: need for injection safety promotion to prevent the spread of blood-borne viruses. Health Promot. Int. 19:95–103.

Kishimoto, T.K., Viswanathan, K., Ganguly, T., Elankumaran, S., Smith, S., Pelzer, K., Lansing, J.C., Sriranganathan, N., Zhao, G., Galcheva-Gargova, Z., Al-Hakim, A., Bailey, G.S., Fraser, B., Roy, S., Rogers-Cotrone, T., Buhse, L., Whary, M., Fox, J., Nasr, M., Dal Pan, G.J., Shriver, Z., Langer, R.S., Venkataraman, G., Austen, K.F.,

Woodcock, J., and Sasisekharan, R. 2008. Contaminated heparin associated with adverse clinical events and activation of the contact system. N. Engl. J. Med. 358:2457–67. 10. 1056/nejmoa0803200.

Kreig, M. 1967. Black Market Medicine. Engelwood Cliffs, NJ: Prentice Hall.

Kron, M.A. 1996. Substandard primaquine phosphate for US Peace Corps personnel. Lancet 348: 1453–4.

Laing, R., Vrakking, H., and Fourie, B. 2004. Quality and stability of TB medicines: let the buyer beware! Int. J. Tuberc. Lung Dis. 8:1043–4.

Laserson, K.F., Kenyon, A.S., Kenyon, T.A., Layloff, T., and Binkin, N.J. 2001. Substandard tuberculosis drugs on the global market and their simple detection. Int. J. Tuberc. Lung Dis. 5: 448–54.

Legris, C. 2005. La détection des médicaments contrefaits par investigation de leur authenticité. Étude pilote sur le marché pharmaceutique illicite de Côte d'Ivoire. Thèse pour le diplôme d'état de docteur en pharmacie soutenus à la Faculté de Pharmacie Nancy I. 2005. Available at: http://www.remed.org/html/theses.html. Accessed 11 April 2006.

Lengeler, C. 2000. Insecticide-treated nets: from social marketing to national programmes. Bulletin of Medicus Mundi Switzerland No. 78, October 2000. Available at: http://www. medicusmundi.ch/mms/services/bulletin/bulletin200003/kap02/06lengeler.html. Accessed 10 May 2007.

Lindegardh, N., Hien, T.T., Farrar, J., Singhasivanon, P., White, N.J., and Day, N.P.J. 2006. A simple and rapid liquid chromatographic assay for evaluation of potentially counterfeit Tamiflu (R). J. Pharm. Biomed. Anal. 42:430–3.

Lipsitch, M., and Levin, B.R. 1997. The population dynamics of antimicrobial chemotherapy. Antimicrob. Agents Chemother. 41:363–73.

Lipsitch, M., and Samore, M.H. 2002. Antimicrobial use and antimicrobial resistance: a population perspective. Emerg. Infect. Dis. 8:347–54.

Lon, C.T., Tsuyuoka, R., Phanouvong, S., Nivanna, N., Socheat, D., Sokhan, C., Blum, N., Christophel, E.M., and Smine, A. 2006. Counterfeit and substandard antimalarial drugs in Cambodia. Trans. R. Soc. Trop. Med. Hyg. 100:1019–24.

Mashelkar, R.A. 2003. Report of The Expert Committee On A Comprehensive Examination Of Drug Regulatory Issues, Including The Problem Of Spurious Drugs. Ministry Of Health and Family Welfare. Government of India. Available at: http://www.cdsco.nic. in/html/Final%20Report%20mashelkar.pdf. Accessed 11 August 2007.

Maurin, J.K., Plucinski, F., Mazurek, A.P., and Fijalek, Z. 2007. The usefulness of simple X-ray powder diffraction analysis for counterfeit control-The Viagra example. J. Pharm. Biomed. Anal. 43:1514–8.

Mulligan, K.J., Brueggemeyer, T.W., Crockett, D.F., and Schepman, J.B. 1996. Analysis of organic volatile impurities as a forensic tool for the examination of bulk pharmaceuticals. J. Chromatogr. B 686:85–95.

Nazerali, H., and Hogerzeil, H.V. 1998. The quality and stability of essential drugs in rural Zimbabwe: controlled longitudinal study. BMJ 317:512–3.

Newton, P.N., Proux, S., Green, M., Smithuis, F., Rozendaal, J., Prakongpan, S., Chotivanich, K., Mayxay, M., Looareesuwan, S., Farrar, J., Nosten, F., and White, N.J. 2001. Fake artesunate in southeast Asia. Lancet 357:1948–50.

Newton, P.N., Rozendaal, J., Green, M., and White, N.J. 2002. Murder by fake drugs – time for international action. BMJ 324: 800–1.

Newton, P.N., Dondorp, A., Green, M., Mayxay, M., and White, N.J. 2003. Counterfeit artesunate antimalarials in SE Asia. Lancet 362:169.

Newton, P.N., Green, M.D., Fernández, F.M., Day, N.P.J., and White, N.J. 2006a. Counterfeit anti-infective medicines. Lancet Inf. Dis. 6: 602–13.

Newton, P.N., McGready, R., Fernández, F.M., Green, M.D., Sunjio, M., Bruneton, C., Phanouvong, S, Millet, P., Whitty, C.J., Talisuna, A.O., Proux, S., Christophel, E.M., Malenga, G., Singhasivanon, P., Bojang, K., Kaur, H., Palmer, K., Day, N.P.J., Greenwood,

B.M., Nosten, F., and White, N.J. 2006b. Manslaughter by fake artesunate in Asia -will Africa be next ? PLoS Med. 3:e197

Newton, P.N., Cockburn, R., and White, N.J. 2007. E-letter to submit in response to "The global threat of counterfeit drugs: why industry and governments must communicate." PLoS Med.. http://medicine.plosjournals.org/perlserv/?request = read-response&doi = 10.1371/journal.pmed.0020100#r1765.

Newton, P.N., Fernández, F.M., Plançon, A., Mildenhall, D.C., Green, M.D., Ziyong, L., Christophel, E.M., Phanouvong, S., Howells, S., McIntosh, E., Laurin, P., Blum, N., Hampton, C.Y., Faure, K., Nyadong, L., Soong, S.W.R., Santoso, B., Zhiguang, W., Newton, J., and Palmer, K. 2008. A collaborative epidemiological investigation into the criminal fake artesunate trade in South East Asia. PLoS Med. 5: e32

Nyadong, L., Green, M., De Jesus, V., Newton, P.N., and Fernandez, F.M. 2007. Reactive desorption electrospray ionization linear ion trap mass spectrometry of latest-generation counterfeit antimalarials via non-covalent complex formation. Anal. Chem. 79:2150–57.

Ogwal-Okeng, J.W., Okello, D.D.O., and Odyek, O. 1998. Quality of oral and parenteral chloroquine in Kampala. East Afr. Med. J. 75:692–4.

Okeke, I.N., Lamikanra, A., and Edelman, R. 1999. Socioeconomic and behavioral factors leading to acquired bacterial resistance to antibiotics in developing countries. Emerg. Inf. Dis. 5:18–27.

Okeke, I.N., Klugman, K.P., Bhutta, Z.A., Duse, A.G., Jenkins, P., O'Brien, T.F., Pablos-Mendez, A., and Laxminarayan, R. 2005a. Antimicrobial resistance in developing countries. Part II: strategies for containment. Lancet Infect. Dis. 5:568–80.

Okeke, I.N., Laxminarayan, R., Bhutta, Z.A., Duse, A.G., Jenkins, P., O'Brien, T.F., Pablos-Mendez, A., and Klugman, K.P. 2005b. Antimicrobial resistance in developing countries. Part I: recent trends and current status. Lancet Infect. Dis. 5:481–93.

Payne, D. 1988. Did medicated salt hasten the spread of chloroquine resistance in *Plasmodium falciparum*? Parasitol. Today 4: 112–5.

Petralanda, I. 1995. Quality of antimalarial drugs and resistance to *Plasmodium vivax* in Amazonian region. Lancet 345:1433.

Phanouvong, S. 2003. Registration, inspection and testing: how to prioritise. SEAM Conference, Da es salaam, Tanzania. Dec 10–12th, 2003. http://www.uspdqi.org/pubs/other/USPDQISEAMConference.pdf Accessed 9 December 2005.

The Pharmacy and Poisons Board, Republic of Kenya Ministry of Health. 2007. Available at: http://www.pharmacyboardkenya.org. Accessed 22 August 2007.

Prazuck, T., Falconi, I., Morineau, G., Bricard-Pacaud, V., Lecomte, A., and Ballereau, F. 2002. Quality control of antibiotics before the implementation of an STD program in Northern Myanmar. Sex. Transm. Dis. 29:624–7.

Price, R.N., and Nosten, F. 2001. Drug resistant falciparum malaria: clinical consequences and strategies for prevention. Drug. Resist. Update 4:187–96.

Primo-Carpenter, J. 2004. A review of drug quality in Asia with focus on anti-infectives. USP. Available at: http://www.uspdqi.org/pubs/other/ANEReview.pdf. Accessed 9 August 2007.

Primo-Carpenter, J., and McGinnis, M. 2007. A matrix of drug quality reports on USAID-assisted countries by the USP Drug Quality and Information Program, 2007. http://www.uspdqi.org/pubs/other/GHC-DrugQualityMatrix.pdf. Accessed 9 August 2007.

Rago, L. 2002. Counterfeit drugs: threat to public health. Global Forum Pharmaceutical AntiCounterfeiting, Geneva, Switzerland, 22–25th September 2002.

Rao, K.V., Eidus, L., Evans, C., Kailasam, S., Radhakrishna, S., Somasundaram, P.R., Stott, H., Subbammal, S., and Tripathy, S.P. 1968. Deterioration of cycloserine in the tropics. Bull. World Health Organ. 39:781–9.

Raufu, A. 2002. Influx of fake drugs to Nigeria worries health experts. BMJ 324:698.

Raufu, A. 2003. India agrees to help Nigeria tackle the import of fake drugs. BMJ 326:1234.

Reidenberg, M.M., and Conner, B.A. 2001. Counterfeit and substandard drugs. Clin. Pharmacol. Ther. 69: 189–93.

Ricci, C., Nyadong, L., Fernández, F.M., Newton, P.N., and Kazarian, S. 2007. Combined Fourier transform infrared imaging and desorption electrospray ionization linear ion trap mass spectrometry for the analysis of counterfeit antimalarial tablets. Anal. Bioanal. Chem. 387:551–9.

Risha, P.G., Shewiyo, D., Msami, A, Masuki, G., Vergote, G., Vervaet, C., and Remon, J.P. 2002. In vitro evaluation of the quality of essential drugs on the Tanzanian market. Trop. Med. Int. Hlth. 7:701–7.

Rookkapan, K., Chongsuvivatwong, V., Kasiwong, S., Pariyawatee, S., Kasetcharoen, Y., and Pungrassami, P. 2005. Deteriorated tuberculosis drugs and management system problems in lower southern Thailand. Int. J. Tuberc. Lung Dis. 9:654–60.

Rozendaal, J. 2000. Fake antimalarials circulating in Cambodia. Mekong Malar. Forum 7:62–9.

Sanderson, K. 2006. Fake pesticides pose threat – flood of counterfeit chemicals is harming people and industry. Nature. 5th November 2006. doi:10.1038/news061030-14.

Saywell, T., and McManus, J. 2002. What's in that pill? Far East. Econ. Rev. Feb 21st 2002: 34–40.

Scafi S.H.F., and Pasquini, C. (2001). Identification of counterfeit drugs using near-infrared spectroscopy. Analyst 126:2218–24.

Schwartz, E., Regev-Yochay, G., and Kurnik, D. 2000. Short report: a consideration of primaquine dose adjustment for radical cure of *Plasmodium vivax* malaria. Am. J. Trop. Med. Hyg. 62:393–5.

Shakoor, O., Taylor, R.B., and Behrens, R.H. 1997. Assessment of the incidence of substandard drugs in developing countries. Trop. Med. Int. Hlth. 2:839–45.

Shen, H., Carter, J.F., Brereton, R.G., and Eckers, C. 2003. Discrimination between tablet production methods using pyrolysis-gas chromatography-mass spectrometry and pattern recognition. Analyst 128:287–92.

Shwe, T.N., Nyein, M.M., Yi, W., and Mon, A, 2002. Blood culture isolates from children admitted to Medical Unit III, Yangon Children's Hospital, 1998. Southeast Asian J. Trop. Med. Pub. Health. 33:764–71.

Smine, A., Diouf, K., and Blumm, N.L. 2002. Antimalarial drug quality in Senegal. USP DQI. Available at: http://www.uspdqi.org/projects/senegalreportenglish.pdf. Accessed 23 October 2006.

Stepniewska, K., and White, N.J. 2008. The pharmacokinetic determinants of the window of selection for antimalarial drug resistance. Antimicrob. Agents Chemother. 52:1589–96.

Sunenshine, R.H., Tan, E.T., Terashita, D.M., Jensen, B.J., Kacica, M.A., Sickbert-Bennett, E.E., Noble-Wang, J.A., Palmieri, M.J., Bopp, D.J., Jernigan, D.B., Kazakova, S., Bresnitz, E.A., Tan, C.G., and McDonald, L.C. 2007. A multistate outbreak of *Serratia marcescens* bloodstream infection associated with contaminated intravenous magnesium sulfate from a compounding pharmacy. Clin. Inf. Dis. 45:527–33.

Tan. T.Q., Mason, E.O. Jr, and Kaplan, S.L. 1993. Penicillin-resistant systemic pneumococcal infections in children: a retrospective case-control study. Pediatrics 92:761–7.

Taylor, R.B., Shakoor, O., and Behrens, R.H. 1995. Drug quality, a contributor to drug resistance? Lancet. 346:122.

Taylor, R.B., Shakoor, O., Behrens, R.H., Everard, M., Low, A.S., Wangboonskul, J., Reid, R.G., and Kolawole, J.A. 2001. Pharmacopoeial quality of drugs supplied by Nigerian pharmacies. Lancet 357:1933–6.

ten Ham, M. 1992. Counterfeit drugs: implications for health. Adv. Drug. React. Toxicol. Rev. 11: 59–65.

Thomas, J.K., Forrest, A., Bhavnani, S.M., Hyatt, J.M., Cheng, A., Ballow, C.H., and Schentag, J.J. 1998. Pharmacodynamic evaluation of factors associated with the development of bacterial resistance in acutely ill patients during therapy. Antimicrob.. Agents Chemother. 42:521–7.

Tillson, A.H., and Eisenberg, W.V. 1954. Tables for the identification of N.F.X crystalline substances by the microscopic-crystallographic method. J. Pharm. Sci. 43: 760–7.

USP. 2004. Fake antimalarials found in Yunnan Province, China, 2004. Available at: http://www.uspdqi.org/pubs/other/FakeAntimalarialsinChina.pdf. Accessed 9 December 2005.

White, N.J. 1999. Antimalarial drug resistance and combination chemotherapy. Phil. Trans. R. Soc. Lond. B 354: 739–49.

White, N.J., and Pongtavornpinyo, W. 2003. The de novo selection of drug-resistant malaria parasites. Proc. R. Soc. Lond. B Biol. Sci. 270:545–54.

White, N.J. 2004. Antimalarial drug resistance. J. Clin. Invest. 113:1084–92.

Whitney, C.G., Farley, M.M., Hadler, J., Harrison, L.H., Bennett, N.M., Lynfield, R., Reingold, A., Cieslak, P.R., Pilishvili, T., Jackson, D., Facklam, R.R., Jorgensen, J.H., Schuchat, A. and Active Bacterial Core Surveillance of the Emerging Infections Program Network. 2003. Decline in invasive pneumococcal disease after the introduction of protein-polysaccharide conjugate vaccine. N. Engl. J. Med. 348: 1737–46.

Wold, S., Esbensen, K., and Geladi, P. 1987. Principal component analysis. Chemom. Intell. Lab. Syst. 2:37–52.

Wolff, J.C., Thomson, L.A., and Eckers, C. 2003. Identification of the 'wrong' active pharmaceutical ingredient in a counterfeit Halfan[TM] drug product using accurate mass electrospray ionisation mass spectrometry, accurate mass tandem mass spectrometry and liquid chromatography/mass spectrometry. Rapid Comm. Mass Spect. 17: 215–21.

Wondemagegnehu, E. 1999. Counterfeit and substandard drugs in Myanmar and Viet Nam. WHO Report. WHO/EDM/QSM/99.3. Geneva: WHO.

World Health Organisation. 1999. Counterfeit drugs – guidelines for the development of measures to combat counterfeit drugs. WHO/EDM/QSM/99.1. Geneva: WHO.

World Health Organisation 2000. World Health Organisation Counterfeit drug reports: 1999–October 2000. Available at: www.who.int/medicines/library/pnewslet/pn300cfd.html. Accessed 17 June 2003.

WHO 2003. Counterfeit triple antiretroviral combination product (Ginovir 3D) detected in Cote d' Ivoire. WHO QSM/MC/IEA.110, Nov. 28, 2003. Available at: http://www.who.int/medicines/publications/drugalerts/DrugAlert110.pdf. Accessed 4 April 2008.

World Health Organisation. 2005. Counterfeit and substandard drugs. Frequently asked questions. Available at: http://www.who.int/medicines/services/counterfeit/faqs/16/en/index.html. Accessed 1 December 2005.

Yeung, S., and White, N.J. 2005. How do patients use antimalarial drugs? A review of the evidence. Trop. Med. Int. Health 10:121–38.

Part V
Strategies to Contain Antimicrobial Resistance

Chapter 25
Containment of Antimicrobial Resistance in Developing Countries and Lessons Learned

Aníbal de J. Sosa

Abstract Developing countries are not an exemption when we describe the gravity of antimicrobial resistance (AMR) and its impact in morbidity and mortality. Emergence of AMR in poor-resource countries is a complex issue that transcends all ethnic groups, races, classes, etc.; however, human beliefs, practices, and behavior are very similar and AMR has lot to do with its growing risk in treatment failures. This situation is not indicative that resistance is a health-care problem until data mining shows inexplicable increasing death rates with many of the most prevalent infectious diseases. In the meantime, emergence of multiresistance microorganisms continues to rise. Multidrug-resistant malaria and tuberculosis are the most frequent situations where treatment failures translate into high rates of morbidity and mortality with the concomitant effects on increasing human suffering, treatment costs, and public health crisis. Often though, policy makers and clinicians do not think outside the box and neglect to investigate antimicrobial resistance as a cause of treatment failures. A good example of it is seen in acute respiratory infections – particularly in children under 5 years of age.

We now confront an alarming situation when pharmaceuticals have forsaken the development of new antimicrobials for considering them of low profitability. This and the occurrence of MDR organisms have brought back old therapies previously abandoned. The availability and access to antimicrobial agents in the developing world are certainly much less than in industrialized countries. On the contrary, access and availability of substandard and counterfeit antimicrobial drugs are quite easy and spread out in poor-resource countries. Most of the developing countries lack effective policies and regulations in place to guarantee effective and timely access to essential medications. This, in turn translates to the public seeking easy solutions with inadequate treatment schemes.

A. de J. Sosa (✉)
Alliance for the Prudent Use of Antibiotics, Boston, MA, USA
e-mail: anibal.sosa@tufts.edu

A. de J. Sosa et al. (eds.), *Antimicrobial Resistance in Developing Countries,*
DOI 10.1007/978-0-387-89370-9_25, © Springer Science+Business Media, LLC 2010

447

25.1 Introduction

This chapter attempts to address lessons learned in developing countries where the author has created "grassroot groups" for the Alliance for the Prudent Use of Antibiotics under a global network of more than 60 chapters. Their main role has been to advocate and contribute with a multidisciplinary and multisectorial approach to foster interventions that are country specific. In 2005, APUA published a *Global Advisory on Antibiotic Resistance* Data (GAARD) highlighting the potential resistant threats (APUA 2005).

25.2 The World Health Organization Global Strategy for Containment of Antimicrobial Resistance

In September 2001, the World Health Organization (WHO) introduced the *Global Strategy for Containment of Antimicrobial Resistance* (WHO 2001). Initially, it would invite member states to adopt it in an effort to reduce the emergence of antimicrobial resistance. This plan was very ambitious and unrealistic knowing that it proposed 14 priority interventions and more than 53 other comprehensive interventions. These interventions could be adopted and executed in rich countries, but not in poor-resource ones, where other pressing priorities demand immediate attention with the unfortunate result of neglecting other important ones. Equally demanding in developing countries could be the result of unexpected catastrophes such as flooding, cyclones, tsunamis, etc.

In 2002, the WHO gathered 34 regional advisors and other health professionals from a wide range of agencies worldwide and held a workshop on the implementation of this strategy. Their task was to find out how the "plan" has been accepted and implemented by member countries. After deliberations, the group proposed a Framework for Implementation of the Global Strategy (Chalker 2002). Based on this, seven research projects were funded by the United States Agency for International Development (USAID). These projects were located in *India, South Africa*, and *Ghana*. The goal was to develop a model cost-effective methodology to contain AMR and show the association between antimicrobial use and the emergence of resistance.

In February 2007, in an effort to revamp efforts toward rational use of medicines, the WHO Sixties General Health Assembly reactivated the language described "from the discussions at the Fifty-eighth World Health Assembly of rational use of medicines by prescribers and patients in the context of the threat of antimicrobial resistance to global health security and the adoption of resolution WHA58.27 on improving the containment of antimicrobial resistance." (WHO 2007)

25.3 Root Causes of AMR

AMR is global and is causing numerous deaths every day. Factors associated with it are well documented and known, but unfortunately the root causes of it continue to be ignored. Unjustified use of antimicrobials by consumers is the main root cause of AMR. (Nordberg et al. 2005). To learn why consumers are promoting AMR, we need to understand the lack of access to health care and health-care-seeking behaviors seen in consumers of disenfranchised countries where its economic and social environment is fragmented and often corrupted.

Factors influencing the development of antimicrobial resistance have been previously described (Okeke et al. 1999; Okeke 2002; Planta 2007; Shears 2007). A more in-depth description of the root causes and poverty is presented in this book by Okeke.

25.4 Detection and Surveillance of Antimicrobial Resistance

Antimicrobial resistance is a phenomenon that deserves ongoing monitoring. In learning its trends for a particular human pathogen and antimicrobial agent, we could be able to devise a strategy to contain further transmission. Knowing how resistance spreads, it is imperative for clinicians and policy makers whom it should work together to come up with effective interventions that address the emergence of pathogens in the community and hospitals (Hindler and Stelling 2007; Morrissey et al. 2007; Simonsen et al. 2004).

Generally speaking detecting a resistant pathogen in the microbiology laboratory requires a careful phenotypic characterization. Occasionally, a clinician might perform a Gram stain from a specimen (blood, pus, spinal fluid, etc.) and get an idea if the culprit is a Gram-positive or Gram-negative organism. This information will allow him to empirically choose and begin therapy. In the next 48–72 h, results from a culture might corroborate this initial assumption and assist in what antimicrobial agent is better used to eradicate this pathogen (NCCLS 2003). A much faster diagnosis is possible nowadays by using readily available techniques for some invasive organisms. Also, new molecular tests are also available to detect the type of resistance determinant present in the microorganism (e.g., *mec-* and *vanA*-resistant genes in methicillin-resistant *Staphylococcus aureus*) (Courvalin 2006). Microbiology tests can also demonstrate in vitro and in vivo the acquisition and transmission of the *vanA*, *vanH*, *vanX*, and *vanY* genes from *Enterococcus feacalis* to methicillin-resistant *S. aureus* (Bush 2004). This type of finding is crucial for clinicians and microbiologists to understand and also to monitor resistance.

Antimicrobial resistance surveillance has certainly evolved in the past decades. Initially, pharmaceutical laboratories invested in several networks such as RESISTNET (Mendes et al. 2001), a surveillance program for bacterial resistance against several antimicrobial agents initiated in Latin American countries by Pfizer Pharmaceuticals; or a network of countries such as EARSS (European Antimicrobial Resistance Surveillance System) supported by the European Commission's Directorate-General for Health and Consumer Affairs (DG SANCO) and the Dutch Ministry of Health, Welfare and Sport (Bronzwaer et al. 1999); or regions such as the Pan American Health Organization Surveillance System for Latin America and the Caribbean where 21 countries have learned trends of resistance among key human pathogens. Most of these existing surveillance systems are using WHONET,[1] which allows the analysis of resistance patterns of key bacteria and its trends over time, among many other uses. Unfortunately, most poor-resource country governments cannot afford to set up AMR surveillance, and hence, AMR data is not available to guide policy and clinical decision makers (Masterton 2008).

25.5 Impact on Morbidity and Mortality of Antimicrobial Resistance in Developing Countries

Why are we so concerned about resistance in developing countries? In the last 20 years, only four new antibiotics have been developed: *daptomyicn, linezolid, quinupristin–dalfopristin*, and *tigecycline*. Unfortunately, resistance has already emerged in all of them. In addition, four of the largest world pharmaceuticals (Bayer, Bristol-Myers Squibb, Eli Lilly, and Roche) have abandoned research of new antimicrobial molecules for being considered of "low profit". Pharmaceuticals prefer to invest in other more lucrative drugs such as those used to improve lifestyle (e.g., viagra, cialis, etc.), or in chronic diseases such as diabetes, hypertension, arthritis, etc. (Poupard 2006; Vicente et al. 2006; Sosa 2007a) (Table 25.1).

When a resistance organism is detected in an infectious disease process, the chance of treatment failure is much greater. Physicians are often left with no effective treatment options, and choose less effective and even more toxic antimicrobials in an attempt to save patient's life. In other instances, physicians opt to use compounds which are no longer in use such as the colistin to treat multiresistant strains of *Acinetobacter baumannii*. Colistin is no longer available in many countries (Michalopoulos and Falagas 2008).

As resistance grows in particular settings, its impact on morbidity and mortality rates also grow. More often, we find resistant *Streptococcus pneumoniae* infections producing invasive clinical conditions with the expected deleterious outcome (Cotton et al. 1992; Roca et al. 2006).

Developing countries are inequitably affected by leading infectious disease killers. Not only do we blame emerging infectious disease organisms, but

---

[1] WHONET is Windows-based database software developed for the management of microbiology laboratory data and the analysis of antimicrobial susceptibility test results. See http://www.who.int/drugresistance/whonetsoftware/en/

Table 25.1 Discovery of antibiotics and emergence of resistance

Antibiotic	Discovered	Introduced	Emergence of resistance
Penicillin	1940	1943	1940 (methicilin, 1965)
Streptomycin	1944	1947	1947, 1956
Cloramphenicol	1947	1949	1970
Tetracycline	1948	1952	1956
Erithromycin	1952	1955	1956
Gentamicin	1963	1967	1970
Vancomicin	1956	1972	1987
TMP/SMX	1969	1973	1979
Ciprofloxacin	1980	1987	1992
Linezolid	1990	2000	2001
Telithromycin	1997	2001	2003
Quinupristin-Dalfopristin	1999	2000	2000
Daptomycin	1980	2003	2004
Tigecycline	1999	2005	Reported in vitro

June 2001. Modified by A. Sosa from Julian E. Davies in "Antibiotic Resistance: Origins, Evolution, Selection and Spread," SB Levy, J Goode, D Chadwick, eds., 1997. © John Wiley & Sons Limited. Reproduced with permission.

resistant ones affecting populations where public health infrastructure is fragmented, vaccines are not readily available, existence of substandard and counterfeit drugs, lack of access to health-care services, and poor heath-care-seeking behavior (Sosa 2004). All of these factors coincide in a poor-resource country in the detriment of people's health and their well-being (Hart and Kariuki 1998; Okeke et al. 1999, 2005a, b; Okeke and Sosa 2003).

25.6 Regional Approaches to Contain AMR: Africa, Asia, and Latin America and the Caribbean

In the past 6 years, the Alliance for the Prudent Use of Antibiotics (APUA) has been working worldwide with other international and local partners in Africa, Asia, Europe, and the Latin America and the Caribbean. Resources and efforts have been invested in local assessments of the AMR situation, infrastructure, capacity building, surveillance, policy and regulation, research, training and education, and public awareness campaigns.

25.6.1 Africa

Commissioned by the Rational Pharmaceutical Management program of Management Sciences for Health and with USAID funding, APUA has developed chapters in Zambia and Ethiopia increasing its number of African chapters to 10 (Sosa and Stelling 2004; Sosa 2005, 2006b; Goredema et al. 2006;

Lawrence et al. 2007). Later, and with funding from the Izumi Foundation, APUA developed chapters in Namibia, The Gambia, and Tanzania and in July 2008 APUA begun developing a chapter in Mozambique (Sosa 2006a; Sosa 2007b; Sosa 2007c; Sosa 2008).

Initially in the process of chapter development, we identified key opinion leaders to assemble a core group who will take the task of assessing the antimicrobial resistance problem in the country and existing resources and manpower. In collecting local information, we often find out that it is fragmented and in many instances unreliable given the circumstances of which findings were collected and analyzed. Nevertheless, chapter stakeholders gather information on key pathogens responsible for threatening infectious diseases. At the same time, we engage them in surveying prescribers in an effort to learn their behavior practices, norms, and attitudes. Other partners are able to assess drug procurement systems as well as sale and dispensing mechanisms. Prescription compliance is not needed in almost all of the African countries to get an antimicrobial agent. Existence of unsanctioned drug sellers is evidently everywhere (Sosa 2004).

We also *have assessed* infrastructure for AMR surveillance. In doing so, we have assisted targeted countries in setting up WHONET and training their laboratory staff. In collaboration with Links Media (www.linksmedia.net), we trained journalists in Ethiopia on antimicrobial resistance issues and set up an electronic listserv for health journalists on antimicrobial use and resistance (Lawrence et al. 2007). This is a forum for them to exchange opinions on the subject, receive up-to-date AMR information, training opportunities, and events.

In most of the African countries where we work with, we have designed a national symposium on AMR to highlight local AMR data, resources, and current interventions in an effort to gain momentum and emphasize public awareness. A press conference is held at the end of symposium.

25.6.2 Asia

The Asian-Pacific Research Foundation for Infectious Diseases (ARFID) initiated efforts in the 1990s. By 1996, the Asian Network for Surveillance of Resistant Pathogens (ANSORP[2]) was born. Since then, ANSORP was established as a regional research network. Representatives were from South Korea, China, Hong Kong, Thailand, Chinese Taipei, India, Sri Lanka, Singapore, Malaysia, Indonesia, and Vietnam. ANSORP have also established a regional repository bank, the Asian Bacterial Bank (ABB) which so far has more than 33,000 isolates from 14 countries.

---

[2] http://www.ansorp.org/

25.6.3 Latin American and the Caribbean

The situation in Latin American and the Caribbean is not much different from that of other regions. The antimicrobial resistance surveillance efforts were initiated in 1998 by the Pan American Health Organization (PAHO; PAHO 2004). So far, 21 countries have set up an efficient surveillance network of the following hospital and community pathogens (Tables 25.2 and 25.3).

Table 25.2 Bacteria under surveillance and number of isolates tested, 2003

Hospital	# isolates tested	Community	# isolates tested
Enterococcus spp.	5,800	Salmonella spp.	4,800
Klebsiella pneumoniae	12,400	Shigella spp.	5,600
Acinetobacter spp.	8,700	Staphylococcus aureus	2,800
Pseudomonas aeruginosa	14,000	Escherichia coli	30,000
Staphylococcus aureus	22,000	Campylobacter spp.	1,700
Escherichia coli	23,000	Streptococcus pneumoniae	2,497
Enterobacter cloacae	4,800	H. Influenzae	887

Source: Birth of a Public Surveillance System: PAHO Combats the Spread of Antimicrobial Resistance in Latin America. APUA Newsletter 2006, Vol. 24. No.1.

Table 25.3 S. pneumoniae: percentage of resistant strains, 2003

Country	Number of strains	Percentage of resistant isolates (all ages)						
		OXA*	PEN	VAN	ERI	OFX	CHL	SXT
Argentina	921	22	14	0	9	0.2	4	28
Brazil	865	30	8	0	5	0	1.5	48
Chile	737	26	7	0	17	0	2	29
Mexico	134	0	64	0	22	0	–	5
Paraguay**	88	32	17	0	6	0	2	49
Peru	32	50	34	0	6	0	12	56

* 1 μ disk <19 mm; **<5 years.
Source: Birth of a Public Surveillance System: PAHO Combats the Spread of Antimicrobial Resistance in Latin America. APUA Newsletter 2006, Vol. 24. No.1.

25.7 The Problem: What Do We Know?

- *Country national assessment on AMR*
  Most countries in the developing world do not have a national assessment on AMR. Some with the assistance of the APUA local chapters have initiated the collection of data from research studies previously conducted in the country and from ongoing surveillance AMR data collected from clinical specimens.
- *Capacity building of staff and infrastructure to deal with AMR*
  We have encountered serious deficiencies in the infrastructure and capacity of the microbiology laboratories. Often, there is not a standardized protocol

for phenotypic characterization of isolates, lack of quality control, and storage conditions of media and disks are undesirable as well as unexisting protocol for antibiotic susceptibility testing. Most of the staff are not appropriately trained or supervised. When staff are trained, they are taken away by other in-country international organizations which offer better pay and benefits. Brain drain is another cause for not retaining the qualified staff (Lucas 2005). This emigration of skilled workers to more industrialized areas is pervasive in least developed countries (LDCs).

- *Local research capabilities to deal with AMR*
  Often, most of microbiology laboratories lack the appropriate infrastructure to perform the required work to identify an organism and for further testing of susceptibility to antibiotics. Culture media and reagents do not meet the specified requirements. Protocols for research studies often do not meet the international standards when studying human subjects.

- *The consumer's point of view*
  Consumers in general are unaware of the dangers of antimicrobial resistance. Often, they lack health-care-seeking behaviors for themselves and their families. They trust their local healers and unsanctioned drug dealers for those ailments that require a comprehensive medical care. They purchase incomplete treatment dosage and very often, they do not comply or adhere to the treatment indications. Once they begin to feel better, they stop the medication. Leftover drugs are kept at home in the event a similar illness develops later. Consumers are not properly informed of their diseases, the treatment, or their evolution and prognosis. They are not prepared to inquire or challenge their doctor or to negotiate their treatment options. Whatever the doctor says, goes. At home, often in some cultures the matriarch exercises absolute power in the decision-making process of the entire family (Okeke et al. 1999; Nordberg et al. 2005; Ezechukwu et al. 2005; Sosa 2004).

- *Substandard and counterfeit drugs*
  Frequently, traditional healers and unsanctioned dealers sell drugs or medicinal preparations of questionable composition. Sometimes they are more of a placebo or "dummy drugs" containing only sugar or inactive ingredients. In some countries, traditional healers override a doctor's indication. This custom is deeply rooted in people's belief system. The WHO has created the International Medical Products Anti-Counterfeiting Taskforce (IMPACT) to deal with the increasing growth of dangerous health products (Okeke et al. et al. 2005; Kelesidis et al. 2007; Lancet 2008).

25.8 What Have We Learned?

AMR per se is not a public health priority unless is attached to an excessive morbidity and mortality caused by an infectious disease. Classical example is seen in acute respiratory infections (ARIs) as the leading cause of death in children under 5 years of age. ARIs are often caused by penicillin-resistant

S. pneumoniae. Countries such as Mexico, South Africa, and Mozambique have a 50% penicillin-resistant *S. pneumoniae* (PRSP) while most of the countries in South America PRSP range between 20 and 35%, with the exception of Bolivia and Ecuador where resistance prevalence is 5–10%. In this case, policy makers and clinicians must come together to undertake AMR as a priority public health issue that requires specific strategies to reduce overuse of penicillin in acute respiratory infections not requiring an antibiotic such as those caused by respiratory syncytial viruses (RSV). Also, by learning the prevalence of pneumococcus serotypes, vaccination can then be instituted to also reduce infections caused by most common serotypes. Also, research studies could define how much of the child population is harboring resistant serotypes as colonizers (Janoff et al. 1993; Rusen et al. 1997; Kellner et al. 1998; Abdel-Haq et al. 1999; Leibovitz et al. 1999; Harper 1999; Joloba et al. 2001; Garcia-Rodriguez and Fresnadillo Martinez 2002; Bhutta 2008; Cardoso et al. 2008).

25.9 What Has Worked?

When efforts are coordinated and planned accordingly, available resources can be more effectively used. These efforts should be comprehensive and cost-effective. Governments should lead efforts and work in conjunction with professional associations and international donors to contain antimicrobial resistance. More detailed solutions are provided under Section 25.11.

In some countries, the WHO *Integrated Management of Childhood Illness, IMCI*, is implemented for the care of children under 5 to reduce the burden of disease and disability among the population in this age group. "It is an evidence-based, syndrome approach to case management that supports the rational, effective, and affordable use of drugs and diagnostic tools" (Benguigui and Stein 2006). Health-care workers (HCWs) and other child caretakers are trained to provide *oral antibiotics* in those situations when the child presents recognized non-severe symptoms and signs of bacterial infection. In 2005, the WHO released new evidence and recommendations for the use of oral amoxicillin or cotrimoxazole in children with pneumonia in the communities of non-HIV countries. A multicenter, prospective observational study conducted in three countries in Latin America revealed that IV penicillin remains the drug of choice in children with severe pneumonia in hospital settings caused by *S. pneumoniae* with penicillin MIC <2 µg/ml (Cardoso et al. 2008; Pemba et al. 2008). Another multicenter randomized controlled trial (RCT) conducted in eight countries has shown that oral amoxicillin is as effective as injectable penicillin for the treatment of severe pneumonia in hospital settings (Rojas and Granados 2006). A reduction of up to 36% in ARI-related mortality was observed in communities where these standardized case management guidelines were implemented.

With the progressing growth of HIV-infected children in Africa, pneumococcal disease risk increases 20–40-fold, and antibiotic resistance makes treatment difficult and expensive (Madhi et al. 2000a, b; Levine et al. 2006).

With the advent and use of pneumococcal conjugate vaccine, most serious pneumococcal disease can be prevented in 50–80% of cases and significantly reduce the frequency of antibiotic-resistant infections (Levine et al.. 2006; Madhi et al. 2007).

25.10 What Has Not Worked?

caducity
= the quality of being transitory or perishable

Several strategies used in developing countries have shown to be ineffective or have not provided the desired outcome. For instance, most of the developing countries have enacted regulation and laws about the use of prescription to obtain antibiotics; however, this law is neither enforced nor monitored. A success story is the one being implemented in Chile (Bavestrello et al. 2002). Dispensers (pharmacies, drug outlets, *boticas*, etc.) do not require surrender of it; or when it is asked, they return original prescription to the buyer. This situation facilitates the reiterated use of prescriptions in future similar ailments for the buyer, relatives, or friends and the prescription has no caducity. The lesson learned is that regulatory measures by themselves will not solve the problem. It implicates a change of behavior from the consumers, dispensers, and prescribers' point of view. It also implicates a dedicated budget to monitor and enforce regulation by practicing surprise and routine visits to pharmacies/ drug outlets to require proof of sales and related prescriptions (Sosa 2004). Because resistance trends take time (often years) to show any changes, policy makers and clinicians do not favor this type of effort.

Some countries have developed standard treatment guidelines (STGs) to guide the use of antimicrobial agents in the community by general practitioners. Often times these STGs are adapted from existing ones from another country and are not periodically updated as needed. Public health offices distribute them through several channels, but their deprived budget does not allow the appropriate introduction, training, and follow-up of their impact in prescribing practices.

Experience gathered by APUA chapters in developing countries shows that advocacy efforts are not sustained or followed-up. Most times, chapter members devote their time, effort, and energy on a voluntary basis which plays a very negative effect due to the competing needs of these volunteers in order for them to professionally survive and provide the sustention for their families. When funding is provided to chapters to assist in their operational and programmatic activities, its viability and sustainability are expected with better outcomes.

Some countries have instituted training of unsanctioned dealers so that distributions of antimicrobial agents meet the desired indication. One problem is that these antimicrobial agents might be a substandard quality or counterfeit.

25.11 The Solution

- *Role of leadership and advocacy*
  Leadership and advocacy are essential to keep the momentum of AMR as priority public health issue. When reputable key opinion leaders make themselves visible, they are heard and able to influence public health decisions. Advocacy needs to be ongoing and consistent with a clear message to those who make the decisions and held responsible for their impact. Key opinion leaders need to closely work with the media (press, TV, radio) to educate them and keep them informed with accurate information.
- *Role of policy and regulation*
  Policy and regulation have its place in the design of an AMR containment strategy, but it needs to be tailored to country needs and priorities. It cannot be transplanted from another country or region. It needs to be in agreement with the level of education of people being served so that acceptance and compliance can be achieved. It also needs to have a reasonable budget allocated through the expected lifespan of the policy (Sosa 2004).
- *Role of providers' education and training*
  Continuing education is essential for any profession. Health-care providers need to understand and learn the changing needs in prescribing practices, trends, and prevalence of antimicrobial resistance in the community and hospital settings, evidence-based scientific and medical information, and consumer's knowledge, attitude, and practices. Providers must invest in learning communication skills to better inform and educate their consumers (McNulty 2001; Steiner et al. 2004; Blakely et al. 2006).
- *Role of media*
  Media communication professionals are key in the process of conveying medical and scientific information in lay language so that community as a whole is able to not only understand the message but also absorb it, digest it, and begin change of behavior to improve health-seeking behavior (Sosa 2006b).
- *Role of public awareness campaigns*
  AMR is a public health issue that needs to be introduced to the public by experts in media communication who can clearly dissect the topic and make it part of the person's well-being (Nweneka et al. 2008). The design of public awareness campaigns merits a solid economic investment from government authorities. These campaigns should be periodically assessed in their impact and changes made accordingly (McNulty 2001; Steiner et al. 2004).
- *Role of APUA country chapters as local grassroot network of resources and expertise*
  Founded in 1981, APUA (www.apua.org) is a non-profit, international organization dedicated to promoting proper antibiotic use and curbing antibiotic resistance worldwide. APUA's mission is to promote global public health by raising public awareness through education and research projects on antibiotic resistance and the proper use of antibiotics. Headquartered in Boston, APUA supports a network of affiliated chapters in more than 60

countries throughout the world and a scientific membership representing over 100 countries. APUA thus provides a unique global network to support country-based activities to control and monitor antibiotic resistance tailored to local needs and realities, and to facilitate the exchange of objective, up-to-date information among scientists, health-care providers, consumers, and policy makers worldwide. APUA brings expertise in infectious disease medicine, microbiology, pathology, clinical pharmacology, and antibiotic resistance surveillance (Sosa 2004, 2005, 2006a,b 2007b,c, 2008).

APUA chapters serve the following vital functions:

- raising awareness about the problem of resistance within a country and about the dangers of incorrect antibiotic usage and faulty prescriptions;
- communicating information on proper antibiotic usage;
- fostering related research and educational projects;
- providing a multidisciplinary approach to interventions;
- fostering scientifically sound solutions;
- affording a local platform for input and feedback into global planning efforts; and
- providing local leaders with regular international networking opportunities to enhance their knowledge and effective at the country level.

25.12 Conclusion

Antimicrobial resistance is growing worldwide in an exaggerated pace. Its impact will result in more deaths, less treatment options and more expensive therapies, it will threaten the effectiveness of other public health programs (Malaria, TB, HIV) and it will threaten national security and global stability.

References

Abdel-Haq, N., W. Abuhammour, B. Asmar, R. Thomas, S. Dabbagh and R. Gonzalez (1999). "Nasopharyngeal colonization with Streptococcus pneumoniae in children receiving trimethoprim-sulfamethoxazole prophylaxis." *Pediatr Infect Dis J* 18: 647–649.
APUA (2005). "Executive summary: global antimicrobial resistance alerts (GAARD) and implications." *Clin Infect Dis* 41(s4): S221–S223.
Bavestrello, L., A. Cabello and D. Casanova (2002). "Impacto de medidas regulatorias en la tendencia de consumo comunitario de antibióticos en Chile." *Rev. méd. Chile* 130(11): 1265–1272.
Benguigui, Y. and F. Stein (2006). "Integrated management of childhood illness: an emphasis on the management of infectious diseases." *Semin Pediatr Infect Dis* 17(2): 80–98.
Bhutta, Z. (2008). "A major burden on children." *BMJ* 336(7650): 948–949.
Blakely, J., R. Sinkowitz-Cochran and W. Jarvis (2006). "Infectious diseases physicians' preferences for continuing medical education on antimicrobial resistance and other general topics." *Infect Control Hosp Epidemiol* 27(8): 873–875.

Bronzwaer, S., W. Goettsch, B. Olsson-Liljequist, M. Wale, A. Vatopoulos and M. Sprenger (1999). "European antimicrobial resistance surveillance system (EARSS): objectives and organisation." *Euro Surveill* 4(4): 66.

Bush, K. (2004). "Editorial commentary: vancomycin resistant staphylococcus aureus in the clinic: not quite Armageddon." *Clin Infect Dis* 38(8): 1056–1057.

Cardoso, M., C. Nascimento-Carvalho, F. Ferrero, E. Berezin, R. Ruvinsky, P. Camargos, C. Sant'Anna, M. Brandileone, M. de Fatima P March, J. Feris-Iglesias, R. Maggi, Y. Benguigui and the CARIBE Group (2008). "Penicillin-resistant pneumococcus and risk of treatment failure in pneumonia." *Arch Dis Child* 93(3): 221–225.

Chalker, J. (2002). Trip report: WHO Workshop for regional Advisers on the Implementation of the Global Strategy to Contain Antimicrobial resistance. Arlington, Management Sciences for Health: 30.

Cotton, M., P. Burger and W. Bodenstein (1992). "Bacteraemia in children in the southwestern Cape. A hospital-based survey." *S Afr Med J* 81: 87–90.

Courvalin, P. (2006). "Vancomycin resistance in gram-positive cocci." *Clin Infect Dis* 42(s1): S25–S34.

Ezechukwu CC, I. Egbuonu and J. O. Chukwuka (2005). "Drug treatment of common childhood symptoms in Nnewi: what mothers do?" *Niger J Clin Pract* 8(9): 1–3.

Garcia-Rodriguez, J. and M. Fresnadillo Martinez (2002). "Dynamics of nasopharyngeal colonization by potential respiratory pathogens." *J Antimicrob Chemother* 50(Suppl S2): 59–73.

Goredema, W., N. Nelson, O. Hazemba, M. Sanchez and A. Sosa (2006). A Call-to-Action National Workshop on Antimicrobial Resistance Containment; Adama, Ethiopia. Trip Report. Arlington, VA, Management Sciences for Health & Alliance for the Prudent Use of Antibiotics: 50.

Harper, M. (1999). "Nasopharyngeal colonization with pathogens causing otitis media: how does this information help us?" *Pediatr Infect Dis J* 18: 1120–1124.

Hart, C. A. and S. Kariuki (1998). "Antimicrobial resistance in developing countries." *BMJ* 317(7159): 647–650.

Hindler, J. and J. Stelling (2007). "Medical microbiology: analysis and presentation of cumulative antibiograms: a new consensus guideline from the clinical and laboratory standards institute." *Clin Infect Dis* 44(6): 867–873.

Janoff, E., J. O'Brien, P. Thompson, J. Ehret, G. Meiklejohn, G. Duvall and J. Douglas (1993). "Streptococcus pneumoniae colonization, bacteremia, and immune response among persons with human immunodeficiency virus infection." *J Infect Dis* 167: 49–56.

Joloba, M. L., S. Bajaksouzian, E. Palavecino, C. Whalen and M. R. Jacobs (2001). "High prevalence of carriage of antibiotic-resistant Streptococcus pneumoniae in children in Kampala Uganda." *Int J Antimicrob Agents* 17(5): 395–400.

Kelesidis, T., I. Kelesidis, P. I. Rafailidis and M. E. Falagas (2007). "Counterfeit or substandard antimicrobial drugs: a review of the scientific evidence." *J. Antimicrob. Chemother.* 60(2): 214–236.

Kellner, J., A. McGeer, M. Cetron, D. Low, J. Butler, A. Matlow, J. Talbot and E. Ford-Jones (1998). "The use of Streptococcus pneumoniae nasopharyngeal isolates from healthy children to predict features of invasive disease." *Pediatr Infect Dis J* 17: 279–286.

Lancet (2008). "Combating counterfeit drugs." *Lancet* 371(9624): 1551.

Lawrence, J., M. Sanchez and A. Sosa (2007). Ethiopian Journalist, Spokesperson, and Advocate Training on Antimicrobial Resistance: Increasing Knowledge, Developing Skills, and Exploring Strategies. Trip Report. Gaithersburg, MD, Links Media & Alliance for the Prudent Use of Antibiotics (APUA): 61.

Leibovitz, E., C. Dragomir, S. Sfartz, N. Porat, P. Yagupsky, S. Jica, L. Florescu and R. Dagan (1999). "Nasopharyngeal carriage of multidrug-resistant Streptococcus pneumoniae in institutionalized HIV-infected and HIV-negative children in northeastern Romania." *Int J Infect Dis* 3: 211–215.

Levine, O. S., K. L. O'Brien, M. Knoll, R. A. Adegbola, S. Black, T. Cherian, R. Dagan, D. Goldblatt, A. Grange, B. Greenwood, T. Hennessy, K. P. Klugman, S. A. Madhi, K. Mulholland, H. Nohynek, M. Santosham, S. K. Saha, J. A. Scott, S. Sow, C. G. Whitney and F. Cutts (2006). "Pneumococcal vaccination in developing countries." *Lancet* 367(9526): 1880–1882.

Lucas, A. O. (2005). "Human resources for health in Africa." *BMJ* 331(7524): 1037–1038.

Madhi, S., P. Adrian, L. Kuwanda, C. Cutland, W. Albrich and K. Klugman (2007). "Long-term effect of pneumococcal conjugate vaccine on nasopharyngeal colonization by strep-tococcus pneumoniae – and associated interactions with *Staphylococcus aureus* and *Haemophilus influenzae* colonization – in HIV-infected and HIV-uninfected children." *J Infect Dis* 196: 1662–1666.

Madhi, S., K. Petersen, A. Madhi, M. Khoosal and K. Klugman (2000a). "Increased disease burden and antibiotic resistance of bacteria causing severe community-acquired lower respiratory tract infections in human immunodeficiency virus type 1-infected children." *Clin Infect Dis* 31: 170–176.

Madhi, S., K. Petersen, A. Madhi, A. Wasas and K. Klugman (2000b). "Impact of human immunodeficiency virus type 1 on the disease spectrum of Streptococcus pneumoniae in South African children." *Pediatr Infect Dis J* 19(12): 1141–1147.

Masterton, R. (2008). "The importance and future of antimicrobial surveillance studies." *Clin Infect Dis* 47(s1): S21–S31.

McNulty, C. A. M. (2001). "Optimising antibiotic prescribing in primary care." *Int J Antimicrob Agents* 18(4): 329–333.

Mendes, C. M. F., S. I. Sinto and C. P. Oplustil (2001). "In vitro susceptibility of gram-positive cocci isolated from skin and respiratory tract to azithromycin and twelve other antimicrobial agents", *Braz J Infect Dis* 5:269–276.

Michalopoulos, A. and M. E. Falagas (2008). "Colistin and polymyxin B in critical care." *Crit Care Clin Hosp Acquired Infect Antimicrob Therapy* 24(2): 377–391.

Morrissey, I., A. Colclough and J. Northwood (2007). "TARGETed surveillance: susceptibility of Streptococcus pneumoniae isolated from community-acquired respiratory tract infec-tions in 2003 to fluoroquinolones and other agents." *Int J Antimicrob Agents* 30(4): 345–351.

NCCLS (2003). "Performance Standards for Antimicrobial Disk Susceptibility Tests; Approved Standard—Eighth Edition." NCCLS document M2-A8. Wayne, PA.

Nordberg, P., C. Stålsby-Lundborg and G. Tomson (2005). "Consumers and providers – could they make better use of antibiotics?" *Int J Risk Safety Med* 17(3–4): 117–125.

Nweneka, C., N. Tapha-Sosseh and A. Sosa (2008). Curbing the Menace of Antimicrobial Resistance in Developing Countries. Banjul, Alliance for the Prudent Use of Antibiotics (APUA): 12.

Okeke, I. (2002). "Special challenges to the study of antimicrobial resistance in Africa." *APUA Newslett* 20(4): 1–3.

Okeke, I., A. Lamikanra and R. Edelman (1999). "Socioeconomic and behavioral factors leading to acquired bacterial resistance to antibiotics in developing countries." *Emerg Infect Dis* 5(1): 18–27.

Okeke, I. and A. Sosa (2003). "Antibiotic resistance in Africa – discerning the enemy and plotting a defence." *Africa Health*. 25(3): 10–15.

Okeke, I. N., K. P. Klugman, Z. A. Bhutta, A. G. Duse, P. Jenkins, T. F. O'Brien, A. Pablos-Mendez and R. Laxminarayan (2005a). "Antimicrobial resistance in developing countries. Part II: strategies for containment." *Lancet Infecti Dis* 5(9): 568–580.

Okeke, I. N., R. Laxminarayan, Z. A. Bhutta, A. G. Duse, P. Jenkins, T. F. O'Brien, A. Pablos-Mendez and K. P. Klugman (2005b). "Antimicrobial resistance in developing countries. Part I: recent trends and current status." *Lancet Infect Dis* 5(8): 481–493.

PAHO (2004). Annual Report of the Monitoring/Surveillance Network for Resistance to Antibiotics. Brasilia, Brazil, Pan American Health Organization: 115.

Pemba, L., S. Charalambous, A. von Gottberg, B. Magadla, V. Moloi, O. Seabi, A. Wasas, K. P. Klugman, R. E. Chaisson, K. Fielding, G. J. Churchyard and A. D. Grant (2008). "Impact of cotrimoxazole on non-susceptibility to antibiotics in Streptococcus pneumoniae carriage isolates among HIV-infected mineworkers in South Africa." *J Infect* 56(3): 171–178.

Planta, M. B. (2007). "The role of poverty in antimicrobial resistance." *J Am Board Fam Med* 20(6): 533–539.

Poupard, J. A. (2006). "Is the pharmaceutical industry responding to the challenge of increasing bacterial resistance?" *Clin Microbiol Newslett* 28(2): 13–15.

Roca, A., B. Sigaúque, I. Ll. Quintó, X. Mandomando, M. Vallès, F. Espasa, J. Abacassamo, E. Sacarlal, A. Macete, M. Nhacolo, M. Levine and P. Alonso (2006). "Invasive pneumococcal disease in children <5 years of age in rural Mozambique." *Trop Med Int Health* 11(9): 1422–1431.

Rojas, M.X. and C. Granados C. (2006). "Oral antibiotics versus parenteral antibiotics for severe pneumonia in children. 2006, Issue 2. Art. No.: CD004979." Retrieved 2.

Rusen, I., L. Fraser-Roberts, L. Slaney, J. Ombette, M. Lovgren, P. Datta, J. Ndinya-Achola, J. Talbot, N. Nagelkerke, F. Plummer and J. Embree (1997). "Nasopharyngeal pneumococcal colonization among Kenyan children: antibiotic resistance, strain types and associations with human immunodeficiency virus type 1 infection." *Pediatr Infect Dis J* 16: 656–662.

Shears, P. (2007). "Poverty and infection in the developing world: healthcare-related infections and infection control in the tropics." *J Hosp Infect* 67(3): 217–224.

Simonsen, G., J. Tapsall, B. Allengrazi, E. Talbot and S. Lazzari (2004). "The antimicrobial resistance containment and surveillance approach – a public health tool." *Bull World Health Organ* 82(12): 928–934.

Sosa, A. (2004). Antibiotic policies in developing countries. *Antibiotic Policies: Theory and Practice.* Ian M. Gould and J. W. M. van der Meer, Springer: New York, 766.

Sosa, A. (2005). Trip Report: The APUA-Zambia Chapter as the Local Champion in the Advocacy for Antimicrobial Resistance Country-Level Implementation Pilot in Zambia. Boston, MA, Alliance for the Prudent Use of Antibiotics (APUA): 47.

Sosa, A. (2006a). Trip Report: Development of APUA-Namibia Chapter. Trip Report. Boston, MA, Alliance for the Prudent Use of Antibiotics (APUA): 17.

Sosa, A. (2006b). Trip Report: Ethiopia Country Visit: Assessment of Antimicrobial Resistance Interest and Development of the APUA-Ethiopia Chapter. Boston, MA, Alliance for the Prudent Use of Antibiotics (APUA). 17.

Sosa, A. (2007a). The Threat of Antibiotic-resistant Bacteria and the Development of New Antibiotics. *Antimicrobial Resistance in Bacteria.* C. Amabiles-Cuevas. Mexico, D.F., Horizon Bioscience, 202.

Sosa, A. (2007b). Trip Report: APUA-Tanzania Chapter Inauguration. Trip Report. Boston, MA, Alliance for the Prudent Use of Antibiotics (APUA): November 9–16.

Sosa, A. (2007c). Trip Report: Development of APUA-Gambia Country Chapter – Trip Report. Boston, MA, Alliance for the Prudent Use of Antibiotics (APUA), 44.

Sosa, A. (2008). Trip Report: Exploratory Visit for the Development of APUA-Mozambique Chapter. Boston, MA, Alliance for the Prudent Use of Antibiotics (APUA), 25.

Sosa, A. and J. Stelling (2004). Trip Report: APUA-Zambia Chapter Development. Trip Report. Boston, Alliance for the Prudent Use of Antibiotics (APUA).

Steiner, E., L. Saddler and L. Fagnan (2004). "Promoting appropriate antibiotic use: teaching doctors, teaching patients." *Californian J Health Promot* 2 (Special Issue: Oregon): 22–30.

Vicente, V., J. Hodgson, O. Massidda, T. Tonjum, B. Henriques-Normark and E. Ron (2006). "The fallacies of hope: will we discover new antibiotics to combat pathogenic bacteria in time?" *FEMS Microbiol Rev* 30(6): 841–852.

WHO (2001). Global Strategy for Containment of Antimicrobial Resistance. Geneva, World Health Organization: 171.

WHO (2007). Improving the containment of antimicrobial resistance. *WHA 58.27.*

Chapter 26
Surveillance of Antibiotic Resistance in Developing Countries: Needs, Constraints and Realities

E. Vlieghe, A.M. Bal, and I.M. Gould

Abstract Surveillance for antimicrobial resistance is important in terms of defining prevention and treatment strategies. While surveillance is routinely and efficiently carried out in developed countries, several major constraints often make it impossible for developing countries to carry out major surveillance programmes in this field. This chapter reviews the problems faced by the developing world in carrying out effective surveillance for antibiotic resistance and briefly describes the available data.

26.1 Introduction

Surveillance data for antibiotic resistance are needed to define or update guidelines for empirical and directed treatment. They may guide appropriate drug supplies and identify the need for implementation of infection control measures. They allow assessment of the resistance problem at local, national and international levels, detect changes in resistance rates and alert the emergence and spread of new resistance types. Conversely, the lack of surveillance makes it impossible to monitor the effects of interventions.

However, obtaining accurate and analysable resistance data is no easy task. The collection and interpretation of such data at a global level is difficult because of reporting bias, differences in study designs and methods used, differences in bacterial strains in various parts of the world, availability of different antibiotics and their patterns of usage and impact of varying patient populations. In economically developed parts of the world, a framework already exists wherein data are easy to collect and interpret. This does not imply that they have necessarily succeeded in controlling the spread of resistant bacterial infections. In many other regions of the world, the extent of the

I.M. Gould (✉)
Department of Medical Microbiology, Aberdeen Royal Infirmary, Foresterhill,
Aberdeen, Scotland
e-mail: i.m.gould@abdn.ac.uk

A. de J. Sosa et al. (eds.), *Antimicrobial Resistance in Developing Countries*,
DOI 10.1007/978-0-387-89370-9_26, © Springer Science+Business Media, LLC 2010

problem is largely unknown due to scattered reports and inconsistent methods of surveillance. In the following paragraphs, we highlight some key problems of resistance surveillance in bacterial pathogens in developing countries and summarize recent antibiotic resistance data pertaining to major pathogens from Asia and Africa.

26.2 The Microbiological Gap

In 2001, the WHO released a strategy for antimicrobial resistance containment (WHO global strategy for containment of antimicrobial resistance). The strategy was based on community education in relation to transmission of infections, educating the prescribers on appropriate antibiotic use, formulating guidelines for use of antimicrobials, establishing infection control programmes in hospitals, monitoring antibiotic usage in hospitals and developing microbiology laboratory support. In addition, it suggested that individual governments develop effective programmes (e.g. establishment of task force projects) and take part in international collaborations in order to tackle the problems of growing resistance. Unfortunately, many countries, particularly in the developing world, have been unable to give priority to this issue. Contributing factors include lack of well-functioning local laboratories, the lack of trained laboratory staff and the fact that antimicrobial resistance is not always perceived as a major health problem at the national and international levels (Archibald and Reller 2001; Petti et al. 2006). While clinical microbiology shares resource limitations (including training, quality assurance, supplies and equipment) with other laboratory disciplines, e.g. clinical chemistry, its differences can set it apart from accountability. Inappropriate aid can provide expensive new equipment but not the resources to run it properly. Individual patient-directed bacteriologic analysis and antimicrobial susceptibility tests are expensive, thereby limiting the numbers of samples submitted from those with relapsing and protracted infections leading to skewed resistance patterns.

Minimum essential, sustainable microbiology should be incorporated into future development projects as a highly efficient way of generating crucial public health antimicrobial resistance data. Ongoing support, continuing education and quality assurance are the essence for further well functioning.

Of all low-resources regions, sub-Saharan Africa is probably the least described region with regard to antimicrobial resistance and its surveillance. For instance, only one or two tertiary level care centres in Nigeria and Ghana keep a complete record of antibiotic susceptibility data over a prolonged period (Okeke et al. 2007). A large number of the African laboratories that perform antimicrobial susceptibility testing (AST) do not collate, store or disseminate their resistance data. Antibiotics tested are often those that are available rather than those that are appropriate.

26.2.1 Definitions of Surveillance

Surveillance has many possible definitions, depending on the level studied (international, regional, national or local), on the case-finding method (active or passive), and on the topic (focused or comprehensive).

Active surveillance data at national level (i.e. a systematic, prospective and multicentre collection of quality-assured microbiological culture and resistance data) are scarce in developing countries. If available, they are performed in the setting of other programmes such as those for sexually transmitted infections, tuberculosis or interventions that are put in place during outbreaks of enteric pathogens (*Vibrio cholerae*, *Shigella* spp., *Salmonella* spp). Active screening for pathogens or for the detection of resistance is rarely carried out.

26.2.2 Basic Needs for Improving Surveillance Capability

The characteristics of good laboratory methodology for AST are affordability and simplicity with acceptable precision and accuracy. An educational infrastructure (e.g. university curriculum in medical laboratory science) is desirable, a competent workforce of laboratory staff essential. Benches and running water are of the essence, as much as a reliable supply system, quality assurance and power supply. Good and regular Internet connectivity is highly desirable but in many places it might be limited and irregular. The data generated from unique samples need to be stored in user-friendly and reliable databases.

With the laboratory capabilities in place, the supply of specimen for culture needs to be assured by education of the clinicians on correct indications and methods while taking cultures with minimal risk of contamination. Similarly, clinicians need to be provided with adequate and sterile specimen containers (e.g. blood culture bottles). In diagnostic bacteriology, blood cultures stand out because of their high clinical relevance and their unequivocal sampling criteria and interpretation. In sub-Saharan Africa, bloodstream infections are a major cause of mortality, especially in pediatric patients. In addition, the use, indications and sampling of blood cultures can easily be trained, and their laboratory work-up can be carried out in a centralized facility.

Besides the need to test susceptibility to the correct antibiotics for clinical use in the individual patient, the results of surveillance should be integrated in the local antibiotic policy. An effective policy needs proper antibiotic stewardship, with regular audit and updating based on changing susceptibility patterns.

26.2.3 WHONET

WHONET is a free windows-based software developed by the WHO Collaborating Centre for Surveillance of Antimicrobial Resistance, Boston, Massachusetts, for antimicrobial resistance surveillance and now in use across the world (Stelling and O'Brien 1997). It is robust because of its inbuilt quality control system which allows detection of common laboratory errors. Its modular configuration allows for customization for various purposes such as clinical, epidemiological, veterinary, molecular, pharmacological and infection control. It can provide analysis of susceptibilities, with line listings, histograms and scatter plots, and has built-in interpretative guidelines for most standardized testing methodologies, e.g. CLSI. Its data structure is compatible with major databases, spreadsheets, statistical and word softwares and is available in most commonly spoken languages. BacLink allows transfer of data from existing computer systems. WHONET has interface capabilities to automated systems, e.g. Vitek. It is very flexible in data presentations, allows combination of files from different laboratories, encryption, multifile presentation and good graphical functions. It is used now in over 80 countries around the world, managing data from over 1000 clinical, public health, veterinary and food laboratories.

26.3 Problems in Organizing Surveillance

While lack of funding would appear to be the most easily identifiable problem, the unique cultural and social environment in developing areas of the world coupled with the existing political climate can also play a role in determining the outcome of network projects particularly those that involve a group of countries. Some of the more direct causes of failure with regard to an organized surveillance structure are discussed below.

26.3.1 Planning and Development

Surveillance requires an effective planning and a clear understanding of the goals. While infections are a major cause of poor health in most developing countries, the response from governments is often limited to finding short-term temporary solutions that provide quick relief to people while very little is put in place to address the issues over a longer time. Painstaking and sustained data collection is thus seen as a waste of time as immediate benefits are uncertain. The emergence of corporate health care industry with simultaneous collapse of the public-funded health care system in countries such as India has severely affected the resources that the government could have otherwise utilized in carrying out large-scale surveillance studies.

26.3.2 Impact of Funding

Lack of funding severely affects the capacity to conduct surveillance studies and also affects the outcome. For example, most studies in relation to antibiotic susceptibility data focus on susceptibility and resistance rates derived from traditional methods, e.g. disk diffusion tests rather than the determination of minimum inhibitory concentrations (MIC) of various antibiotics for which the necessary equipments are often beyond the scope of most approved budgets. Thus, lack of funding not only restricts the extent of surveillance but also affects its quality. Health sector does not often get a priority in terms of budget allocation from governments in developing countries, and consequently, the implementation of national programmes suffers (Vaughan and Walt 1984).

26.3.3 Manpower and Human Resources

Trained manpower is often lacking in most developing regions of the world. While these areas still manage to recruit planners and highly skilled workers, there is a shortage of typical medium-skilled motivated individuals required for carrying out the work. Internal migration of the population from rural to urban areas makes it nearly impossible to recruit people where the need to conduct studies is most important. The extreme vertical hierarchy in many developing regions of the world has a negative impact on surveillance studies because the field workers do not necessarily feel a viable and important part of the system but instead carry on with the job that is assigned. This affects the motivation over long term.

26.4 Available Data

Not withstanding the quality assurance difficulties of surveillance in developing countries, the restricted data available suggest that resistance problems are just as great as in industrialized countries. Central and South America will be covered in another chapter of this book.

26.4.1 Asia

Some of the wider transnational surveillance networks operational in Asia include the SENTRY, ANSORP and DOMI programmes. The SENTRY antimicrobial surveillance programme for the Asia-Pacific region and South Africa was initiated in 1998 as part of a worldwide surveillance system with an aim to study the resistance patterns for both community- and hospital-acquired infections. It includes data from 17 sentinel centres in eight countries (Australia, Hong Kong, China, Japan, the Philippines, Singapore, Taiwan and South Africa).

The Asian Network for Surveillance of Resistant Pathogens (ANSORP) was initiated in 1996 by the Korean Professor Jae Hoon-Song. Starting with surveillance of antibiotic resistance in *Streptococcus pneumoniae*, ANSORP is currently in phase V of its programme, with the aim to study the trends in community-acquired MRSA and hospital-acquired pneumonia across Asia while continuing to gather further data on *S. pneumoniae*. ANSORP operates across several countries in Asia (Song et al. 2004).

The DOMI (Diseases of the Most Impoverished) Typhoid Study Group, a surveillance programme in India, Pakistan, Indonesia, China and Vietnam, is primarily concerned with the use of typhoid vaccine in these endemic regions of Asia, but it also contributes towards gathering data in relation to disease burden and antibiotic resistance in *Salmonella typhi* (Ochiai et al. 2007).

Surveillance for antimicrobial resistance in individual countries is often carried out in collaboration with international agencies, e.g. the Invasive Bacterial Infection Surveillance (IBIS) study in India (supported by USAID through the International Clinical Epidemiology Network). More affluent Asian countries such as Singapore have their own national antimicrobial resistance surveillance programme in place. The programme in Singapore based on WHONET was initiated in 2006 and was the first of its kind in the country involving six public sector hospitals including two tertiary centres (Hsu et al. 2007). Similarly, the KONSAR (Korean Nationwide Surveillance of Antimicrobial Resistance) programme in Korea became operational in 1997 when a systematic attempt was made to carry out active surveillance for antibiotic resistance (Lee et al. 2006).

26.4.2 *Africa*

In contrast to the surveillance activities in Latin America and in some Asian countries, sub-Saharan Africa is underserved in terms of antimicrobial resistance data. South Africa has been involved in supranational surveillance programmes or surveys so far such as the SENTRY study (Bell et al 2002a,b, 2003), the Alexander Project (Felmingham et al. 2005), the PEARLS study (Bouchillon et al. 2004) and the SARISA study (Zinn et al. 2004). Most of them were commercially funded and worked with a central laboratory.

Several African countries, mostly former French colonies, have Pasteur Institutes (e.g. Cameroun, Central African Republic) integrated in the worldwide network of the Pasteur Institutes and associated institutes. Often they are among the few well-functioning laboratories in the country and act as national reference laboratories. However, most other antibiotic resistance data in sub-Saharan Africa are derived from passive surveillance at the local laboratories. Those studies are mainly in urban, monocentric, teaching hospital settings, hence studying a biased population for a limited period of time. Unfortunately, not all national surveillance data are published in international journals. Those studies most often present a retrospective collection of strains from a variety of samples. Methods, standards and quality assurance are often not or incompletely mentioned. Removal of doubly entered data is often not clearly stated. Those methodical problems however are to a certain extent a worldwide problem, and not unique for the situation in sub-Saharan Africa.

In many studies, the antibiotic panels used do not comply with international recommendations or even contain inappropriate antibiotics; often the same panel is used for Gram-positive and Gram-negative bacteria alike. Some publications, likely those that are funded by pharmaceutical companies, mention only the susceptibility of a single antibiotic (Obi et al. 1998). Correlating clinical data (e.g. on antibiotic pre-treatment or mode of acquisition) are rarely explored. Until now, only a few African countries have adopted the WHONET system for their data management (Stelling and O'Brien 1997; Blomberg et al. 2004).

In spite of the methodical limitations and weaknesses, those locally derived antibiotic resistance data can give an orienting estimation of the problem of antimicrobial resistance and on the important role local laboratories can play, as discussed by Van den Ende (2007) and Brink et al. (2007). The latter described the first results of a surveillance network of 12 South African laboratories studying Gram-negative bacteria and *Staphylococcus aureus* from blood cultures. In spite of methodological and technical limitations [e.g. testing for extended spectrum β lactamase (ESBL) was not performed in all laboratories], the results show extensive resistance among all Gram-negative bacteria, with 84% ampicillin resistance in *Escherichia coli* and ESBL rates varying from 5% among *E. coli* to 26% among *Klebsiella* spp. MRSA rates were 36%.

Blomberg et al. (2004) described the first hospital surveillance data 18 months after the instalment of the WHONET software in a major Tanzanian referral hospital in 1998. They found extensive resistance among all Gram-negative bacteria for all commonly used first-line drugs (ampicillin, tetracycline, cotrimoxazole) with third-generation cephalosporins and fluoroquinolones as possible treatment options. MRSA and penicillin resistance in *S. pneumoniae* remained surprisingly low at 2 and 4%, respectively.

Surveillance in North Africa is available through the EU-funded ARMed (Borg et al. 2006) and other collaboration networks. Surveys at national level describing nosocomial and community-acquired infections have also been reported (Srikantiah et al. 2006).

26.5 Resistance Data on Key Pathogens

26.5.1 S. aureus

Asian rates for meticillin resistance among *S. aureus* are generally very high, between 35.3% in Singapore (Hsu et al. 2007) and 68% in Korea (Lee et al. 2006). Published data on community-acquired MRSA in lower income Asian countries are scarce. Very high rates of nosocomially acquired MRSA (up to

87.5% of nosocomial *S. aureus* infections) were noted in Indian intensive care units. These alarming rates of resistance in this study could be explained by the high-risk environment of the intensive care coupled with indiscriminate use of antibiotics (Mehta et al. 2007).

In Africa, data from a local surveillance network of 14 hospitals in KwaZulu-Natal (Shittu and Lin 2006) confirm a quite high MRSA prevalence of 26.9% MRSA, with 80% of those multidrug resistant. A study in eight African hospitals (Kesah et al. 2003) from 1996 to 1997 confirmed these figures for the sub-Saharan countries.

Data from *S. aureus* bacteraemia studies in Zimbabwe (Mukonyora et al. 1985) and Senegal (Seydi et al. 2004) show very high rates of meticillin resistance among the invasive strains (28 and 51–72%, respectively), with the former dating back to 1985. Although many of those data are retrospective in nature and have methodical limitations, they remain alarming and warrant persistent surveillance. In contrast, resistance rates in nasal carriage are generally low (Lamikanra et al. 1985; Aires De Sousa et al. 2000; Amir et al. 1995).

26.5.2 S. pneumoniae

Asian resistance data from the ANSORP study group showed very high resistance rates to penicillin resistance in *S. pneumoniae* (Lee et al. 2006). South African surveillance data from invasive strains can be found in SENTRY publications (Bell et al. 2002a), but also in national surveillance reports (Liebowitz et al. 2003; Huebner et al. 2000) and show a dramatic increase in penicillin (and macrolide) resistance in the last two decades, up to 30% intermediate and 46% high-level resistance to penicillin. However, such high rate of increasing resistance was not seen in the Gambia in the same period (Adegbola et al. 2006).

Variable information is available from surveys on nasopharyngeal carriage, mainly in sick children from many different countries (Hill et al. 2006; Feikin et al. 2003; Denno et al. 2002; Kacou-N'douba et al. 2004).

26.5.3 Enteric Pathogens

26.5.3.1 *Salmonella*, *Shigella* and *Vibrio* spp.

Typhoid surveillance in five Asian countries (China, India, Indonesia, Pakistan and Vietnam) showed an annual incidence of typhoid fever ranging from 24.2 (Vietnam) to 493.5 (India) per 100,000 population (Ochiai et al. 2008). A staggering 23% isolates were multidrug resistant (resistant to ampicillin, cotrimoxazole and chloramphenicol). While incidence of typhoid fever varied significantly between study sites, the methodology was not uniform. In the two sites with very high incidence rates (India and Pakistan), regular home visits could have influenced health seeking behaviour. Nonetheless, the high incidence paints a more accurate picture of disease burden. Also, because only one blood culture was taken in order to ascertain bacteraemia, the true incidence could be much higher, given the sensitivity of blood culture is not 100%.

Population-based surveillance data on *S. typhi* from Egypt reveal that an alarming 30% of strains (26 out of 89) were multidrug resistant (Srikantiah et al. 2006). The annual incidence in this report was 59 cases per 100,000 population per year. Many local reports from a variety of other African countries have been published. Most studies describe high rates of resistance for first-line antibiotics such as chloramphenicol (Nkuo-Akenji et al. 2001) while ciprofloxacin resistance was found to be low (Kariuki et al. 1999). Evolution towards multidrug-resistant non-typhoid Salmonellae is variable (Kariuki et al. 2005, 2006) and is mainly seen in HIV/AIDS patients.

Shigella species have been extensively described during outbreaks in Central and Eastern Africa, mainly in DRC and Rwanda. Several publications describe the historical evolution of resistance patterns towards multidrug resistance under the antibiotic pressure of the treatment used during previous recent outbreaks (Malengreau et al. 1983; Frost et al. 1982). von Seidlein et al. (2006) describe a similar phenomenon in several Asian countries. In this multi-centre surveillance comprising more than 600,000 people, *Shigella* was isolated from 2927 out of 56,958 diarrhoeal cases giving an overall isolation rate of 5%. Among the isolates identified as *Shigella flexneri*, majority were resistant to ampicillin (83.7%) and cotrimoxazole (75.4%) while a small but significant minority (0.9%) were multidrug resistant (resistant to ampicillin, cotrimoxazole, nalidixic acid and ciprofloxacin). These first multicentre surveillance data on shigellosis from this region provide a useful benchmark for future studies.

Most available data on *V. cholerae* are scattered outbreak reports from Western, Central and Eastern Africa. Most show also an increasing pattern of combined resistance for many first-line drugs, including tetracycline and cotrimoxazole (Urassa et al. 2000; Guévart et al. 2006) which is often plasmid mediated and linked to the increased consumption of antibiotics during and outside epidemics.

26.5.4 E. coli *and Other Enterobacteriaceae*

The Asian surveillance studies describe overall high resistance rates among Enterobacteriaceae and non-fermenters to third-generation cephalosporins and carbapenems (20–50%). A recent surveillance study from India showed that roughly 8% of *E. coli* strains causing urinary tract infections were resistant to three or more antibiotics commonly used to treat such infections (Mathai et al. 2008).

Resistance data of mainly nosocomial samples have been published from the SENTRY (Hirakata et al. 2005; Bell et al. 2002b, 2003) and PEARLS studies

(Bouchillon et al. 2004), showing alarming resistance rates, especially in *Klebsiella* and *Enterobacter* species, with rates of ESBL detection up to 28 and 36%, respectively, leaving limited treatment options for the empiric treatment of Gram-negative sepsis.

Local surveillance data of drug resistance in enteric *E. coli* in healthy Nigerian students were studied and followed over a 10-year period by Okeke et al. (2000), showing a significant increase in resistance for ampicillin, tetracycline and chloramphenicol in this "reservoir". A survey in asymptomatic bacteriuria among pregnant woman in Tanzania (Blomberg et al. 2005) showed relatively low resistance figures among *E. coli* and other Gram-negative bacteria. Multiple smaller studies, mainly on molecular resistance mechanisms of urinary or enteric *E. coli*, have been published.

26.6 Conclusion

Genuine surveillance data in developing countries remain scarce. The available data however suggest an important problem of extensive and combined resistance, especially among Gram-negative bacteria, the enteric pathogens and *S. aureus*. Resistance data on *S. pneumoniae* is also matter of concern. Further confirmation by genuine surveillance is urgent; this is however labour and cost intensive. Possible solutions include cohort surveillance in sentinel laboratories of particular regions and for selected key organisms. Data derived from cohort surveillance should be integrated in treatment guidelines and regularly updated. Partnerships with laboratories in resource-rich countries can be helpful for local capacity building through teaching and providing feedback and performing more expensive or technically complex investigations on selected strains.

But what are the benefits of carrying out surveillance? As can be noted from the KONSAR programme, almost a decade of surveillance had very little impact on antibiotic resistance. Several factors could contribute towards a lack of observed benefit of these surveillance reports. In most countries, antibiotic use remains indiscriminate and surveillance reports only push clinicians towards using antibiotics that are commonly perceived to be "stronger". There is a need to have robust policies on antibiotic usage while surveillance data are used as an educational tool rather than a marketing exercise for newer antibiotics. In addition, organizations that carry out surveillance studies need to be better coordinated. Fortunately, surveillance programmes across nations are beginning to take some shape. For instance, the ANSORP was able to gather data from several countries and its findings on macrolide resistance were recently published (Song et al. 2004).

In summary, surveillance data for antimicrobial resistance in developing countries are highly needed, but they are not to remain a collector's item. They are most useful whenever interventions to contain resistance are possible, through programmes for infection control, improving antibiotic prescribing behaviour and setting up standard treatment guidelines.

References

Adegbola, R.A., Hill, P.C., Secka, O., et al. 2006. Serotype and antimicrobial susceptibility patterns of isolates of Streptococcus pneumoniae causing invasive disease in The Gambia 1996–2003. Trop Med Intl Health 11: 1128–1135.

Aires De Sousa, M., Santos Sanches, I., Ferro, M.L., De Lencastre, H. 2000. Epidemiological study of staphylococcal colonization and cross-infection in two West African Hospitals. Microb Drug Resist 6: 133–141.

Amir, M., Paul, J., Batchelor, B. et al. 1995. Nasopharyngeal carriage of Staphylococcus aureus and carriage of tetracycline-resistant strains associated with HIV-seropositivity. Eur J Clin Microbiol Infect Dis 14: 34–40.

Archibald, L.K., Reller, L.B. 2001. Clinical microbiology in developing countries. Emerg Infect Dis 7: 302–305.

Bell, J.M., Turnidge, J.D., Jones, R.N. et al. 2002a. Antimicrobial resistance trends in community-acquired respiratory tract pathogens in the Western Pacific Region and South Africa: report from the SENTRY antimicrobial surveillance program, (1998–1999) including an in vitro evaluation of BMS284756. Int J Antimicrob Agents 19: 125–132.

Bell, J.M., Turnidge, J.D., Gales, A.C. et al. 2002b. Prevalence of extended spectrum beta-lactamase (ESBL)-producing clinical isolates in the Asia-Pacific region and South Africa: regional results from SENTRY Antimicrobial Surveillance Program (1998–99). Diagn Microbiol Infect Dis 42: 193–198.

Bell, J.M., Turnidge, J.D., Jones, R.N. et al. 2003. Prevalence of extended-spectrum beta-lactamase-producing Enterobacter cloacae in the Asia-Pacific region: results from the SENTRY Antimicrobial Surveillance Program, 1998–2001. javascript:AL_get(this, 'jour', 'Antimicrob Agents Chemother.'); Antimicrob Agents Chemother 47: 3989–3993.

Blomberg, B., Mwakagile, D.S., Urassa, W.K., et al. 2004. Surveillance of antimicrobial resistance at a tertiary hospital in Tanzania. BMC Public Health 4: 45.

Blomberg, B., Olsen, B.E., Hinderaker, S.G. et al. 2005. Antimicrobial resistance in urinary bacterial isolates from pregnant women in rural Tanzania: implications for public health. Scand J Infect Dis 37: 262–268.

Borg, M.A., Scicluna, E., de Kraker, M., et al. 2006. Antibiotic resistance in the southeastern Mediterranean – preliminary results from the ARMed project. Euro Surveill 11: 164–167.

Bouchillon, S.K., Johnson, B.M., Hoban, D.J. et al. 2004. Determining incidence of extended spectrum beta-lactamase producing Enterobacteriaceae, vancomycin-resistant Enterococcus faecium and methicillin-resistant Staphylococcus aureus in 38 centres from 17 countries: the PEARLS study 2001–2002. Int J Antimicrob Agents 24: 119–124.

Brink, A., Moolman, J., da Silva, M.C., et al. 2007. Antimicrobial susceptibility profile of selected bacteraemic pathogens from private institutions in South Africa. S Afr Med J 97: 273–279.

Denno, D.M., Frimpong, E., Gregory, M., Steele, R.W. 2002. Nasopharyngeal carriage and susceptibility patterns of Streptococcus pneumoniae in Kumasi, Ghana. West Afr J Med 21: 233–236.

Feikin, D.R., Davis, M., Nwanyanwu, O.C. et al. 2003. Antibiotic resistance and serotype distribution of Streptococcus pneumoniae colonizing rural Malawian children. Pediatr Infect Dis J 22: 564–567.

Felmingham, D., White, A.R., Jacobs, M.R. et al. 2005. The Alexander Project: the benefits from a decade of surveillance. J Antimicrob Chemother 56 Suppl 2: ii3–ii21.

Frost, J.A., Rowe, B., Vandepitte, J. 1982. Acquisition of trimethoprim resistance in epidemic strains of Shigella dysenteriae type 1 from Zaire. Lancet 8278: 963.

Guévart, E., Solle, J., Mouangue, A. et al. 2006. Antibiotic susceptibility of Vibrio cholerae 01: evolution after prolonged curative and preventive use during the 2004 cholera epidemics in Douala (Cameroon). Med Mal Infect 36: 329–334.

Hill, P.C., Akisanya, A., Sankareh, K. et al. 2006. Nasopharyngeal carriage of Streptococcus pneumoniae in Gambian villagers. Clin Infect Dis 43: 673–769.

Hirakata, Y., Matsuda, J., Miyazaki, Y. et al. 2005. Regional variation in the prevalence of extended-spectrum beta-lactamase-producing clinical isolates in the Asia-Pacific region (SENTRY 1998–2002). Diagn Microbiol Infect Dis 52: 323–329.

Hsu, L.Y., Tan, T.Y., Jureen, R. et al. 2007. Antimicrobial drug resistance in Singapore hospitals. Emerg Infect Dis 13: 1944–1947.

Huebner, R.E., Wasas, A.D., Klugman, K.P. 2000. Trends in antimicrobial resistance and serotype distribution of blood and cerebrospinal fluid isolates of Streptococcus pneumoniae in South Africa, 1991–1998. Int J Infect Dis 4: 214–218.

Kacou-N'douba, A., Guessennd-Kouadio, N., Kouassi-M'bengue, A., Dosso, M. 2004. Evolution of Streptococcus pneumoniae antibiotic resistance in Abidjan: update on nasopharyngeal carriage, from 1997 to 2001 Med Mal Infect 34: 83–85.

Kariuki, S., Cheesbrough, J., Mavridis, A.K., Hart, C.A. 1999. Typing of salmonella enterica serotype paratyphi C isolates from various countries by plasmid profiles and pulsed-field gel electrophoresis. J Clin Microbiol 37: 2058–2060.

Kariuki, S., Revathi, G., Kariuki, N. et al. 2005. Increasing prevalence of multidrug-resistant non-typhoidal salmonellae, Kenya, 1994–2003. Int J Antimicrob Agents 25: 38–43.

Kariuki, S., Revathi, G., Kiiru, J., Lowe, B., Berkley, J.A., Hart, C.A. 2006. Decreasing prevalence of antimicrobial resistance in non-typhoidal Salmonella isolated from children with bacteraemia in a rural district hospital, Kenya. Int J Antimicrob Agents 28:166–171.

Kesah, C., Ben Redjeb, S., Odugbemi, T.O. et al. 2003. Prevalence of methicillin-resistant Staphylococcus aureus in eight African hospitals and Malta. Clin Microbiol Infect 9: 153–156.

Lamikanra, A., Paul, B.D., Akinwole, O.B., Paul, M.O. 1985. Nasal carriage of Staphylococcus aureus in a population of healthy Nigerian students. J Med Microbiol 19: 211–216.

Lee, K., Lim, C.H., Cho, J.H. et al. 2006. High prevalence of ceftazidime-resistant Klebsiella pneumoniae and increase of imipenem-resistant Pseudomonas aeruginosa and Acinetobacter spp. in Korea: a KONSAR program in 2004. Yonsei Med J 47: 634–645.

Liebowitz, L.D., Slabbert, M., Huisamen, A. 2003. National surveillance programme on susceptibility patterns of respiratory pathogens in South Africa: moxifloxacin compared with eight other antimicrobial agents. J Clin Pathol 56: 344–347.

Malengreau, M., Molima-Kaba, Gillieaux, M., de Feyter, M., Kyele-Duibone, Mukolo-Ndjolo. 1983. Outbreak of Shigella dysentery in eastern Zaire, 1980–1982. Ann Soc Belge Med Trop 63: 59–67.

Mathai, E., Chandy, S., Thomas, K., et al. 2008. Antimicrobial resistance surveillance among commensal Escherichia coli in rural and urban areas in Southern India. Trop Med Int Health13: 41–45.

Mehta, A., Rosenthal, V.D., Mehta, Y., et al. 2007. Device-associated nosocomial infection rates in intensive care units of seven Indian cities. Findings of the International Nosocomial Infection Control Consortium (INICC). J Hosp Infect 67: 168–174.

Mukonyora, M., Mabiza, E., Gould, I.M. 1985. Staphylococcal bacteraemia in Zimbabwe 1983. J Infect 10: 233–239.

Nkuo-Akenji, T.K., Ntemgwa, M.L., Ndip, R.N. 2001. Asymptomatic salmonellosis and drug susceptibility in the Buea District, Cameroon. Cent Afr J Med 47: 254–257.

Obi, C.L., Makandiramba, B., Robertson, V., Tswana, S.A., Moyo, S.R., Nziramasanga, P. 1998. In-vitro activity of piperacillin and tazobactam combination against clinically significant bacteria. East Afr Med J 75: 162–165.

Ochiai, R.L., Acosta, C.J., Agtini, M. et al. 2007. The use of typhoid vaccines in Asia: the DOMI experience. Clin Infect Dis 45 Suppl 1: S34–S38.

Ochiai, R.L., Acosta, C.J., Danovaro-Holliday, M.C., et al. 2008. A study of typhoid fever in five Asian countries: disease burden and implications for controls. Bull World Health Organ 86: 260–268.

Okeke, I.N., Fayinka, S.T., Lamikanra, A. 2000. Antibiotic resistance in Escherichia coli from Nigerian students, 1986–1998. Emerg Infect Dis. 6: 393–396.

Okeke, I.N., Aboderin, O.A., Byarugaba, D.K., Ojo, K.K., Opintan, J.A. 2007. Growing problem of multidrug-resistant enteric pathogens in Africa. Emerg Infect Dis 13: 1640–1646.

Petti, C.A., Polage, C.R., Quinn, T.C., Ronald, A.R., Sande, M.A. 2006. Laboratory medicine in Africa: a barrier to effective health care. Clin Infect Dis 42: 377–382.

Seydi, M., Sow, A.I., Soumare, M. et al. 2004. Staphylococcus aureus bacteremia in the Dakar Fann University Hospital. Med Mal Infect 34: 210–215.

Shittu, A.O., Lin, J. 2006. Antimicrobial susceptibility patterns and characterization of clinical isolates of Staphylococcus aureus in KwaZulu-Natal province, South Africa. BMC Infect Dis 6: 125.

Song, J.H., Sung, S.I., Ko, K.S. et al. 2004. High prevalence of antimicrobial resistance among clinical Streptococcus pneumoniae isolates in Asia (an ANSORP study). Antimicrob Agents Chemother 48: 2101–2107.

Srikantiah, P., Girgis, F.Y., Luby, S.P., et al. 2006. Population-based surveillance of typhoid fever in Egypt. Am J Trop Med Hyg 74: 114–119.

Stelling, J.M., O'Brien, T.F. 1997. Surveillance of antimicrobial resistance: the WHONET program. Clin Infect Dis 24 Suppl 1: S157–S168.

Urassa, W.K., Mhando, Y.B., Mhalu, F.S., Mjaonga, S.J. 2000. Antimicrobial susceptibility pattern of Vibrio cholerae O1 strains during two cholera outbreaks in Dar es Salaam, Tanzania. East Afr Med J 77: 350–353.

Van den Ende, J. 2007. Antibiotic resistance in hospitals – are surveillance data of any value? S Afr Med J 2007 97: 264–266.

Vaughan, J.P., Walt, G. 1984. Implementing primary health care: some problems of creating national programmes. Trop Doct 14: 108–113.

Von Seidlein, L., Kim, D.R., Ali, M. et al. 2006. A multicentre study of Shigella diarrhoea in six Asian countries: disease burden, clinical manifestations, and microbiology. PLoS Med 3: e353.

WHO Global Strategy for Containment of Antimicrobial Resistance. WHO/CDS/CSR/DRS/2001.2.

Zinn, C.S., Westh, H., Rosdahl, V.T. et al. 2004. An international multicenter study of antimicrobial resistance and typing of hospital Staphylococcus aureus isolates from 21 laboratories in 19 countries or states. Microb Drug Resist 10: 160–168.

Chapter 27
Vaccines: A Cost-Effective Strategy to Contain Antimicrobial Resistance

Richard A. Adegbola and Debasish Saha

Abstract There is rapid spread of infectious disease globally. Emergence of antimicrobial resistance has made it difficult to contain the spread and the associated escalating costs of treatment. Rational use of antibiotics and an effective infection control strategy in health-care facilities can reduce the growing spread of infectious disease dramatically but these are often not possible in developing countries due to overburdened health-care structure. Nevertheless the infectious disease has to be contained.

The immune system in the human body is responsible for protection against disease. Once vaccines are introduced into the body they mimic natural infection and stimulate immune responses, which are directed to act against invading organisms. The protection thus conferred is often long lasting. Most vaccines act in this way against acute infections. Advanced molecular techniques can also be used to produce vaccines against pathogens causing chronic infections.

The success of the smallpox vaccine has been a history-making event and the World Health Organization's Expanded Programme of Immunization (EPI) has become a hallmark of disease containment with a bare minimum cost of $2 per person against six major infectious diseases. There is evidence from an US study that a savings of $53.2 billion can be met by immunizing children of a specific year birth cohort while the expenditure for the vaccination programme will only be $5.1 billion in terms of direct and indirect (societal) costs. Success stories have also been recorded with vaccines against invasive pneumococcal and Hib diseases and that of yellow fever in terms of lives saved.

Vaccine and antibiotics are not competitors but are equally important for the containment of the infectious disease. Antibiotic is a rescue measure while the vaccine can be seen as a long-term remedy. The ultimate goal is to reduce disease burden and the use of antibiotics. An effective vaccine can

R.A. Adegbola (✉)
Medical Research Council Laboratories (UK), Fajara, Banjul, The Gambia, West Africa
e-mail: radegbola@mrc.gm

A. de J. Sosa et al. (eds.), *Antimicrobial Resistance in Developing Countries*, 477
DOI 10.1007/978-0-387-89370-9_27, © Springer Science+Business Media, LLC 2010

almost stand-alone as a cost-effective containment strategy against antimicrobial resistance and should augment the fight against infection in a cost-effective way.

27.1 Introduction

The emergence of resistance to antimicrobials and the spread of infectious disease depend on geopolitics, economy, population structure and the social and cultural stigma. It is rather difficult to find a common approach towards achieving the goal of containing antibiotic resistance, but increasing antibiotic resistance and the spread of infectious diseases must be urgently contained.

It is generally perceived that rational use of antibiotics is the only way to stop the emergence of resistance. A good prescriber's guideline, judicious use, initiation of combination therapy and introduction of vaccine can bring down resistance substantially (Wilton et al. 2002).

Resistance containment strategies are reviewed in other chapters. The most immediate action that a health-care facility can embark on is an infection control strategy to slowdown progression of resistance. In an over-populous health-care centre, patient isolation, use of sterile gloves, disposable syringe and keeping the environment of the area clean can reduce the transmission of resistant organism (Nicolle 2001). The accuracy of the diagnostic tests, ability of the laboratory personnel to interpret data and dissemination of available data to the prescribers also play a vital role in the containment of resistance. Pharmaceutical companies are gradually moving away from research and development of newer antibiotics because of their short market life. Newer generation antibiotics become resistant within a short period of time, well before the cost of development is recovered. Efforts may be underway to rejuvenate antibiotic development by pharmaceutical industries but the process is unduly long and expensive (Williams and Heymann 1998).

Interestingly, resistance containment strategies are not new and most of them are in practice in at least some developed countries. An integrated approach to improve the public health infrastructure and ensure resistance containment is needed at health-care facilities and in the community. Unfortunately, developing countries often do not have the resources to set up such an infrastructure, upgrade technical skills, improve the basic supplies to the health-care facilities and sustain relevant programmes. In the interim, antibiotic use continues and resistance will continue to emerge.

This relentless emergence of resistance and the lack of an integrated approach are quietly leading us to a post-antibiotic era. An alternative "vaccine era" could be cost-effective and sustainable irrespective of developed or developing country status. This chapter focuses on the mechanisms of action of vaccines, the use of vaccines in containment of some deadly diseases and their overall cost-effectiveness.

27.2 Vaccines and Infectious Disease

Vaccines act on the immune system. They are produced from micro-organisms that can inflict a disease and are typically prepared from the pathogen or are versions of the pathogen that are weakened or attenuated through laboratory techniques. Once the vaccine is inside the human body, the defence system is activated against the particular disease. Vaccines induce immune responses directed at containing and eliminating the causative agent of the disease. The emergence of antibiotic resistance and the attendant escalating costs of treatment have most probably pushed public health researchers to look for more cost-effective disease-control interventions, vaccines being the most favourable option.

27.2.1 Vaccines Against Viral Infections and Their Cost-Effectiveness

Infections with signs that do not clearly point to bacterial origin are often thought to be viral. Since specific anti-viral treatments are often not available, vaccines that mimic closely to the natural infections are the principal means for control. Vaccines have been proved to be effective against acute, self-limiting infections and can confer long-lasting immunity. The same may not be true for chronic viral infections but with the advent of molecular biology there is a strong possibility of developing vaccines against chronic viral infections like hepatitis C virus (HCV), human papilloma virus (HPV) and HIV (Berzofsky et al. 2004). The major viral vaccination success story is the eradication of smallpox by the year 1977. The WHO spent more than $300 million in 11 years, an amount that has been repaid many times over – probably each year – in human lives saved (Ehreth 2003).

Polio vaccines were discovered by Salk (1947) and then by Sabin (1957). The polio vaccines were made widely available by 1963. Industrialized countries started using polio vaccines and in 1994 the polio was declared eradicated from the Americas. Eradication saved the USA alone from an annual expenditure of $230 million while the global savings from the polio eradication can be estimated as $1.5 billion from the treatment and other associated costs (Satcher 1998). Although there are ongoing challenges in the last phase, the polio eradication programme can also be considered a successful case study in infectious disease containment with resultant economic gains (Sabin 1984; Thompson et al. 2006).

Measles vaccine was licensed in the early 1960s and the disease is probably in existence from 7th century A.D.. Respiratory tract ailments following measles require an effective antibiotic to be instituted without delay to prevent any case fatality. Vaccination is the only way to prevent the disease and its deadly complications. With mass campaigns and target group

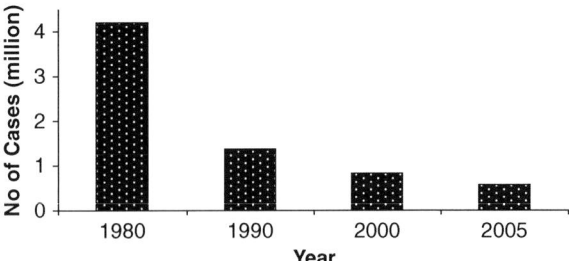

Fig. 27.1 Global decline in measles incidence following EPI

vaccinations, few countries have been able to eradicate the measles completely while the others have contained it (Fig. 27.1) (MMWR 1997). Apart from averting 3.2 million cases and 16,000 deaths until the year 2020, it has been estimated that vaccination strategy would save $208 million in treatment costs (Acharya et al. 2002).

Vaccination against hepatitis B is a rather late inclusion in the immunization schedule in most of the developing countries. There are still countries that are reluctant to include it for economic reasons even though the cost of the vaccine would be about $1. The benefit of yellow fever vaccination through the Expanded Programme on Immunization clearly outweighs the cost of treatment (Monath and Nasidi 1993). The disease can be contained efficiently through mass coverage of vaccination and with a cost less than $1 per person per year (Tomori 2002).

27.2.2 Vaccines Against Bacterial Infections and Their Cost-Effectiveness

Unlike viral infections, bacterial infections were diagnosed clinically or in laboratories and antibiotics were the natural choice for their cure. The first bacterial disease against which a whole cell vaccine was used, with limited success, was plague (Williamson 2001). Vaccination for bacterial infections such as diphtheria, tetanus, pertussis and tuberculosis followed a major breakthrough in infectious disease containment.

Pertussis is a childhood disease that brings about enormous direct cost on the hospitalization and resulting complications. Pertussis vaccination is usually given along with diphtheria and tetanus. It has been shown in a simulated birth cohort of 4.1 million children from birth to 15 years that with this combined vaccination 3 million cases and 28,000 deaths could be averted with a societal cost saving of over $45 million (Ekwueme et al. 2000).

Immunization against diphtheria also started in the mid-19th century in industrialized countries and the decline in the diphtheria cases in those countries motivated its adoption into childhood immunizations in many developing countries. The introduction of vaccination resulted in a dramatic reduction of cases in the early 1980s before a resurgence in Russia and the neighbouring countries (Galazka and Robertson 1995; Galazka et al. 1995). Following aggressive control through vaccination, the incidence of diphtheria was successfully reduced (Dittmann et al. 2000).

Neonatal tetanus can be prevented through active immunization of the mothers during their pregnancy and the immunity conferred to the child lasts for several years. An intervention package that includes tetanus vaccination can prevent 41–72% of the neonatal deaths worldwide (Darmstadt et al. 2005). The treatment of tetanus requires sophisticated artificial respirator, neuromuscular blocking agents and a selective antibiotic, which are not available in majority of the health-care settings in the developing countries (Daud et al. 1981). Figure 27.2 shows the decline in the incidence of cases due to combination vaccination against diphtheria, pertussis and both maternal and neonatal tetanus (WHO 2006).

The introduction effective *Haemophilus influenzae* type b (Hib) vaccine can prevent 400,000–700,000 global deaths due to Hib pneumonia or meningitis mostly in the age group under 5. The number of lives saved through vaccination can easily justify the cost of the vaccine (WHO 2007). In countries like the Gambia, Hib vaccine prevented 21% of all-cause radiological pneumonia in children (Mulholland and Adegbola 1998; Peltola 2000a).

Tuberculosis (TB) is mainly a problem of the impoverished countries. The problem is escalated by the emergence of multidrug-resistant tuberculosis (MDR-TB) and the emerging extensively drug-resistant tuberculosis (XDR-TB). The close interaction of TB with HIV/AIDS has worsened the situation in sub-Saharan Africa and Asia and TB is considered to be one of the most frequent causes of death (Dolin et al. 1994). Global incidence of TB is rising by about 1% per annum despite a wide coverage of BCG immunization. Improved technology is being applied to identify potential new TB vaccine candidates some of which are now undergoing clinical evaluation (Fruth and Young 2004; Ibanga et al. 2006; Orme 2005).

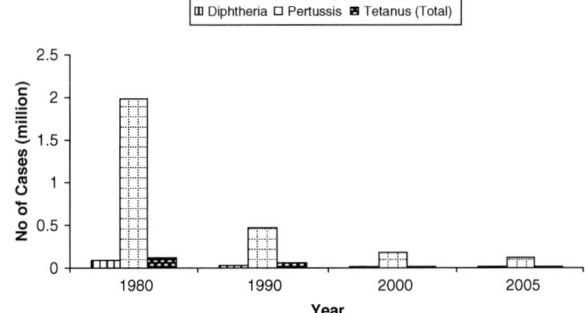

Fig. 27.2 The reported incidence of diphtheria, pertussis and tetanus from the WHO member countries

27.3 Vaccine for the Expanded Programme of Immunization

Until recently, early vaccines were mainly used by affluent countries. In the early- and mid-20th century there was rapid development of vaccine production following which World Health Organization (WHO) in the year 1974 undertook a massive program to control six major childhood diseases namely tetanus, diphtheria, pertussis, tuberculosis, poliomyelitis and measles through effective vaccination using EPI. During the early 1990s, hepatitis B was incorporated into the EPI while yellow fever was recommended for the countries where it is endemic. WHO is now periodically setting time-bound targets to control each of these diseases. EPI became the trademark in vaccine dosing, safety, efficacy, coverage and cost-effectiveness (Le Gonidec 1985). EPI was successful in combating disease that caused enormous suffering to the human being at a very marginal cost. The common EPI protocol adopted by the member countries of WHO does not cost more than $2 per individual as opposed to the treatment that would have placed enormous social and economic burden on the individual as well as the society (Table 27.1).

Information is lacking regarding the cost-effectiveness of vaccine initiatives in most of the developing countries. In a study conducted in the USA, there was evidence of savings of $53.2 billion by immunizing children of a specific year birth cohort while the expenditure for the vaccination programme was only $5.1 billion in terms of both direct and indirect (societal) costs (Zhou et al. 2005). A series of analysis was also conducted at Slovak Republic and reported the cost-effectiveness of routine and regular vaccination through EPI (Hudeckova and Straka 2000).

Table 27.1 Outcome of major diseases and cost of vaccine against them

Disease	Age affected	Outcome	Vaccine available	Cost (US $/dose)
Tuberculosis	All	High case fatality requires active treatment with multiple antibiotics for at least 6 months. MDR is a growing problem	BCG	0.096
Poliomyelitis	Children	Lifelong disability	Polio	0.152
Diphtheria, pertussis and tetanus	All	High case fatality requires intensive treatment and antibiotics	DTP	0.210
Measles	Children	High case fatality	Measles	0.500
Hepatitis B	All	Requires a long term and expensive treatment. Major cause of hepatic carcinoma	Hepatitis B	0.270

27.4 Newer Vaccines

Approximately 2.0 of the 10.6 million deaths that occur in young children each year are due to pneumonia (Wardlaw et al. 2006). About half of these deaths occur among children in sub-Saharan Africa and another 30% occur in the countries of south Asia (Bryce et al. 2005). In developing countries 11–20 million children with severe pneumonia are known to require hospital admission (Rudan et al. 2004). *Streptococcus pneumoniae* has been identified as a main bacterial cause of pneumonia (Shann 1986) and other pneumococcal diseases particularly in very young children where a polysaccharide vaccine is not very effective. Recent estimates from WHO indicate that pneumococcal infections are responsible for 1.6 million deaths each year, half of them in the age group under 5 and the highest risk is among the children less than 2 years old and in elderly person (WHO, 2007). Emergence of penicillin- and multiple drug-resistant pneumococci has compounded the problems of managing children with pneumococcal disease. Apart from preventing invasive infections like meningitis, bacteraemia and pneumonia due to *S. pneumoniae*, a conjugate pneumococcal vaccine was able to reduce at least more than a million of estimated 20 millions of episode of otitis media in the USA and thereby reducing both the medical and societal cost indicating that vaccination is much more cost-effective than antibiotic treatment (Black et al. 2000; Capra et al. 2000; Eskola et al. 2001; Ray et al. 2002). An effective infant vaccination programme can prevent 78% of the potential meningitis cases, 69% of the potential pneumonia cases and 8% of the potential otitis media episodes in a birth cohort and lower the total cost of pneumococcal disease by about $750 million in a year in the USA (Black et al. 2000). In countries where pneumococcal disease is caused by only few serotypes, a conjugate vaccine is highly efficacious and can reduce rates of hospital admission, thus reducing both the direct and indirect cost of treatment and should be considered to reduce the disease burden and contain antibiotic resistance (Adegbola et al. 2006; Cutts et al. 2005).

Proper vaccination strategy can reduce meningococcal disease substantially and can be an effective strategy to prevent an impending epidemic even in the less developed countries (Peltola 2000b). Routine immunization with meningococcal vaccine is within the cost-effective reach of the public health interventions especially in countries where the disease is endemic (Miller and Shahab 2005). A simulation of epidemics of meningococcal disease proved that a preventive vaccination strategy is more cost-effective in averting a possible adverse outcome (Parent du Chatelet et al. 2001).

27.5 Vaccine for the Future

Apart from pneumonia there are an estimated 5 billion cases of morbidity and 5 million cases of mortality due to enteric infections, particularly in developing countries and in terms of severity and incidence they are more to affect children in age group under 5 (Wiedermann and Kollaritsch 2006). Most industrialized countries have been able to contain these infections through improvement of water supply and sanitation, whereas the people living in the underdeveloped countries and those travelling to these areas still suffer from these infections (Bloom and Canning 2000; Reiff et al. 1996). Majority of the enteric diseases are caused by bacteria but in very young children diarrhoea is mainly due to rotavirus and highly affected by seasonality. Early studies in rotavirus vaccination showed cost saving promises (Carlin et al. 1999; Takala et al. 1998). There are difficulties with inducing immunity against a wide range of pathogens causing enteric infections but there are some effective vaccines available and research is underway to have clearer definition of the aetiology and epidemiology of enteric infections with a view to finding newer and more appropriate vaccines that can prevent diarrhoea due to viral or bacterial aetiology (Dougan et al. 2002; Kollaritsch and Wiedermann 2007; Levine and Noriega 1995; Svennerholm and Steele 2004).

The number of childhood vaccinations is increasing and there is a growing interest from parents to have their children vaccinated with maximum possible number of shots rather than face some dire consequences of illness (Fig. 27.3).

It is quite an understandable argument that most of the currently available effective vaccines are against viral diseases and it is often not obvious that these

Fig. 27.3 A busy and crowded immunization centre at the Gambia

can be cost-effective in containing antibiotic resistance. It must also be appreciated that most viral infections prompt antimicrobial treatment to counter secondary infections and any post-disease complications. It will be unfortunate if we do not foresee the long-term cost-effectiveness that is brought about by disease containment and reduction in the use of antibiotics, thereby reducing the selective pressure and the hefty budget required for research and development of newer antibiotics.

27.6 Why Choose Vaccines?

Emergencies are always unpredictable and it often becomes difficult to plan for health emergencies to ensure adequate and timely response. Diseases that kill millions each year require immediate resolution, which may be conferred through antibiotics, as well as lasting measures to prevent recurrence, and here vaccines would certainly fulfil the charge.

Introduction of an effective vaccine helped in the eradication of smallpox; others have been contained over the period while research is still underway aimed at controlling or eradicating other dreadful viral diseases. It is evident from the WHO (2006) that affluent countries are well ahead on the vaccine coverage but it is also reassuring to see that developing countries of Southeast Asia and sub-Saharan Africa are also catching up (Table 27.2). Along with this, the industrialized countries took initiative to introduce vaccines against some diseases that caused outbreaks with a high mortality with overwhelming success. Presently, HIV/AIDS, tuberculosis and malaria are a major global public health problem. Multidrug-resistant tuberculosis, failure of malaria to respond

Table 27.2 EPI coverage (%) against major diseases by WHO member countries in 2005

Regions/countries	BCG	DTP	OPV/IPV	MCV	Hep B
AFR	77	67	69	65	39
AMR	96	92	91	92	85
EMR	84	82	84	82	74
EUR	91	95	95	93	76
SEAR	80	66	65	65	27
WPR	87	87	87	87	76
Developed	83	96	94	92	63
Developing	83	75	76	75	57
Least developed	82	76	76	72	41
Economy in transition	95	97	97	97	95

AFR-WHO African Region, AMR-WHO Region of the Americas, EMR-WHO Eastern Mediterranean Region, EUR-WHO European Region, SEAR-WHO South-East Asia Region, WPR-WHO Western Pacific Region

BCG-Bacille Calmette-Guérin; vaccine against tuberculosis, DTP-diphtheria, tetanus, pertussis vaccine (coverage indicates three doses were completed), OPV-oral polio vaccine IPV-inactivated polio vaccine, MCV-measles-containing vaccine, HepB-hepatitis B vaccine.

to conventional drugs and failure of combination therapy against HIV/AIDS have made vaccines important candidates for cost-effective protection against these killer diseases.

27.7 Vaccines: The Cost

A true cost-effectiveness study for vaccines is probably not possible as it will require a non-vaccinated or placebo-receiving comparison group, which is unethical. Most of cost-effectiveness studies are based on epidemiological models or simulations of the disease in question but all of these clearly showed the benefit of vaccination in terms of both direct and indirect costs (Table 27.3) (Hudeãková et al. 2004).

The cost-effectiveness of a vaccine depends on the combined information on the incidence of the disease, its sequelae if not treated, the safety and efficacy of the vaccine, and its cost from production to program administration and the alternative intervention available for the disease in question.

Cost-effectiveness for any intervention has become an important issue in national health budgeting. When the intervention is able to give the maximum benefit at an acceptable cost, the effectiveness is attained. But there is no clear margin of the cost of an effective intervention. Acceptability depends on the willingness to pay for the health benefits, which do not have any defined threshold. Cost-effectiveness is a mode of economic evaluation and for any health-related intervention gains are measured in the number of illnesses prevented, workdays saved due to illness and quality of life years gained. Vaccines may not provide an immediate result to an acute infectious state and may look as an unworthy expenditure but importance should be given not only to the immediate cost but also to the relative reduction of mortality and morbidity due to vaccine introduction and compared with other available alternatives in a define set of population.

Table 27.3 The changes in disease and financial cost at Slovakia in pre-vaccination (model situation) and post-vaccination era (1951–2002)

Disease	Cases in the pre-vaccination period (/100000)	Cases in the post-vaccination period (/100000)	Changes in incidence of disease (%)	Change in the treatment cost (%)	Cost saving (%)
Diphtheria	66.5	0.0	100	100	11.1
Tetanus	1.2	0.1	95.8	95.7	0.2
Pertussis	56.7	0.7	98.4	98.8	7.7
Poliomyelitis	7.4	0.0	100	100	1.1
Measles	381.9	0.0	100	100	39.5
Rubella	309.9	0.1	99.9	99.9	19.4
Mumps	417.8	0.2	99.9	99.9	21.0
All	X	X	99.9	99.9	100

27.8 Vaccines – A Renewed Hope

The EPI has ushered a new era in preventive medicine. Without the EPI, most preventable diseases will re-emerge and continue to sicken the children and increase mortality, health problems and chronic disability. But EPI has its drawbacks with the reporting of adverse events, cold chain requirements and less than optimal implementation in many countries. Renewed efforts are being made through global donor programmes such as the Advanced Market Commitment and the plan of Global Alliance for Vaccines and Immunization to make new and possibly expensive vaccines available to those who would not have otherwise had access to them in developing countries.

The future appears even more promising; more and more countries are incorporating newer vaccines or rescheduling programmes in order to achieve a wide range of coverage against some of the most deadly diseases. The immediate costs of the vaccine seem to be high but vaccination costs should not be taken in isolation but must be seen in the context of other alternatives and the growing menace of antimicrobial resistance. A vaccine should be able to decrease the morbidity and mortality and eventually eliminate the disease in a cost-effective way.

27.9 Conclusion

Antibiotics and vaccines are not competitors. Antibiotics should be considered as the rescue measure while vaccines are a primary mode of prevention of infectious disease and should augment the fight against infection. Vaccines are not an infectious disease panacea. There are issues with escape mutants, capsular switching and serotype replacement diseases and, in some situations antimicrobial use will still be required. Despite the possible challenges with vaccine use, prior vaccination provides protection against shedding of pathogens. It may also raise the threshold load of pathogens required for infection and prevent pathogenic organisms from acquiring resistance from the environment. Moreover, indirect protective effects of some vaccines resulting from herd immunity in individuals who have not been vaccinated represent additional benefits and another indication of cost-effectiveness of a vaccination program.

Vaccines act primarily as a protective measure against infectious disease. If vaccination can prevent an infectious disease, antibiotic use will be drastically reduced. Lower antibiotic use will reduce the pressure on the pathogen to transform in a manner that will promote resistance. Vaccine development needs continued efforts and resources and technical skills from developed countries can be combined with the present target sites and the health-care set-up in the least- or under-developed countries. The focal point of research then becomes unidirectional. Among the strategies available, the vaccine can almost stand-alone as a cost-effective containment strategy against antibiotic resistance.

References

Acharya A, Diaz-Ortega JL, Tambini G, de Quadros C and Arita I. 2002. "Cost-effectiveness of measles elimination in Latin America and the Caribbean: a prospective analysis." *Vaccine* 20: 3332–3341.

Adegbola RA, Hill PC, Secka O, Ikumapayi UN, Lahai G, Greenwood BM and Corrah T. 2006. "Serotype and antimicrobial susceptibility patterns of isolates of Streptococcus pneumoniae causing invasive disease in The Gambia 1996–2003." *Trop Med Int Health* 11: 1128–1135.

Berzofsky JA, Ahlers JD, Janik J, Morris J, Oh S, Terabe M and Belyakov IM. 2004. "Progress on new vaccine strategies against chronic viral infections." *J Clin Invest* 114: 450–462.

Black S, Lieu TA, Ray GT, Capra A and Shinefield HR. 2000. "Assessing costs and cost effectiveness of pneumococcal disease and vaccination within Kaiser Permanente." *Vaccine* 19 Suppl 1: S83–S86.

Black S, Shinefield H, Fireman B, Lewis E, Ray P, Hansen JR, Elvin L, Ensor KM, Hackell J, Siber G, Malinoski F, Madore D, Chang I, Kohberger R, Watson W, Austrian R and Edwards K. 2000. "Efficacy, safety and immunogenicity of heptavalent pneumococcal conjugate vaccine in children. Northern California Kaiser Permanente Vaccine Study Center Group." *Pediatr Infect Dis J* 19:187–195.

Bloom DE and Canning D. 2000. "Policy forum: public health. The health and wealth of nations." *Science* 287:1207–1209.

Bryce J, Black RE, Walker N, Bhutta ZA, Lawn JE and Steketee RW. 2005. "Can the world afford to save the lives of 6 million children each year?" *Lancet* 365: 2193–2200.

Capra AM, Lieu TA, Black SB, Shinefield HR, Martin KE and Klein JO. 2000. "Costs of otitis media in a managed care population." *Pediatr Infect Dis J* 19: 354–355.

Carlin JB, Jackson T, Lane L, Bishop RF and Barnes GL. 1999. "Cost effectiveness of rotavirus vaccination in Australia." *Aust N Z J Public Health* 23:611–616

Cutts FT, Zaman SM, Enwere G, Jaffar S, Levine OS, Okoko JB, Oluwalana C, Vaughan A, Obaro SK, Leach A, McAdam KP, Biney E, Saaka M, Onwuchekwa U, Yallop F, Pierce NF, Greenwood BM and Adegbola RA. 2005. "Efficacy of nine-valent pneumococcal conjugate vaccine against pneumonia and invasive pneumococcal disease in The Gambia: randomised, double-blind, placebo-controlled trial." *Lancet* 365: 1139–1146.

Darmstadt GL, Bhutta ZA, Cousens S, Adam T, Walker N and de Bernis L. 2005. "Evidence-based, cost-effective interventions: how many newborn babies can we save?" *Lancet* 365: 977–988.

Daud S, Mohammad T and Ahmad A. 1981. "Tetanus neonatorum (a preliminary report of assessment of different therapeutic regimens)." *J Trop Pediatr* 27:308–311.

Dittmann S, Wharton M, Vitek C, Ciotti M, Galazka A, Guichard S, Hardy I, Kartoglu U, Koyama S, Kreysler J, Martin B, Mercer D, Ronne T, Roure C, Steinglass R, Strebel P, Sutter R and Trostle M. 2000. "Successful control of epidemic diphtheria in the states of the Former Union of Soviet Socialist Republics: lessons learned." *J Infect Dis* 181 Suppl 1:S10–S22.

Dolin PJ, Raviglione MC and Kochi A. 1994. "Global tuberculosis incidence and mortality during 1990–2000." *Bull World Health Organ* 72:213–220.

Dougan G, Huett A and Clare S. 2002. "Vaccines against human enteric bacterial pathogens." *Br Med Bull* 62:113–123.

Ehreth J. 2003. "The global value of vaccination." *Vaccine* 21:596–600.

Ekwueme DU, Strebel PM, Hadler SC, Meltzer MI, Allen JW and Livengood JR. 2000. "Economic evaluation of use of diphtheria, tetanus, and acellular pertussis vaccine or diphtheria, tetanus, and whole-cell pertussis vaccine in the United States, 1997." *Arch Pediatr Adolesc Med* 154:797–803.

Eskola J, Kilpi T, Palmu A, Jokinen J, Haapakoski J, Herva E, Takala A, Kayhty H, Karma P, Kohberger R, Siber G and Makela PH. 2001. "Efficacy of a pneumococcal conjugate vaccine against acute otitis media." *N Engl J Med* 344:403–409.

Fruth U and Young D. 2004. "Prospects for new TB vaccines: Stop TB Working Group on TB Vaccine Development." *Int J Tuberc Lung Dis* 8:151–155.

Galazka AM and Robertson SE. 1995. "Diphtheria: changing patterns in the developing world and the industrialized world." *Eur J Epidemiol* 11:107–117.

Galazka AM, Robertson SE and Oblapenko GP. 1995. "Resurgence of diphtheria." *Eur J Epidemiol* 11:95–105.

Hudeãková H, Szilágyiová, M, Nováková, E, Huboãan, P and Oleár V. 2004. "Public health importance of vaccination against selected communicable diseases." *Acta Medica Martiniana* 4:34–37.

Hudeckova H and Straka S. 2000. Health and economic benefits of compulsory regular vaccination in the Slovak Republic. I. Methods. *Epidemiol Mikrobiol Imunol* 49:24–27.

Ibanga HB, Brookes RH, Hill PC, Owiafe PK, Fletcher HA, Lienhardt C, Hill AV, Adegbola RA and McShane H. 2006. "Early clinical trials with a new tuberculosis vaccine, MVA85A, in tuberculosis-endemic countries: issues in study design." *Lancet Infect Dis* 6:522–528.

Kollaritsch H and Wiedermann U. 2007. "Examples for vaccines against diarrheal diseases – rotavirus and traveller's diarrhea". *Wien Med Wochenschr* 157:102–106.

Le Gonidec G. 1985. "The expanded WHO vaccination program". *Ann Inst Pasteur Immunol* 136D:167–174.

Levine MM and Noriega F. 1995. "A review of the current status of enteric vaccines." *P N G Med J* 38:325–331.

Miller MA and Shahab CK. 2005. "Review of the cost effectiveness of immunisation strategies for the control of epidemic meningococcal meningitis." *Pharmacoeconomics* 23:333–343.

MMWR. 1997. "Measles eradication: recommendations from a meeting cosponsored by the World Health Organization, the Pan American Health Organization, and CDC." *MMWR Recomm Rep* 46:1–20.

Monath TP and Nasidi A. 1993. "Should yellow fever vaccine be included in the expanded program of immunization in Africa? A cost-effectiveness analysis for Nigeria." *Am J Trop Med Hyg* 48:274–299.

Mulholland EK and Adegbola RA. 1998. "The Gambian *Haemophilus influenzae* type b vaccine trial: what does it tell us about the burden of *Haemophilus influenzae* type b disease?" *Pediatr Infect Dis J* 17:S123–S125.

Nicolle LE. 2001. "Infection control programmes to contain antimicrobial resistance." WHO/CDS/CSR/DRS/2001.7.

Orme IM. 2005. "Tuberculosis vaccines: current progress." *Drugs* 65:2437–2444.

Parent du Chatelet I, Gessner BD and da Silva A. 2001. "Comparison of cost-effectiveness of preventive and reactive mass immunization campaigns against meningococcal meningitis in West Africa: a theoretical modeling analysis." *Vaccine* 19:3420–3431.

Peltola H. 2000a. "Worldwide *Haemophilus influenzae* type b disease at the beginning of the 21st century: global analysis of the disease burden 25 years after the use of the polysaccharide vaccine and a decade after the advent of conjugates." *Clin Microbiol Rev* 13: 302–317.

Peltola H. 2000b. "Emergency or routine vaccination against meningococcal disease in Africa?" *Lancet* 355:3.

Ray GT, Butler JC, Black SB, Shinefield HR, Fireman BH and Lieu TA. 2002. "Observed costs and health care use of children in a randomized controlled trial of pneumococcal conjugate vaccine." *Pediatr Infect Dis J* 21:361–365.

Reiff FM, Roses M, Venczel L, Quick R and Witt VM. 1996. "Low-cost safe water for the world: a practical interim solution." *J Public Health Policy* 17:389–408.

Rudan I, Tomaskovic L, Boschi-Pinto C and Campbell H. 2004. "Global estimate of the incidence of clinical pneumonia among children under five years of age." *Bull World Health Organ* 82:895–903.

Sabin AB. 1984. "Strategies for elimination of poliomyelitis in different parts of the world with use of oral poliovirus vaccine." *Rev Infect Dis* 6 Suppl 2:S391–S396.

Satcher D. 1998. "Testimony on Eradication of Polio and Control or Elimination of Measles" Testimony presented before the Senate Committee on Appropriations, Subcommittee on Labor, Health and Human Services, Education and Related Agencies. September 23, 1998.

Shann F. 1986. "Etiology of severe pneumonia in children in developing countries." *Pediatr Infect Dis* 5:247–252.

Svennerholm AM and Steele D. 2004. "Microbial-gut interactions in health and disease. Progress in enteric vaccine development." *Best Pract Res Clin Gastroenterol* 18:421–445.

Takala AK, Koskenniemi E, Joensuu J, Makela M and Vesikari T. 1998. "Economic evaluation of rotavirus vaccinations in Finland: randomized, double-blind, placebo-controlled trial of tetravalent rhesus rotavirus vaccine." *Clin Infect Dis* 27:272–282.

Thompson KM, Duintjer Tebbens RJ, Pallansch MA, Kew OM, Sutter RW, Aylward RB, Watkins M, Gary H, Alexander JP, Venczel L, Johnson D, Caceres VM, Sangrujee N, Jafari H and Cochi SL. 2006. "Development and consideration of global policies for managing the future risks of poliovirus outbreaks: insights and lessons learned through modeling." *Risk Anal* 26:1571–1580.

Tomori O. 2002. "Yellow fever in Africa: public health impact and prospects for control in the 21st century." *Biomedica* 22:178–210.

Wardlaw T, Salama P, Johansson EW and Mason E. 2006. "Pneumonia: the leading killer of children." Lancet 368:1048–1050.

WHO. 2006. "WHO vaccine-preventable diseases: monitoring system.2006 global summary." WHO/IBV/2006.

WHO. 2007. "The evolving vaccine pipeline." www.who.int/immunization_delivery/new_vaccines/Evolving-vaccine-pipeline.pdf accessed on 27th March, 2008 at 1310 hrs.

Wiedermann U and Kollaritsch H. 2006. "Vaccines against traveler's diarrhoea and rotavirus disease – a review." *Wien Klin Wochenschr* 118:2–8.

Williams RJ and Heymann DL. 1998. "Containment of antibiotic resistance." *Science* 279: 1153–1154.

Williamson ED. 2001. "Plague vaccine research and development." *J Appl Microbiol* 91:606–608.

Wilton P, Smith R, Coast J and Millar M. 2002. "Strategies to contain the emergence of antimicrobial resistance: a systematic review of effectiveness and cost-effectiveness." *J Health Serv Res Policy* 7:111–117.

Zhou F, Santoli J, Messonnier ML, Yusuf HR, Shefer A, Chu SY, Rodewald L and Harpaz R. 2005. "Economic evaluation of the 7-vaccine routine childhood immunization schedule in the United States, 2001." *Arch Pediatr Adolesc Med* 159:1136–1144.

Chapter 28
Teaching Appropriate Antibiotic Use in Developing Countries

Celia M. Alpuche Aranda and Luis Romano Mazzotti

Abstract The use of antimicrobial drugs has saved countless lives and reduced the morbidity of infectious diseases. However, the growing threat from resistant microorganisms calls for cost-effective interventions to prevent the emergence of new resistant strains and the spread of existing ones. One approach to reduce the incidence of infections due to antibiotic-resistant organisms is to control the inappropriate use of antibiotics in both the hospital and community settings. In order to achieve this goal, it is necessary to identify the factors involved in physician's antibiotic prescription patterns and to elaborate educational interventions that can be adapted to different clinical scenarios. Several studies in developed countries have shown the potential benefit of these interventions. However, developing countries pose a special situation as they lack adequate pharmacological surveillance systems, antimicrobial drugs are widely available to the public (even without prescription), and continuous medical education programs for physicians are non-existent. The implementation of educational programs directed to the judicious use of antimicrobial drugs might probe to be the most efficient intervention in developing countries in the worldwide battle against antimicrobial resistance.

28.1 Introduction

The discovery of antimicrobial drugs in the first part of the nineteenth century is considered as a major milestone in medical history; however, the remarkable success of antimicrobial drugs generated a misconception in the late 1960s and early 1970s that infectious diseases had been conquered. This belief has misled generations of physicians and patients causing them to overuse these drugs. One

C.M. Alpuche Aranda (✉)
Dirección General Adjunta. Instituto de Diagnóstico y Referencia Epidemiológicos (InDRE). Secretaria de Salud, Mexico
e-mail: calpuche@salud.gob.mx

A. de J. Sosa et al. (eds.), *Antimicrobial Resistance in Developing Countries*,
DOI 10.1007/978-0-387-89370-9_28, © Springer Science+Business Media, LLC 2010

of the inevitable consequences of this practice was the appearance and spread of antimicrobial-resistant bacteria.

Fleming, during his Nobel Prize lecture, warned us: "It is not difficult to make microbes resistant to penicillin in the laboratory by exposing them to concentrations not sufficient to kill them, and the same thing has occasionally happened in the body... Moral: If you use penicillin, use enough" (Fleming 1945).

It seems likely that this problem will only get worse in the future. Despite the critical need for new antimicrobial agents, the development of these agents is declining, the biggest pharmaceutical companies are curtailing their antiinfective research programs (Spellberg et al. 2004), and a future with a limited or obsolete antibiotic armamentarium against pan-resistant bacteria is no longer a remote possibility. We might be facing, in a worst-case scenario, the dawn of a second pre-antibiotic era.

28.2 Inappropriate Antibiotic Use

Despite the association between the increased use of antibiotics and the spreading of resistant organisms in the community, the use of these drugs continues to grow worldwide. Unfortunately, most of the information regarding antimicrobial resistance and intervention strategies comes from developing countries. One of the reasons for this is that most reports from developing countries do not get published in peer-reviewed journals and end up in non-indexed publications (Keiser et al. 2004).

In the United States, an office-based study suggested an increase of 48% in antibiotic prescribing for children between 1980 and 1992 (McCaig et al. 2003). The Centers for Disease Control and Prevention (CDC) estimates that more than 100 million courses of antibiotics are prescribed each year, and approximately 50% of those prescriptions are unnecessary (usually indicated for viral or spontaneously resolving bacterial infections) (Dowell et al. 1998). The National Hospital Ambulatory Medical Care Survey demonstrated that almost 50% of common cold and up to 80% of acute bronchitis consults are treated with antimicrobial drugs.

Another phenomenon that has been observed in the last decade is the increase in the prescription of broad-spectrum antibiotics that were originally indicated for the treatment of severe infections. For example, fluoroquinolone use has increased 300% between 1995 and 2000. Approximately 42% of these prescriptions were not justified (Linder et al. 2005). The social and economic cost of the misuse of antibiotic drugs is surpassed by the costs that are generated from the treatment of multidrug-resistant (MDR) bacterial infections. For example, the cost of an ICU patient with MDR *Acinetobacter* is $237,195, compared with $54,504 of non-infected patients (Niederman 2001).

The burden of the inappropriate use of antimicrobial drugs in developing countries cannot be estimated since most of them do not have pharmacological surveillance programs. Current inferences about antibiotic prescription patterns in developing countries are based on a small number of reports generated by microbiology laboratories in urban areas and there their data do not usually represent the true situation of their countries. In many African, Asian, and Latin American countries, antibiotics are available on demand from hospitals, pharmacies, drugstores, convenience stores, and even from illegal drug dealers. In rural Bangladesh, for example, 95% of drugs consumed for 1 month by more than 2000 study participants came from local pharmacies (Okeke et al. 1999).

In Mexico, a survey of 1659 households in a periurban community in Mexico City revealed that antibiotics were used in 37% diarrheal episodes although only in 5% of all episodes was this therapy indicated. Patients seen by a physician were six times more likely to be treated with an antibiotic compared to those who did not consult a physician (Bojalil and Calva 1994). Using data from the same population, these authors determined that the main reasons for drug use were acute respiratory tract ailments and gastroenteritis. Approximately two–thirds of individuals using an antibiotic said they had used it for less than 5 days and 72% of the purchases were for insufficient quantities of drugs (Calva and Bojalil 1996).

Gamboa-Salcedo et al. (2006) assessed the outpatient management of 260 children with diarrhea in Mexico City. They found that 64% of patients that were seen by a physician received antibiotic treatment. The main reasons for prescribing an antibiotic were the following: >5 bowel movements (27.3%), >2 vomits (26.7%), and illness duration (23%). Fever and bloody stools were recognized as reasons for prescribing an antibiotic only in 7.3 and 3.6% of cases, respectively. In Pakistan, Nizami et al. (1996) observed the antibiotic prescribing practices of general physicians and pediatricians for childhood diarrhea. Sixty percent of GPs and 50% of pediatricians prescribed antibacterials; cotrimoxazole was the most frequently prescribed antibacterial by both types of practitioners. These findings support the use of physician- and community-oriented educational interventions in order to achieve a more rational use of antibiotics.

Several studies have analyzed the reasons why physicians prescribe antibiotics. First, there is the patient point of view: patients are accustomed to receiving antibiotics, so there is an increased demand for antibiotics treatment, since these drugs will make them feel better. Second, the physician must deal with the patient's pressure along with the fear of missing a severe infection, losing the patient to other physicians who will prescribe antibiotics, not having enough time to discuss the treatment, or not knowing the natural history of common viral and bacterial infections (Weissman and Besser 2004; Scott et al. 2001; Mangione-Smith et al. 1999; Colgan and Powers 2001; Bauchner et al. 1999; Avorn and Solomon 2000).

28.3 Intervention Studies

Since it might be difficult to distinguish between bacterial and viral infections, physicians must be aware of elements that might help them elaborate a presumptive diagnosis and decide which patients might benefit from antibiotic therapy. The knowledge of the natural history of some infectious diseases and whether antibiotics will affect the course of the illness must be reinforced since viral and many common bacterial infections are self-limited and there is no additional benefit from antimicrobial treatment. For example, antibiotic therapy has little effect on the course of acute bronchitis caused by *Mycoplasma pneumoniae* or *Chlamydophila pneumoniae* (Gonzales and Sande 2000). Due to the complexity of this problem, the medical community has adopted a number of interventions directed to physicians, nurses, pharmacist, and the public.

In 1995, the CDC launched the Campaign for Appropriate Antibiotic Use in the Community (Emmer and Besser 2002). This campaign targeted the five respiratory conditions that account for more than 75% of all office-based prescriptions for all ages combined: otitis media, sinusitis, pharyngitis, bronchitis, and the common cold. In collaboration with the American Academy of Pediatrics (AAP) and the American Academy of Family Physicians, the CDC developed six principles for appropriate use of antibiotics for pediatric upper respiratory tract infections (Dowell et al. 1998). They also produced health education materials for both parents and providers to promote appropriate use of antibiotics use. The objective of these materials was to stimulate discussion between patients and providers and change the current social perspective toward antibiotic drugs (Table 28.1).

Acute otitis media (AOM) is one of the most common infections for which antibiotics are prescribed in the pediatric population, resulting in more than 20 million antibiotic prescriptions and a cost of more than $3 billion each year (Powers 2007). The prescription rate of antimicrobials for this infection varies between 31% in the Netherlands and 98% in the United States, Australia, and New Zealand (Froom et al. 1990). In 2004, the AAP released its new guidelines

Table 28.1 CDC tips for practicing physicians

When parents ask for antibiotics to treat viral infections:
- Explain that unnecessary use of antibiotics can be harmful
- Share the facts
- Build cooperation and trust
- Encourage active management of the illness
- Be confident with the recommendation to use alternative treatments

Create an office environment to promote the reduction in antibiotic use:
- Talk about antibiotic use at 4- and 12-month-old child visits
- Start the educational process in the waiting room
- Involve office personnel in the educational process
- Use the CDC/AAP pamphlets and principles to support your treatment decisions

for the management of OMA including a subset of patients that might be carefully observed with no antimicrobial treatment. All these efforts to improve the clinical practice are obscured by the lack of continuous medical education. It is not unusual that these kinds of guidelines are totally unknown to general physicians, and even to infectious diseases specialists (Ibia et al. 2003).

Vernacchio et al. (2007) compared physicians' practice with the AAP acute otitis media guideline's recommendations among 299 physicians (77% pediatricians). Although the "watch-and-wait" option for low-risk children was considered to be acceptable by 83%, it was only used in 15% of AOM cases during the past 3 months. Parental reluctance was the most common reason for rejecting the observation option. Also, this group found that the antibiotic suggestions for treating OMA are not being followed. Only 57% reported the use of high-dose amoxicillin (80–90 mg/kg daily) for treatment of non-severe AOM, 13% used high-dose amoxicillin plus clavulanate for severe AOM, 43% used high-dose amoxicillin plus clavulanate for patients with AOM who had not responded to initial high-dose amoxicillin, and 17% used intramuscular ceftriaxone for patients who had not responded to high-dose amoxicillin plus clavulanate.

It is important to emphasize that all these educational and intervention strategies must be implemented in different social environments, in both a national and a regional basis. It is important to use local data when developing these programs, since the epidemiologic and social conditions might differ between communities.

Our modern lifestyle has caused that both physicians and patients have less time to discuss important issues, such as the true therapeutic activity of antibiotics, their limitations, and the risks of their inappropriate use. It has been observed that physicians who spend less time with their patients are likely to prescribe more antibiotics than those who spend more time with their patients (Hutchinson and Foley 1999).

On the other hand, several studies have analyzed the correlation between patients' degree of satisfaction and their treatment expectations; the results show that most patients would be satisfied with a non-antibiotic treatment as long as their physician explained thoroughly their diseases, treatment options, and the reasons for the decision to withhold antibiotics (Colgan and Powers 2001; Ong et al. 2007; Barden et al. 1998).

28.4 Medical Students and Residents

One of the target populations where intervention studies can benefit the most is the physicians in training. Unfortunately, this intervention has not been generalized enough. Fakih et al. (2003) surveyed 182 resident physicians of 11 primary care programs in Michigan regarding the management of upper respiratory infections and antimicrobial resistance. Although 91% agreed that overuse of antibiotics has a major role in increasing antimicrobial resistance, 20% answered that antibiotics are useful in the treatment of common cold, more than 30% would prescribe antibiotics when the diagnostic is not certain, and only 21% knew that there is no resistance to penicillin for *Streptococcus pyogenes*.

Ibia et al. (2005) surveyed senior medical students in 21 medical schools in New England and the Mid-Atlantic States to evaluate their knowledge of and compliance with principles of judicious antimicrobial use, as defined by the Centers for Disease Control and others. Ninety-nine percent of the students were aware of the increase in antimicrobial resistance, but almost 50% had read none of the principles, and only 2.9% had read all six. Thirty percent answered that they would treat a 4-year-old child with pharyngitis (with no positive culture) with antibiotics, and 60% thought that an 18-month-old toddler with purulent rhinitis and wheezing should be treated with antimicrobial drug.

In a similar study, Nambiar surveyed the knowledge of these six principles in pediatric residents of 12 teaching hospitals of the Mid-Atlantic States. While 50% of third- and fourth-year residents (R3–R4) knew these principles, only 16% of first-year residents (R1) were aware of them. When asked about their management on a child with purulent rhinitis and a fever of 38.8°C, 63% of R1 answered that they would start antibiotics, compared with 47% of R3–R4 (Nambiar et al. 2002).

In Costa Rica, Mora et al. (2002) reviewed 500 charts in a tertiary care teaching hospital to assess use, misuse, and abuse of antibiotics. Of the 500 patients, 175 (35%) did not received antibiotics, no appropriate cultures were obtained in 45 patients (14%), no record or justification for the prescription was documented in 130 patients (40%), no cultures were obtained before modifying therapy in 80 patients (46%), no planned duration of therapy was stated in 180 patients (55%), and an incorrect weight-based dosage was prescribed in 23 patients (8%).

Apisarnthanarak et al. (2006) evaluated the impact of education and an antibiotic control program on antibiotic prescribing practices in a tertiary care teaching hospital in Thailand. After the intervention, there was a 24% reduction in the rate of antibiotic prescription and the incidence of inappropriate antibiotic use was significantly reduced. Rates of use of third-generation cephalosporins and glycopeptides were significantly reduced but use of cefazolin and fluoroquinolones increased. Significant reductions in the incidence of infections due to methicillin-resistant *Staphylococcus aureus*, extended-spectrum beta-lactamase-producing *Escherichia coli* and *Klebsiella pneumoniae*, and third-generation, cephalosporin-resistant *Acinetobacter baumannii* were also observed. The authors calculated a total cost savings of $32,231 during the 1-year study period.

In Mexico City, Pineda et al. (2007) evaluated the knowledge on appropriate use of antibiotics among pediatric residents in a tertiary care teaching hospital. These pediatric residents were surveyed using a questionnaire that included very basic knowledge of antibiotics and practice experiences. This questionnaire was fully validated before the interview. The questions were as simple as "How do beta-lactam antibiotics work?" "Does cefotaxime belong to third-generation cephalosporins?" "What is the difference between bacteriostatic and bactericidal antibiotics?" "What is the use of antibiotics in upper respiratory tract infections and diarrhea?" among others. We expected that a longer time in the pediatric residency program would correlate with increased knowledge about rational antibiotic prescription; however, third-year residents did not achieve a significant superior grade when compared with first-year residents and the most surprising finding was that they were not able to reach 65% of the correct answers significantly lower compared to the positive control group that included infectious disease (ID) fellows (Fig. 28.1).

A significant difference was found when the grades of the ID fellows were compared with the grades of pediatric residents and other subspecialty fellows ($p = <0.001$, one-way ANOVA). No difference was detected between pediatric residents and non-ID fellows or between surgical and non-surgical residents. Although the grade achieved by the ID fellows control group (78%) seems low, we must point out that this interview took place in the first 3 months of residency year and those ID fellows with the lower grades were the ones in the beginning of the fellowship. This study clearly demonstrated that general knowledge regarding antibiotic use is not reaching good standards during regular medical school training, because these residents were trained in different medical schools around the country. It is obvious that this Pediatric Residency Program requires to introduce a well-defined antibiotic use course in the curricula with continuous evaluation during the 3 years of training (Table 28.2).

Medical schools all over the world need to extend and improve the curriculum for microbiology and infectious disease. In the Universidad Nacional Autonoma de Mexico (UNAM) School of Medicine, a rational use of antibiotics course has been designed by infectious diseases specialists (personal communication) and is given every year to third- and fourth-year medical students as an optional course but it is included as part of the curriculum to gain credits. In the 4 years that this course has been taking place, the initial evaluation of the students clearly shows a very limited knowledge regarding antibiotic

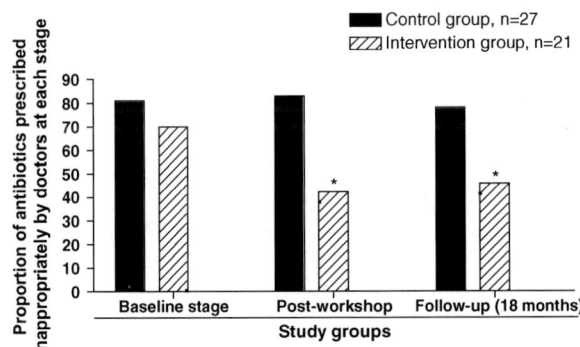

Fig. 28.1 Effect of the intervention on physicians' practice. *$p < 0.05$. Adapted from Perez-Cuevas

498 C.M. Alpuche Aranda and L. Romano Mazzotti

Table 28.2 Mean grades of pediatric residents on an appropriate use of antibiotics questionnaire

Group	N	Mean grade (IC 90%)
First-year residents	44	60.45 (55.89–65.02)
Second-year residents	37	57.7 (52.17–63.24)
Third-year residents	31	62.9 (57.17–68.64)
Control group (ID fellows)	13	77.3 (74.14–80.48)

Adapted from Pineda et al.

characteristics and use, and even more worrisome is that these students completed their microbiology and pharmacology courses just 1 or 2 years before.

The agenda of this rational use of antibiotics course includes a description of the antibiotic family (chemistry) and an explanation of the evidence for the antibiotic treatment of selected diseases. A review on antimicrobial resistance mechanisms, clinical impact, and regional data is also covered. Infectious diseases experts in different fields are invited to each class not only to give a lecture but also to explain their own and international experience regarding antibiotic prescription behavior in the specific topic and how this could be linked to antibacterial resistance. At the end of each class, a message on appropriate use of antimicrobial drugs is always underscored. In the last class, the students are surveyed using a questionnaire very similar to the one used in the pediatric residency study mentioned above and the obtained results highlight the increased comprehensive view of rational antibiotic use. Even more, the general opinion among these students is very homogeneous that they were able to understand the importance to reinforce the need of "excellence" in clinical diagnosis supported by scientific evidence in order to define the best treatment, in this case antibiotic treatment, for their patients. Unfortunately, there is no further analysis in order to determine this course's impact on this group of students over time.

28.5 Community Interventions

The possibility to change antibiotic prescription behavior patterns using specific training on rational use of antibiotics as target group interventions is one of the options that have shown an impact in a short period of time. Several interventions have targeted both physicians and general public to increase awareness of the adequate use of antimicrobial drugs.

Perez-Cuevas et al. conducted an educational intervention in 18 primary care facilities in Mexico City to improve prescribing practices for rhinopharyngitis. Using a workshop and a managerial peer review committee, the authors documented a decrease in the percentage of patients receiving antibiotics from 85.2 to 48%; the physicians' inappropriate use of antibiotics went from 70%

(baseline) to 42.3% immediately after the intervention, and 45.8% at the 18-month follow-up (Perez-Cuevas et al. 1992).

Two major things can be learnt from this study: first, that almost 80% of the physicians were using inappropriate antibiotics for rhinopharyngitis and that, after the intense intervention, antibiotic use decreased significantly; however, almost 50% of these physicians maintained their behavior to use them; second, it is evident that the intervention must be sustained for longer periods of time to be able to change prescription behavior patterns.

In Mexico, Gutierrez et al. (1994) used an intervention strategy designed to decrease drug prescription in the treatment of acute diarrhea aimed at family medicine practitioners. The authors evaluated 20 physicians who received the complete intervention (study group) and 20 physicians who received none (control group). The treatment behaviors of the study and control groups were similar on baseline but differed significantly in the post-workshop evaluation. The study group showed a reduction in the use of antibiotics (from 78.8 to 39.3%) for children younger than 5 years old with acute diarrhea. In the mid-term evaluation, the use of antibiotics by the study group decreased to 27.6%. In the long-term evaluation, persistent positive prescribing behavior was still present in the study group, with a significant difference compared to the control group, where no modification was found in the prescribing behavior throughout the study.

In a similar study conducted in Indonesia, Santoso et al. (1996) investigated the impacts of two different methods of educational intervention, i.e., a small-group face-to-face intervention and a formal seminar for prescribers, on prescribing practice in acute diarrhea. The results showed that both interventions were equally effective in improving the levels of knowledge of prescribers about the appropriate management of acute diarrhea. They were also partially effective in improving the appropriate use of drugs, reducing the use of non-rehydration medications. There was a highly significant reduction of antimicrobial usage after either small-group face-to-face intervention or formal seminar, and the former caused significantly greater reduction than the latter. The authors concluded that the small-group face-to-face intervention is as effective as, and less costly than, the seminar, making it a feasible option for improving antimicrobial prescribing patterns in developing countries.

Bexell et al. (1996) explored the impact of continuing education seminars on the quality of patient management and rational drug use in Zambia. In the intervention health centers, the average number of drugs per patient decreased from 2.3 to 1.9 ($p = 0.005$) and the proportion of patients managed with nonpharmacological treatment increased from 1 to 13.2%. More drugs were correctly chosen in the intervention health centers compared to control health centers. The proportion of patients prescribed antibiotics decreased and the proportion of patients adequately managed increased in the intervention health centers.

Hennessy et al. (2002) were able to diminish the antibiotic prescription rate in Alaskan rural villages by 35% after an appropriate use of antibiotics education program focused on both the community and the physicians. Rubin et al. (2005)

conducted a multifaceted intervention to improve antimicrobial prescribing for upper respiratory tract infections (URTIs) in a small rural community in Utah. The intervention involved patient education materials, a media campaign to increase public awareness, small group sessions involving physicians, and physician use of URTI algorithms. After 6 months, the percentage of patients in the community who received antibiotics for URTIs during the intervention period was 15.6% less than that for the baseline period. There was a decrease of 56% on antibiotic prescriptions for acute bronchitis and a 13.4% decrease on macrolides prescriptions. Among hospitalized patients, there was no significant decrease in the number of patients with URTI who were prescribed an antibiotic, although there was also a decrease in macrolide use (11.2%).

In Australia, Dollman et al. (2005) conducted a community-based intervention to reduce antibiotic use for upper respiratory tract infections. These investigators used consumer information on antibiotic use for URTIs (including a local media campaign) and education of health professionals (including sessions with general practitioners at the four practices in the study area). After 4 months of intervention, the prescription rates for the six most common antibiotics prescribed for URTI (amoxicillin, amoxicillin/clavulanic acid, cefaclor, doxycycline, erythromycin, and roxithromycin) decreased by 32%. These kinds of interventions have been able to reduce the number of prescription of certain types of antibiotics.

In the United States, during the 1992–2000 period, there was a reduction in the prescription rates of amoxicillin–clavulanic acid, cephalosporins, and erythromycin, but an increase was noted in macrolides, such as azithromycin and clarithromycin, and fluoroquinolones. In patients under 15 years of age, there was an increase of 69% in the prescription of amoxicillin–clavulanic acid (McCaig et al. 2003). In Manitoba, Canada, a decrease of almost 30% in antibiotic prescriptions was estimated in the 1996–2000 period, but the rate of broad-spectrum macrolide prescriptions increased 12-fold in the same period (Kozyrskyj et al. 2004).

As shown by the data above, a decrease in the prescription rates of some antimicrobial drugs has been accompanied by an increase in others, usually with a broader spectrum, which has a negative impact on the antimicrobial resistance phenomenon.

A large-scale multinational intervention, the Integrated Management of Childhood Illness (IMCI), was developed by WHO and UNICEF to provide effective and affordable interventions to reduce child mortality and improve child health and development. Gouws et al. (2004) assessed the effect of IMCI case management training on the use of antimicrobial drugs among health-care workers treating young children at first-level facilities in Brazil, Uganda, and Tanzania. Overall, children seen by IMCI-trained, health-care workers were less likely to receive unnecessary antibiotics in all three countries and more likely to receive correct prescriptions for antimicrobial drugs when needed. Caregivers whose children were seen by workers not trained in IMCI received little or no information about how to administer antibiotics.

28.6 Conclusion

The antimicrobial resistance that emerged in the last 50 years will continue growing in the future. The number of multiresistant bacteria will expand and some antibiotics will become obsolete. The patients who cannot be treated will have to be quarantined causing social unrest, affecting whole communities and even international travel and commerce. The impact on developing countries will be even more severe due to the economic burden of their health-care systems.

The only effective way to achieve a real change is by means of an integral and structured approach, focused on rational use of antibiotics, involving health administrations, health-care agencies, scientists, physicians, the public, and the agricultural and pharmaceutical industry (Table 28.3).

The Alliance for Prudent Use of Antibiotics (APUA) represents an international force devoted to improve antimicrobial use at all levels and in a worldwide basis (Barza et al. 2002).

One of the most important recommendations made by this organization refers to the ban of nontherapeutic use of antimicrobials in agriculture as an attempt to lower the burden of antimicrobial resistance in the environment with consequent benefits to human and animal health. Some European countries, Japan, and Australia have already adopted these recommendations, but their impact will not be noticeable until more countries, including developing ones, join this global effort.

Infectious diseases must be approached with a more "ecological" point of view, focusing on the host–parasite relationship and prevention, instead of the disease-based approach commonly used ("just kill the bug").

One of the basic strategies is to teach appropriate use of antibiotics and it has to be done since early time in training such as medical school and residency or fellowship programs. However, these interventions must also include general practitioners, especially in rural areas of developing countries where physicians are required to work with scarce resources and often ignore basic concepts of antimicrobial prescription (Okeke et al. 1999). It is also obvious that we need to approach this strategy in a different way and be able to develop a deep sense of the importance of appropriate antibiotic use. As the rest of the world, developing countries must include in their agenda the rational use of antibiotic strategies as part of the battle against antibacterial resistance problem.

Table 28.3 Components and strategies for appropriate use of antibiotics

Form a coalition of diverse partners

Target changes at multiple levels – individuals, groups, and organizations or institutions

Educate providers

Educate patients

Educate the public

References

American Academy of Pediatrics Subcommittee on Management of Acute Otitis Media. Diagnosis and management of acute otitis media. Pediatrics 2004;113:1451–65.

Apisarnthanarak A, Danchaivijitr S, Khawcharoenporn T, et al. Effectiveness of education and an antibiotic-control program in a tertiary care hospital in Thailand. Clin Infect Dis 2006;42:768–75

Avorn J, Solomon DH. Cultural and economic factors that (mis)shape antibiotic use: the nonpharmacologic basis of therapeutics. Ann Intern Med 2000;133:128–35.

Barden LS, Dowell SF, Schwartz B, Lackey C. Current attitudes regarding use of antimicrobial agents: results from physician's and parents' focus group discussions. Clin Pediatr (Phila) 1998;37:665–71.

Barza M, Gorbach S, DeVincent SJ. Introduction. Clin Infect Dis 2002;34 Suppl 3:S71–2.

Bauchner H, Pelton SI, Klein JO. Parents, physicians, and antibiotic use. Pediatrics 1999;103:395–401.

Bexell A, Lwando E, von Hofsten B, Tembo S, Eriksson B, Diwan VK. Improving drug use through continuing education: a randomized controlled trial in Zambia. J Clin Epidemiol 1996;49:355–7

Bojalil R, Calva JJ. Antibiotic misuse in diarrhea. A household survey in a Mexican community. J Clin Epidemiol 1994;47:147–56

Calva J, Bojalil R. Antibiotic use in a periurban community in Mexico: a household and drugstore survey. Soc Sci Med 1996;42:1121–8

CDC Centers for Disease Control and Prevention Practice Tips. Accessed at http://www.cdc.gov/drugresistance/community/files/ads/Practpac.pdf

Colgan R, Powers JH. Appropriate antimicrobial prescribing: approaches that limit antibiotic resistance. Am Fam Physician 2001;64:999–1004

Dollman WB, LeBlanc VT, Stevens L, O'Connor PJ, Turnidge JD. A community-based intervention to reduce antibiotic use for upper respiratory tract infections in regional South Australia. Med J Aust 2005;182:617–20.

Dowell SF, Marcy SM, Phillips WR, Gerber MA, Schwartz B. Principles of judicious use of antimicrobial agents for pediatric upper respiratory tract infections. Pediatrics 1998;101:163–5.

Emmer CL, Besser RE. Combating antimicrobial resistance: intervention programs to promote appropriate antibiotic use. Infect Med 2002;19:160–73.

Fakih MG, Hilu RC, Savoy-Moore RT, Saravolatz LD. Do resident physicians use antibiotics appropriately in treating upper respiratory infections? A survey of 11 programs. Clin Infect Dis 2003;37:853–6.

Fleming A. 1945. Penicillin. Nobel Lecture. Accessed at http://nobelprize.org/medicine/laureates/1945/fleming-lecture.pdf

Froom J, Culpepper L, Grob P, et al. Diagnosis and antibiotic treatment of acute otitis media: report from International Primary Care Network. BMJ 1990;300:582–6.

Gamboa-Salcedo T, Gutierrez-Camacho C, Mota-Hernandez F. Manejo de la diarrea en el hogar: errores y aciertos. Gac Med Mex 2006;142:309–14

Gonzales R, Sande MA. Uncomplicated acute bronchitis. Ann Intern Med 2000;133:981–91.

Gouws E, Bryce J, Habicht JP, et al. Improving antimicrobial use among health workers in first-level facilities: results from the multi-country evaluation of the Integrated Management of Childhood Illness strategy. Bull World Health Organ 2004;82:509–15.

Gutierrez G, Guiscafre H, Bronfman M, Walsh J, Martinez H, Munoz O. Changing physician prescribing patterns: evaluation of an educational strategy for acute diarrhea in Mexico City. Med Care 1994;32:436–46.

Hennessy TW, Petersen KM, Bruden D, et al. Changes in antibiotic-prescribing practices and carriage of penicillin-resistant Streptococcus pneumoniae: a controlled intervention trial in rural Alaska. Clin Infect Dis 2002;34:1543–50.

Hutchinson JM, Foley RN. Method of physician remuneration and rates of antibiotic prescription. CMAJ 1999;160:1013–7.

Ibia E, Sheridan M, Schwartz R. Knowledge of the principles of judicious antibiotic use for upper respiratory infections: a survey of senior medical students. South Med J 2005; 98:889–95.

Ibia EO, Schwartz RH, Freij BJ, Sheridan MJ. Judicious antibiotic use for purulent rhinorrhea: do pediatric infectious disease specialists practice what they preach? Clin Pediatr (Phila) 2003;42:443–6.

Keiser J, Utzinger J, Tanner M, Singer BH. Representation of authors and editors from countries with different human development indexes in the leading literature on tropical medicine: survey of current evidence. BMJ 2004;328:1229–32

Kozyrskyj AL, Carrie AG, Mazowita GB, Lix LM, Klassen TP, Law BJ. Decrease in antibiotic use among children in the 1990s: not all antibiotics, not all children. CMAJ 2004;171:133–8.

Linder JA, Huang ES, Steinman MA, Gonzales R, Stafford RS. Fluoroquinolone prescribing in the United States: 1995 to 2002. Am J Med 2005;118:259–68.

Mangione-Smith R, McGlynn EA, Elliott MN, Krogstad P, Brook RH. The relationship between perceived parental expectations and pediatrician antimicrobial prescribing behavior. Pediatrics 1999;103:711–8.

McCaig LF, Besser RE, Hughes JM. Antimicrobial drug prescription in ambulatory care settings, United States, 1992–2000. Emerg Infect Dis 2003;9:432–7.

Mora Y, Avila-Aguero ML, Umana MA, Jimenez AL, Paris MM, Faingezicht I. Epidemiological observations of the judicious use of antibiotics in a pediatric teaching hospital. Int J Infect Dis 2002;6:74–7.

Nambiar S, Schwartz RH, Sheridan MJ. Antibiotic use for upper respiratory tract infections: how well do pediatric residents do? Arch Pediatr Adolesc Med 2002;156:621–4.

National Center for Health Statistics-National Hospital Ambulatory Medical Care Survey. Accessed at http://www.cdc.gov/nchs/about/major/ahcd/ahcd1.htm

Niederman MS. Impact of antibiotic resistance on clinical outcomes and the cost of care. Crit Care Med 2001;29:N114–20.

Nizami SQ, Khan IA, Bhutta ZA. Drug prescribing practices of general practitioners and paediatricians for childhood diarrhoea in Karachi, Pakistan. Soc Sci Med 1996;42:1133–9.

Okeke IN, Lamikanra A, Edelman R. Socioeconomic and behavioral factors leading to acquired bacterial resistance to antibiotics in developing countries. Emerg Infect Dis 1999;5:18–27

Ong S, Nakase J, Moran GJ, Karras DJ, Kuehnert MJ, Talan DA. Antibiotic use for emergency department patients with upper respiratory infections: prescribing practices, patient expectations, and patient satisfaction. Ann Emerg Med 2007;50:213–20.

Perez-Cuevas R, Guiscafre H, Munoz O, et al. Improving physician prescribing patterns to treat rhinopharyngitis. Intervention strategies in two health systems of Mexico. Soc Sci Med 1996;42:1185–94.

Pineda M, Romano L, Pacheco A, Pavia N, Santos JI, Alpuche CM. 2007. Encuesta sobre conocimientos básicos de antibióticos y su uso clínico en residentes del Hospital Infantil de México Federico Gómez. UNAM. Mexico

Powers JH. Diagnosis and treatment of acute otitis media: evaluating the evidence. Infect Dis Clin North Am 2007;21:409–26, vi.

Rubin MA, Bateman K, Alder S, Donnelly S, Stoddard GJ, Samore MH. A multifaceted intervention to improve antimicrobial prescribing for upper respiratory tract infections in a small rural community. Clin Infect Dis 2005;40:546–53.

Santoso B, Suryawati S, Prawaitasari JE. Small group intervention vs formal seminar for improving appropriate drug use. Soc Sci Med 1996;42:1163–8.

Scott JG, Cohen D, DiCicco-Bloom B, Orzano AJ, Jaen CR, Crabtree BF. Antibiotic use in acute respiratory infections and the ways patients pressure physicians for a prescription. J Fam Pract 2001;50:853–8.

Spellberg B, Powers JH, Brass EP, Miller LG, Edwards JE, Jr. Trends in antimicrobial drug development: implications for the future. Clin Infect Dis 2004;38:1279–86.

Vernacchio L, Vezina RM, Mitchell AA. Management of acute otitis media by primary care physicians: trends since the release of the 2004 American Academy of Pediatrics/American Academy of Family Physicians clinical practice guideline. Pediatrics 2007;120:281–7.

Weissman J, Besser RE. Promoting appropriate antibiotic use for pediatric patients: a social ecological framework. Semin Pediatr Infect Dis 2004;15:41–51.

Chapter 29
Containing Global Antibiotic Resistance: Ethical Drug Promotion in the Developing World

Catherine Olivier, Bryn Williams-Jones, Béatrice Doizé, and Vural Ozdemir

Abstract The introduction of antimicrobial agents was a breakthrough health intervention that helped save millions of lives around the world and that provided a sense of control on the part of clinicians over host–pathogen interactions. Yet despite the concrete advances in prevention and treatment of infectious diseases, there has been a parallel surge in resistance to antimicrobials that is seriously compromising the gains made over the past century. Acknowledging the underlying mechanisms such as inappropriate use of antibiotics in humans and the agricultural applications of antibiotics for growth promotion and prophylaxis is a first and essential step to contain global antimicrobial resistance. However, it is also critical to consider in parallel the broad social, economic and political drivers and ethical significance of antimicrobial promotion in developing countries. Moreover, these socio-ethical factors constitute tangible targets against which public policy interventions can be developed to remedy growing concerns over the spread of antimicrobial resistance. In this chapter, we focus on drug promotion practices in the developing world that have important repercussions on physician prescribing habits, antimicrobial use and development of resistance. Other social factors that are endemic in developing countries – such as inadequate resources in health and education systems – can amplify erratic or suboptimal drug use. The public health consequences of, and ethical responsibilities associated with, drug promotion are substantially larger when viewed through the prism of the realities faced by persons living in developing countries. In our world of increasing urbanization and globalization, the actions of pharmaceutical companies have a global impact. Thus, their social responsibilities extend beyond developing

B. Williams-Jones (✉)
Programmes de bioéthique, Département de médecine sociale et préventive,
Faculté de médecine, Université de Montréal, Montréal, Québec, Canada
e-mail: bryn.williams-jones@umontreal.ca

V. Ozdemir (✉)
Bioethics Programs, Department of Social and Preventive Medicine, Faculty
of Medicine, University of Montréal, Montréal, Québec, Canada
e-mail: vural.ozdemir@umontreal.ca

A. de J. Sosa et al. (eds.), *Antimicrobial Resistance in Developing Countries*,
DOI 10.1007/978-0-387-89370-9_29, © Springer Science+Business Media, LLC 2010

medications for the populations of the developing world, to include socio-ethical and public health consideration for populations worldwide. Focusing on the fact that 'what happens in the developing world *does not* stay in the developing world', we propose that ethical considerations should be included as an integral part of the framework in the evaluation of appropriateness of drug promotion practices.

29.1 Introduction

The introduction of antimicrobial agents (antibiotics and related therapeutics for medicinal use) early in the twentieth century markedly improved the clinical management of infectious diseases, such as syphilis and tuberculosis, caused by microbes (a broad term including bacteria, fungi, parasites and viruses) that were previously untreatable and often fatal. The widespread implementation of population-level immunization programmes, improved sanitation, access to safe nutrition and other public health measures saved millions of lives around the world – this success also helped engender a sense of control on the part of clinicians and public health professionals over host–pathogen interactions. And for hitherto fatal infections that have resisted outright cure, such as HIV/AIDS, the advent of second- and third-generation antimicrobials has transformed them into treatable chronic diseases, at least in developed countries. Yet despite these concrete advances in prevention and treatment of infectious diseases, a parallel surge in resistance to antimicrobials is seriously compromising the gains made over the past century; and this resurgence has been directly associated with the overuse (or misuse) of antimicrobials.

An analysis of published data on the efficacy of drugs in major therapeutic classes indicates that only about 50% of patients respond to their medications (Spear et al. 2001). Seen in this light, resistance to antimicrobials becomes an important compounding factor in the difficulties faced in treating common complex diseases. When a patient fails to respond to non-microbial treatment such as a cardiovascular or an anticancer drug, the adverse medical consequences are often limited to the patient. But for patients with common pathogenic infections, treatment failures (e.g. due to inappropriate antibiotic use) are accompanied by the development of resistant strains of microbes, which are then spread to other patients or the healthy public, who then become carriers of these resistant strains. Antimicrobial resistance also extends the duration of infection and the time interval during which microbes can be transmitted to others.

Methicillin-resistant *Staphylococcus aureus,* penicillin-resistant *Streptococcus pneumoniae* and multi-resistant *Mycobacterium tuberculosis*, to name only a few of the many antimicrobial-resistant microbes, pose serious ongoing challenges to biomedicine and public health. While substantial progress has been made in discerning the underlying biological, genetic and environmental causes of the emergence of antimicrobial resistance, there has been much less attention to important social factors such as socio-economic disparities and the impact of drug development and delivery strategies (in both developed and developing countries). The inappropriate use of antibiotics and the underlying mechanisms of antimicrobial resistance have broad social and ethical significance that transcends individual patients or specific communities who suffer from treatment failures.

Attention to the ethical issues and social factors surrounding antimicrobial resistance is necessary, not only because this can help generate more nuanced models of biological susceptibility to antimicrobial resistance but because these factors constitute tangible targets against which public policy interventions can be directed. In this chapter, we focus on one such significant social factor (and the associated ethical issues): drug promotion practices in the developing world. We examine the ways in which marketing practices contribute to inappropriate drug use, and thereby to antimicrobial resistance, and propose a means by which to evaluate whether a marketing practice is ethical in a developing world context. We conclude by emphasizing that 'what happens in the developing world *does not* stay in the developing world'. With the globalization of travel, education, commerce and health care, antimicrobial resistance can be spread easily between developed and developing countries; antimicrobial resistance is thus a global problem and requires global and equitable solutions. An important first step in developing such solutions is for health professionals, policy makers and civil society to better understand the drivers and interests of the pharmaceutical industry and recognize the diversity of current drug promotion practices around the world and their contribution to antimicrobial resistance. With such knowledge, it then becomes possible to start building practical mechanisms (e.g. policy, regulation, commercial incentives) that can positively shape or influence drug promotion practices, in both developing and developed countries, so that the public health benefits of antimicrobial drugs are not so quickly overshadowed by the spectre of antimicrobial resistance.

29.2 Resistance to Antimicrobials: Biology and Inappropriate Drug Use

Selective pressure from antimicrobial agents inevitably leads to resistance development in a small subpopulation of microbes, which can subsequently spread by virtue of the rapid replication rates of microorganisms, and the efficient transfer of antimicrobial resistance genes via the process of conjugation whereby plasmids (carrying resistance genes) are exchanged among microbes. Although many social, financial and behavioural factors contribute to the emergence and dissemination of antimicrobial resistance in pathogens, inappropriate use of antibiotics constitutes one of the most important factors in this biological process (Hart and Kariuki 1998; Okeke et al. 1999, 2005; Byarugaba 2004).

The emergence of resistance in microbes can be precipitated most obviously by overuse of antimicrobials. Inappropriate antimicrobial use can easily develop as a downstream consequence of excessive drug promotion activities in vulnerable global populations who may be devoid of information channels to make informed decisions on benefits and risks of antimicrobials. Hence, it is not surprising that the use and prescription of antimicrobials, which escalated significantly until the mid-1990s, was immediately followed by an increase in antimicrobial resistance worldwide (McCaig et al. 1998, 2002, 2003; Garcia-Rey et al. 2002; Gaur and English 2006). Such extensive antimicrobial use has been precipitated (and justified by some local authorities) by the high prevalence of infectious diseases in certain developing countries. It is interesting to note that many antimicrobials (e.g. penicillin) have become victims of their initial success (i.e. efficacy) – they were widely overused and not always with a sound rational indication of their prescription. It is now estimated that 50% of antimicrobial prescriptions are associated with inappropriate use, either because of unnecessary length of treatment, wrong choice of prescription or dosage regimen, or use in persons without a discernible sign of infection (Usluer et al. 2005).

A relatively less appreciated form of inappropriate drug use, and one that can also precipitate treatment resistance, is underuse or erratic/inconsistent self-administration. This can result from 'drug holidays' in patients who stop taking their medication after an initial favourable therapeutic response or acute side effects, only to return to their physicians' office with a recurring infection by a more resistant strain of microbe. Such erratic use or underuse of antimicrobials is likely to be more predominant in the context of developing countries, where patients often lack both the financial resources and general knowledge about pharmaceutical use to adhere to their antimicrobial treatment regimen (Hart and Kariuki 1998; Okeke et al. 1999). Other social factors that are widespread in developing countries – such as low-quality generic or counterfeit antimicrobials, ineffective distribution and inappropriate storage conditions of pharmaceuticals – can also contribute to erratic or suboptimal drug use. And these factors can have even more drastic consequences when compounded by aggressive drug promotion practices. Hence, the public health consequences of and ethical responsibilities associated with drug promotion are substantially larger when viewed through the prism of the realities faced by persons living in developing countries.

29.3 Drug Promotion and Inappropriate Drug Use in Developing Countries: Ethical Considerations

29.3.1 Building Ethical Frameworks Based on Resource Availability

In the United States and New Zealand, there is political agreement (although accompanied with some controversy) on whether the benefits of direct-to-consumer advertising (DTCA) for prescription pharmaceuticals (patient education and empowerment) (Deshpande et al. 2004) outweigh the costs (misinforming consumers, over-consumption) (Findlay 2001). Nonetheless, both countries allow DTCA for prescription pharmaceuticals and medical products (Hoek and Gendall 2002). By contrast, in Canada, the United Kingdom and most other European countries, DTCA of prescription drugs is much more tightly controlled (e.g. only the drug name or the targeted disease can appear in advertising, not both), although there has been pressure to lift these restrictions (Jaderberg 2002).

The ethics of drug promotion practices remain hotly contested in North America and Europe, and are even becoming a global public concern, as evidenced by the fact that the 2007 World Consumer Rights Day was dedicated to the theme of unethical drug promotion (WCRD 2007). In the developing world, there have been concerns about inappropriate drug promotion and the impact on public health for at least two decades. In 1988, the World Health Organization (WHO) produced a guideline for drug promotion entitled *Ethical Criteria for Drug Promotion* (WHO 1988) to regulate practices worldwide. But as an international non-governmental organization, the WHO cannot make laws or enforce regulations. Hence, these recommendations are essentially voluntary, until such time that they are incorporated into national and local regulations and legislation. Nevertheless, WHO recommendations carry important political and ethical weight for member countries and thus can play an important role in shaping global standards.

The aim of the WHO ethical criteria was to improve health care and regulate drug promotion practices to ensure the rational use of medicinal drugs (with particular attention to promotion, advertising, drug representatives, free samples and medical education). But how does one classify a drug promotion practice as unethical? The WHO defines drug promotion as 'all informational and *persuasive* activities by manufacturers and distributors, the effect of which is to induce the prescription, supply, purchase and/or use of medicinal drugs' (WHO 1988). Hence, unethical drug promotion encompasses practices that deviate from these normative criteria, e.g. by misleading patients and doctors as to the utility or the efficacy of a particular drug, and that can then lead to inappropriate use of medicines, which can have a large socio-ethical significance for the global society (and not solely for affected individuals). Yet while it may be easy to identify egregious violations as unethical (e.g. lying or blatant misrepresentation of scientific data), it is much more challenging to parse the grey zone between positively and justifiably promoting a drug and engaging in unethical marketing practices, such as misrepresentation or fear mongering (we will develop some aspects of this issue in more detail below).

For our purposes, it is important to note that drug promotion practices in a developing country context raise several distinct socio-ethical issues. In developing countries, the sparse resources allocated to health care and education limit physician involvement in continuous medical education. Thus, pharmaceutical promotion often becomes the major source of information on new drugs, and ways to use them, for the majority of physicians in many countries (both developed and developing) (Bhutta 1996; Vancelik et al. 2007). When countries have poor or inadequate economic resources in health care and education, and also lack adequate regulatory measures to control medicinal drug promotion (or lack sufficient compliance if some form of regulation exists), their populations become even more vulnerable to aggressive drug promotion practices by the large pharmaceutical companies (Dal-Pizzol 2002).

Developing country physicians with a large volume of daily patient appointments often do not have adequate time or resources for meaningful education of and communication with patients on how to use their medications and avoid adverse drug/food interactions. While increasing the prescription volume of medicines (and in particular, antimicrobials to treat the many infectious diseases endemic in developing countries), the excessive promotion of medicines can lead to a gap between drug prescription and physician practices designed to educate patients on the rational and appropriate use of their medicines. Such nuances (and burdens) of clinical practice in developing country settings with low physician/patient ratios and limited resources can make otherwise ethical drug promotion practices unethical in the context of a developing country – it is important to acknowledge this 'normative flexibility' in ethical thresholds, which results from contextual differences between developed and developing countries.

As noted above, the increased or inappropriate use of antimicrobials eventually leads to treatment resistance, initiating a vicious cycle whereby more potent second- and third-generation agents are prescribed. However, newer antimicrobials are often under patent protection and markedly more expensive than first-line (older) generic treatments. These cost differences become further accentuated in developing countries, which often do not have the economic capacity to include these medications in their already limited public health programmes, thereby widening existing gaps in access to health care and antimicrobial agents. Hence, countries wanting to offer universal programmes to their populations for much needed antimicrobial drugs (e.g. antiretroviral programmes in India, Uganda, South Africa and Brazil) have, at times, had to sidestep international economic agreements and regulations (particularly with regard to drug patents) (Yamey 2001; Mutume 2001; BBC News 2005, 2007). These isolated yet impressive initiatives have succeeded in increasing access to medications, facilitated patients' adherence to their antimicrobial treatment (because the medications are available and affordable), thus reducing an important factor in resistance emergence in pathogens (i.e. suboptimal adherence) (Oyugi et al. 2004; Kumarasamy et al. 2005).

29.3.2 Moving Beyond the Resource Availability Criteria in Defining Ethical Thresholds

Infectious diseases are responsible for 41.9% of deaths in the developing world (WHO Global Burden of Diseases statistics), and about one in three deaths worldwide (Yoshikawa 2002). Developing countries thus constitute an important market for companies selling antimicrobials drugs. Indeed, a study conducted in India between July 1995 and June 1996 demonstrated that the majority of drug promotional material provided to physicians at the Guru Teg Bahadur Hospital in Delhi concerned antimicrobial agents (Lal 1998), suggesting that pharmaceutical companies quickly adapt to local and regional health-care markets. At the same time, this study also showed that 60% of the promotional material encountered in Delhi did not conform with the afore-mentioned WHO guidelines for drug promotion (WHO 1988). Similar studies in Latin America and South Asia further support the case that there is poor compliance with the WHO ethical guidelines for medicinal drug promotion in developing countries (Dal-Pizzol 2002; Islam and Farah 2007).

A particular concern with global drug promotion practices is the absence of concordance between the content of promotional messages concerning a particular drug and the evidence base in the scientific literature (Mintzes 2006). A 2004 study in Germany, for example, suggested that 94% of the drug advertisements encountered were not scientifically based (Tuffs 2004; WCRD 2007). A high prevalence of misleading drug promotion claims has also been found in certain developing countries. According to a study in Bangladesh, 15–35% of the promotional pamphlets provided to physicians were misleading, with 50% of these pamphlets citing evidence that substantially lacked scientific support (Islam and Farah 2007). In many instances, the promotional material not only does not rely on scientific studies but also promotes the dissemination of results of non-scientific studies as scientific 'evidence' (Menkes 2006). Such unethical promotion practices clearly undermine pharmaceutical companies' claims (and responsibilities) to be supporting patient empowerment through education. But there is also arguably a potential negative impact on company shareholders – and a failure of full corporate accountability – when the true efficacy and utility of drugs is misrepresented to the general public (Dukes 2002). As seen with a number of major drug recalls in recent years, unsubstantiated (or at least overly optimistic) scientific claims of significant benefit and limited risk can result in extremely costly drug withdrawals and class action lawsuits, which render shareholders financially vulnerable. Drug promotion practices that overstate the benefits and utility of medicinal drugs are thus bad for both patients and shareholders.

Another consideration for developing countries is that the self-medication of antimicrobials is a common practice, since these medications can often be purchased without a prescription and their sales are poorly regulated by local governments (Kunin 1993; Gassman 1999). In the face of these realities, drug promotion activities could lead to a greater coercive influence on consumers, whether they be healthy persons taking antibiotics for unjustified (e.g. fear of disease) reasons or patients with infectious diseases. Taken together, we suggest that the adverse downstream consequences of aggressive drug promotion can be more deleterious in a developing world context. This also means that the ethical standards and stringency by which drug promotion practices are evaluated need to be different (i.e. higher ethical standards and lower ethical thresholds) in the case of developing world populations that are vulnerable in terms of both economic access to antimicrobials and access to information and educational resources to objectively interpret the drug promotion material provided by drug manufacturers.

The case of the anti-malarial drug *artemisinin* illustrates well the extent to which drug distribution and promotion can impact the prescription of an antimicrobial and then contribute to the emergence of drug resistance. Artemisinin-combination therapies (ACTs) – which involve the use of multiple anti-malarial drugs – have been proven to rapidly eliminate symptoms associated with malaria infections (3 days after treatment begins) while reducing chances of resistance emergence in pathogens (due to the multi-drug approach), and its adoption for malaria treatment was initially strongly recommended by the WHO (2006). But because of their relatively high efficacy (although not as good as ACTs) and lower costs, artemisinin *monotherapy* (individual drug approach) has been used and/or promoted by certain physicians and manufacturers (e.g. local manufacturers in China and Sanofi-Aventis) in developing countries, subsequently increasing the potential for development of drug resistance. Given that malaria kills over a million people a year worldwide and infects individuals on every continent (although predominantly in sub-Saharan Africa) (Benkimoun 2005), the promotion of artemisinin as a monotherapy to physicians in developing countries could at first glance seem ethical for its immediate benefits – these drugs do work against malaria and are more affordable individually than the more complex multi-drug ACTs. But their use as monotherapies is ultimately short-sighted from a public health perspective, and their promotion unethical, because it contributes to the development of drug resistance and the failure of other artemisinin derivatives to then be able to effectively treat malaria infections. Conscious of the risks and dangers of pathogen resistance to artemisinin derivatives, in 2006 the WHO called for a halt to monotherapy usage for these drugs (MSNBC 2006) and recently underpinned the emergence of such resistance in Asia (WHO 2007).

29.4 Mapping the 'Control Points' in Drug Promotion Practices in Health Care: The 'Drug Representative' as a Key Mediator

The pharmaceutical companies' investment in drug promotion has consistently increased over the years, from 15 to 20% of sales income in 1987 (Lexchin 1987) to 35% of sales income according to a 1997 analysis (Mintzes 1998). The expenditures, solely for drug promotion in the United States for 2002, amount to $21 billion, which reflects the importance that large pharmaceutical companies attach to these strategies (Norris et al. 2005). The major strategies used in drug promotion by the pharmaceutical companies can be classified as promotion directed at health professionals (e.g. physicians, nurses and pharmacists or associated employees), promotion through social factors (e.g. political lobbying, interest groups or education funds) or promotion by targeting consumers through DTCA (WCRD 2007).

As mentioned earlier, DTCA for prescribed drugs is permitted in only two developed countries (the United States and New Zealand) and is restricted in Canada and Europe. By contrast, a lack of regulations concerning medicinal drug promotion in numerous developing countries means that this type of drug promotion may be more common (and pervasive) in the latter (Mintzes 1998). Nonetheless, the dominant focus of drug promotion, worldwide, remains the physician and accounts for close to 84% of all promotional efforts (Brodie and Levitt 2002; Mastin et al. 2007). In consequence, physicians themselves have arguably become the most important transmitter of drug promotion information to patients and consumers.

The pharmaceutical representative/physician relation is one that has raised numerous discussions in the clinical ethics literature. Even though there are various perceptions in the medical community on the influence that pharmaceutical gifts may have on the prescribing habits of physicians (Hodges 1995; Rogers et al. 2004), most authors agree that there is important concern for conflict of interests and bias with regard to physician behaviour (Wazana 2000; Katz et al. 2003). In developing countries, where previously the number of pharmaceutical representatives had been relatively low, there has been a dramatic increase over the last few years (e.g. in Turkey, in recent years, there was a 2.5-fold increase) (Semin et al. 2007). Yet while this increased number of representatives might enable an overall increase in visits to physicians, this new situation has been unexpectedly associated with more difficult working conditions for representatives, due in large part to reduced time with physicians.

Shorter visits with physicians is due to the fact that pharmaceutical representatives choose to visit physicians with higher patient volumes in order to optimize their returns from a single visit (Vancelik et al. 2007), with the end result being that physicians get less information from these visits, which can in turn negatively influence downstream prescribing to patients. Needless to say, this curious phenomenon also provokes an obvious concentration of informed physicians within specific geographic regions, reducing access to needed information in more remote areas, thereby further enlarging the current gap in access to health as well as health-related information between rural and urban populations worldwide.

Another major concern associated with the representative/physician interaction is the impact that financial incentives have on drug prescription. As has been widely noted in the medical and ethics literatures, incentives do not have to be overt or large to be effective! Paid lectures, free roundtable lunches, conference invitations, free drug samples or assortment of coffee mugs and pens, medical student tuitions or distribution fee compensations all contribute to building brand recognition with physicians, which can then shape, in often very subtle ways, physician preferences and thus the number of prescriptions for particular medications. Although such incentives are increasingly being considered unethical and unacceptable in many contexts (particularly in

Western teaching hospitals and medical schools), their occurrence is widespread in the developing countries, and their effects are certain to be amplified. In this context, physicians are often poorly paid, operate outside coordinated state health-care systems and have little or no access to continuing medical education, so access to information on pharmaceutical drugs and to the drugs themselves will be much more limited. In these difficult circumstances, even the best intentioned physician will be hard-pressed to avoid being unduly influenced by drug promotion activities (whether these come via a drug representative or through other avenues such as the media).

Regulations of and guidelines for drug promotion activities by pharmaceutical representatives (among others) are beginning to be implemented in a number of developed countries (Harris 2007), in order to control the spread of unnecessary and inefficient drug prescribing; similar efforts in developing countries are of utmost importance, if progress is to be made towards positively influencing the appropriate prescribing of antimicrobials, so as to mitigate the emergence of resistance in microorganisms. Particular attention should also be given to production and distribution (by government and industry) of comprehensible and accurate patient information that can be easily accessed through doctors' clinics.

29.5 Antimicrobial Use in the Food Industry and Adverse Effects on Human Health

Tremendous amounts of antimicrobials are used annually in food production (e.g. chicken, cattle, pigs); unfortunately, no unbiased estimates of their use are available and the published data differ markedly; and unlike the case of drug promotion for human medicines, there appear to be only limited discussions in the veterinary and agricultural sciences literature of the impact of drug promotion on veterinary prescribing practices and farmer utilization.

The WHO has estimated that half of the worldwide use of antibiotics can be associated with agricultural and fish farming methods (WHO 1998). A direct link between agricultural use of antibiotics and the appearance of resistant pathogens in humans has been proposed, but not yet demonstrated conclusively (Rollin 2001). So while experts recognize that the most serious problems of antimicrobial resistance in people are due to the widespread use of antimicrobials in human medicine, the effect of antimicrobial resistance in bacteria of animal origin on human health is still the subject of ongoing debate. Nonetheless, a major concern is that the unrestricted use in food animals (for growth promotion and prophylactic purposes) of antimicrobials that are also important in treating human infectious diseases will exacerbate the occurrence of antibiotic resistances in humans (Phillips et al. 2004; Aarestrup 2005).

The WHO has recommended that such 'dual-use' antimicrobials not be used in food animals, unless their safety is demonstrated by comprehensive risk-assessment analyses (WHO 2001). Along these lines, the European Union has banned the use of antimicrobials at sub-therapeutic levels for growth promotion in pigs, but such use continues on a large scale in other developed countries, under the supervision of health and agriculture authorities. In developing countries, antimicrobials are not used very often as growth promoters for food animals, but they are widely and commonly used for prophylaxis and therapy (Mitema et al. 2001; Muriuki et al. 2001; Dipeolu and Alonge 2002).

International guidelines for appropriate use of antimicrobials in animals have been published (Anthony et al. 2001; Nicholls et al. 2001), and in developed countries, surveillance programmes to monitor antimicrobial use and track the emergence of antimicrobial resistance in food animals have been well established. However, in developing countries, very little is being done to control antimicrobial use in food animals, and there is very little data on actual antimicrobial use in agriculture, or the emergence of drug-resistant pathogens. Nonetheless, a few studies in Asia and Africa are beginning to document the situation. For example, a study in Thailand revealed that antimicrobials listed as critically important for human medicine are frequently used in small-scale broiler farms (Na Iampang et al. 2007). In Kenya and Nigeria, high antimicrobial residues have been observed in meat (Mitema et al. 2001; Dipeolu and Alonge 2002; Lu et al. 2002). In Uganda, between 1990 and 2000, a 1% per year increase of resistance in bacterial isolates from bovine mastitis was observed (Byarugaba 2004).

In the developed countries of North America and Europe, farms are primarily large scale and closed systems (the family farm will often include hundreds or even thousands of animals), while in developing countries, many farms are very small (a few dozen animals) and commonly associated with household subsistence. This lifestyle has as a consequence that a large proportion of the population in developing countries (including those living in urban centres) lives in close contact with food animals (Simango and Rukure 1991), thus increasing the chances of microorganism transmission from food animals to humans through the food chain and animal handling (Rolland et al. 1985; Corpet 1988). This situation was clearly illustrated in a 2002 study of an outbreak of sepsis in pigs in China, which showed that a multi-drug-resistant strain of *Enterococcus faecium* was transferred to humans, leading to several deaths (Lu et al. 2002). Similarly, it was also observed that resistant staphylococci conventionally associated with animals were often present in septic wounds in Nigerian hospital patients (Kolawole and Shittu 1997).

Antimicrobials can also be introduced into the environment through the spraying of fruit trees for disease prophylaxis (Vidaver 2002), the application of antibiotic-containing animal manure on croplands (Barza and Gorbach 2002) and the use for fish farming of waste from farm animals that had received antimicrobials as growth promoters (Petersen et al. 2002). These applications contribute to the selection of resistant bacteria. Such injudicious use of antimicrobials in animals and agriculture, and its consequences for public health, underpins the importance of limiting the inappropriate promotion and application of antimicrobials in agricultural and veterinary contexts. There is thus a pressing need for more responsible and controlled use of antimicrobials in veterinary and agricultural methods in both developed and developing countries (Anthony et al. 2001).

29.6 Overcoming the Link Between Promotion of Antimicrobial Agents and Emergence of Antimicrobial Resistance in Pathogens

By not following the proposed WHO ethical criteria and guidelines, companies may be able to maximize their returns on the promoted drugs while avoiding the 'regulatory burden' that such regulation and ethical guidelines would engender. In the particular case of the promotion of antimicrobials, the consequences of ignoring these ethical guidelines are reflected in the increased or inadequate use of antimicrobial agents, and thus the emergence and spread of drug-resistant pathogens (Mintzes 2006). But not all blame should be laid at the feet of the pharmaceutical industry; governments are equally guilty for not regulating drug promotion (i.e. by not setting clear and enforceable rules of the game for industry) and for insufficiently supporting objective and independent medical and public health education programmes. A move in the right direction would be for policy makers and industry to address the gaps separating the content of promotional messages (i.e. scientific accuracy) and the target population (i.e. physicians and patients), in order to reach an acceptable equilibrium between the promotion of antimicrobial agents and the emergence of antimicrobial resistance.

It is important to recognize that in the developing country context, the recipient of antimicrobial drug promotion (physician/veterinarian or patient/ farmer) is very likely already favourable to their use because of the high incidence of infectious diseases in these regions of the world. Thus, part of the pharmaceutical advertisers' job is already accomplished even before the promotional material is distributed to the recipients, ensuring greater success of promotional campaigns. Given that individuals in many developing countries can often obtain antimicrobial drugs without prescription or even medical advice (Okeke et al. 1999), pharmaceutical companies arguably have a social responsibility to facilitate the provision of drugs with appropriate information to ensure correct use.

The global problem of antimicrobial-resistant pathogens, however, shifts the boundaries of corporate social responsibility of the pharmaceutical companies. An essential element in this particular problem is that the qualitative aspect (the emergence of resistance) and the quantitative aspect (the spread and extent of resistance) are separate in both a temporal and medical manner. As was seen with the anti-malarial drug *artemisinin*, promoting antimicrobials as if they were like any other individual therapy on the market (e.g. gastrointestinal acid reflux drugs) without attention to the consequences for the broader social environment (including animals and people) can contribute to both the emergence of resistance in pathogens and their spread. But if these promotional messages are shaped so as to take into consideration the specific social, economic and political realities of developing countries, then they could be an important means of mitigating the spread of significant regional and global health problems, such as antimicrobial resistance.

The transmission of antimicrobial-resistant strains between individuals is further exacerbated by the high urbanization rate in many developing countries, but also worldwide (Dugger 2007). Even in developed countries, the spread of resistant microbial strains has been shown to occur not only through hospital settings but also via community-acquired infections (Bancroft 2007). In the United States, a study on methicillin-resistant *S. aureus* (MRSA) demonstrated that more than one in four invasive infections could be associated with community-onset infections (Klevens et al. 2007). The public health implications of human concentration on antimicrobial resistance emergence and transmission of strains affecting human populations are correlated with the poor sanitation conditions and high poverty levels in many urban areas. Thus, recipients of promotional messages about antimicrobials must be aware of the risks of (1) the emergence of resistance in the pathogen and (2) the potential dissemination of resistant strains in the patient/consumer's environment. We suggest that promotional messages should therefore include these broader repercussions, information on the prevalence of existing resistance to the particular antimicrobial molecule in the drug, as well as information required by the WHO ethics guidelines.

Drug advertising to physicians in developed countries is often less informative than that provided to their counterparts in developed countries. This reduced information quality is also accompanied by limited access to information on cost-effectiveness of drugs (Waud 1992; Donald 1999). Thus, one could imagine the mandatory inclusion of information on price range comparisons (or even alternate generic drugs) in the promotional materials for health professionals and patients, although companies would obviously argue that such requirements unreasonably undermine economic competitiveness. More important, though, is access to scientific information regarding antimicrobial efficacy, safety and risk of resistance, which can contribute to limiting the improper prescription or inappropriate use of antimicrobial agents. It is imperative that such information be included in the promotion material distributed to physicians, pharmacists and other health-care employees who are likely to prescribe or provide individuals with antimicrobial drugs in the developing world. In a context where local governments may not have the financial and human resource capacity to provide such information, there is clearly a place for international organizations (governmental and non-governmental), and the pharmaceutical industry, to help with this important knowledge transfer and dissemination.

Increased surveillance of resistance emergence in pathogens responsible for infectious diseases has been identified as a top priority in containing the emergence and spread of antimicrobial-resistant pathogens (Fidler 1998; Richet et al. 2001). Such surveillance is clearly within the remit and responsibility of national and international agencies (local health agencies aided by the WHO and centres for disease control in developed countries) (Menkes 1997; Fidler 1998), and not the pharmaceutical industry, whose business is drug development and production. But the way in which drugs are produced and marketed can have important consequences for surveillance and containment of antimicrobial resistance.

The common response to antimicrobial resistance from industry (and the biomedical sciences more generally) has been to develop new and often substantially more expensive antimicrobials that could be used as replacement treatments for pathogens that had become drug resistant. But in recent years, pharmaceutical companies have all but given up on the development of new antimicrobial agents; this is clearly illustrated by the 56% decrease in the number of antimicrobial agents approved by the FDA between 1998 and 2002, when compared to the previous decade (Spellberg et al. 2004). The pharmaceutical companies attribute this reduction in antimicrobial development to the low investment returns provided by new antimicrobials, which will inevitably provoke the emergence of resistance in pathogens and thus limit their potential usefulness; they also point to the failure of genomics technologies to provide new targets for antimicrobial drugs (Projan 2003; Wenzel 2004). And while smaller biotechnology companies have shown a willingness to enter into antimicrobial drug development, their efforts are insufficient to make a noticeable impact on the production of new drugs to keep pace with ongoing emergence of antimicrobial resistance (Talbot et al. 2006).

The antimicrobials being promoted in the developing world today have already encountered resistance in pathogens; thus their increased distribution following successful drug promotion to physicians and other health professionals can only contribute to the spread of resistance in the population and the worsening of already existing epidemics worldwide. Given this fatal result of antimicrobial agents' promotion, it is essential that pharmaceutical companies, governmental organizations and local frontline physicians adopt an evidence-based ethical system for the promotion of antimicrobials in the developing world.

29.7 Essential Medicines Library: An Empirical Resource to Evaluate the Accuracy of Drug Promotional Material in Developing Countries

In 2001, the WHO inaugurated a 'Global Strategy for Containment of Antimicrobial Resistance', which is built upon several existing WHO programmes to promote the rational use of medicines, most notably the Model Lists of

Essential Drugs. According to the WHO, while spending on pharmaceuticals represents less than one-fifth of total public and private health spending in most developed countries, it represents 15–30% of health spending in transitional economies and 25–66% in developing countries. In most low-income countries, pharmaceuticals are the largest public expenditure on health after personnel costs and the largest household health expenditure; the expense of serious family illness, including drugs, is a major cause of household impoverishment (WHO).

Rising costs of drugs are a universal occurrence, but the issues surrounding access to pharmacotherapy and adverse effects of aggressive drug promotion practices are more acutely felt in developing countries. Hence, the WHO Essential Drugs List, which was first published in 1977, sought to provide safe and effective treatments for communicable and chronic diseases affecting the vast majority of the world's population. The Model List of Essential Medicines is updated every 2 years by an expert committee to include drugs that will satisfy priority health-care needs. The selection criteria take into account, for instance, public health relevance, evidence on efficacy and safety and comparative cost-effectiveness (Quick 2003; Hogerzeil 2004; Ozdemir et al. 2006). We submit that the concept of essential medicines is also an effective (and often neglected) empirically based tool to safeguard the rational use of medicines and to ensure the objective interpretation of drug promotion in developing countries. It would be anticipated that enhancing awareness of the Essential Medicines Library among citizens, clinicians and health-care decision makers in developing countries could help prevent potential abuses by or misuse of drug promotion activities by different stakeholders.

29.8 Conclusions and 'Points to Consider'

Infectious diseases are an important contributor to the mortality and morbidity rates in developing countries. The development and promotion of antimicrobial drugs can thus be perceived as an essential tool in the fight against epidemics. Unfortunately, history has shown that shortly after the introduction of new antimicrobials, there is an emergence of resistance in pathogens, worsening the occurrence of infectious diseases through the spread of resistant pathogen strains in the population. This resistant pathogen phenomenon complicates enormously the task of national and international public health organizations. Physicians may thus reasonably feel unequipped when facing infectious disease occurrence and resistance.

In developing countries, the sparse resources available for and allocated to continuous education for health-care professionals necessarily limit the capacity of these professionals to critically judge pharmaceutical drug promotion material. The promotional material provided by the pharmaceutical companies through medical journals, Internet sites and pharmaceutical representatives often constitutes the primary (even only) source of information for physicians and associated health professionals. Recognizing this, promotional messages should aim at reducing mismatches between the content and the target populations' needs. In order to address these mismatches properly, an evidenced-based ethical framework should be implemented by pharmaceutical companies, governments and health-care professionals to evaluate appropriate drug promotion. Such a framework would ensure that promotional messages (1) acknowledge the epidemiological, social and financial realities of the target population; (2) conform with the WHO ethical criteria for drug promotion; (3) improve accessibility to objective and independent scientific information on the part of the recipients of promotional material (health professionals and patients); and (4) present evidence from the scientific literature and provide continuing medical and public education on antimicrobial resistance.

While some may question whether companies should have social responsibilities beyond the pure maximization of profits, scholars within the business ethics community have argued convincingly that companies do have important social responsibilities that reach beyond mere adherence to laws and regulatory policies (Heath 2006). For our purposes, we maintain that drug manufacturers have an ethical obligation to identify both the present and the future health-care needs of the population in a more global manner so as to better respond to these pressing health challenges. As citizens, scientists and patients, there is a need to recognize that pharmaceutical research and promotion, and the scientific enterprise more generally, are embedded in a socio-cultural, economic and political agenda driven by multiple stakeholders who may (or may not) share the same ethical values (Ozdemir and Godard 2007; Master and Ozdemir 2008; Ozdemir et al. 2008). As a way forward, and similar to the already widely accepted concept of good clinical practice (GCP), the pharmaceutical industry should consider adopting an evidence-based ethical drug promotion for antimicrobial agents geared towards the vulnerable and importantly affected populations of the developing world. In our world of increasing urbanization and globalization, the actions of pharmaceutical companies have regional, national and global impact: drug promotion practices in one country affect the behaviour of (and impact on) consumers and health professionals around the world, a fact that is most noticeable in the case of promotion of antimicrobial drugs. Thus, the social responsibilities of pharmaceutical companies arguably go beyond providing charity to the populations of the developing world (e.g. through drug donation programmes) to include objective considerations over the risks and benefits of the drug promotional activities worldwide. To return to our original assertion that 'what happens in the developing world does not remain in the developing world', it should be clear that as global citizens, the pharmaceutical industry and governments bear a responsibility to contain and, if possible, overcome the development of antimicrobial resistance through ethical promotion of medicines.

Acknowledgments Supported in part by an ethics operating seed grant from the Canadian Institutes of Health Research and a career scientist award in ethics, science and society research from the Fonds de la recherche en santé du Québec to Vural Ozdemir. We thank Anne-Marie Dion for constructive critique and expert advice on pharmaceutical advertising and promotion. Catherine Olivier was supported by a fellowship from the Centre de Recherche en Éthique at the University of Montreal, and from the Canadian Institutes of Health Research. The ideas expressed represent the personal views of the authors only.

References

Aarestrup, F. M. 2005. Veterinary drug usage and antimicrobial resistance in bacteria of animal origin. Basic Clin Pharmacol Toxicol 96: 271–281.

Anthony, F., Acar, J., Franklin, A., Gupta, R., Nicholls, T., Tamura, Y., et al. 2001. Antimicrobial resistance: responsible and prudent use of antimicrobial agents in veterinary medicine. Rev Sci Tech 20: 829–839.

Bancroft, E. A. 2007. Antimicrobial resistance: it's not just for hospitals. JAMA 298: 1803–1804.

Barza, M. and Gorbach, S. 2002. The need to improve microbial use in agriculture: ecology and human health consequences. Clin Infect Dis 34: S71–S144.

BBC News. 2005. Brazil may break Aids drug patent. Brazil has threatened to break the patent on an anti-Aids drug in order to make a cheaper generic version. http://news.bbc.co.uk/go/pr/fr/-/2/hi/americas/4621735.stm.

BBC News. 2007. Uganda opens first HIV drug plant. A factory that will produce treatments for HIV/Aids is opening in Uganda, the first of its kind in the country http://news.bbc.co.uk/2/hi/africa/7033162.stm.

Benkimoun, P. 2005. Le paludisme, un tueur du tiers-monde. Le Monde. Paris.

Bhutta, T. I. 1996. Deception by design: pharmaceutical promotion in the Third World. Br Med J 313:60.

Brodie, M. and Levitt, L. 2002. Drug advertising: the right or wrong prescription for our ailments? Nat Rev Drug Discov 1: 916–920.

Byarugaba, D. K. 2004. A view on antimicrobial resistance in developing countries and responsible risk factors. Int J Antimicrob Agents 24: 105–110.

Corpet, D. E. 1988. Antibiotic resistance from food. N Engl J Med 318: 1206–1207.

Dal-Pizzol, F. 2002. Drug advertisements in less-developed countries. Lancet 359: 1439–1440.

Deshpande, A., Menon, A., Perri, M., 3rd and Zinkhan, G. 2004. Direct-to-consumer advertising and its utility in health care decision making: a consumer perspective. J Health Commun 9: 499–513.

Dipeolu, M. and Alonge, D. 2002. Residues of streptomycin antibiotic in meat sold for human consumption in some states in Nigeria. Arch Zootec 51: 477–480.

Donald, A. 1999. Technology transfer: the problem with "trickle down" theory. Br Med J 319: 1298–1299.

Dugger, C. W. 2007. U.N. predicts urban population explosion. The New York Times. New York.

Dukes, G. M. N. 2002. Accountability of the pharmaceutical industry. Lancet 360: 1682–1684.

Fidler, D. P. 1998. Legal issues associated with antimicrobial drug resistance. Emerg Infect Dis 4: 169–177.

Findlay, S. D. 2001. Direct-to-consumer promotion of prescription drugs. Economic implications for patients, payers and providers. Pharmacoeconomics 19: 109–119.

Garcia-Rey, C., Aguilar, L., Baquero, F., Casal, J. and Martin, J. E. 2002. Pharmacoepidemiological analysis of provincial differences between consumption of macrolides and rates of erythromycin resistance among Streptococcus pyogenes isolates in Spain. J Clin Microbiol 40: 2959–2963.

Gassman, N. 1999. Social and economic value: two sides of the self-medication coin. Responsible self-medication: a challenge to consumers and industry. The World Self-Medication Industry (WSMI) 13th General Assembly, The Association of the European Self-Medication Industry (AESGP) 35th Annual Meeting, Self-care – a vital element of health policy in the information age. Berlin.

Gaur, A. H. and English, B. K. 2006. The judicious use of antibiotics – an investment towards optimized health care. Indian J Pediatr 73: 343–350.

Harris, G. 2007. Minnesota limit on gifts to doctors may catch on. The New York Times. New York.

Hart, C. A. and Kariuki, S. 1998. Antimicrobial resistance in developing countries. Br Med J 317: 647–650.

Heath, J. 2006. Business ethics without stakeholders. Bus Ethics Q 16: 533–557.

Hodges, B. 1995. Interactions with the pharmaceutical industry: experiences and attitudes of psychiatry residents, interns and clerks. CMAJ 153: 553–559.

Hoek, J. and Gendall, P. 2002. Direct-to-consumer advertising down under: an alternative perspective and regulatory framework. J Public Policy Mark 21: 202–212.

Hogerzeil, H. V. 2004. The concept of essential medicines: lessons for rich countries. 329: 1169–1172.

Islam, M. S. and Farah, S. S. 2007. Misleading promotion of drugs in Bangladesh: evidence from drug promotional brochures distributed to general practitioners by the pharmaceutical companies. J Public Health (Oxf) 29: 212–213.

Jaderberg, M. 2002. The pharmaceutical industry – a key partner in providing information to the patient and a key player in the European debate on direct to consumer communication. Int J Med Mark 2: 179–183.

Katz, D., Caplan, A. L. and Merz, J. F. 2003. All gifts large and small. Am J Bioeth 3: 39–46.

Klevens, R. M., Morrison, M. A., Nadle, J., Petit, S., Gershman, K., Ray, S., et al. 2007. Invasive methicillin-resistant Staphylococcus aureus infections in the United States. JAMA 298: 1763–1771.

Kolawole, D. and Shittu, A. 1997. Unusual recovery of animal staphylococci from septic wounds of hospital patients in Ile-Ife, Nigeria. Lett Appl Microbiol 24: 87–90.

Kumarasamy, N., Safren, S. A., Raminani, S. R., Pickard, R., James, R., Krishnan, A. K., et al. 2005. Barriers and facilitators to antiretroviral medication adherence among patients with HIV in Chennai, India: a qualitative study. AIDS Patient Care STDS 19: 526–537.

Kunin, C. M. 1993. Resistance to antimicrobial drugs – a worldwide calamity. Ann Intern Med 118: 557–561.

Lal, A. 1998. Information contents of drug advertisements: an Indian experience. Ann Pharmacother 32: 1234–1238.

Lexchin, J. 1987. Advertisement scrutiny. Lancet 1: 1323–1324.

Lu, H. Z., Weng, X. H., Li, H., Yin, Y. K., Pang, M. Y. and Tang, Y. W. 2002. Enterococcus faecium-related outbreak with molecular evidence of transmission from pigs to humans. J Clin Microbiol 40: 913–917.

Master, Z. and Ozdemir, V. 2008. Selling translational research: is science a value-neutral autonomous enterprise? Am J Bioeth 8(3): 52–54

Mastin, T., Andsager, J. L., Choi, J. and Lee, K. 2007. Health disparities and direct-to-consumer prescription drug advertising: a content analysis of targeted magazine genres, 1992–2002. Health Commun 22: 49–58.

McCaig, L. F., Besser, R. E. and Hughes, J. M. 2002. Trends in antimicrobial prescribing rates for children and adolescents. JAMA 287: 3096–3102.

McCaig, L. F., Besser, R. E. and Hughes, J. M. 2003. Antimicrobial drug prescription in ambulatory care settings, United States, 1992–2000. Emerg Infect Dis 9: 432–437.

McCaig, L. F., Hooker, R. S., Sekscenski, E. S. and Woodwell, D. A. 1998. Physician assistants and nurse practitioners in hospital outpatient departments, 1993–1994. Public Health Rep 113: 75–82.

Menkes, D. B. 1997. Hazardous drugs in developing countries. Br Med J 315: 1557–1558.

Menkes, D. B. 2006. Calling the piper's tune. Primary Care Commun Psychiatry 11: 147–149.

Mintzes, B. 1998. Blurring the boundaries: new trends in drug promotion. In: Promotion Targeting Consumers. Amsterdam: HAI-Europe www.haiweb.org/pubs/blurring/blurring. intro.html Dated Accessed January 20, 2009.

Mintzes, B. 2006. Disease mongering in drug promotion: do governments have a regulatory role? PLoS Med 3: e198.

Mitema, E. S., Kikuvi, G. M., Wegener, H. C. and Stohr, K. 2001. An assessment of antimicrobial consumption in food producing animals in Kenya. J Vet Pharmacol Ther 24: 385–390.

MSNBC. 2006. WHO calls for halt on malaria treatment. Agency fears improper use may lead to drug resistance. Associated Press.

Muriuki, F. K., Ogara, W. O., Njeruh, F. M. and Mitema, E. S. 2001. Tetracycline residue levels in cattle meat from Nairobi slaughter house in Kenya. J Vet Sci 2: 97–101.

Mutume, G. 2001. Health and intellectual property; poor nations and drug firms tussle over WTO patent provisions. Africa Recover 15(1): 14–15.

Na Iampang, K., Chongsuvivatwong, V. and Kitikoon, V. 2007. Pattern and determinant of antibiotics used on broiler farms in Songkhla province, southern Thailand. Trop Anim Health Prod 39: 355–361.

Nicholls, T., Acar, J., Anthony, F., Franklin, A., Gupta, R., Tamura, Y., et al. 2001. Antimicrobial resistance: monitoring the quantities of antimicrobials used in animal husbandry. Rev Sci Tech. 20: 841–847.

Norris, P., Herxheimer, A., Lexchin, J. and Mansfield, P. 2005. Drug promotion. What we know, what we have yet to learn. Reviews of materials in the WHO/HAI database on drug promotion. Amsterdam: WHO/HAI.

Okeke, I. N., Klugman, K. P., Bhutta, Z. A., Duse, A. G., Jenkins, P., O'Brien, T. F., et al. 2005. Antimicrobial resistance in developing countries. Part II: strategies for containment. Lancet Infect Dis 5: 568–580.

Okeke, I. N., Lamikanra, A. and Edelman, R. 1999. Socioeconomic and behavioral factors leading to acquired bacterial resistance to antibiotics in developing countries. Emerg Infect Dis 5: 18–27.

Oyugi, J. H., Byakika-Tusiime, J., Charlebois, E. D., Kityo, C., Mugerwa, R., Mugyenyi, P., et al. 2004. Multiple validated measures of adherence indicate high levels of adherence to generic HIV antiretroviral therapy in a resource-limited setting. J Acquir Immune Defic Syndr 36: 1100–1102.

Ozdemir, V., Aklillu, E., Mee, S., Bertilsson, L., Albers, L., Graham, J. E., et al. 2006. Pharmacogenetics for off-patent antipsychotics: reframing the risk for tardive dyskinesia and access to essential medicines. Expert Opin Pharmacother 7: 119–133.

Ozdemir, V. and Godard, B. 2007. Evidence-based management of nutrigenomics expectations and ELSIs. Pharmacogenomics 8:1051–1062.

Ozdemir, V., Graham, J. E. and Godard, B. 2008. Race as a variable in pharmacogenomics research: From empirical ethics to publication standards. Pharmacogenet Genomics 18(10): 837–841.

Petersen, A., Andersen, J., Kaewmak, T., Somsiri, T. and Dalsgaard, A. 2002. Impact of integrated fish farming on antimicrobial resistance in a pond environment. Appl Environ Microbiol 68: 6036–6042.

Phillips, I., Casewell, M., Cox, T., De Groot, B., Friis, C., Jones, R., et al. 2004. Does the use of antibiotics in food animals pose a risk to human health? A critical review of published data. J Antimicrob Chemother 53: 28–52.

Projan, S. J. 2003. Why is big Pharma getting out of antibacterial drug discovery? Curr Opin Microbiol 6: 427–430.

Quick, J. D. 2003. Ensuring access to essential medicines in the developing countries: a framework for action. Clin Pharmacol Ther. 73: 279–283.

Richet, H. M., Mohammed, J., McDonald, L. C. and Jarvis, W. R. 2001. Building communication networks: international network for the study and prevention of emerging antimicrobial resistance. Emerg Infect Dis 7: 319–322.

Rogers, W. A., Mansfield, P. R., Braunack-Mayer, A. J. and Jureidini, J. N. 2004. The ethics of pharmaceutical industry relationships with medical students. Med J Aust 180: 411–414.

Rolland, R. M., Hausfater, G., Marshall, B. and Levy, S. B. 1985. Antibiotic-resistant bacteria in wild primates: increased prevalence in baboons feeding on human refuse. Appl Environ Microbiol 49: 791–794.

Rollin, B. 2001. Ethics, science and antimicrobial resistance. J. Agric Environ Ethics 14: 29–37.

Semin, S., Aras, S. and Guldal, D. 2007. Direct-to-consumer advertising of pharmaceuticals: developed countries experiences and Turkey. Health Expect 10: 4–15.

Simango, C. and Rukure, G. 1991. Potential sources of campylobacter species in the homes of farmworkers in Zimbabwe. J Trop Med Hyg 94: 388–392.

Spear, B. B., Heath-Chiozzi, M. and Huff, J. 2001. Clinical application of pharmacogenetics. Trends Mol Med 7: 201–204.

Spellberg, B., Powers, J. H., Brass, E. P., Miller, L. G. and Edwards, J. E., Jr. 2004. Trends in antimicrobial drug development: implications for the future. Clin Infect Dis 38: 1279–1286.

Talbot, G. H., Bradley, J., Edwards, J. E., Jr., Gilbert, D., Scheld, M. and Bartlett, J. G. 2006. Bad bugs need drugs: an update on the development pipeline from the Antimicrobial Availability Task Force of the Infectious Diseases Society of America. Clin Infect Dis 42: 657–668.

Tuffs, A. 2004. Only 6% of drug advertising material is supported by evidence. Br Med J 328: 485.

Usluer, G., Ozgunes, I. and Leblebicioglu, H. 2005. A multicenter point-prevalence study: antimicrobial prescription frequencies in hospitalized patients in Turkey. Ann Clin Microbiol Antimicrob 4: 16.

Vancelik, S., Beyhun, N. E., Acemoglu, H. and Calikoglu, O. 2007. Impact of pharmaceutical promotion on prescribing decisions of general practitioners in Eastern Turkey. BMC Public Health 7: 122.

Vidaver, A. K. 2002. Uses of antimicrobials in plant agriculture. Clin Infect Dis 34 Suppl 3: S107–S110.

Waud, D. 1992. Pharmaceutical promotions – a free lunch? N Engl J Med. 327: 351–353.

Wazana, A. 2000. Physicians and the pharmaceutical industry: is a gift ever just a gift? JAMA 283: 373–380.

World Consumer Rights Day (WCRD). 2007. Unethical Drug Promotion.

Wenzel, R. P. 2004. The antibiotic pipeline – challenges, costs, and values. N Engl J Med 351: 523–526.

World Health Organization. (WHO) http://www.who.int/medicines/en.

World Health Organization. (WHO). 1988. Criteria for medicinal drug promotion. Geneva: World Health Organization.

World Health Organization. (WHO). 1998. Emerging and other communicable diseases: antimicrobial resistance. Geneva: World Health Organization.

World Health Organization. (WHO). 2001. Pharmaceutical promotion. IN WHO global strategy for containment of antimicrobial resistance. Executive Summary, World Health Organization (Ed.). 51–53. Geneva: World Health Organization.

WHO. 2006. Facts on acts (artemisinin-based combination therapies). Geneva: World Health Organization.

WHO. 2007. Resistance to artemisinin derivatives along the Thai-Cambodian border. Weekly Epidemiol Rec 82(41): 360.

Yamey, G. 2001. US trade action threatens Brazilian AIDS programme. Br Med J 322: 383.

Yoshikawa, T. T. 2002. Antimicrobial resistance and aging: beginning of the end of the antibiotic era? J Am Geriatr Soc 50: S226–S229.

Chapter 30
News Media Reporting of Antimicrobial Resistance in Latin America and India

Marisabel Sánchez and Satya Sivaraman

Abstract To provide an insight into current media environments, types of coverage, reasons for such coverage, and potential improvements in news media coverage of antimicrobial resistance (AMR) in Latin America and India. To better understand the potential influences of news reporting of antimicrobial resistance it is important to gain an insight into the current trends in mass media in Latin America and India. As elsewhere, the mass media permeates daily life.

30.1 Analysis of the Mass Media in Latin America

Latin America comprised of 18 Spanish-speaking countries—Mexico, most of Central and South America, Cuba, the Dominican Republic, and Puerto Rico in the Caribbean—has an exceptionally dynamic and complex mass media. Despite sharing similar languages and cultural heritage, the region cannot be considered a homogenous media market.

30.1.1 Radio and Television

Radio continues to be the dominant medium in Latin America, with the greatest market penetration and opportunities for promoting open dialogue with citizens. A myriad of low-power radio transmitters reach only a small area and are operated locally to serve remote rural areas and indigenous communities. These stations are highly commercial, community oriented, and provide a pivotal platform for dialogue and community mobilization on important issues. Currently, there is an impressive proliferation of web links and live Internet broadcast to Latin American radio stations. Although many of the web-linked stations are not representative of Latin American radio stations as a whole,

M. Sánchez (✉)
Links Media, LLC, Gaithersburg, MD, USA
e-mail: msanchez@linksmedia.net

A. de J. Sosa et al. (eds.), *Antimicrobial Resistance in Developing Countries*,
DOI 10.1007/978-0-387-89370-9_30, © Springer Science+Business Media, LLC 2010

many do provide a taste of the trends in local and regional popular programming. International radio broadcasting services also provide news programming. Although their audience is fairly small—around 9 million or 3.2% of the general radio listening audience—these services communicate significant political, economic, religious, social, and cultural opinions. In general, radio listeners in Latin America tune in for the news. Unfortunately, most news is largely repeated reports from newspapers and news agencies. Critical social and political issues are often times not treated with the depth and analysis required. Many news shows only focus on one side of the issue or on only one aspect of the story. In studying the state of radio news in Latin America, organizations such as the Fundación Nuevo Periodismo Iberoamericano have found a general lack of formal training of radio journalists.

Television viewing in Latin American averages about 3 h and 14 min per person. Fiction, drama, and entertainment, as Spanish-language *telenovelas*, games, variety, celebrity gossip entertainment, and reality programming, such as the Spanish version of "Big Brother" are very popular. Latin Americans watch an average of three to four channels regularly. Most television stations show news and talk shows only in the morning.

In most Latin American countries, such as in Puerto Rico, the television market is mainly commercial and highly competitive. Current changes in Venezuela have brought restrictions in favor of government-controlled media. Mexico, on the other hand, has relaxed media ownership regulations in the past few years, thus increasing market competition for advertising revenue.

30.1.2 The Internet

Public Internet kiosks and cybercafés provide increasingly affordable access to the web as well as economic opportunities for entrepreneurs. Popular forms of mass media include electronic and print media—broadcasting for radio and television, discs and tapes, film, recorded music, video games, software publishing, web sites, blogs and podcasts, newspapers, books, magazines, and popular literature. All these forms of mass media shape content, quality, depth, and frequency in which news is reported. After years of slow growth, the Internet is moving to a faster beat. Brazil, Mexico, Argentina, and Chile are driving this growth. E-commerce and online advertising are benefiting from the expansion. Internet World Stats reported close to 50 million Internet users in Latin America in 2006. Many organizations in Latin America are turning to the Internet as a channel for promoting development and political reform to addressing problems such as the environment, diseases, economic competitiveness, and political pluralism. However, some critics are pointing at the Internet as yet another form of exclusion that could further deepen the economic and social divide that already exists in the region.

30.1.3 Newspapers

There is a vast range of size and resources in newspapers in Latin America. Clarín of Argentina has a daily circulation of 700,000 and more than 1 million on Sundays is the largest Spanish-language daily in the world. Brazil's Folha de Sao Paulo sells 650,000 copies daily and 1.2 million copies on Sundays. These newspapers have larger circulation than many US newspapers. There are many multilingual newspapers such as El Regional, a Guatemalan weekly published in Spanish, and five Maya languages.

As in the United States and western Europe, Latin America's newspaper circulation has been falling, largely due to the Internet. Newsrooms are also changing as many Latin American newspapers are considering unifying their Internet and print edition newsrooms to save on costs and improve on revenues and circulation. Most online newspapers rely on their print counterparts for content and on a small pool of mostly untrained and inexperienced journalists. Online journalists focus on writing and editing and do very little reporting. It is uncertain whether newspaper owners will opt for keeping their more experienced and trained journalists.

In Latin America, advertising messages about medications, primarily in print media, are key sources of information and education for most consumers. Unfortunately, too many of these messages are unclear, simplistic, and potentially misleading. The bombardment of advertising by pharmaceutical companies during the past 10 years has increased at fast rate both in volume and frequency. An example of such aggressive promotion was documented by the Asociación Mexicana para la Defensa del Consumidor (AMEDEC), which found 45 advertisements of one medication in two large circulation newspapers in Mexico in a period of 2 days. Advertising of medications, not only in print media but also in all mass media, has become a major source of revenues. The implication of such publicity on how journalists report on issues related to medications is largely unknown, as few studies have documented the impact. What is evident is that pharmaceutical companies are the primary sources of information about medications including antibiotics.

30.2 Analysis of News Reporting of AMR

A 2005 gray literature review showed that coverage related to AMR was limited, deficient, and rarely based on scientific evidence locally generated. Information regarding determinant factors of AMR was limited to a few articles. Most notably absent was information about consumers' perception of risk of arbitrary use of antimicrobials, their expectations, and motivations. Additionally, there was lack of news coverage about AMR surveillance, antibiotic prescribing and dispensing practices, and legal regulations[1]. The literature

---

[1] M. Sanchez, B. Kubiak. A Review of Gray and Popular Literature Regarding Antimicrobial Use 1989-2005 Coverage: Determining the Determinants for Assessment: Meeting of South

review also found that most articles focused on the economics of antimicrobials such as the high cost of medications, limited purchasing power, or financial resources for acquisition of medications by consumers and the government, as well as easy access to less expensive antibiotics from informal and illegal markets. In Paraguay, for instance, the majority of print news coverage was related to this issue, with over 60% of the coverage focusing on the economic aspects. Access to health care and medications was the second most widely covered issue related to AMR. News reports about shortages of primary care physicians and pharmacists in rural areas, severe shortages of medications, environmental conditions affecting access to quality medications, and social issues such as exclusion of poor consumers were the primary topics covered.

An interesting finding in the literature review was the fact that although some Latin American countries, as in the case of Paraguay, recognize in their Constitution, health as a fundamental right of their citizens, there is absence of consumer protection laws against the illegal sale and distribution of medications, including antibiotics. Additionally, the news media reported about the need for advocacy of enforcement of laws against counterfeits and the sale of antibiotics without prescriptions. Other important news reports, though limited, focused on overall distrust in the health-care system including pharmacies and providers. Case in point was the massive coverage of expired medications in public health facilities through various newspapers in Peru, Paraguay, and Bolivia found in the literature review.

Although the quantity and frequency of news media coverage about antimicrobial resistance and its determinant factors have not been sufficient, a number of some of relevant issues have been raised. For example, the high levels of self-medication by consumers and the problem of abandoning treatment, as well as poor management of medications in health facilities, lack of government regulations and enforcement, and the problem of counterfeited drugs have been given some attention primarily in newspapers. Raids and confiscation of illegally sold medications including antibiotics were also reported on television and radio news programs.

In Latin America—where the present threat of infectious diseases and antimicrobial resistance continue to be a significant public health peril—the United States Agency for International Development (USAID) has funded the South American Infectious Disease Initiative (SAIDI) since 2005. Through this mechanism, efforts to develop multisector strategies to contain the advance of AMR have been undertaken in three countries: Bolivia, Paraguay, and Peru. These initiatives include working with the news media to educate them about how to effectively cover and report issues related to AMR. Significant improvement has been made in the quality and quantity of news media reporting of AMR due to the efforts of the initiative, but only in the three countries currently

American Infectious Disease Initiative (SAIDI), March 2005. Prepared for U.S. Agency for International Development Global Health Division. Contract GHS-I-10-03-00037-00. Prepared by Links Media, Gaithersburg MD USA.

involved in the project. Coverage generated by the initiative has resulted in depth news reporting of issues such as rational use of medications related to prescribing physicians, dispensers, and consumers; the importance of enacting and enforcing regulation of the sale of antibiotics; the need for quality production and management of antibiotics including good storage practices; and the necessity and value of the adoption of standard treatment guidelines by physicians and medical schools in the public and private sector. Interestingly, trained journalists have been advocates in their newsrooms about the need for greater attention to AMR-related stories because they see the connection between morbidity and mortality and therapeutic failure.

Unfortunately, stories are not an easy sell in the newsrooms unless they involve significant numbers of deaths and illness related to failure of medications. On the other hand, the interest exists to learn more about AMR and the linkages to tuberculosis and upper respiratory infections.

In-depth and specialized journalism, particularly of health issues while limited, is still alive and well, but not without significant challenges and major obstacles. Anyone seriously considering working with the media to promote issues such as AMR must consider these facts.

In Latin America, journalists who conduct in-depth investigate reporting—particularly dealing with issues such as corruption, drug trafficking, and other crimes face serious challenges from death threats, bribes and extortion, and lawsuits. In many countries, journalists must also contend with laws that make libel a criminal offence, and use a very broad standard to define libel as disrespectful expressions directed to a public official even if fully verifiable.

Latin America still has many great investigative journalists who are interested in in-depth reporting and analysis of issues such as those required in AMR. Unfortunately, far too many journalists are not able to get their stories published. The considerable time and effort required for in-depth reporting is generally not profitable for the newspaper or the news organization involved. Convincing publishers and editors in Latin America to embrace in-depth investigative journalism is very difficult. Countries with a strong tradition of in-depth journalism such as Argentina, Colombia, and Peru have shown a notable decline in reporting. In many Latin American countries, despite all odds, investigative journalism continues to gain strength both in quantity, quality of newsgathering and reporting, and networks of sources. The once promising emergence of alternative channels via the Internet may still render some hope for the future.

Science and health reporting has its own part to play in the media networks of these countries; however, this is not necessarily true in Latin America, where data are highly curtailed, and scientific and health journalists are forced to make great efforts to overcome this situation. Consequently, journalists searching for going beyond the mere role of providing information find themselves unable to do so because there is limited information and few well-trained and credible sources willing to provide unbiased scientific evidence. On the other hand, there are scientists who are eager to share information but lack the training to be able to provide data in a manner that is understood by journalists with limited backgrounds in reporting technical, scientific, and health issues.

Scientific and health journalists can contribute to the promotion of an environment that leads to greater understanding of health issues such as AMR and solutions that affect Latin America, provided they conceive their roles as specialized journalists determined to use all the communication means and technologies available to offer information, analysis, and opinions on this activity. Journalists and organizations interested in improving understanding and promoting action toward AMR containment should support the creation of capacity building activities that makes sense in the context of coherent research, communication, and educational policies. They should also seek to ensure the development and dissemination of national and local data on AMR and its determinant factors as well as the proper training of credible scientific sources to provide journalists with scientific evidence of the problem and the solutions.

The scientific community interested in improved news reporting of AMR should also consider promoting the participation of the public in decisions affecting the promotion and management of AMR containment activity. Discussions on development or the adoption of specific interventions and policy should include public debates sparked by news media reporting. Journalists can play a role in promoting debate and provide experts, potential users, and policy makers information, analysis, and opinion on the AMR. Unfortunately, such an environment does not currently exist, though the environment in some Latin American countries such as Paraguay and Peru is clamouring for such dialogue.

News media coverage of issues related to AMR is of vital importance in order to gain a greater share of the limited budgetary resources earmarked for health care and the use of these resources aimed at containing the spread of resistance. Journalists should also contribute in the fight to search for more resources and credible sources by providing information, education, and a democratic approach when it comes to making decisions about containment of AMR.

30.3 Recommendations and Priorities for Improved News Reporting of AMR in Latin America

The possibilities of news coverage will increase if its purpose is clearly understood and access to information is consistent, frequent, and timely. News reports should not just provide information, but explain and analyze issues related to AMR, its impact and solutions, particularly in regard to policy issues that have a tremendous impact on the country's ability to contain the spread of AMR. AMR news reporting should provide credible and accurate information from different and multiple sources, it is not acceptable to continue to report based on one source only. In order to reach in-depth understanding journalists should reach out to the members of the academic and business communities, private and public hospitals, scientific societies such as the Infectious Diseases Societies, Alliance for Prudent Use of Antibiotics, pediatric societies.

Journalists should consider providing information, analysis, and opinions on AMR containment strategies as well as required actions to support and sustain containment efforts.

Journalists in Latin America interested in AMR issues should also address a number of priority issues, including the problems and needs that are common to the entire region, the current characteristics of AMR, and solutions from a variety of sectors and systems, at the local, national, and regional levels. Journalists should also report on issues related to social and government responsibility, community participation, enactment and enforcement of laws that support containment, and resources allocated to supporting the systems that support containment.

The priorities proposed here are based on the specific problems of a region at a particular moment in time. The current problems will gradually change; some will be overcome and others will likely spring up. This reflection on priorities is merely aimed at encouraging discussion about improving the environment, commitment, and access to information, in order to enhance the quality and quantity of news reporting about AMR-related issues.

30.3.1 Key Recommendations

- Improved training for journalists and media representatives on AMR issues related to drug management, quality of medications, surveillance, rational use of antibiotics, policy and regulations, and resources
- Better training for spokespersons
- Materials for journalists, including development of press kits and high quality, verified briefing content
- More influx of news and information
- Multisource reporting and information collection
- Consistent feeding of news articles, studies, and advances related to AMR to the media
- Access to trained and credible sources of information
- Collaboration with scientific societies, academia, agencies, and NGOs to help ensure access to information, and credible sources that can provide perspective and explanation of technical topics

30.4 Analysis of the Mass Media in India

The history of the Indian media is closely intertwined with the struggle for India's freedom from British rule when newspapers—both in English and in the vernacular languages—were started by nationalist businessmen or activists espousing their cause.

After India's independence in 1947, the state-run and private media focused on promoting the developmental policies of the new Indian government and the myriad needs of a vast and poor country. Since the 1980s, with the liberalization of the Indian economy and its subsequent rapid growth, the Indian media has undergone a transformation in terms of quantity and quality.

In 2006, the media in India—print, radio, and television—had a business turnover on the order of US $6 billion, exceeding 1% of the country's gross domestic product and matching the size of many of the country's domestic industries. The driving forces behind this expansion include rising levels of literacy and education, an increasingly integrated national market, growing consumption, and a burgeoning middle class.

Facilitating the process are improved access to and falling costs of the means or instruments of communication. The Indian media has obtained access to both new technology as well as editorial content in recent years through ties with foreign counterparts.

Thanks to the rapid growth of television, the media's influence is also no longer confined to literate or educated Indians alone. Television plays a key role in setting national and regional agendas, and in shaping social values and political choices, including electoral outcomes.

Such rapid growth in reach and economic clout has, however, not necessarily meant any improvement in the overall quality of the output of the Indian media. In contrast to the earlier phase of the media's growth, when responding to national developmental needs through high-quality journalism was the main motive, the current phase of media expansion is driven largely by pressure from advertisers, the market, and the media business's own search for higher and higher profits.

The rapid growth in the size of the Indian media business in recent years is, in the main, driven by aggressive marketing strategies. The content of Indian media, including news reporting, is more and more guided by priorities set by the marketing and advertising departments of media companies.

One measure of the success of marketing is the selling of space in the press and airtime in the electronic media, which translates into advertising revenue. Advertising revenue for the Indian media as a whole expanded sevenfold between 1991 and 2004.

Spending on advertising in the Indian media has expanded exponentially, from less than Rs 18 billion in 1991–1992 to more than Rs 130 billion in 2004. Spending on the press has risen from Rs 10.7 billion to Rs 60.4 billion, while that on television has risen from Rs 3.90 to Rs 58 billion. The 1991 ratio of television to press in spending on advertising was 23:63. In 1999, television overtook the press. The ratio soon stabilized at around 45:47. Television still leads the press in reach—a viewership of 79 million versus a readership of 47 million. However, radio is witnessing a revival of sorts—its reach expanding from 16 to 26 million homes.

30.4.1 Newspapers

Advertisements on average occupy more than 20% of Indian newspaper space (and as much as 30% in northern India). This figure tells only one part of the story. For the bigger or leading papers, the ratio is 60% or more advertising to reporting. For the English-language press, this ratio is 65%-plus advertising to reporting.

In recent years, the press has become even more dependent on advertising for its very survival. One of the main reasons for this is the low issue price of Indian newspapers. Since circulation revenues can only meet a fraction of this cost, newspapers have become dependent on advertising to make up the bulk of the expenses necessary for survival. This has extremely unhealthy consequences for media independence and diversity. Growing dependence on advertising means less and less autonomy from the corporate interests that buy advertising space or time. Attracting advertisements is a function primarily of circulation, and secondarily of the editorial position of a publication on a range of issues such as the economy and social and political policies.

Newspapers are finding it increasingly difficult to garner the increasingly high circulations necessary to attract advertising. This is especially so because of the growing competition in advertising for mass-consumption products, mainly from television. Only newspapers with relatively large circulations can survive in the marketplace. The others face a bleak future.

30.4.2 Television

Thanks to round-the-clock news channels, Indian television has now emerged as a primary source of news for many people. In a month-long 2006 study of prime time news bulletins in 24-h news channels the Centre for Media Studies found that malaria, tuberculosis, disability, cancer, or sexually transmitted diseases did not figure in any of the six Hindi-language news channels monitored. These channels were Zee News, Doordarshan, Star News, Sahara, NDTV, and Aajtak.

Even other health-related themes such as immunization, diabetes, and mental health hardly figured in the entire month's bulletin. Only 3% of all news items in the news bulletins of these channels could be categorized as health related. What got mentioned relating to health, mostly by the way of commercial advertisements, were medicine, diet/nutrition, colds, avian influenza, allergies, and HIV/AIDS.

One-third of all that is to do with health in the news channels, mostly by way of advertisements, was about "care and treatment;" only 15% of health reports were to do with "precaution and prevention." News channels promote health, if at all, mostly by way of commercial claims to do with cough syrup, headache pills, vitamin tablets, digestive medicines, ulcer gels, and hand and mouth wash liquids.

Of 15 stories on the Zee News channel during the month in the area of health, 10 were to do with doctors' strikes and 2 were on alleged neglect of hospitals. The case was similar with Sahara TV. The other stories in this regard were to do with bird flu or were about yoga. There was only one story on tuberculosis in the entire month by one news channel, which was to do with a government policy intention to eradicate tuberculosis in the country.

30.4.3 The Internet

The increased access to information and resources through the Internet is helping health journalists improve their knowledge and ability to carry out accurate reporting. However, the dearth of Internet connections within India means that sustaining independent web sites (without corporate or institutional sponsorship) devoted to health issues is quite difficult for want of readership or advertising support.

30.4.4 Scientific Reporting

Several popular science magazines and journals have been launched in the Indian market from time to time but very few of them have survived, lacking adequate readership or advertising support. In recent years, *Science Today*, *Science Age*, *Bulletin of Sciences*, *Research and Industry*, and Indian editions of foreign science magazines like *Vigyan* (*Scientific American*) and *World Scientist* (*La Recherche*) have ceased publication.

Though there has been something of a revival of coverage of science and health issues in Indian newspapers in the past decade and some have introduced regular science columns, it is still rare to find publications with their own science editor or a science reporter. Most of them are dependent on press handouts from research institutions, companies, or lobby groups, or they reprint articles from foreign journals.

Another major challenge facing science communication in India is writing in non-English languages. The country has 18 recognized regional languages and much of the latest scientific information is generally available in English. The quality of translation of science-related material is not yet satisfactory.

30.5 Problems and Issues in Reporting

The Indian media has traditionally accorded low priority to science-related journalism and health journalism in particular. Instead, top priority has been given only to reporting political and economic news.

Although India produces a large number of science graduates every year and has earned a global reputation as a scientific and technological power, there is a paucity of good science journalists in general and health journalists in particular in the Indian media. There are several reasons for this seemingly paradoxical situation:

- Journalism was not a well-paid profession, so qualified personnel often shied away from joining the media.
- Due to rapid advances and narrow specialization in various fields, journalists and newspaper editors alike find themselves unable to keep abreast of developments within a particular scientific or technical subject area.
- Science articles published in the newspapers/magazines, with a few notable exceptions, are poorly written in prosaic style, using technical jargon and an excess of avoidable statistics. Science articles in the Indian-language media are mere translations of those that appear in the English-language media, while the latter itself lacks originality and lifts ideas from foreign journals.

Singh contends that "popular science writing in India is still shackled by complacency and over dependence on foreign sources; they are unfortunately used for plagiarism."

Not only in print but in broadcast media also, misleading scientific information, a continuous decay of creativity in presentation, distortion in translation, inconsistency in organizing the contents, lapses in the use of language, and many more deviations can be seen frequently.

30.6 Misuse or Simplification of Data for Lay Audiences

According to Dr. Anoop Mishra, director of the Department of Diabetes and Metabolic Diseases at Fortis Hospitals, New Delhi, there are a lot of sensational reports in the media from time to time about "miracle cures" that are often ordinary new treatments, or reports on "new diseases," which are sometimes existing diseases not diagnosed earlier. Sometimes behind a so-called "blockbuster" drug, hailed as a panacea, lurks a dangerous adverse effect that is underplayed by the scientists and under-reported by the media.

Dr. Mishra also notes that many Indian health reporters lack knowledge in complex health-related issues, have insufficient training in health reporting, and do not make the necessary inquiries to obtain analytical comments from unbiased experts. As a result, discovery of one or two cases of an infectious disease is reported prominently as almost an epidemic, unusual but routine surgeries are named one of a kind; incorrect, misleading treatment information is given as expert advice; and statistically insignificant research findings are termed highly significant.

At times, Dr. Mishra says, a treatment is claimed to be extremely effective when only a few cases have been treated with mixed results. Many such studies, which are actively doled out to reporters, are wrongly executed, hastily analyzed, and conclusions dramatized. Untrained journalists are easy prey for such scientists and physicians.

30.7 Indian Media Reporting on Antibiotic and Antimicrobial Resistance

Reporting on the subject of antibiotic resistance in India is additionally difficult for the simple reason that not enough work has been done by various state or non-state institutions in studying the phenomenon and developing a proper understanding of its extent and impact.

For example, there has been no collation of existing national-level data on the occurrence and burden of antibiotic resistance. At hospitals around the country, there is no uniformity of testing procedures, lack of proper diagnostic facilities, and no emphasis on recording data about antibiotic resistance in a systematic way. The understanding of how widespread antibiotic resistance is at the community level is even weaker.

Given the large number of drug manufacturers in India, many of them operating in an unregulated manner from ad hoc production facilities, there is no proper data even on the overall volume and consumption of antibiotics.

Further, in a developing country context, antibiotic resistance has not yet been recognized as a serious threat to public health especially because of the perception that there are far more important health concerns—from malaria to HIV/AIDS—that need to be tackled first. The lack of importance given to antibiotic resistance in the overall context of health issues in India also means a lack of people within the medical establishment with the expertise to speak on the subject with proper data and insight.

A recurring problem in the Indian context is that of mistrust between members of the medical profession and journalists, with the former perceiving the latter as being too prone to sensational reporting and misrepresentation of facts. Among the suggestions made by the Indian Science Writers Association, for example, is to create a science media center, including a centralized web site to facilitate media access to the latest research reports and other developments in the field.

As in other developing countries, the issue of antibiotic resistance in India is complex. In urban and semi-urban centers, there is a growing trend of such microbial resistance due to overuse and abuse of antibiotics. In the vast and populated rural areas the problem is often about the complete lack of access to any antibiotics whatsoever.

What is required is a balanced approach toward containing antibiotic resistance, and not a rash approach that either promotes or denounces antibiotics blindly. For the media, which prefers simplistic and sensational sound bites, conveying such complex messages is not very attractive.

One consequence of this is that while the media might occasionally touch upon the subject—typically with a scary story about so-called "super bugs" in hospital intensive care units—it would not persist with coverage that requires it to go into the nuances and convey the various aspects of the problem to its audience.

Thanks to pressure on competing media outfits to increase newspaper circulation—or capture eyeballs in the case of television and web sites—there has been a conscious "dumbing down" of news coverage through sensational reporting, an unhealthy obsession with personalities and celebrities, and reliance on hearsay and scientifically unverified reports.

In the context of health reporting what this means is that the media tends to look for sensational angles to every health issue with minor outbreaks of infectious diseases being dubbed as "epidemics" or a new viral strain called a "mysterious killer bug," and so on. Minus the hype it is often difficult for media persons also to sell the stories to their own editors, and this sets off a vicious cycle whereby people working on health issues also tend to exaggerate the facts in order to win media attention.

30.8 Trends and Future Directions

With health figuring higher and higher on the Indian public agenda, particularly in urban areas, there is an increase in the space and time devoted to health issues in the Indian media.

A trend in recent years on Indian television has been to feature such issues by way of regular programs, outside news bulletins. Zee News has a health-related program called *"Total Tandrusti"* every Wednesday, Doordarshan News channel has *"Total Health"* on Sundays, while New Delhi Television has *"Sahat Ka Yog"* Monday to Friday in the mornings. Popular magazines devoted to health issues have also made their appearance in the Indian media market.

All these are also prompting a growing recognition within various Indian institutions working on public health about the need for regular training of all those involved with health communication work. The combination of increasing demand for health journalism with institutional support for training health journalists bodes well for the future of this genre of reporting.

Index

A. de J. Sosa et al. (eds.), *Antimicrobial Resistance in Developing Countries*,
DOI 10.1007/978-0-387-89370-9, © Springer Science+Business Media, LLC 2010